THE BIOCHEMISTRY
OF VIRUSES

THE BIOCHEMISTRY
OF VIRUSES

Edited by

Hilton B. Levy, Ph.D.

NATIONAL INSTITUTES OF HEALTH
BETHESDA, MARYLAND
DEPARTMENT OF HEALTH, EDUCATION, AND WELFARE

MARCEL DEKKER   New York and London   1969

A/576.64

MARCEL DEKKER, INC.
95 Madison Avenue, New York, New York 10016

United Kingdom edition published by
MARCEL DEKKER LTD.
14 Craufurd Rise, Maidenhead, Berkshire, England

LIBRARY OF CONGRESS CATALOG CARD NUMBER 75-90149

PRINTED IN THE UNITED STATES OF AMERICA

To the National Institutes of Health,
without whose support
most of the work reported here could not have been done

PREFACE

For several years, there has been an increasing association of molecular biology with virology. The restricted size of the virus genome offers unusual opportunities for studying the interrelationships between nucleic acids and protein synthesis. A great deal of research in molecular biology has been done using viruses, and much basic virological research in recent years has utilized the methods of molecular biology. In one sense this area of overlapping research is the subject of this book. Each of the contributors has attempted to summarize in a critical way the state of knowledge in his area of interest, addressing himself to an audience just slightly less knowledgeable than he in that particular area. Thus, the book is largely aimed at the research scientist or advanced graduate student. Chapters on virus chemistry and virus structure have been added because of their inherent interest and because they tend to act as a framework relevant to all the viruses. A chapter on interferon has been included because of its relationship to the molecular biology of virus replication and because of the editor's personal association with this field. The inclusion of a chapter on DNA tumor viruses was planned, but its cancellation at the last minute became necessary. It is hoped that the book may serve as a reference source and as a textbook in advanced courses in virus biochemistry.

CONTRIBUTORS TO THIS VOLUME

Bader, John P., *Chemistry Branch, National Cancer Institute, Bethesda, Maryland*

Baltimore, David,† *Department of Biology, Massachusetts Institute of Technology, Cambridge, Massachusetts*

Baron, Samuel, *U.S. Department of Health, Education, and Welfare, National Institute of Allergy and Infectious Diseases, Laboratory of Viral Diseases, Bethesda, Maryland*

Buckler, Charles E., *U.S. Department of Health, Education, and Welfare, National Institute of Allergy and Infectious Diseases, Laboratory of Viral Diseases, Bethesda, Maryland*

Cooper, Peter D., *Department of Cell Biology and of Microbiology and Immunology, Albert Einstein College of Medicine, New York, New York and Department of Microbiology, John Curtin School of Medical Research, Australian National University, Canberra, Australia*

Drzeniek, R., *Institut für Virologie, Justus Liebig-Universität, Giessen, Germany*

Gauntt, C. J., *The Wistar Institute of Anatomy and Biology, Philadelphia, Pennsylvania*

Ginsberg, Harold S., *Department of Microbiology, School of Medicine, University of Pennsylvania, Philadelphia, Pennsylvania*

† This work was begun at the Salk Institute for Biological Studies, San Diego, California.

ix

Graham, A. F., *The Wistar Institute of Anatomy and Biology, Philadelphia, Pennsylvania*

Green, Maurice, *Institute for Molecular Virology, Saint Louis University School of Medicine, Saint Louis, Missouri*

Levy, Hilton B., *U.S. Department of Health, Education, and Welfare, National Institute of Allergy and Infectious Diseases, Laboratory of Viral Diseases, Bethesda, Maryland*

McAuslan, B. R., *Roche Institute for Molecular Biology, Nutley, New Jersey*

Mathews, Christopher K., *Department of Biochemistry, University of Arizona College of Medicine, Tucson, Arizona*

Mattern, Carl F. T., *Laboratory of Viral Diseases, National Institute of Allergy and Infectious Diseases, National Institutes of Health, Bethesda, Maryland*

Roizman, Bernard, *Department of Microbiology, The University of Chicago, Chicago, Illinois*

Rott, R., *Institut für Virologie, Justus Liebig-Universität, Giessen, Germany*

Scholtissek, C., *Institut für Virologie, Justus Liebig-Universität, Giessen, Germany*

CONTENTS

4. The Genetic Analysis of Poliovirus

177

Peter D. Cooper

5. Myxoviruses

219

C. Scholtissek, R. Drzeniek, and R. Rott

6. The Reoviruses   259

C. J. Gauntt and A. F. Graham

7. RNA Tumor Viruses   293

John P. Bader

8. Biochemistry of Adenovirus Infection   329

Harold S. Ginsberg

9. The Biochemistry of Poxvirus Replication   361

B. R. McAuslan

10. Herpesviruses   415

Bernard Roizman

11. DNA-Containing Bacteriophage   483

Christopher K. Mathews

12. Biochemistry of Interferon   579

Hilton B. Levy, Samuel Baron, and Charles E. Buckler

1

CHEMICAL COMPOSITION
OF ANIMAL VIRUSES

MAURICE GREEN

INSTITUTE FOR MOLECULAR VIROLOGY
SAINT LOUIS UNIVERSITY SCHOOL OF MEDICINE
SAINT LOUIS, MISSOURI

1-1. Introduction

The remarkable advances during the past fifteen years in understanding cell function in molecular terms is due in large measure to the study of viruses and virus–cell systems. Purified virus is one of the few sources of homogeneous nucleic acid molecules for physical and chemical studies. Virus-infected cells provide important experimental systems for studying some of the difficult problems in cell and molecular biology, such as DNA replication, transcription of DNA to mRNA, and translation of mRNA molecules to specific proteins. However, most of these studies have utilized bacteriophage and bacterial cells; much less is known about the synthesis, function, and regulation of macromolecules in mammalian cells. The increased use of animal viruses to study mammalian cell function suggests that mammalian cell biology similarly will experience a rapid growth of new knowledge, especially in view of (a) the improved methodology for the large scale growth of mammalian cells in culture, (b) the development of methods for the cultivation and purification of milligram quantities of many animal viruses, and (c) the availability of increasingly sophisticated methods for studying macromolecules.

This chapter summarizes our current knowledge of the chemical properties of animal viruses. About 600 animal viruses are known, containing from five to several hundred genes, and replicating and maturing at different sites within the cell. The large number of animal viruses makes it difficult to assimilate new knowledge about viruses, but this difficulty is minimized upon recognition that most animal viruses can be classified into major virus groups, each possessing similar chemical, physical and biological properties. The identification of a virus within a group permits reasonable deductions concerning the chemical composition of the virus, the size of the viral genome, the mechanism of virus replication, the intracellular site of virus synthesis, and virus-induced alterations of cellular metabolism and enzyme synthesis.

A. GENERAL PROPERTIES OF VIRIONS

The virion is the mature, extracellular virus particle whereas the term "virus" is used to describe all phases of the virus life cycle (*1*). Animal virions range from 20–300 mμ in diameter, representing more than a 1000-fold difference in mass between the smallest and largest virions. Despite this large size range, virions have unique common properties which clearly distinguish them from other forms of life. As defined by Lwoff (2), viruses (a) contain DNA or RNA but not both, (b) multiply as viral nucleic acid, and (c) do not possess enzymes of energy metabolism. Excluded from the realm of viruses are the large intracellular Rickettsiae, the psittacosis-lymphogranuloma venereum group, the smaller bacterium, and the pleuropneumonia-like organisms, all of which contain both DNA and RNA, reproduce by binary fission, and possess an independent metabolism.

All virions contain protein and either DNA or RNA. The simplest virions possess only nucleic acid and protein while the most complex virions may possess lipids, polysaccharides, and several minor constituents. The numerous enzymes involved in cellular energy metabolism are absent from virions. The enzymes of macromolecular biosynthesis, many of which were discovered only during the past 10–15 years, are assumed to be absent from most virions, but this has been tested rarely. Recently the large poxviruses, which replicate exclusively in the cytoplasm of the cell, have been shown to contain RNA polymerase activity (3, 4).

B. MAJOR GROUPS OF ANIMAL VIRUSES

The continuing discovery of new viruses and the rapid accumulation of information on the biological, chemical, and physical properties of animal virions have provided the stimulus and necessity for classifying animal viruses. Eight major animal virus groups were proposed by the International Subcommittee on Virus Nomenclature (5), based on the chemical and structural features of the virion (6). Recently, alternative schemes have been suggested (7) and a universal system of nomenclature was proposed (8) which included viruses infecting animals, bacteria, plants, and insects. But considerable disagreement on virus nomenclature exists among virologists and no changes in classification have been approved. Since the classification of viruses by major animal virus groups is in common use, it is retained here.

Table 1-1 lists the eight original virus groups and four additional groups proposed recently (9, 10) based on four fundamental properties of the virion (8): (a) the chemical nature of the viral genetic material, (b) the symmetry of the nucleocapsid, (c) the presence or absence of an envelope, and (d) the number of capsomeres for viruses

TABLE 1-1
Classification and Properties of Animal Virus Groups[a]

			Properties of virion		
Virus group	Nucleic acid	Capsid symmetry	Presence of envelope	No. of capsomeres or (diameter of helix, mμ)	Diameter of virion, mμ
Picodnavirus[b]	DNA	Cubic	Naked	12 or 32	18–24
Papovavirus	DNA	Cubic	Naked	72	40–55
Adenovirus	DNA	Cubic	Naked	252	70–90
Herpesvirus	DNA	Cubic	Envelope	162	120–180
Poxvirus	DNA	Unknown	Envelope	—	150–300
Picornavirus	RNA	Cubic	Naked	32	17–30
Reovirus	RNA	Cubic	Naked	92	74
Arbovirus[c]	RNA	Unknown	Envelope	—	40–50
Myxovirus	RNA	Helical	Envelope	(9 mμ)	80–200
Paramyxovirus	RNA	Helical	Envelope	(18 mμ)	100–300
Rhabdovirus[b]	RNA	Helical	Envelope	(18 mμ)	60 × 225
Leukemiavirus[b] (RNA tumor viruses)	RNA	Unknown	Envelope	—	~100

[a] Taken from Green (6) and from Melnick (9) and Melnick and McCombs (10).
[b] All group names but these have been approved by the International Subcommittee on Virus Nomenclature (5).
[c] The physical and chemical properties of only several members of this large group (over 200 members) have been studied.

with cubic symmetry and the diameter of the nucleocapsid for viruses with helical symmetry.

1. Chemical Nature of Viral Nucleic Acid

Virions are classified according to the type of nucleic acid they contain, either DNA or RNA.

2. Symmetry of the Nucleocapsid

The virion contains viral nucleic acid enclosed in a protein shell(s) termed the "capsid" which protects the viral genetic material; the capsid with its enclosed nucleic acid is termed the "nucleocapsid." Capsids are constructed of protein subunits organized either as icosahedral shells (cubic symmetry) or helical tubes (helical symmetry). The fine structure and symmetry of the virion is determined by electron microscopy of negatively stained virus particles (11) (see Chap.

2). The second criterion for classifying virions is the symmetry of the viral capsid, whether cubic or helical.

3. *Presence or Absence of an Envelope*

Virions are further classified as either (a) "naked" nucleocapsids or (b) enveloped nucleocapsids. The presence of a lipid-containing envelope is determined by electron microscopy and is correlated with the loss of viral infectivity upon treatment with ether. The lipid portion of the envelope is derived from cytoplasmic or nuclear membrane during the final stages of virus maturation but the protein component may be virus coded (*12*).

4. *Number of Capsomeres in Cubic Viruses and Diameter of Nucleocapsid in Helical Viruses*

The capsids of virions possessing cubic symmetry contain morphological components called capsomeres seen by electron microscopy. Each capsomere consists of one or several protein subunits. The number of capsomeres characteristic for each virus group possessing cubic symmetry is a criterion for classification.

The nucleocapsids of virions possessing helical symmetry consist of multiple protein subunits helically arranged about the viral nucleic acid molecule. The diameter of the nucleocapsid characteristic for each virus group with helical symmetry is used as a criterion for classification.

Table 1-1 summarizes the properties of the 12 major virus groups. A partial list of the about 100 members of the five DNA virus groups, the picodnaviruses, papovaviruses, adenoviruses, herpesviruses, and poxviruses is presented in Table 1-2A. A partial list of the about 500 members of the RNA virus groups, the picornaviruses, reoviruses, arboviruses, myxoviruses, paramyxoviruses, rhabdoviruses, and leukemiaviruses is presented in Table 1-2B.

Before discussing the results of chemical studies on virions, the methods used in the isolation, purification, and characterization of virions and viral nucleic acids are described briefly; the subject was reviewed in detail in 1965 by Schwerdt (*13*).

C. METHODS FOR THE ISOLATION AND PURIFICATION OF VIRIONS

Purification of Virions

The purification of most animal virions no longer presents a serious problem. Mammalian cells, suitable for the propagation of animal

TABLE 1-2
Members of Major Groups of Animal Viruses[a]
(partial list)

A. DNA Viruses

Picodnavirus group[b] (8 members)
  H-1, H-3 (hamster)
  RV (Kilham rat virus), X-14 (rat)
  Adeno-satellite viruses (human, simian, murine)
Papovavirus group (7 members)
  Shope rabbit papilloma, human papilloma (wart), bovine papilloma,
  Polyoma (murine), simian vacuolating virus (SV40)
  K virus (murine)
Adenovirus group (55 members)
  Human (31 types)
  Simian (18 types)
  Bovine (3 types)
  Canine, avian, and murine
Herpesvirus group (12 members)
  Herpesvirus (human), varicella-zoster (chickenpox) (human)
  Cytomegalovirus (human and other species)
  B (Old world simian), Marmoset (New world simian)
  Pseudorabies (porcine)
  Bovine rhinotracheitis
  Infectious laryngotracheitis (fowl)
  Equine abortion, LK (equine)
Poxvirus group (24 members)
  Vaccinia, cowpox, rabbitpox, ectromelia, variola (smallpox), alastrim
  Myxoma, fibroma
  Fowlpox, canarypox, turkeypox, pigeonpox
  Molluscum contagiosum, goatpox, sheeppox
  Yaba monkey tumor virus, swinepox

B. RNA Viruses

Picornavirus group (about 200 members)
  Picornaviruses of human origin
    Enteroviruses:
      Poliovirus (3 types)
      Coxsackie virus A (23 types)
      Coxsackie virus B (6 types)
      ECHO virus (30 types)
    Rhinovirus (about 100 types)
  Picornaviruses of lower animals
    Foot-and-mouth disease, murine encephalomyelitis, Mengo,
    encephalomyocarditis

TABLE 1-2 *(Continued)*

Reovirus group (3 members)
    Types 1, 2, and 3
Arbovirus group (about 200 members)
    Subgroup A: Equine encephalitis—Eastern equine encephalitis, Western equine
            encephalitis, Venezuelan, Semliki forest, Sindbis, chikungunya, and
            13 other named viruses
    Subgroup B: St. Louis encephalitis, Dengue, yellow fever, Japanese encephalitis,
            and 28 other named viruses.
    Subgroup C: 9 named viruses
    Other groups
Myxovirus group (5 members)
    Influenza A, B, and C
    Fowl plague
    Swine influenza
Paramyxovirus group (13 members)
    Mumps; Newcastle disease virus; parainfluenza 1, 2, 3, and 4; Sendai; measles,
    distemper; respiratory syncytial; rubella(?)
Rhabdovirus group[b] (2 members)
    Rabies virus (human and lower animals)
    Vesicular stomatitis virus (bovine)
Leukemiavirus group[b,c] (a minimum of 60 members)
    Avian leukosis–sarcoma viruses
    Murine leukemia–sarcoma viruses
    Murine mammary tumor virus (Bittner)

---

[a] Taken from Green (*6*) and from Melnick (*9, 10*).

[b] All group names but these have been approved by the International Subcommittee on Virus Nomenclature (*5*).

[c] Refers to leukemiaviruses of avian and murine species; the etiology of human leukemia is not known.

viruses, can be derived from various tissues and species, and grown in culture, often on a large scale. Reliable plaque assays for measuring virus infectivity are available for monitoring virus growth and purification. Many different procedures for purifying animal virions have been described. Several high-resolution separation techniques, such as equilibrium density-gradient sedimentation and zone sedimentation, virtually guarantee the isolation of many animal viruses in a highly purified state. Several procedures most commonly used for purifying virions are described briefly.

    a. *Differential Centrifugation.* Viruses that are not inactivated or aggregated by high centrifugal forces can be purified by alternating cycles of low- and high-speed centrifugation. Lysates of infected cells are centrifuged first at low speed to remove cellular debris, then the supernatant fluid is centrifuged at high speeds to deposit virions as a pellet; the small molecular weight contaminants remain in the supernatant fluid.

b. *Fluorocarbon Treatment.* Homogenization of infected cell preparations in the presence of a fluorocarbon, such as trichlorotrifluoroethane, will often selectively denature cell proteins at the fluorocarbon–aqueous solvent interface; viruses generally are not inactivated and remain in the aqueous phase.

c. *Treatment with Proteolytic Enzymes and Nucleases.* Cellular proteins and nucleic acids in a partially purified virus preparation can be digested by added proteolytic enzymes and nucleases. But some virions are inactivated by proteolytic enzymes and the usefulness and effect on the virion of enzyme treatments must be evaluated first.

d. *Differential Precipitation.* The precipitation of virions or cellular constituents can often be accomplished with various concentrations of inorganic salts or organic solvents. The optimal concentrations of salt or solvent must first be determined empirically for each virus–cell system.

e. *Equilibrium Density-Gradient Sedimentation and Zone Sedimentation.* Virus purification can be achieved most effectively by separating components on the basis of buoyant density (equilibrium density-gradient sedimentation) or velocity sedimentation (zone sedimentation). Equilibrium density-gradient sedimentation involves the centrifugation of the virus in a density-gradient solution formed with cesium chloride, rubidium chloride, sucrose, glycerol, potassium tartrate, or potassium citrate. The density gradient may be generated during centrifugation of the virus in a solution of cesium chloride or rubidium chloride, or the virus preparation may be layered on a preformed density gradient prior to centrifugation. At equilibrium, virus particles form a band (or zone) at the position in the gradient corresponding to the buoyant density of the virion. Animal virions possess buoyant densities which differ from those of most cell macromolecules, thus extensive purification is accomplished by equilibrium density-gradient sedimentation.

Zone sedimentation exploits the relatively large size of virions which sediment more rapidly than most cellular macromolecules. The virus preparation is layered over a preformed stabilizing density gradient and centrifuged for a period sufficient to resolve the zones of virion and cellular macromolecules. When large amounts (greater than about

100 μg) of virus particles are purified by equilibrium density-gradient or zone sedimentation, the virion band is visible as a light scattering region.

f. *Miscellaneous Procedures.* Viruses have been purified by a variety of additional procedures including adsorption and ion-exchange chromatography and partition in two-phase liquid polymer systems. Viruses such as myxoviruses which agglutinate erythrocytes can be purified substantially by adsorption to and elution from erythrocytes.

D. CHARACTERIZATION OF THE VIRION

Virus preparations should be of high purity for physical and chemical measurements to be meaningful. Criteria for purity generally involve tests for homogeneity of the virion by several chemical and physical properties including: (a) uniformity in size and shape by electron microscopy; (b) a single band by zone sedimentation or a sharp boundary by velocity sedimentation with the analytical ultracentrifuge; (c) coincident distribution of infectivity and other virion properties, such as optical density or radioactivity of labeled virus after zone sedimentation or equilibrium density-gradient sedimentation; (d) the absence of detectable host cell protein in nonenveloped virions by immunological analysis; (e) the absence of cell protein and nucleic acids as judged by the absence of radioactivity in purified virus prepared from infected cells to which uninfected cells containing labeled protein, DNA, or RNA have been added; (f) the presence of only RNA or DNA, and a constant chemical composition for different virus preparations, especially when purified by different procedures.

1. Optical Properties

Direct observation of negatively stained virions by electron microscopy (11) is a routine and valuable procedure for studying virion morphology, fine structure, size, and homogeneity. Virus particle counts by electron microscopy can be compared with the concentration of "infectious" particles. Light scattering, X-ray scattering, and X-ray diffraction measurements require specialized facilities and generally are not used for the routine characterization of virions.

2. Hydrodynamic Properties

Measurements of the movement of virus particles in solution such as sedimentation and diffusion provide information on the size and homogeneity of the virion (14). Electrophoresis, less commonly employed in virus research, provides data on the surface charge and homogeneity of the virion.

3. Chemical Properties

Relatively small amounts (1 to 5 μg) of DNA, RNA, protein, and phosphorus can be quantitatively measured routinely in virion preparations with errors of $\pm 3\%$. Lipid and polysaccharide analyses are more difficult, requiring larger amounts of material.

4. Immunological Properties

Immunodiffusion and immunoelectrophoresis analyses of virus preparations using antiserum to purified virus provide information on the number of virion proteins. Contamination of virus preparations with host-cell proteins can be detected with antiserum to host cells.

5. Viral Protein

Purified virus may be characterized with regard to the number and types of polypeptides, their amino acid composition, molecular weight, amino and carboxyl terminal amino acids, and amino acid sequence. Very few animal virions have been studied with regard to these properties.

6. Viral Nucleic Acid

The genetic substance of virions has been studied in greater detail than other components. The properties of viral nucleic acids most amenable to investigation are (a) base composition, (b) molecular weight, (c) number of strands per molecule, and (d) composition. In several instances, nucleotide sequence homology relationships between viral DNA molecules have been determined.

E. CHARACTERIZATION OF VIRAL NUCLEIC ACIDS

Viral nucleic acid is extracted from purified virion preparations by several procedures including treatment with sodium dodecyl sulfate (SDS) to disrupt the virion and/or extraction with phenol to denature and remove viral protein. Prior treatment with a proteolytic enzyme such as pronase and papain to digest viral proteins is necessary to obtain high yields of viral DNA from adenoviruses, herpesviruses, and poxviruses.

1. DNA Base Composition

The base composition of viral DNA is determined commonly by four procedures: (a) acid hydrolysis of DNA to purine and pyrimidine bases which are resolved by paper chromatography and quantitated by uv spectrophotometry; (b) enzymatic hydrolysis of viral [^{32}P]-DNA to [^{32}P]-deoxyribonucleotides which are resolved by paper electrophoresis and quantitated by radioactivity; (c) calculation of percent guanine (G) + cytosine (C) of duplex DNA from the thermal denaturation temperature (T_m) (i.e., the temperature corresponding to the midpoint of the absorbance rise when native DNA is denatured by heating in solution); or (d) calculation of percent G + C from the buoyant density of duplex DNA. A linear relationship exists between the base composition of duplex DNA molecules and both their buoyant density (15) and T_m values (16) with only a few exceptions. Methods (c) and (d) are most commonly used now because they are accurate and require only several micrograms of DNA for each analysis. In duplex DNA molecules, since the molar percentage adenine (A) and thymine (T) are equal as are those of G and C, the one value, percent G + C conveniently expresses the molar percentage of all four DNA bases. All animal virion DNA's but one are duplex molecules.

2. RNA Base Composition

The base composition of viral RNA is usually determined by (a) acid hydrolysis to free bases or nucleosides and quantitation by uv spectrophotometry, or (b) more conveniently and accurately by alkaline hydrolysis of viral [^{32}P]-RNA to [^{32}P]-ribonucleotides which are resolved by paper electrophoresis and quantitated by radioactivity.

3. Molecular Weight and Conformation of Viral DNA

The molecular weight of duplex DNA is determined conveniently and accurately by measurements of sedimentation coefficient (S) or molecular length (17). Three methods are used to estimate S of nucleic acid molecules: (a) boundary sedimentation in the analytical ultracentrifuge, (b) zone sedimentation in the analytical ultracentrifuge, and (c) zone sedimentation of labeled DNA through stabilizing density gradients in the preparative ultracentrifuge. In (a) and (b), the molecular weight of duplex DNA can be calculated from the relationship established between S and molecular weight (18). In (c), S and the corresponding molecular weight are calculated by comparison with cosedimented DNA markers of known S and molecular weight.

Length measurements on viral DNA are made by electron microscopy, most often by the Kleinschmidt procedure (19), in which DNA molecules are spread on a protein film, deposited on carbon grids, and their lengths determined. The molecular weight of the DNA molecule is calculated from its measured length based on the mass per unit length of the duplex DNA (1.92×10^6 daltons/μ) (17). (One dalton is the weight of a hydrogen atom.)

Circular duplex DNA molecules are identified, and their lengths determined by electron microscopy, from which the molecular weight is calculated. Sedimentation analysis does not ordinarily distinguish between linear and circular DNA duplexes. But circular duplex viral DNA such as those of the papovaviruses can be detected by conversion to a rapidly sedimenting form, a "supercoil," upon denaturation (20).

The molecular weight of circular single-stranded DNA molecules is estimated from electron microscopic length measurements and that of linear single-stranded DNA molecules from S since the relationship between S values of single-stranded linear DNA molecules and their molecular weight is known (18).

4. Molecular Weight and Conformation of Viral RNA

All animal virus RNA's studied to date are single stranded except reovirus RNA which is double stranded. Molecular weight estimations of single-stranded RNA are less reliable than those of DNA because (a) the sedimentation coefficient of RNA is markedly dependent upon salt concentration to an extent which may vary for different RNA's, (b) empirical equations relating molecular weight and S are of unknown reliability for different RNA's, (c) electron microscopic length measurements of single-stranded RNA molecules and the relationship between length and molecular weight are not as yet well established.

5. Homology between Viral Nucleic Acids

Nucleic acid hybridization (or annealing) refers to the specific interaction in vitro between two polynucleotide strands by complementary base pairing. This reaction provides one of the few methods available for studying nucleotide sequence homology between viral nucleic acids. The degree of hybrid formation between the relatively simple viral nucleic acids appears to be highly specific and measurements agree well with genetic data (*20a*). Three types of hybridization reactions are in use, DNA–RNA, DNA–DNA, and RNA–RNA hybridization (*20b*).

1-2. Chemistry of DNA Animal Virions

The chemistry of DNA animal virions was reviewed in detail in 1965 (*21*). Since then it was recognized that the papovavirus group included two diverse groups of virions, the new group was termed picodnaviruses (*22*). Five animal virus groups that contain DNA are known, from the smallest to the largest: picodnaviruses, papovaviruses, adenoviruses, herpesviruses, and poxviruses.

The chemical composition of the members of DNA virus groups that have been analyzed are summarized in Table 1-3. As far as we know, the picodnaviruses, papovaviruses, and adenoviruses consist only of DNA and protein. The larger enveloped herpesviruses and poxviruses possess additional constituents.

Particle weight estimations of DNA animal virions are tabulated in Table 1-4. Virions range in size from 11×10^6 daltons for the picodna-

TABLE 1-3
Chemical Composition of DNA Animal Virions[a]

Virus group	DNA, %	Protein, %	Lipid, %	Others
Picodnavirus	19–25[b]	75–81	None	None
Papovavirus	12[c]	88[c]	None	None
Adenovirus	12–13[d]	87	None	None
Herpesvirus	Present	Present	Present	None
Poxvirus	5–6[e]	89	5.6	Cu, FAD, biotin

[a] Chemical analysis has been performed on only several virions within each group, except for the 31 human adenoviruses all of which have been analyzed (*23*).

[b] Only adeno-satellite viruses types 1 and 4 were analyzed (*24*).

[c] Only SV40 and Shope papilloma viruses were analyzed (*26, 27*).

[d] The 31 human adenoviruses (*23*) and the simian SV15 virus (*24*) were analyzed.

[e] Only the vaccinia virus was analyzed (*25*).

TABLE 1-4
Particle Weight of DNA Animal Virions

Virus group	Virion	Particle weight, daltons $\times 10^6$	Method
Picodnavirus	Adeno-satellite type 4[a]	11.1	DNA content and molecular weight[b]
Papovavirus	SV40[c]	17	Sedimentation coefficient and diffusion coefficient[d]
	Shope papilloma[e]	40	DNA content and molecular weight[b]
Adenovirus	Human adenovirion 3, 7, 12, 18, 2, 4, and 21[f]	175	DNA content and molecular weight[b]
Herpesvirus	—	—	—
Poxvirus	Vaccinia[g]	3200	DNA content and molecular weight[b]

[a] Estimated from data of Parks et al. (24).

[b] The molecular weight of viral DNA divided by the fractional DNA content of the virion provides a reliable estimate of the particle weight of the virion assuming that each virion contains one DNA molecule.

[c] From Ref. 26.

[d] Calculated from the sedimentation coefficient and diffusion coefficient of the virion using the Svedberg equation (14).

[e] Estimated from the DNA content of the virion (27) and the molecular weight of viral DNA (28).

[f] Estimated from the DNA content of the virion (23) and the molecular weight of viral DNA (29).

[g] Estimated from the DNA content of the virion (25) and the molecular weight of viral DNA (30).

viruses to 3×10^9 daltons for the poxviruses. The particle weight of SV40 virus is based on the sedimentation coefficient and diffusion coefficient of the virion (26). The particle weights of the other DNA animal virions were calculated from the DNA content of the virion and the molecular weight of viral DNA.

The molecular weight and conformation of the DNA molecules extracted from DNA animal virions, as determined from sedimentation analysis, molecular-length measurements, and band-width analysis, are summarized in Table 1-5. All DNA viruses thus far analyzed possess one molecule of DNA per virion. Viral DNA ranges in size from $2–5 \times 10^6$ daltons for the small picodnaviruses and papovaviruses to $160–240 \times 10^6$ daltons for the large poxviruses. The poxvirus genome may code for 160 to 500 different proteins, assuming that the size of

TABLE 1-5

Molecular Weight and Conformation of Animal Virus DNA Molecules

Virus group and virion	$S_{20,w}$, svedbergs	Length, μ	Conformation	Molecular weight, daltons $\times 10^6$
Picodnavirus group				
Adeno-satellite type 1	15.5 (31)	—	Linear duplex	3.5 (S)[a]
Adeno-satellite type 4	15.7 (24)	1.5 (24)	Linear duplex	3.2 (S), 2.9 (L)[a]
Minute virus of mice			Single stranded (32)	—
Kilham rat virus	12.9 (38)	0.83 (38)	Linear duplex	2.0 (S), 1.6–2.1 (L)
Papovavirus group				
Polyoma	21[b] (33)	1.56 (34)	Circular duplex	2.9–3.4 (S)
SV40	21[b] (53)	1.40 (53)	Circular duplex	2.3–2.5 (S) (L)
Rabbit kidney vacuolating		1.44 (34a)	Circular duplex	2.8 (L)
Shope papilloma	28[b] (35)	2.5 (28)	Circular duplex	5.0 (L)
Human papilloma	28[b] (36)	—	Circular duplex	5.3 (S)
Bovine papilloma		2.5 (37)	Circular duplex	4.9 (L)
Adenovirus group				
Human types 2, 3, 4, 7, 12, 18, 21	30–32 (29)	11–13 (29)	Linear duplex	20–25 (S) (L)
Herpesvirus group				
Herpes simplex	44 (39)	—	Linear duplex	68 (S)
Pseudorabies	44 (39)	—	Linear duplex	68 (S), 70 (B)[a] (40)
Human cytomegalovirus		—	Linear duplex	64 (B)[a] (41)
Infectious bovine rhinotracheitis	39 (39)	—	Linear duplex	54 (S)
Equine abortion	52 (42)	48 (42)	Linear duplex	94 (S), 92 (L), 92 (B) (42)
LK	49 (39)	—	Linear duplex	84 (S)
Poxvirus group				
Vaccinia		87 (30), 78 (43)	Linear duplex	150–170 (L)
Fowlpox	80 (44)	100 (45)	Linear duplex	200–240 (S), 200 (L)

[a] Abbreviations: (S) calculated from sedimentation coefficient; (L) calculated from molecular length determined by electron microscopy; (B) from band width determined by equilibrium density-gradient centrifugation.
[b] S value of circular duplex.

the average viral gene is 0.5–1 $\times$ 10^6 daltons. For example, 10^6 daltons of DNA contains about 1500 nucleotide pairs which would code for 500 amino acids (coding ratio 3:1), corresponding to a protein of molecular weight about 50,000; 0.5 $\times$ 10^6 daltons of DNA would code for a protein of molecular weight about 25,000. All animal virus DNA's so far studied are duplex molecules except that of the minute virus of mice which is single stranded (32). Papovavirus DNA's are circular molecules, the other four DNA virus groups possess linear DNA molecules. The circular DNA's of the papovaviruses are infectious, as is the recently reported linear DNA molecule of Kilham rat virus (a picodnavirus); the DNA's of all other animal virions have not been isolated in an infectious state.

The base compositions of animal virus DNA's range from 35% G + C for the poxviruses to 74% G + C for some of the herpesviruses (Table 1-6).

The amino acid composition of only a few animal virions is known (Table 1-7). Of the DNA viruses, the amino acid composition of the Shope papilloma virus and five human adenoviruses of "highly oncogenic," "weakly oncogenic," and "non-oncogenic" nature have been determined. All the common amino acids are present, no distinctive features were found. The amino acid content of DNA and RNA animal

TABLE 1-6
Base Composition of Animal Virus DNA Molecules

Virion	Molar per cent G + C
Picodnavirus group	
Adeno-satellite types 1 and 4	54–62 (24, 31)
Minute virus of mice (single stranded)	48 (32)
Kilham rat virus	44 (38)
Papovavirus group	
Polyoma, SV40, Shope papilloma, canine, bovine, human papilloma, rabbit kidney vacuolating	41–49 (21, 34a, 68)
Adenovirus group	
Human adenovirus types 1–31	48–61 (23)
Simian adenovirus SA7 and SV15	58–60 (46, 47)
Herpesvirus group	
Equine abortion, LK, and human cytomegalovirus	56–58 (21)
Herpes simplex, pseudorabies, and infectious bovine rhinotracheitis	68–74 (21)
Poxvirus group	
Vaccinia, rabbitpox, cowpox, ectromelia, and fowlpox	35–40 (21)

Amino Acid Composition of DNA and RNA Animal Virions

Moles per 100 moles of amino acid recovered

Amino acids	Adenovirus type[a]					KB cells[a]	Shope papilloma[b]	Polio-virus[c]	Encephalo-myocarditis[d]	Maus-Elber-feld[e]	Foot-and mouth disease[f]
	2	4	7	12	18						
Aspartic	11.8	11.7	11.7	11.5	11.2	9.4	10.8	11.9	9.3	10.4	10.3
Threonine	6.9	7.3	7.9	7.2	6.5	5.5	5.6	9.1	8.6	9.9	10.7
Serine	6.7	7.4	7.3	7.0	6.9	5.9	6.3	7.0	7.9	8.3	6.4
Glutamic	9.0	7.3	8.3	8.8	8.8	12.0	12.0	7.7	8.7	8.0	8.6
Proline	7.2	6.2	6.1	7.4	6.4	5.2	6.4	7.2	7.5	8.0	5.9
Glycine	7.8	7.6	7.8	8.4	8.0	7.4	5.9	6.8	8.7	8.4	8.1
Alanine	9.0	10.0	8.8	9.6	9.4	8.1	6.1	7.8	7.1	7.9	8.7
Valine	6.1	6.1	5.9	5.8	6.2	7.2	6.3	7.2	6.8	6.0	6.7
1/2-cystine	0.3	0.5	0.3	0.5	0.6	1.1	3.3	0.8	1.5	1.1	0.8
Methionine	2.3	3.4	3.4	2.6	2.7	2.4	1.9	1.5	1.9	1.8	1.5
Isoleucine	3.4	3.4	3.6	3.3	3.5	5.2	3.9	4.8	3.9	4.0	3.2
Leucine	7.4	6.8	6.9	7.1	7.4	8.9	7.9	8.5	8.3	7.7	7.4
Tyrosine	4.4	4.9	4.8	4.1	4.5	3.0	4.2	3.9	3.8	4.2	4.9
Phenylalanine	3.8	3.5	3.7	3.7	3.7	3.9	3.9	4.4	5.1	5.6	3.8
Ammonia[g]	21.4	18.8	18.1	22.0	22.7	9.8	—	—	—	—	—
Lysine	4.4	4.2	4.5	4.1	4.7	7.6	7.1	4.7	4.5	3.8	4.6
Histidine	1.6	1.5	1.5	1.6	1.7	2.1	1.9	2.4	1.4	1.9	3.4
Arginine	7.9	8.3	7.6	7.3	7.9	5.2	5.2	4.7	3.6	3.5	4.0
Tryptophane	1.16	—	1.43	1.37	—	0.45	1.2	—	—	—	1.0

[a] From Ref. 48.
[b] Calculated from Ref. 27.
[c] From Ref. 49.
[d] From Ref. 50.
[e] From Ref. 51.
[f] From Ref. 52.
[g] Ammonia is not included in the total.

TABLE 1-8

Number of Polypeptide Components in
RNA and DNA Animal Virions

Virus group	Virion	Number of polypeptides
Papovaviruses	SV40	3 (*53*)
	Polyoma	1 (*53a*)
Adenoviruses	Ad 2	10 (*54*)
Poxvirus	Vaccinia	17 (*55*)
Picornaviruses	Poliovirus	4 (*56*)
	Maus-Elberfeld	3–4 (*57*)
	Encephalomyocarditis	3–4 (*57–59*)
	Mengovirus	3–4 (*57*)
	Foot-and-mouth disease	1 (*52*)
Arbovirus	Sindbis	2 (*12*)
Myxovirus	Influenza A and B	3 (*60, 61*)
Leukemiavirus	Mouse leukemia virus (Rauscher)	2 (*136*)
	Rous sarcoma virus + Rous associated virus	4–6 (*196*)
	Avian myeloblastosis	4–6 (*196*)
	Avian myeloblastosis	6–8 (*195*)

virions are similar in general, except for the higher arginine content of the adenoviruses.

Estimates of the number of different polypeptide components comprising animal virions has usually been analyzed by polyacrylamide gel electrophoresis of dissociated virus particles (Table 1-8). There is one polypeptide in the small polyoma virion and a minimum of 17 polypeptides in the large vaccinia virion.

The chemical and physical properties of individual virions and their constituents are described below. Newer findings are discussed in more detail than the older literature which are fully described in reviews (*21, 62, 63*). For further details, the reader is referred to chapters on specific virus groups.

A. PICODNAVIRUSES

The name picodnavirus (pico = small, DNA, virus) was suggested (*22*) for this group of small DNA viruses (diameter 18 to 24 mμ) by analogy with the small RNA viruses, the picornaviruses. Included in this group are the hamster osteolytic viruses (H-1 and H-3), Kilham rat virus, X-14 rat virus, and the adeno-satellite viruses of human, simian, and murine origin. The adeno-satellite viruses are defective

because they replicate only within cells infected with adenoviruses
(*64*); four serotypes have been reported (*65*), two of which infect
man (*66*).

Of the picodnaviruses, only adeno-satellite types 1 and 4 have been
chemically characterized. The virus was grown in monkey kidney
cells and purified by fluorocarbon treatment and CsCl equilibrium
density-gradient centrifugation. Adeno-satellite type 4 contains 27%
DNA and 73% protein and type 1 contains 19% DNA and 81%
protein (*24*), assuming, as is most likely the case, that DNA and
protein are the sole constituents of the virion. The total phosphorous
content of types 1 and 4 is accounted for by the DNA content and
no RNA was detected by the orcinol reaction.

Very little is known about the structure of picodnaviruses. They
exhibit cubic symmetry but the number of capsomeres is difficult to
determine, different investigators estimating either 12 or 32 capsomeres.
Based on the DNA content of the virion (27%) and the molecular
weight of the viral DNA (3.0×10^6), the particle weight of adeno-
satellite type 4 is estimated at 11.1×10^6 daltons (Table 1-4).

DNA was extracted from purified adeno-satellite types 1 and 4 by
treatment with a proteolytic enzyme, sodium dodecyl sulfate (SDS),
followed by phenol extraction (*24, 31*). The complementary base
pattern of type 1 DNA (A = T and G = C) and the sharp melting
profiles of types 1 and 4 DNA's provide strong evidence for a duplex
DNA structure. The per cent G + C of type 1 DNA is 54 ([32]P-
nucleotide analyses), 55 (based on T_m), and 58 (based on buoyant
density) while that of type 4 DNA is 58 (T_m) and 62 (buoyant
density). The lack of agreement between per cent G + C calculated
from T_m and buoyant density is unusual and occurs also with the
DNA's of the "weakly" oncogenic human adenoviruses (*23*).

The DNA's of adeno-satellite types 1 and 4 are linear duplex molecules of molecular weight $2.9–3.5 \times 10^6$ daltons, as shown by electron
microscopy of native DNA and analytical zone sedimentation of native and alkali-denatured DNA (Table 1-5). Since papovaviruses
possess DNA genomes in the same size range as those of adeno-satellite viruses and are not defective, it is probable that the defectiveness
of the adeno-satellite viruses is a function of the type of information
in the genome and not its size. In contrast, the tobacco necrosis satellite virus is defective presumably because of the small size of its RNA
(molecular weight of 400,000) which would code for a maximum of
two proteins (*67*).

The minute virus of mice (MVM) was reported to possess a single-

stranded DNA molecule (*32*) based on two findings: (a) an increase
in ultraviolet absorption (hyperchromicity) occurred upon reaction
of purified virus with formaldehyde; the same characteristic is observed with the single-stranded DNA of bacteriophage ϕX174, presumably because of the reaction of formaldehyde with unpaired, single-
stranded DNA; (b) duplex DNA molecules are separated by alkali
denaturation into their constituent single strands which possess a
higher buoyant density than native DNA; but MVM DNA has the
same buoyant density before and after alkali denaturation, suggesting that it is single stranded in the native state. The size and conformation (whether linear or circular) of MVM DNA is not known.
MVM is the only animal virion found to possess a single-stranded
DNA molecule; the further study of MVM DNA and its replication
should be of considerable interest.

The DNA of Kilham rat virus is a linear duplex molecule with
a molecular weight of 2×10^6 (*38*), sufficient genetic information to
code for 2 to 4 proteins, thus being the smallest duplex DNA molecule
known (Table 1-5). Viral DNA contains 43 to 45% G + C (*38*), and
is infectious, being the only infectious linear animal virus DNA known.

B. PAPOVAVIRUSES

This group of small DNA viruses includes the highly oncogenic
mouse polyoma and simian SV40 viruses, the wart-producing papilloma viruses of the rabbit (Shope papilloma), human, bovine, canine,
and other species, and the nontumorigenic mouse K virus.

Most viruses in this group have been purified during the past several
years. Based on chemical analysis, SV40 (*26*) and Shope papilloma
(*27*) virions contain 12–12.5% DNA and 88% protein; the Shope
papilloma virion contains no detectable polyamines (*27*).

Papovavirions are icosahedral particles containing 72 capsomeres
and form two distinct subgroups on the basis of the size of the virion
and viral DNA: (a) human, bovine, canine, and Shope papilloma
virions have diameters of 55 mμ, and (b) polyoma and SV40 virions
have diameters of 41–45 mμ (*26, 68*). These relatively small differences
in diameter reflect twofold differences in particle weight. The sedimentation coefficients of Shope papilloma, canine papilloma, and
human papilloma virions are 295S–300S (*68*) while that of SV40 is
219S (*26*). The SV40 virion has a molecular weight of 17.3×10^6 calculated from a sedimentation coefficient of 219S, a diffusion coefficient

of 0.90×10^{-7} cm²/sec, and a partial specific volume of 0.658 (*26*)
(Table 1-4). This value derived from hydrodynamic measurements
agrees with that calculated from the DNA content of the virion
(12.5%) and the molecular weight of viral DNA (2.3×10^6). The
Shope papilloma virion has a molecular weight of about 40×10^6,
twice that of SV40, as estimated from the DNA content of the virion
(12%) and the molecular weight of viral DNA (5×10^6) (Table 1-4).

Circular, superhelical duplex DNA molecules are found in many
papovavirions, including polyoma, SV40, human papilloma, Shope
papilloma, and bovine papilloma (Table 1-5). Prominent features of
these circular viral DNA molecules are (a) conversion to a rapidly
sedimenting "supercoil" by denaturation, (b) the rapid renaturation
of denatured DNA, and (c) the twisted circular appearance in electron
micrographs. Viral DNA preparations extracted from papovavirions
also contain smaller amounts of slower sedimenting "nicked" circular
forms (one strand has a discontinuity). A third DNA form, observed
frequently in polyoma DNA preparations, is cell DNA fragments
enclosed in some virions (*69, 70*) by an unknown mechanism which
is of interest for understanding the assembly of virions. The incorporation of cellular DNA in polyoma virions suggests the need to critically
analyze DNA molecules of other virions for cellular components which
may represent mistakes in virus assembly.

The molecular weight of polyoma DNA is 2.9 to 3.4×10^6 based on
its sedimentation coefficient and that of SV40 DNA is 2.3 to 2.5×10^6
based on S and molecular length (Table 1-5). Human papilloma and
Shope papilloma DNA's have molecular weights close to 5×10^6,
about twice that of polyoma and SV40 DNA's (Table 1-5). The base
composition of papovavirus DNA's range from 41 to 49% G + C
(Table 1-6), values similar to those of mammalian cell DNA which
contains 42 to 44% G + C.

Nucleotide sequence relationships between several papovavirus
DNA's have been studied by DNA–RNA hybridization measurements.
No nucleotide sequence homology was detected between human papilloma and Shope rabbit papilloma when viral DNA was annealed with
radioactive synthetic viral complementary RNA made on the heterologous viral DNA template with RNA polymerase (*36*). Likewise no
homology was detected between Shope papilloma and polyoma viruses
(*35*) nor between SV40 and polyoma viruses (*70a*). In contrast, the
adenoviruses share from 10–25% of their nucleotide sequences (see
next section).

The SV40 virion contains three different polypeptides of the same average molecular weight, 16,000 (53), two are components of the viral capsid while the third appears to be associated with viral DNA (71). The amino acid composition and amino and carboxyl end groups have not been reported.

C. Adenoviruses

The members of this large group of medium-sized, ether-resistant DNA viruses share several properties including a near-identical fine structure, common group-specific antigen(s), and common DNA nucleotide sequences. At least 55 different adenovirus types are known based on the specificity of neutralization and hemagglutination inhibition (72). These may be divided into six subgroups, 31 human (types 1 to 31), 18 simian, 3 bovine, and one each of canine, murine, and avian adenoviruses (Table 1-2A).

1. Chemical Composition of Adenoviruses

Of the adenovirus group, only the human adenoviruses and simian SV-15 have been chemically characterized. The 31 human adenoviruses were isolated in highly purified form and their chemical and physical properties studied by Green and Piña (89–92, 23). The DNA, protein, and phosphorus content of Ad 1 to 31 and the dry weight of Ad 3, 5, 6, 7, 10, 15, 21, 24, 26, and 28 were determined simultaneously (93). The sum of the DNA and protein weights was equal to the dry weight of the virus, supporting the conclusion that human adenoviruses consist of only DNA and protein as major constituents. The amount of phosphorus agreed with the DNA content indicating that RNA and phospholipid are absent. A negative orcinol reaction and the absence of uracil in viral DNA hydrolysates provided further evidence for the absence of RNA from adenovirions.

The 31 human adenovirions range in DNA content from 11.6–13.5% DNA as shown by chemical analysis and buoyant density measurements (23); SV-15 contains 13.7% DNA (24). Ad 2, 4, and 21 have the highest DNA content of the human adenovirions (13–13.5%), the "highly oncogenic" A group, Ad 12, 18, and 31, have the lowest DNA content (11.6–12.1%), whereas the remaining 25 adenovirions have intermediate DNA contents. The smaller molecular weight of Ad 12 DNA (90) accounts for the 9% lower relative DNA content of Ad 12 compared to Ad 2. This difference represents about 2×10^6 daltons of

DNA which could code for four proteins of average molecular weight, 25,000. The significance of the lower DNA content of the highly oncogenic adenoviruses is not known.

2. Oncogenicity of Adenoviruses

The properties of oncogenic human adenoviruses are of particular interest since they are the first "human" viruses shown to possess neoplastic properties, i.e., they are carcinogenic in newborn rodents (73, 74). Oncogenic adenoviruses of simian (75), bovine (76), and avian (77) origin also have been described.

As summarized in Table 1-9 and 1-10, there are three distinctly different groups of oncogenic and transforming human adenoviruses (78, 79): "highly oncogenic" group A, which includes Ad 12, 18, and 31; "weakly oncogenic" group B, which includes Ad 3, 7, 11, 14, 16, and 21; and "nononcogenic" but transforming group C, which includes

TABLE 1-9

Oncogenic and Transforming Human Adenoviruses: Base Composition of Viral DNA and Relationship between Virus-Specific RNA's in Transformed Cells

Adenovirus			Viral DNA, % G + C[a]	Viral DNA–mRNA, % homology[b]
Group	Members	Oncogenicity		
A	Ad 12, 18, 31	"Highly oncogenic"[c] in newborn hamsters	48–49	30–60
B	Ad 3, 7, 11, 14, 16, 21	"Weakly oncogenic"[c] in newborn hamsters (all but Ad 11)	49–52	40–100
C	Ad 1, 2, 5, 6	"Nononcogenic" in newborn hamsters but morphologically transform rat embryo cells in vitro[d]	57–59	90–100

[a] Piña and Green (23).

[b] Hybridization of virus-specific RNA from adenovirus induced tumor and transformed cells with heterologous DNA in this group, % of homologous hybridization (80–83).

[c] Highly oncogenic adenoviruses induce tumors in a large proportion of newborn hamsters within two months after injection with purified virus; "weakly oncogenic" in a small proportion after 4–18 months.

[d] Freeman et al. (79).

TABLE 1-10

DNA–DNA Homology Relationships between Adenovirus DNA's

DNA pair hybridized	% homology
Group A $\times$ Group A (Ad 12, 18, 31)	80–85[a,b]
Group B $\times$ Group B (Ad 3, 7, 11, 14, 16, 21)	70–100[c]
Group C $\times$ Group C (Ad 1, 2, 5, 6)	85–95[b]
Group A $\times$ Group B	10–25[c]
Group A $\times$ Group C (Ad 12 only)	20–24[b]
Group B $\times$ Group C (Ad 7 only)	24–26[b]
Simian SA-7 $\times$ Group A (Ad 12 only)	10–12[b]
Simian SA-7 $\times$ Group B (Ad 7 only)	19[b]
Simian SA-7 $\times$ Group C (Ad 2 only)	23[b]
Simian SV-15 $\times$ Group A (Ad 12 only)	9–11[b]
Simian SV-15 $\times$ Group B (Ad 7 only)	16[b]
Simian SV-15 $\times$ Group C (Ad 2 only)	19[b]

[a] Lacy and Green (86).
[b] Piña and Green (47).
[c] Lacy and Green (87, 88).

Ad 1, 2, 5, and 6. Group A and B adenoviruses induce tumors in newborn rodents, whereas group C adenoviruses have not yet done so. Group C adenoviruses transform rat embryo cells *in vitro* with regard to morphological properties but not neoplastic properties since group C transformed cells do not induce tumors when injected into newborn rats (79). In contrast, cells transformed by group A and B adenoviruses induce tumors in newborn or adult animals. Cells transformed by group A, B, and C adenoviruses are free of infectious virus but synthesize relatively large amounts of virus-specific RNA molecules transcribed from a small fraction of the viral genome (80–83).

The members of group A are closely related, sharing 80–85% of their DNA nucleotide sequences as determined by DNA-DNA hybridization measurements; group B members share 70–100%, and group C members share 85–95% of their nucleotide sequences (Table 1-10). Although the virions within each group are very closely related, those in different groups share only 10–26% of their nucleotide sequences (A $\times$ B, A $\times$ C, B $\times$ C) (Table 1-10).

Recently the relationships between the DNA's of oncogenic simian adenovirus SA-7 and nononcogenic simian adenovirus SV-15 and those of the group A, B, and C adenoviruses were determined (Table 1-10) (47). Between 9–23% of the nucleotide sequences of the simian adenoviruses are shared with the human adenoviruses but oncogenic

SA-7 is no more closely related to the oncogenic human adenoviruses than is nononcogenic SV-15. These DNA–DNA homology measurements suggest that a basic core of genetic information, amounting to 10–26% of the viral genome, is shared by both human and simian adenoviruses. In contrast, no homology was found between the DNA's of the different papovaviruses (*35, 36, 70a*).

3. *Adenovirus DNA*

The 31 human adenovirus DNA's were prepared in 60–80% yields as single homogeneous molecules by digestion of purified virions with papain, treatment with SDS, and deproteinization with phenol (*23*). Without prior treatment with a proteolytic enzyme, viral DNA remains associated with a protein component(s) (*92*) during phenol extraction or equilibrium sedimentation in cesium chloride gradients. Recent studies have suggested that viral DNA may be associated with arginine-rich internal proteins (*94, 95*).

The 31 human adenovirus DNA's are double stranded and range in $G + C$ content from 48–61% (*23*) as shown by T_m and buoyant density measurements and direct chemical analysis of viral DNA. Good agreement was found between these three measurements except for the "weakly oncogenic" group B adenoviruses which possess DNA molecules with higher buoyant densities than predicted from their base composition. The oncogenic human adenoviruses (group A and B) have 48–52% $G + C$ (Table 1-9), values closer to mammalian cell DNA (42–44% $G + C$) than those of the nononcogenic human adenoviruses which possess 55–61% $G + C$. The oncogenic papovaviruses have a similar low $G + C$ content, 41–49% (Table 1-6). These findings suggest that DNA tumor viruses may possess DNA regions homologous to segments of the host-cell chromosomes by analogy with temperate bacteriophage which have DNA base compositions similar to that of their host cell and share nucleotide sequences with host-cell DNA (*84, 85*).

There is now an exception to the correlation between low $G + C$ content and oncogenicity of DNA viruses: The DNA of oncogenic simian adenovirus SA-7 contains a much higher $G + C$ content, 58–60% (*46, 47*) than that of the oncogenic human adenoviruses and papovaviruses. However, this exception is not too surprising, for if a low $G + C$ content is required for oncogenesis more than likely only a small part of the relatively large adenovirus genome of molecular weight 23×10^6 would be involved. This interpretation is strengthened by studies on the base composition of purified virus-specific RNA's from adenovirus tumor and transformed cells (82, 83) which indicate that only certain adenovirus DNA regions with a low G + C content are transcribed in adenovirus-transformed cells.

Since the oncogenic papovavirus DNA's are circular duplexes, it was important to establish whether this might be a common feature of oncogenic DNA viruses. The intact DNA molecules isolated from six different human adenoviruses, "highly oncogenic" Ad 12 and 18, "weakly oncogenic" Ad 7 and 21, and "nononcogenic" Ad 2 and 4 were shown to be linear molecules, 11 to 13 μ in length, as determined by protein-film electron microscopy (29) (Table 1-5). The molecular weights calculated from the sedimentation coefficients of native DNA agreed with the observed molecular lengths (Table 1-5). Further, the sedimentation rates of alkali-denatured DNA (conditions under which single strands of linear duplex DNA are separated) predict molecular weights that are one-half those of the duplex adenovirus DNA molecules. Sedimentation under alkaline conditions revealed no detectable "supercoil"—that compact structure formed by the denaturation of an uninterrupted circular helix. Thus three independent measurements provide conclusive evidence that the DNA's isolated from the human adenoviruses are double-stranded, uninterrupted linear molecules of molecular weight 20 to 25 $\times$ 10^6 daltons. Although the DNA molecules isolated from the adenovirions are linear molecules, the conformation of DNA within the virion or within the cell during virus replication is not known.

Two remarkable properties of some bacteriophage DNA molecules, (a) circular permutation and (b) terminal redundancy, have been discovered recently. (a) The DNA molecules extracted from bacteriophage T2, T4, and P22 are circularly permuted (17), i.e., they consist of a population of molecules, each with the same linear sequence of nucleotides but beginning and ending at different nucleotides. Upon denaturation and annealing, a circularly permuted population of DNA molecules will form circular molecules; helical regions will form between strands originating from different duplexes. (b) The DNA molecules extracted from bacteriophage T2, T3, T7, and P22 are terminally redundant, i.e., the ends of each molecule contain the same sequence for a region amounting to about 1–3% of the length of the molecule. After partial digestion with exonuclease III which removes nucleotides stepwise from the 3' end of both DNA strands, followed by annealing, terminally redundant molecules form circular molecules.

Of animal virus DNA's, only the adenovirus nucleic acids have been examined with regard to circular permutation or terminal redundancy. In preliminary experiments, adenovirus DNA molecules were tested for terminal redundancy by partial digestion with exonuclease III followed by annealing. Only 1–2% of DNA molecules were rendered circular in the case of Ad 2, 4, 7, and 18, and none in the case of Ad 12 and 21 (29). The significance of this low incidence of artificial circles is uncertain because 10–15% of bacteriophage T2 DNA molecules will form circles under similar conditions (96).

Preliminary experiments testing adenovirus DNA's for circular permutation also have yielded inconclusive results: Four per cent of Ad 2 DNA molecules formed circles after denaturation and annealing under conditions where 34% of P22 molecules formed circles (97).

4. Persistent Viral Genes in Transformed Cells

The distant relationship (10–26% homology) between the group A, B, and C adenoviruses does not *a priori* rule out the possession of common genes responsible for cell transformation since the shared sequences may amount to from 2 to 6 genes. But recent studies virtually eliminate this possibility. Labeled virus-specific RNA molecules isolated from tumor and transformed cells induced by group A, B, and C adenoviruses were annealed with the DNA's of the group A, B, and C adenoviruses (80–83) (Table 1-9): group A virus-specific RNA's cross hybridized with group A DNA's with 30–60% efficiency but not with group B and C DNA's; similarly group B and C virus-specific RNA's hybridized within groups to the extent of 40–100 and 90–100%, respectively, but not between groups. Thus, comparative hybridization measurements did not detect ribonucleotide sequences common to the virus-specific RNA molecules synthesized in cells transformed by the A, B, and C adenoviruses, leading to the conclusion that different viral genes are involved in transformation by viruses of these three groups.

5. Adenovirus Proteins

The icosahedral adenovirion consists of 240 nonvertex capsomeres called "hexons" and 12 vertex capsomeres called "pentons," each penton consisting of a base and a "fiber" (98). The following considerations suggest that additional protein(s) are components of the virion: (a) Viral DNA remains associated with a protein(s) after disruption of the virion with SDS unless the virion is digested first with a proteolytic enzyme (92). (b) About 18% of the total protein of acetone-disrupted Ad 2 virions remains associated with viral DNA (94). (c) Serological evidence for a possible internal protein has been obtained (95). (d) The molecular weight of the hexon has been estimated at 400,000 (72) and the molecular weight of adenovirus DNA is 20–25 × 10⁶ (29). The sum of the 240 capsomeres and the viral DNA molecule is therefore about 120 × 10⁶ daltons (the 12 pentons contribute less than 5% to the mass of the virion). This accounts for only 70% of the mass of the virion which is estimated at 175 × 10⁶ daltons (Table 1-4). Thus 30% of the virion is unaccounted for, part or all of which may represent the proposed "internal protein(s)" of the adenovirion.

The amino acid compositions of purified Ad 2, 4, 7, 12, and 18 (48) (Table 1-7) are almost identical, and grossly similar to those of other animal virions except that the arginine content of adenovirions is twice that of other animal virions. Furthermore, the Ad 2 hexon, recently obtained in highly purified form (99), has an amino acid composition very similar to that reported for the intact virion (48) with one notable exception: The arginine content of the hexon is only 60% of the intact virion. Since the hexon comprises 95% of the proteins of the viral capsid, the lower arginine content of the hexon compared to the intact virion suggests the presence of an arginine-rich internal protein(s).

The N-terminal amino acid end groups of Ad 2, 4, 5, 6, 18, 21, and 27 are alanine and glycine, in molar ratios ranging from 2.5 to 5.5 alanine:glycine (94). Treatment with acetone disrupts Ad 2 into two fractions, one of which is associated with viral DNA during sedimentation through sucrose gradients and during electrophoresis on cellulose acetate. The DNA-associated protein fraction comprises 18% of the virion protein and contains most of the N-terminal amino acids; the bulk of the viral protein has no free N-terminal amino acids.

Polyacrylamide gel electrophoresis of Ad 2 disrupted by treatment with SDS and mercaptoethanol resolved 10 polypeptides (54); the relationship between the viral polypeptides and the morphological components of the virion is not yet established.

D. HERPESVIRUSES

Members of the herpesvirus group are medium sized, enveloped DNA virions that contain essential lipid (i.e., are ether sensitive) and share several biological and morphological features. The possible association of herpes-like particles (HLP) with various diseases has heightened interest in these viruses. Serological and clinical data have provided reasonably conclusive evidence that a HLP causes infectious mononucleosis (*100*). In addition, HLP have been detected frequently by electron microscopic examination of cell cultures derived from Burkitt's tumors of humans (*101*). Evidence for the possible association of a herpes simplex strain with cervical cancer has been published (*102, 102a*), and HLP has been isolated from frog tumors (as well as normal frog tissues) (*103*). However, because of the frequent occurrence of HLP in "normal" cell cultures derived from various sources, a possible "passenger" role for HLP is difficult to exclude.

The icosahedral herpesvirions consist of a DNA-containing core, surrounded by a capsid consisting of 162 hollow capsomeres and an outer lipid-containing envelope. The core and capsid are assembled in the nucleus but the envelope is derived from the nuclear membrane during the final stages of maturation (*104*). The buoyant density of the herpes simplex virion may vary when grown in different host cells (*105*), probably reflecting differences in the composition of the cell-derived envelope.

The chemistry of the herpesvirus group has been neglected. Most members have been partially purified, including herpes simplex, pseudorabies, equine abortion, infectious bovine rhinotracheitis, human cytomegalovirus, and LK virus. The presence of DNA, protein, and lipid has been demonstrated by several investigators but the amount and type of each component have not been established.

Viral nucleic acid has been extracted from partially purified preparations of most herpesvirions and its DNA nature established by the diphenylamine color reaction (specific for deoxyribose) or by the characteristic buoyant density of DNA. The double-stranded nature of several herpesvirus DNA's has been proven. The molecular weights of herpesvirus DNA's were estimated by three procedures: (a) calculation from the sedimentation coefficient, a reliable method for DNA's of molecular weight less than 40×10^6 but of uncertain application to the large herpesvirus DNA's, (b) calculation from the DNA band width after sedimentation to equilibrium in cesium chloride gradients, a procedure requiring an empirical correction (*17*), and (c) electron microscopic length measurements. Molecular weights ranging from 54–94×10^6 daltons were found for the DNA's of infectious bovine rhinotracheitis, herpes simplex, pseudorabies, LK, human cytomegalovirus, and equine abortion virus (Table 1-5). The best characterized herpesvirus DNA, equine abortion virus DNA, has a molecular weight of 92–94×10^6 daltons derived from S, molecular length, and band-width measurements.

The base compositions of herpesvirus DNA's suggest the existence of distinct subgroups (Table 1-6): (a) herpes simplex, pseudorabies, and infectious bovine rhinotracheitis with an unusually high $G + C$ content, 68–74%, and (b) equine abortion, LK, and human cytomegalovirus with 56–62% $G + C$.

E. POXVIRUSES

These large, complex animal virions possess a common nucleoprotein antigen and multiply exclusively in the cell cytoplasm. Vaccinia virus, the first animal virus to be purified and chemically analyzed (25), contains 89% protein, 5–6% DNA, and 5.6% lipid (1.4% cholesterol, 2.2% phospholipid, and 2.2% neutral fat). Small amounts of copper, flavine-adenine-dinucleotide, and biotin were detected also, but the metabolic functions of these materials, if any, are not known (21).

The origin of the RNA polymerase which transcribes poxvirus DNA within the cell was a mystery until recently. Since the poxvirus genome is transcribed in the cell cytoplasm, it seemed unlikely that the host-cell RNA polymerase which functions normally in the cell nucleus was involved. But the induced synthesis of a virus-coded RNA polymerase was difficult to envision because the induction process itself would require the functioning of RNA polymerase. The answer to this dilemma was remarkably simple. Poxvirus particles possess their own RNA polymerase: DNA-dependent RNA polymerase activity, differing in properties from the host-cell enzyme, has been demonstrated recently in highly purified rabbitpox (3) and vaccinia (4). Purified vaccinia virus also has an enzyme(s) which removes the terminal phosphate from adenosine triphosphate, guanosine triphosphate, uridine triphosphate, and cytidine triphosphate (106); the role of these enzyme activities in virus growth is unknown. These discoveries raise the question of whether other animal virions may possess specific enzymes of macromolecular biosynthesis.

As do all large DNA molecules, poxvirus DNA requires special handling during isolation and experimentation to avoid breakage by shearing forces. Careful analysis has shown that vaccinia virus DNA has a molecular weight of 150–170×10^6 daltons, as determined by electron microscopy (30) and zone sedimentation (30, 43). Fowlpox

DNA which has a molecular weight of $200–240 \times 10^6$ daltons, as shown both by analytical centrifugation and electron microscopy (*44, 45*) (Table 1-5), is the largest viral DNA molecule known. Molecular weights of poxvirus DNA's are consistent with the DNA content of the virion calculated from the fractional DNA content and estimated molecular weight of the virion (*107, 108*).

Poxvirus DNA molecules have similar base compositions, ranging from 35–40% G + C for the DNA's of vaccinia, rabbitpox, cowpox, ectromelia, and fowlpox virions (Table 1-6).

Extensive electron microscopic studies (*109*) have disclosed a very complex structure for the vaccinia virion, still far from completely understood, consisting of a centrally located core containing the viral genome, two lateral bodies on either side of the core, and a protein coat surrounded by an outer membrane. As expected from the complexity of the virion, a large number of different proteins are present, 17 polypeptides being resolved by polyacrylamide gel electrophoresis of the dissociated vaccinia virion (*55*). The location of some polypeptides within the virion has been established (*55*).

1-3. Chemistry of RNA Animal Virions

The RNA virions utilize RNA as their genetic material instead of DNA, the genetic material for most other life forms. The chemistry of RNA animal virions was reviewed in 1965 (*62*). At that time four RNA animal virus groups were recognized, the picornaviruses, reoviruses, arboviruses, and myxoviruses (*6*). Since then three additional groups have been proposed (*9, 10*): (a) the paramyxoviruses, for example, mumps and Newcastle disease virus, which were included initially with the myxoviruses, (b) the newly recognized rhabdoviruses, rabies and vesicular stomatitis virus, and (c) the leukemiaviruses, which include all the RNA tumor viruses.

The chemical composition of members of RNA animal virus groups is summarized in Table 1-11. The picornaviruses and reoviruses contain only RNA and protein. The few arboviruses that have been characterized contain RNA, protein, and lipid. The myxoviruses contain RNA, protein, lipid, and polysaccharide. Adequate chemical analysis is not available for the paramyxoviruses and rhabdoviruses. Of the leukemiaviruses, only avian myeloblastosis virus and the mouse mammary tumor virus have been chemically characterized; they contain RNA, protein, and lipid (*115, 116*).

TABLE 1-11
Chemical Composition of RNA Animal Virions[a]

Virus group	RNA, %	Protein, %	Lipid, %	Others, %
Picornavirus[b]	25–30	70–75	None	None
Reovirus[c]	15	85	None	None
Arbovirus[d]	6	65	29	None
Myxovirus[e]	0.8–0.9	70	18.5	5–7% polysaccharide
Paramyxovirus	Present	Present	Present	None
Rhabdovirus	Present	Present	Present	None
Leukemiavirus	2.2[f]	~60	35	None
Leukemiavirus	0.8[g]	~70	27	None

[a] Chemical analysis has been carried out on only several members of each group; see text for further details.

[b] Range of values for poliovirus, foot-and-mouth disease virus, encephalomyocarditis virus, and Maus-Elberfeld virus is reported. See Ref. 62 for further details.

[c] Value for reovirus type 3 (110).

[d] Value for Sindbis virus (111).

[e] Values for influenza virus (112–114).

[f] Values for avian myeloblastosis virus (115).

[g] Values for mammary tumor virus (116).

Particle size estimates for RNA animal virions are given in Table 1-12. The small picornaviruses range in size from 6–10×10^6 daltons, the reoviruses and arboviruses are about 70×10^6 daltons, the myxoviruses 150×10^6 daltons, and the leukemiaviruses 370–450×10^6 daltons.

Viral RNA molecules have a much narrower range of molecular weights from 2×10^6 daltons for picornavirus RNA (enterovirus subgroup) to 10–15×10^6 daltons for leukemiavirus and reovirus RNA's (Table 1-13) than viral DNA molecules which range from 2–200×10^6 daltons. Picornavirus, arbovirus, and paramyxovirus RNA's are extracted from the virion mainly as single, large, intact RNA molecules, but quite surprisingly reovirus and myxovirus RNA's are isolated in a number of small segments of reproducible size. Although a major 70S RNA species is isolated from leukemiaviruses, recent studies indicate that this consists of aggregated smaller RNA molecules (116a). Whether reovirus and myxovirus RNA molecules possess weak spots or linkers which break during isolation or whether these virions contain multiple RNA species has not yet been established. Of the RNA animal viruses, the RNA molecules of only picornaviruses and arboviruses have been shown to be infectious.

TABLE 1-12
Particle Weight of RNA Animal Virions

Virion group	Virion	Particle weight, daltons $\times 10^6$	Method
Picornavirus	Poliovirus[a]	5.5	S and D[b]
	Maus-Elberfeld[c]	5.7	S andD
	Mengovirus[d]	8.3	S and D
	Encephalomyocarditis[e]	10 (probably too high)	RNA content and molecular weight[f]
Reovirus	Reovirus type 3[g]	70	S and D
Arbovirus	Sindbis[h]	70	S
Myxovirus	Fowl plague[i]	150	S and D
Leukemiavirus	Avian myeloblastosis	450	Mass of virion[j]
	Mammary tumor virus	370	Volume and density of virion[k]

[a] From Ref. *117*.

[b] Calculated from sedimentation (S) and diffusion (D) data using the Svedberg equation.

[c] From Ref. *118*.

[d] From Ref. *119*.

[e] From Ref. *120*.

[f] Calculated from the molecular weight of viral RNA and the fractional RNA content of the virion.

[g] From Ref. *110*.

[h] From Ref. *139* calculated from the sedimentation coefficient using the Svedberg equation, assuming that the virion behaves as a spherical particle during sedimentation.

[i] From Ref. *121*.

[j] From Ref. *115;* calculated from the dry weight of purified virus and the particle concentration.

[k] From Ref. *116;* estimated from the dry weight density and diameter of the virion.

The base compositions of animal virion RNA molecules differ widely (Table 1-14). The major RNA components of the reovirion posses complementary base pairing, i.e., $A = U$, $G = C$, because of their duplex structure. The absence of complementary base pairing in other viral RNA's indicates that these molecules are single stranded.

Of the over 400 RNA animal viruses, the amino acid composition of only poliovirus, encephalomyocarditis, Maus-Elberfeld, and foot-and-mouth disease virions have been reported. As shown in Table 1-7, the amino acid contents are very similar, no distinctive features are evident. The number of protein subunits in RNA virions ranges from one for foot-and-mouth disease virus to six for several leukemia-viruses. (Table 1-8); in contrast DNA animal virions possess 1 to 17 polypeptides, reflecting their wider range of genetic information.

TABLE 1-13
Molecular Weight and Conformation of Animal Virus RNA Molecules

Virus group and virion	$S_{20,w}$ svedbergs	Length, μ	Conformation	Molecular weight, daltons $\times$ 10^6
Picornavirus group				
Poliovirus	28[a] (117)		Single stranded	2.0,[b] 1.7[c]
Maus-Elberfeld	28[a] (118)		Single stranded	2.0,[b] 1.7[c]
Mengovirus			Single stranded	1.7[d] (119)
Rhinovirus			Single stranded	4[e] (122)
Arbovirus				
Sindbis	46[g] (140)		Single stranded	4-5[f] (139)
Japanese encephalitis	45[h] (129)		Single stranded	4.9[g]
Dengue	42 (172)		Single stranded	3.3[h]
Semliki forest			Single stranded	
Reovirus group				
Reovirus type 3	14-15 (128, 142)	1.1 (143)	Double stranded	2.3-2.4 (128, 142)
	12-13	0.6	Double stranded	1.3-1.4
	10.5	0.35	Double stranded	0.8
	2-3 (144, 145)		Single stranded	0.004-0.007 (144, 145)
Myxovirus group				
Influenza	38[g] (146, 147)		Single stranded	2.8[g]
	17-21[g] (61, 146-150)		Single stranded	0.5[g]
	16[g] (61)		Single stranded	
	14[g] (61, 149)		Single stranded	
	7[g] (61, 146)		Single stranded	

Paramyxovirus group				
Newcastle disease virus	57[a] (131)	1.06 (151)	Single stranded	7.5[a]
Sendai	49[a] (132)	1.3–1.4 (152)	Single stranded	7[a]
Simian parainfluenza SV-5	57 (182)	1.02 (181)	Single stranded	7
Leukemiavirus group (see footnote k)				
Rous sarcoma virus + Rous associated virus	71[a] (133)		Single stranded	12[a]
Rous sarcoma virus + Rous associated virus	71[a] (153)		Single stranded	12[a]
Rous sarcoma virus	71[a] (134)		Single stranded	12[a]
Avian myeloblastosis virus	71[a] (133)		Single stranded	12[a]
Avian myeloblastosis virus	67[a] (135)		Single stranded	10[a]
Avian myeloblastosis virus		8.3 (154)	Single stranded	10[i]
Murine leukemia virus–Rauscher	74[a] (136)		Single stranded	13[a]
Murine leukemia virus–Rauscher	73[j] (155)		Single stranded	13[j]
Murine leukemia virus–Rauscher	70[a] (137)		Single stranded	10[a]
Mammary tumor virus	70[a] (156)		Single stranded	12[a]

[a] S determined in 0.02 M phosphate buffer.
[b] Calculated from sedimentation coefficient and viscosity of viral RNA using the formula of Scheraga and Mandelkern (138).
[c] Calculated from the molecular weight of virion (5.5–5.7 × 10⁶) and 30% RNA content.
[d] Calculated from sedimentation coefficient and diffusion coefficient using the Svedberg equation.
[e] Based on electron-microscope length measurement of a rhinovirus ribonucleoprotein.
[f] Calculated from the sedimentation coefficient in dimethyl sulfoxide.
[g] S determined in 0.1 M salt; molecular weight calculated using the formula of Spirin (141).
[h] S determined in 0.1 M salt using ribosomal RNA marker. Molecular weight calculated from S value assuming that the molecular weight varies as the square of S.
[i] Estimated from length of RNA.
[j] S determined in 0.2 M salt; molecular weight calculated from S using the formula of Spirin (141).
[k] Recent studies with RSV and a murine leukemiavirus indicate that the 70S leukemiavirus RNA consists of aggregates of smaller RNA molecules (116a); therefore the sedimentation values and molecular weights of the leukemiavirus RNA molecules are ambiguous at the present time.

TABLE 1-14

Base Composition of Animal Virus RNA Molecules

Virion	Molar per cent				Reference
	A	U	G	C	
Picornavirus group					
Poliovirus (3 types)	29	25	24	22	125
Coxsackie A9	27	25	28	20	124
Coxsackie A10	28	23	28	21	124
Coxsackie B1	29	24	24	23	126
Maus-Elberfeld	25	27	24	24	51
Foot-and-mouth disease	26	22	24	28	123
Reovirus group					
Reovirus type 3 (whole virus RNA)	38	28	18	17	128
Reovirus type 3 (pooled double-stranded fragments)	28	28	22	22	128
Reovirus type 3 (A-rich fragment)	88	10.5	0	1.5	144
Arbovirus group					
Sindbis	30	20	26	25	111
Semliki Forest	27	22	26	24	129a
Dengue	31	21	26	21	129
Myxovirus group					
Influenza A (strains PR-8, MEL, WSE, Swine, and CAM)	23	33	20	24	130
Influenza B (strains LEE, MIL, and ROB)	23	36	18	23	130
Paramyxovirus group					
Newcastle disease virus	24	29	24	23	131
Newcastle disease virus	20	31	25	23	132
Leukemiavirus group					
Rous sarcoma virus + Rous associated virus	25	22	28	24	133
Rous sarcoma virus	25	22	29	24	134
Avian myeloblastosis virus	25	23	29	23	133
Avian myeloblastosis virus	24	23	29	24	135
Murine leukemia virus–Rauscher	26	23	25	27	136
Murine leukemia virus–Rauscher	24	24	26	26	137
Mammary tumor virus	19	29	30	22	116

A. PICORNAVIRUSES

The picornaviruses are the smallest of the RNA animal viruses (diameter 17 to 30 mμ), the name being derived from pico (small) and RNA. Because of the small size of their genome (2×10^6 daltons) viruses such as poliovirus and mengovirus have been widely used to provide experimental systems for analyzing the intracellular translation of small polycistronic viral mRNA molecules. Two subgroups of the human picornaviruses are known (Table 1-2B): (a) the enteroviruses (about 60 members) which include polioviruses, coxsackie viruses, and echoviruses; and (b) the rhinoviruses (about 100 members), the major causes of the common cold. Rhinoviruses and enteroviruses differ in several properties: rhinoviruses are acid sensitive and have a higher buoyant density (1.38–1.41 g/cm³) in cesium chloride gradients than the enteroviruses (1.32–1.35 g/cm³). Recently a rhinovirus was shown to have twice the RNA content of poliovirus, thus explaining the increased buoyant density of the former (*122*).

Poliovirus was the first animal virus to be crystallized (1955), several other picornaviruses were crystallized subsequently. Several members of the chemically simple enterovirus subgroup have been chemically and physically characterized. The RNA content of poliovirus (*157*), encephalomyocarditis (EMC) (*50*), and foot-and-mouth disease (FMDV) (*123*) viruses is 25, 30, and 32%, respectively. The rest is protein; DNA and phospholipid are absent. Coxsackie (*124*), mengovirus (*119*), and Maus-Elberfeld (ME, also called mouse encephalitis) (*118*) viruses also were shown to contain RNA.

The molecular weights of the RNA molecules extracted from poliovirus, ME, FMDV, and mengovirus range from $1.7–2.0 \times 10^6$ daltons (Table 1-13). Rhinovirus RNA is twice this size, based on electron microscopic length measurements of the ribonucleoprotein strand extruded from heat-degraded rhinovirions (*122*).

The RNA base composition has been determined for the three poliovirus types, ME, FMDV, and Coxsackie A9, A10, and B1 (Table 1-14); no particular patterns are evident although differences between viruses can be seen. Variations in base analysis reported by different laboratories for the same virus make it difficult to evaluate small differences in RNA base ratios between different viruses. Because of the limited sensitivity of chemical base composition analysis (±5%), two RNA's may be indistinguishable yet differ by as much as 100 nucleotides per molecule.

Poliovirus, ME, and EMC range in particle weight from $6–10 \times 10^6$ daltons (Table 1-12). Since the molecular weight of viral RNA is about 2×10^6, there are $4–8 \times 10^6$ daltons of capsid protein. Since RNA molecules of small virions do not contain sufficient genetic information to code for a single protein of this size, Crick and Watson in 1956 suggested that the viral capsids of small virions are built of small, identical protein subunits. Several plant viruses were subsequently shown to be constructed from multiple copies of a single polypeptide, for example, tobacco mosaic virus and turnip yellow mosaic virus. As revealed by polyacrylamide electrophoresis of the dissociated viral capsid (Table 1-9), FMDV contains a single polypeptide but poliovirus ME, EMC, and mengovirus contain 3–4 polypeptides. The C-terminal sequences of the FMDV polypeptide is (Val, Asp, Thr, Asn, Phe, Glu, Gly)-Ser-Ala-Leu-Glu (sequence within parenthesis is uncertain) (52).

Poliovirus contains 2 major and 2 minor polypeptides (56). The molecular weight of the viral polypeptides obtained by treatment of the virion with acetic acid followed by performic oxidation (to cleave disulfide bonds) are the same, 27,000 (56), as determined by sedimentation equilibrium. However, the polypeptides prepared by treatment of the virion with SDS have a molecular weight of 42,000 (160); the reason for this discrepancy is not known.

The proteins of ME, EMC, and mengovirus isolated by phenol treatment were reported to consist of two major electrophoretic components, roughly similar in size (26,000 daltons as determined by sedimentation equilibrium) and one to two minor components (57). Recently EMC proteins were resolved into three major components with molecular weights of 29,000 to 33,000 which differed in amino acid composition (58, 59). Two of the polypeptide components in ME (161) differ in amino acid content and peptide sequence and recently ME polypeptides were resolved further into equal amounts of three major components with molecular weights of 25,000, 30,000, and 32,000 (162).

The amino acid composition of poliovirus, EMC, ME, and FMDV are similar (Table 1-7). Minor but significant differences in amino acid composition have been recorded for three plaque-type mutants of EMC (158) and distinct differences in amino acid composition were reported for three immunological types of FMDV (159).

B. REOVIRUSES

The three virions in this group (types 1, 2, and 3) are found virtually in all animal species but their relationship to human disease is not known. Reoviruses are the only known animal virions containing double-stranded RNA. The wound tumor virus, which multiplies in plants and insects, is similar to reovirus in size and fine structure and also contains double-stranded RNA (*127*), but is unrelated antigenically to reovirus. The chemical and physical similarities between the wound tumor virus of plants and the reoviruses, the ubiquitous occurrence of reoviruses, and the lack of association of reoviruses with known diseases have stimulated speculation that reoviruses may play a role in human cancer.

Reoviruses are nonenveloped icosahedral particles, about 74 mμ in diameter, possessing 92 capsomeres. Reovirus type 3 contains 15% RNA and 85% protein (*110*), no DNA (diphenylanine test negative), and no lipid since virus infectivity is resistant to ether treatment. The molecular weight of the reovirion calculated from its sedimentation coefficient and diffusion coefficient (*110*), is 70×10^6 daltons (Table 1-12). The RNA complement of the virion is roughly estimated at 10×10^6 daltons from the molecular weight and RNA content of the virion. But most investigators have failed to extract RNA molecules of this size from the reovirion.

Reovirus RNA is released from the virion by several different procedures as a mixture of segments of three reproducible size classes with sedimentation coefficients of 14–15S (large, L), 12–13S (medium, M), and 10.5S (small, Sm) (*128, 142, 143*) (Table 1-13). Similarly, electron microscopy has revealed three major lengths of 1.1, 0.6, and 0.35 μ (*143*), corresponding to the L, M, and Sm RNA classes. The RNA segments are double stranded as shown by a sharp melting profile, resistance to RNase, relative independence of sedimentation coefficient and ionic strength, and equality of A:U and G:C (*128*). They do not arise from random breakage of the RNA molecules since (a) the size distribution of L, M, and Sm segments are reproducible, (b) no homology was detected between L, M, and Sm segments (*163*), and (c) the reovirus genome is transcribed within the cell as singlestranded RNA segments which hybridize with the homologous doublestranded segments released from the virion but not with heterologous double-stranded segments (*163, 164*).

The molecular weights of the RNA segments were estimated by two procedures: (a) from the sedimentation coefficients assuming that the relationship between S and molecular weight established for doublestranded DNA holds for double-stranded RNA, and (b) from the electron microscopic length measurements based on estimate of the linear density of double-stranded RNA which is not known with precision. The molecular weight of the L, M, and Sm segment classes are $2.3–2.4 \times 10^6$, $1.3–1.4 \times 10^6$, and 8×10^5 daltons (*128, 142*).

Recently, a unique adenine (A)-rich, single-stranded RNA molecule, comprising about 20% of the mass of the viral genome, has been found in the RNA extracted from reovirions (*144, 145*). The base composition of the A-rich component is 88% A, 10.5% U, 1.5% C, and 0% G (*144*). All three reovirus serotypes possess the A-rich components (*145*) and there are about 200 molecules of A-rich RNA per virion, each containing from 12 to 20 nucleotides (*144, 145*). The possibility that A-rich RNA serves as a linker between the double-stranded RNA segments seems unlikely because of the large number of A-rich molecules in each virion. The function and mechanism of synthesis of A-rich RNA are not known.

Electron microscopy of RNA liberated from purified reovirus by short exposure to sodium perchlorate solution detected some molecules of length 6–7 μ, which would correspond to a molecular weight of about $15–17 \times 10^6$ daltons, suggesting that the reovirus RNA segments arise from breakage of the genome at specific sites during extraction of the virion (*143*). The L, M, and Sm double-stranded RNA segments occur in ratios of n, n, and 1.5 n (*128*). If n is 2, then the molecular weight of the genome is 9.6×10^6 daltons, if n is 3, then 14.6×10^6 daltons. The most reasonable expectation is 3. The content of A-rich RNA molecules brings the total RNA complement of the reovirion to about 16×10^6 daltons. However, further work is needed before the structure of the interesting reovirus genome is solved.

Attempts to demonstrate infectious reovirus RNA have failed, perhaps due to the fragility of the reovirus genome inasmuch as an intact RNA molecule may be necessary for infectivity. No published information is available on the number of protein components in the virion.

C. Arboviruses

Only several members of this large group of small (40–50 mμ diameter) enveloped virions have been chemically characterized. Over 200 members have been described including, for example, the viruses of yellow fever, St. Louis encephalitis, and dengue, and three subgroups, A, B, and C, have been formed based on antigenic relationships.

Several arboviruses have been purified, including eastern equine encephalitis (*165*), Semliki forest (*166*), Venezuelan equine encephalitis (*167*), Murray valley encephalitis (*169*), and Sindbis (*111*) viruses. Eastern equine encephalitis was reported in 1943 to contain protein, RNA, phospholipid, and cholesterol (*165*). Murray Valley encephalitis virus, of uncertain purity, was reported to contain 7.8% RNA and 11% lipid (*169*). Sindbis virus, purified and analyzed in 1963 by modern methods, contains 65% protein, 6% RNA, and 29% lipid (22% phospholipid and 7% cholesterol) (*111*). The phospholipid of Sindbis virus is derived largely from pre-existing cellular material (*170*) but viral protein is synthesized *de novo* after infection (*171*). Other arboviruses thus far analyzed are of questionable purity and proof that RNA is the hereditary material rests mainly on the extraction of infectious, RNase-sensitive material from infected cells and tissues (*168*).

The fine structure of the arbovirion is far from understood and whether the virion is constructed with cubic or helical symmetry has not yet been established. Electron microscopy has been performed on only several of the 200 arboviruses, and reveals that the arbovirion consists of an outer lipoprotein envelope and an inner RNA "core" (*173*). Recently this gross structure was established at the biochemical level (*12*). The Sindbis virion sediments at 280S from which a molecular weight of 70×10^6 daltons was calculated using the Svedberg equation, assuming a spherical particle (*139*). Treatment of purified Sindbis virus with 0.2% deoxycholate dissociates the viral envelope, liberating intact cores which were separated from envelope components by zone centrifugation; the core sediments at 140S and envelope components at 10–20S. Sindbis virus has been shown by gel electrophoresis to consist of two major proteins, one is the core protein associated with viral RNA, and the second protein is part of the lipoprotein envelope.

The molecular weights of several arbovirus RNA's, estimated from their sedimentation coefficients (Table 1-13) are about twice that of picornavirus RNA's (enterovirus subgroup). Sindbis RNA has a molecular weight of $4–5 \times 10^6$ daltons based on its S value in dimethylsulfoxide (*139*) which agrees with the value calculated from the size of the virion (70×10^6 daltons) and its RNA content (6%). The RNA's of Japanese encephalitis virus have a sedimentation coefficient of 46S (*140*), Dengue virus, 45S (*129*), and Semliki virus, 42S (*172*), from which molecular weights of $3.3–4.9 \times 10^6$ daltons were estimated (Table 1-13).

The RNA base ratios of highly purified Sindbis virus, a group A arbovirus, are 30, 20, 26, and 25 molar % A, U, G, and C, respectively

(111) (Table 1-14). Similar base ratios are found for the RNA of Semliki forest, another group A virus, and Dengue, a group B arbovirus. These base ratios are distinctly different from those of the other RNA animal virions.

D. MYXOVIRUSES

These large and relatively complex enveloped RNA virions include influenza A, B, and C, fowl plague, and swine influenza viruses. Myxovirions consist of helical ribonucleoprotein, 9 mμ in diameter, inside a lipomucoprotein envelope with projecting spikes, as shown by electron microscopy of negatively stained particles (11).

Influenza virus, analyzed chemically in detail, contains 0.8–0.9% RNA (112–114); no significant DNA (112, 114); 18.5% total lipid consisting of 7% cholesterol and 11.5% phospholipid (cephalin, sphingomyelin, and lecithin) (113); 5 to 7% polysaccharide consisting of galactose, mannose, fucose, and an amino sugar (112, 113); and about 70% protein (Table 1-11). Fowl plague virus has a chemical composition similar to that of influenza virus (174).

Influenza virus has an RNA complement of 2–3 $\times$ 10^6 daltons as calculated from the particle weight of the virion (roughly estimated at 200–300 $\times$ 10^6 daltons based on the diameter of the virus particle) and the fractional RNA content of the virion (0.8–0.9%). Recently RNA species of several different sizes were extracted from the influenza virion (61, 146–150, 175). Two investigators detected a 38S RNA component (146, 147), corresponding to a molecular weight of 2.8 $\times$ 10^6 daltons, as well as 19S–20S and 7S species. The 38S component would correspond to a molecular weight of about 2.8 $\times$ 10^6 daltons and may represent the entire viral genome but this conclusion is unlikely in view of the following: (a) When placed in low salt solution, heated to 70°C, or treated with formaldehyde (147), the 38S RNA was dissociated to 20S RNA, and (b) other investigators using different extraction procedures did not detect the 38S component in influenza virus RNA but found the following components sedimenting at 17S–21S (61, 148–150), 16S (61), 14S (61, 149), and 8–9S (61, 149). The 18S, 16S, 14S, and 9S RNA components are single stranded because they are susceptible to RNase. Further they possess base compositions indistinguishable from that of total viral RNA (61). Five distinct RNA components were demonstrated recently in RNA extracted from influenza virus (175), with S values of 18S and smaller, by polyacrylamide gel electrophoresis, a technique with high resolving power for RNA molecules. The RNA of Newcastle disease virus extracted and analyzed under the identical conditions consists of a single 57S component. The molecular weight of the 18S influenza virus RNA species is estimated at about 0.6×10^6 daltons, by comparison with 18S ribosomal RNA. Thus four 18S RNA molecules would be necessary to account for a viral genome of 2.4×10^6 daltons; the origin of the smaller RNA segments is not known.

Recent experiments suggest that the influenza 38S RNA may be an artifact of the isolation procedure. Over one-third of the RNA extracted from the influenza virion in the presence of 5 mM Mg^{2+} or Ca^{2+} sedimented at 38S whereas no RNA species larger than 18S RNA were detected when extraction was performed in the absence of Ca^{2+} or Mg^{2+} (61); RNA extracted from tobacco mosaic in the presence of Ca^{2+} or Mg^{2+} showed no increase in S value (61). Furthermore, the 38S RNA isolated in the presence of divalent cations is converted to 18S RNA by treatment with EDTA, suggesting that 38S RNA is an aggregate of smaller RNA segments, either formed by or held together by divalent cations (61). Possibly this aggregate is a naturally occurring complex of viral RNA segments or perhaps the 38S component represents an intact viral genome which is cleaved at preferential breaking points during extraction. Further work is needed to clarify the structure of the influenza virus genome.

The RNA base ratios of several influenza A strains are 23, 33, 20, and 24 molar % A, U, G, and C, respectively, while the base ratios of influenza B strains differ slightly from those of influenza A strains (Table 1-14). The high U content of influenza RNA is unique among the RNA virions. The base ratios of RNA extracted from the whole influenza virion and the 18S, 16S, 14S, and 9S fragments are very similar (61).

Disruption of myxovirions by several procedures has disclosed three functional components (174); (a) an inner ribonucleoprotein, termed the "internal antigen," (b) surface hemagglutinin components responsible for the ability of myxovirions to agglutinate erythrocytes, and (c) a virus-specific neuraminidase which cleaves N-acetyl neuraminic acid from polysaccharide components of the cell surface. At least three major protein fractions were resolved by electrophoresis of SDS-disrupted influenza virus strains A and B (60) and three bands were observed by polyacrylamide gel electrophoresis of protein prepared from influenza virus by phenol extraction (61).

The lipomucoprotein envelope of influenza virus (as well as that of other enveloped virions) is derived in part from the cell membrane as shown by (a) electron microscopic observations during influenza virus maturation, (b) the similarity in lipid and polysaccharide composition of influenza virus and the host cell used to propagate the virus (*113*), and (c) the formation of influenza virus phospholipid from ^{32}P-labeled host-cell lipid (*114*).

It has been proposed that the surface spikes of the virion are the hemagglutinin components and that the neuraminidase molecules are stacked in between the hemagglutinin spikes (*177*). The viral neuraminidase has been separated from the virion and several properties described (*176–178*). The enzyme is strain specific since purified neuraminidase from the Asian and PR-8 strains of influenza A virus differ in several properties (*178*). The role of the viral neuraminidase in infection is not known but since antiserum to purified enzyme does not prevent infection (*179*) the enzyme does not appear to be essential to initiate infection. Possible roles for the enzyme in intracellular virus development and in virus release have been suggested.

E. Paramyxoviruses

The paramyxoviruses, for example, Newcastle disease virus (NDV), mumps, parainfluenza 1, 2, 3, and 4, measles, Sendai virus (Table 1-2B), resemble the myxoviruses morphologically consisting of an inner helical nucleocapsid and a spike-covered lipoprotein envelope. Paramyxoviruses were previously included in the myxovirus group but are now grouped separately (*9, 10*) because of several differences in properties: (a) the helical nucleocapsid of paramyxoviruses is 18 mμ in diameter, that of myxoviruses is 9 mμ (*180*), (b) paramyxoviruses possess a single large RNA genome of molecular weight about 7×10^6 while myxoviruses (e.g., influenza virus) appear to possess multiple RNA segments of low molecular weight.

The chemical analysis of paramyxoviruses purified by modern techniques is not available. Recently the inner helical nucleocapsid of simian parainfluenza virus, SV-5 was isolated in milligram quantities from the cytoplasm of infected cells by osmotic shock followed by equilibrium centrifugation in cesium chloride gradients (*181*) and shown to be composed of RNA and protein; tests for DNA and hexose were negative. The RNA content of the SV-5 nucleocapsid is 4.1% (assuming that only RNA and protein are present). The RNA content of the NDV nucleocapsid, released from purified virus by treatment with deoxycholate and purified by zonal centrifugation, is 4.5% (152). The molecular length of the SV-5 nucleocapsid, determined by electron microscopy, is 1.02 μ (181) and that of the NDV nuclocapsid is reported at 1.06 μ (151) and 1.3–1.4 μ (152) by two laboratories. These lengths correspond to molecular weights of about 7×10^6 daltons for viral RNA calculated from the relationship between the length of the tobacco mosaic virus ribonucleoprotein and the molecular weights of its RNA. The size of NDV RNA estimated in this manner agrees with the molecular weight calculated from the sedimentation coefficient of NDV RNA (Table 1-13).

RNA extracted from purified NDV consists of a major, large RNA component which sediments at 57S in 0.1 M NaCl–10^{-3} M EDTA (131) and 49S in 0.01 M sodium acetate–0.05 M NaCl (132), as well as smaller fragments (4S) which most likely are degradation products. Sendai virus RNA has a similar sedimentation coefficient 57S (182). NDV RNA is single stranded as shown by its susceptibility to RNase, noncomplementary base composition, and dependence of S on salt concentration (131). Based on the sedimentation coefficient of NDV RNA, its molecular weight is about 7.5×10^6 daltons (131). Although the NDV RNA molecule is believed to represent the whole viral genome it is not infectious (132); possibly a small RNA segment is removed by RNA extraction or the conditions for demonstrating infectivity are inadequate.

Of the paramyxoviruses, the base composition of only NDV RNA is available. Several laboratories have reported base ratios which differ considerably, perhaps due to inadequately purified virus; two analysis of the purified fast-sedimenting RNA component, in reasonable agreement, are given in Table 1-14. The base ratios of NDV RNA are significantly different from those of the myxoviruses although both influenza RNA and NDV RNA possess a high U content.

There is no information on the protein chemistry of the paramyxoviruses.

F. RHABDOVIRUSES

This group was proposed recently (10) for several virions with distinctive morphological features, not shared by other animal virions, and includes vesicular stomatitis virus (VSV) and rabies virus. The virion is a cylinder, flat on one end and round at the other consisting of an inner helical component and an envelope with surface projections. Similar "bullet-shaped" virions have been described recently for hemorrhagic septicemia virus of rainbow trout and two insect viruses, sigma virus and coccal virus (10).

Very little is known about the chemistry of the two animal rhabdoviruses, VSV and rabies virus. VSV is 175 mμ in length and 68 mμ in diameter, possesses an inner helical nucleocapsid 18 mμ in diameter and fine filamentous projections about 10 mμ in length projecting from the envelope (185–188). Indirect data are available on the chemical nature of VSV. Partially purified ^{32}P-labeled VSV was chemically fractionated, 70% of the radioactivity was associated with phospholipid and 10% with nucleic acid; most of the nucleic acid label was in RNA (183). Additional evidence that RNA is the genetic material of VSV is the inability of 5-fluorodeoxyuridine, an inhibitor of DNA synthesis, to block VSV replication (184).

Rabies virus is similar to VSV in morphology, having dimensions of 180 × 75 mμ, an inner core, and surface projections of 6–7 mμ in length (189). No data is available on the chemical composition of rabies virus.

G. Leukemiaviruses (RNA Tumor Viruses)

Members of this recently proposed (9) group of RNA tumor viruses are similar morphologically, possessing an inner RNA-containing core and an outer lipid-containing envelope. Included in this group (Table 1-2B) are (a) a large number of viruses in the avian leukosis–sarcoma virus complex, (b) the large number of strains (190) of murine leukemia-sarcoma viruses and variants such as the Maloney sarcoma and Harvey sarcoma viruses which induce solid tumors in animals, and (c) the mammary tumor virus of mice.

Representative virions of the three leukemiavirus subgroups have been purified recently including avian myeloblastosis virus (AMV), mammary tumor virus (MTV), the Rauscher murine leukemia virus (MLV), a mixture of Rous sarcoma virus (RSV) and Rous associated virus (RAV), and RSV alone. Virions were purified from the plasma of infected mice (MLV), infected cell cultures (MLV, RSV + RAV, RSV), milk from infected mice (MTV), and myeloblasts in culture (AMV). Only AMV and MTV have been analyzed chemically. AMV contains 2.5% RNA, 35% lipid, about 60% protein, and no significant DNA (115) (Table 1-11). The virion has an RNA complement of about 10×10^6 daltons, calculated from the particle weight of AMV [estimated roughly as 450×10^6 (Table 1-12) from the dry weight of purified virus and particle counts by electron microscopy] and the fractional RNA content. MTV is reported to contain 27% lipid and 0.8% RNA on a dry weight basis; the remainder is presumably protein (116). The RNA value may be low since it predicts 4×10^6 daltons of RNA per virion and recent studies on isolated MTV RNA suggest a molecular weight of about 12×10^6 (156).

Recently the RNA's of several leukemiaviruses were isolated and characterized. Leukemiavirus RNA's are single stranded, as shown by digestion by RNase, dependence of sedimentation rate upon ionic strength, and noncomplementary base ratios. RNA extracted from the Brian high-titer strain of RSV, containing 4–10 times more helper virus RAV than RSV (referred to as RSV + RAV) (191), and from RSV (free of RAV) (134) have sedimentation coefficients of 67S–74S (Table 1-13), corresponding to molecular weights of $10–13 \times 10^6$ daltons. AMV RNA has a molecular weight of 10×10^6 from electron microscopic length measurements and $10–12 \times 10^6$ based on sedimentation analysis. Similar S values and molecular weights are reported for MLV and MTV (Table 1-13). Variable amounts (20–70%) of heterogeneous RNA of much smaller size (about 4S) were present in RNA prepared from each leukemiavirion. Most likely these are degradation products produced by contaminating nucleases acting on damaged virions present in each preparation. However, the fast sedimenting component has not been shown to be infectious and the possibility remains that it does not represent the whole viral genome. Quite surprisingly the 70S RNA of RSV and a murine leukemiavirus was reported recently to consist of aggregates of smaller RNA molecules (116a). Therefore the size and structure of leukemiavirus RNA is uncertain at the present time.

The base composition of leukemiavirus RNA's are shown in Table 1-14. The base ratios of the avian tumor viruses RSV + RAV, RSV, and AMV are indistinguishable, as are their sedimentation coefficients (Table 1-13), suggesting that these viruses are closely related. RSV + RAV, RSV, and AMV base ratios are quite different from those of MLV and MTV (Table 1-14).

AMV (192) and MLV (193) have lipid compositions which are similar qualitatively to those of their host cells, indicating that the lipid of the virion is derived exclusively from the cell.

Relatively little is known about the protein components of the

leukemiavirions. Two antigenic components have been detected (*190*):
(a) a type-specific antigen in the outer envelope and (b) a group-
specific antigen within the virion, probably a constituent of the nucleocapsid. A common group-specific antigen is shared by all avian leukemia–sarcoma viruses and a different group-specific antigen is shared
by the murine leukemia viruses. The AMV group-specific antigen has
been isolated from SDS-disrupted virus by Tiselius electrophoresis
(*194*) and from Tween-ether disrupted virus by electrophoresis on
cellulose acetate and polyacrylamide gels (*195*). It is not known
whether the antigen consists of one or several proteins because the
conditions used to disrupt and analyze the virus may not completely
dissociate its components. Phenol extracted proteins from MLV (*136*),
AMV (*196*), and RSV + RAV (*196*) were resolved by gel electrophoresis into 2, 4–6, and 4–6 protein components, respectively; AMV
disrupted by Tween-ether was reported to give 6–8 protein components
by gel electrophoresis (*195*).

ACKNOWLEDGMENTS

This work was supported by Public Health Service grant AI-01725,
contract PH43-64-928 from the National Institute of Allergy and
Infectious Diseases, and contract PH43-67-692 from the National
Cancer Institute. The author is a Research Career Awardee (5-K6-
AI-4739) of the National Institutes of Health.

REFERENCES

1. D. L. D. Casper, R. Dulbecco, A. Klug, A. Lwoff, M. G. P. Stoker, P.
Tournier, and P. Wildy, *Cold Spring Harbor Symp. Quant. Biol.*, **27**, 49
(1962).
2. A. Lwoff, *J. Gen. Microbiol.*, **17**, 239 (1957).
3. J. R. Kates and B. R. McAuslan, *Proc. Natl. Acad. Sci. U.S.*, **58**, 134 (1967).
4. W. Munyon, E. Paoletti, and J. T. Grace, Jr., *Proc. Natl. Acad. Sci. U.S.*,
58, 2280 (1967).
5. International Subcommittee on Virus Nomenclature, *Intern. Bull. Bacteriol.
Nomenclature and Taxonomy,* **13**, 217 (1963).
6. M. Green, in *Viral and Rickettsial Infections of Man*, 4th ed. (F. L.
Horsfall, Jr., and I. Tamm, eds.), Lippincott, Philadelphia, 1965, Chap. 2,
p. 11.
7. P. Wildy, H. S. Ginsberg, J. Brandes, and J. Maurin, *Progr. Med. Virol.*, **9**,
476 (1967).
8. A. Lwoff and P. Tournier, *Ann. Rev. Microbiol.*, **20**, 45 (1966).
9. J. L. Melnick, *Progr. Med. Virol.*, **9**, 483 (1967).

10. J. L. Melnick and R. M. McCombs, *Progr. Med. Virol.*, **8**, 400 (1966).
11. R. W. Horne and P. Wildy, *Advan. Virus Res.*, **10**, 101 (1963).
12. J. H. Strauss, B. W. Burge, E. R. Pfefferkorn, and J. E. Darnell, *Proc. Natl. Acad. Sci. U.S.*, **59**, 533 (1968).
13. C. E. Schwerdt, in *Viral and Rickettsial Infections of Man*, 4th ed. (F. L. Horsfall, Jr., and I. Tamm, eds.), Lippincott, Philadelphia, 1965, Chap. 3, p. 19.
14. H. K. Schachman and R. C. Williams, in *The Viruses*, Vol. 1 (F. M. Burnet and W. M. Stanley, eds.), Academic Press, New York, 1959, Chap. 3, p. 223.
15. C. L. Schildkraut, J. Marmur, and P. Doty, *J. Mol. Biol.*, **4**, 430 (1962).
16. J. Marmur and P. Doty, *J. Mol. Biol.*, **5**, 109 (1962).
17. C. A. Thomas, Jr., and L. A. MacHattie, *Ann. Rev. Biochem.*, **36**, 485 (1967).
18. F. W. Studier, *J. Mol. Biol.*, **11**, 373 (1965).
19. A. K. Kleinschmidt, D. Lang, and R. K. Zahn, *Z. Naturforsch.*, **16**, 730 (1961).
20. R. Weil and J. Vinograd, *Proc. Natl. Acad. Sci. U.S.*, **50**, 730 (1963).
20a. B. J. McCarthy in *Subviral Carcinogenesis* (Y. Ito, ed.), Nissha Printing Co., Kyoto, Japan, 1966, p. 47–61.
20b. M. Green, K. Fujinaga, and M. Piña, in *Basic Techniques in Virology* (K. Habel and N. Salzman, eds.), Academic Press, New York, in press.
21. M. Green, in *Viral and Rickettsial Infections of Man*, 4th ed. (F. L. Horsfall, Jr., and I. Tamm, eds.), Lippincott, Philadelphia, 1965, Chap. 6, p. 145.
22. H. D. Mayer and J. L. Melnick, *Nature,* **210**, 331 (1966).
23. M. Piña and M. Green, *Proc. Natl. Acad. Sci. U.S.*, **54**, 547 (1965); M. Green and M. Piña, unpublished data.
24. W. Parks, M. Green, M. Piña, and J. Melnick, *J. Virol.*, **1**, 980 (1967).
25. J. E. Smadel and C. L. Hoagland, *Bacteriol. Rev.*, **6**, 79 (1942).
26. M. A. Koch, H. J. Eggers, F. A. Anderer, H. D. Schlumberger, and H. Frank, *Virology*, **32**, 503 (1967).
27. S. J. Kass and C. A. Knight, *Virology*, **27**, 273 (1965).
28. A. K. Kleinschmidt, S. J. Kass, R. C. Williams, and C. A. Knight, *J. Mol. Biol.*, **13**, 749 (1965).
29. M. Green, M. Piña, R. Kimes, P. C. Wensink, L. A. MacHattie, and C. A. Thomas, *Proc. Natl. Acad. Sci. U.S.*, **57**, 1302 (1967).
30. I. Sarov and Y. Becker, *Virology*, **33**, 369 (1967).
31. J. A. Rose, M. D. Hoggan, and A. J. Shatkin, *Proc. Natl. Acad. Sci. U.S.*, **56**, 86 (1966).
32. L. V. Crawford, *Virology,* **29**, 605 (1966).
33. L. V. Crawford, *Virology*, **22**, 149 (1964).
34. W. Stoeckenius, *Proc. Natl. Acad. Sci. U.S.,* **50**, 737 (1963).
34a. L. V. Crawford and E. A. C. Follett, *J. Gen. Virol.,* **1**, 19 (1967).
35. L. V. Crawford, *J. Mol. Biol.*, **8**, 489 (1964).
36. L. V. Crawford, *J. Mol. Biol.*, **13**, 362 (1965).
37. H. Bujard, *J. Virol.*, **1**, 1135 (1967).
38. P. May, A. Niveleau, G. Berger, and C. Brailovsky, *J. Mol. Biol.*, **27**, 603 (1967).
39. W. C. Russell and L. V. Crawford, *Virology*, **22**, 288 (1964).

40. A. S. Kaplan and T. Ben-Porat, *Virology,* **23,** 90 (1964).

41. L. V. Crawford and A. J. Lee, *Virology,* **23,** 105 (1964).

42. R. L. Soehner, G. A. Gentry, and C. C. Randall, *Virology,* **26,** 394 (1965).

43. J. F. McCrea and M. B. Lipman, *J. Virol.,* **1,** 1037 (1967).

44. L. G. Gafford and C. C. Randall, *J. Mol. Biol.,* **26,** 303 (1967).

45. J. M. Hyde, L. G. Gafford, and C. C. Randall, *Virology,* **33,** 112 (1967).

46. D. Axelrod, personal communication.

47. M. Piña and M. Green, *Virology,* **36,** 321 (1968).

48. H. Polasa and M. Green, *Virology,* **31,** 565 (1967).

49. L. Levintow and J. E. Darnell, Jr., *J. Biol. Chem.,* **235,** 70 (1960).

50. P. Faulkner, E. M. Martin, S. Sved, R. C. Valentine, and T. S. Work, *Biochem. J.,* **80,** 597 (1961).

51. R. R. Rueckert and W. Schafer, *Virology,* **26,** 333 (1965).

52. H. L. Bachrach and G. F. Vande Woude, *Virology,* **34,** 282 (1968).

53. F. A. Anderer, H. D. Schlumberger, M. A. Koch, H. Frank, and H. J. Eggers, *Virology,* **32,** 511 (1967).

53a. H. V. Thorne and D. Warden, *J. Gen. Virol.,* **1,** 135 (1967).

54. J. V. Maizel, Jr., *Science,* **151,** 988 (1966).

55. J. A. Holowczak and W. K. Joklik, *Virology,* **33,** 717 (1967).

56. J. V. Maizel, *Biochem. Biophys. Res. Commun.,* **13,** 483 (1963).

57. R. R. Rueckert, *Virology,* **26,** 345 (1965).

58. T. S. Work, *J. Mol. Biol.,* **10,** 544 (1964).

59. A. T. H. Burness and D. S. Walter, *Nature,* **215,** 1350 (1967).

60. W. G. Laver, *J. Mol. Biol.,* **9,** 109 (1964).

61. P. H. Duesberg and W. S. Robinson, *J. Mol. Biol.,* **25,** 383 (1967).

62. F. L. Schaffer and C. E. Schwerdt, in *Viral and Rickettsial Infections of Man,* 4th ed. (F. L. Horsfall, Jr., and I. Tamm, eds.), Lippincott, Philadelphia, 1965, Chap. 5, p. 94.

63. M. Green, *Am. J. Med.,* **38,** 651 (1965).

64. R. W. Atchison, B. Casto, and W. Hammon, *Science,* **149,** 754 (1965).

65. M. D. Hoggan, N. Blacklow, and W. P. Rowe, *Proc. Natl. Acad. Sci. U.S.,* **55,** 1467 (1966).

66. N. R. Blacklow, M. D. Hoggan, and W. P. Rowe, *Proc. Natl. Acad. Sci. U.S.,* **58,** 1410 (1967).

67. M. E. Reichmann, *Proc. Natl. Acad. Sci. U.S.,* **52,** 1009 (1964).

68. L. V. Crawford and E. M. Crawford, *Virology,* **21,** 258 (1963).

69. M. R. Michel, B. Hirt, and R. Weil, *Proc. Natl. Acad. Sci. U.S.,* **58,** 1381 (1967).

70. E. Winocour, *Virology,* **31,** 15 (1967).

70a. E. Winocour, *Virology,* **25,** 276 (1965).

71. F. A. Anderer, M. A. Koch, and H. D. Schlumberger, *Virology,* **34,** 452 (1968).

72. R. W. Schlesinger, *Advan. Virus Res.,* **14,** 1 (1969).

73. J. J. Trentin, Y. Yabe, and G. Taylor, *Science,* **137,** 835 (1962).

74. R. J. Huebner, W. P. Rowe, and W. T. Lane, *Proc. Natl. Acad. Sci. U.S.,* **48,** 2051 (1962).

75. R. N. Hull, I. S. Johnson, C. G. Culbertson, C. B. Reimer, and H. F. Wright, *Science*, **150,**, 1044 (1965).
76. J. H. Darbyshire, *Nature*, **211**, 102 (1966).
77. R. J. Huebner and W. T. Lane, *Science*, **149**, 1108 (1965).
78. R. J. Huebner, in *Perspectives in Virology*, Vol. 5 (M. Pollard, ed.), Academic Press, New York-London, 1967, Chap. 7, p. 147.
79. A. E. Freeman, P. H. Black, E. A. Vanderpool, P. H. Henry, J. B. Austin, and R. J. Huebner, *Proc. Natl. Acad. Sci. U.S.*, **58**, 1205 (1967).
80. K. Fujinaga and M. Green, *Proc. Natl. Acad. Sci. U.S.*, **57**, 806 (1967).
81. K. Fujinaga and M. Green, *J. Virol.*, **1**, 576 (1967).
82. K. Fujinaga and M. Green, *J. Mol. Biol.*, **31**, 63 (1968).
83. K. Fujinaga and M. Green, unpublished data.
84. D. B. Cowie and B. J. McCarthy, *Proc. Natl. Acad. Sci. U.S.*, **50**, 537 (1963).
85. M. H. Green, *Proc. Natl. Acad. Sci. U.S.*, **50**, 1177 (1963).
86. S. Lacy, Sr. and M. Green, *Proc. Natl. Acad. Sci. U.S.*, **52**, 1053 (1964).
87. S. Lacy, Sr. and M. Green, *Science*, **150**, 1296 (1965).
88. S. Lacy, Sr. and M. Green, *J. Gen. Virol.*, **1**, 413 (1967).
89. M. Green and M. Piña, *Virology*, **20**, 199 (1963).
90. M. Green and M. Piña, *Proc. Natl. Acad. Sci. U.S.*, **51**, 1251 (1964).
91. M. Green and M. Piña, unpublished data.
92. M. Green and M. Piña, *Proc. Natl. Acad. Sci. U.S.*, **50**, 44 (1963).
93. M. Green, M. Piña, and R. Kimes, *Virology*, **31**, 562 (1967).
94. W. G. Laver, J. R. Suriano, and M. Green, *J. Virol.*, **1**, 723 (1967).
95. W. C. Russell and B. E. Knight, *J. Gen. Virol.*, **1**, 523 (1967).
96. L. A. MacHattie, D. A. Ritchie, and C. A. Thomas, Jr., *J. Mol. Biol.*, **23**, 355 (1967).
97. M. Green and C. Smith, unpublished data.
98. H. S. Ginsberg, H. G. Pereira, R. C. Valentine, and W. C. Wilcox, *Virology*, **28**, 782 (1966).
99. U. Pettersson, L. Philipson, and S. Hoglund, *Virology*, **33**, 575 (1967).
100. G. Henle, W. Henle, and V. Diehl, *Proc. Natl. Acad. Sci. U.S.*, **59**, 94 (1968).
101. M. Epstein, G. Henle, B. Achong, and Y. Bark, *J. Exptl. Med.*, **121**, 761 (1965).
102. W. Josey, A. Nahmias, and Z. Narb, personal communication.
102a. W. E. Rawls, W. A. F. Tompkins, M. E. Figueroa, and J. L. Melnick, *Science*, **161**, 1255 (1968).
103. M. Mizell, C. W. Stackpole, and S. Halperin, *Proc. Soc. Exptl. Biol. Med.*, **127**, 808 (1968).
104. R. W. Darlington and L. H. Moss, *J. Virol.*, **2**, 48 (1968).
105. P. G. Spear and B. R. Roizman, *Nature*, **214**, 713 (1967).
106. W. Munyon, E. Paoletti, J. Ospina, and J. T. Grace, Jr., *J. Virol.*, **2**, 167 (1968).
107. C. C. Randall, L. G. Gafford, R. W. Darlington, and J. Hyde, *J. Bacteriol.*, **87**, 939 (1964).
108. W. K. Joklik, *Virology*, **18**, 9 (1962).
109. W. K. Joklik, *Bacteriol Proc.*, **30**, 33 (1966).
110. P. J. Gomatos and I. Tamm, *Proc. Natl. Acad. Sci. U.S.*, **49**, 707 (1963).

111. E. R. Pfefferkorn and H. S. Hunter, *Virology,* **20,** 433 (1963).

112. G. L. Ada and B. T. Perry, *Nature,* **175,** 854 (1955).

113. L. H. Frommhagen, C. A. Knight, and N. K. Freeman, *Virology,* **8,** 176 (1959).

114. M. Kates, A. C. Allison, D. A. J. Tyrrell, and A. T. James, *Biochem. Biophys. Acta,* **52,** 455 (1961).

115. R. A. Bonar and J. W. Beard, *J. Natl. Cancer Inst.,* **23,** 183 (1959).

116. M. J. Lyons and D. H. Moore, *J. Natl. Cancer Inst.,* **35,** 549 (1965).

116a. P. H. Duesberg, *Proc. Natl. Acad. Sci. U.S.,* **60,** 1511 (1968).

117. F. A. Anderer and H. Restle, *Z. Naturforsch.,* **19b,** 1026 (1964).

118. P. Hausen and W. Schafer, *Z. Naturforsch.,* **17b,** 15 (1962).

119. D. G. Scraba, C. M. Kay, and J. S. Colter, *J. Mol. Biol.,* **26,** 67 (1967).

120. L. Montagnier and F. K. Sanders, *Nature,* **197,** 1177 (1963).

121. W. Schafer, K. Munk, and O. Armbruster, *Z. Naturforsch.,* **7b,** 29 (1952).

122. S. McGregor and H. D. Mayor, *J. Virol.,* **2,** 149 (1968).

123. H. L. Bachrach, R. Trautman, and S. S. Breese, Jr., *Am. J. Vet. Res.,* **25,** 333 (1964).

124. C. F. T. Mattern, *Virology,* **17,** 520 (1962).

125. F. L. Schaffer, H. F. Moore, and C. E. Schwerdt, *Virology,* **10,** 530 (1960).

126. J. J. Holland, *Proc. Natl. Acad. Sci. U.S.,* **48,** 2044 (1962).

127. P. J. Gomatos and I. Tamm, *Proc. Natl. Acad. Sci. U.S.,* **50,** 878 (1963).

128. A. R. Bellamy, L. Shapiro, J. T. August, and W. K. Joklik, *J. Mol. Biol.,* **29,** 1 (1967).

129. V. Stollar, T. M. Stevens, and R. W. Schlesinger, *Virology,* **30,** 303 (1966).

129a. J. A. Sonnaben, E. M. Martin, and E. Mecs, *Nature,* **213,** 365 (1967).

130. G. L. Ada and B. T. Perry, *J. Gen. Microbiol.* **14,** 623 (1956).

131. P. H. Duesberg and W. S. Robinson, *Proc. Natl. Acad. Sci. U.S.,* **54,** 794 (1965).

132. D. W. Kingsbury, *J. Mol. Biol.,* **18,** 195 (1966).

133. W. S. Robinson and M. A. Baluda, *Proc. Natl. Acad. Sci. U.S.,* **54,** 1686 (1965).

134. H. L. Robinson, *Proc. Natl. Acad. Sci. U.S.,* **57,** 1655 (1967).

135. J. Huppert, F. Lacour, J. Harel, and L. Harel, *Cancer Res.,* **26,** 1561 (1966).

136. P. H. Duesberg and W. S. Robinson, *Proc. Natl. Acad. Sci. U.S.,* **55,** 219 (1966).

137. F. Galibert, C. Bernard, Ph. Chenaille, and M. Boiron, *Nature,* **209,** 680 (1966).

138. H. A. Scheraga and I. Mandelkern, *J. Am. Chem. Soc.,* **75,** 179 (1953).

139. J. H. Strauss, personal communication.

140. A. Igarashi, T. Fukunaga, and K. Fukai, *Biken's J.,* **7,** 111 (1964).

141. A. S. Spirin, *Progr. Nucleic Acid Res.,* **1,** 301 (1963).

142. Y. Watanabe and A. F. Graham, *J. Virol.,* **1,** 665 (1967).

143. T. Dunnebacke and A. Kleinschmidt, *Z. Naturforsch.,* **22b,** 159 (1967).

144. A. R. Bellamy and W. K. Joklik, *Proc. Natl. Acad. Sci. U.S.,* **58,** 1389 (1967).

145. A. J. Shatkin and J. D. Sipe, *Proc. Natl. Acad. Sci. U.S.,* **59,** 246 (1968).

146. H. O. Agrawal and G. Bruening, *Proc. Natl. Acad. Sci. U.S.,* **55,** 818 (1966).

147. M. W. Pons, *Virology,* **31,** 523 (1967).

148. F. Sokol and S. Schramek, *Acta Virol.,* **8,** 193 (1964).

149. D. P. Nayak and M. A. Baluda, *J. Virol.,* **1,** 1217 (1967).

150. P. Davies and R. D. Barry, *Nature,* **211,** 384 (1966).

151. R. W. Compans and P. W. Choppin, *Virology,* **33,** 344 (1967).

152. D. W. Kingsbury and R. W. Darlington, *J. Virol.,* **2,** 248 (1968).

153. J. Harel, J. Huppert, F. Lacour, and L. Harel, *Compt. Rend.,* **261,** 4559 (1965).

154. N. Granboulan, J. Huppert, and F. Lacour, *J. Mol. Biol.,* **16,** 571 (1966).

155. P. T. Mora, V. W. McFarland, and S. W. Luborsky, *Proc. Natl. Acad. Sci. U.S.,* **55,** 438 (1966).

156. P. H. Duesberg and P. B. Blair, *Proc. Natl. Acad. Sci. U.S.,* **55,** 1490 (1966).

157. C. E. Schwerdt and F. L. Schaffer, *Ann. N.Y. Acad. Sci.,* **61,** 740 (1955).

158. M. A. Moscarello and M. E. Kaighn, *Biochim. Biophys. Acta.,* **90,** 161 (1964).

159. H. L. Bachrach and J. Polatnick, *Proc. Soc. Exptl. Biol. Med.,* **124,** 465 (1967).

160. A. Boeye, *Virology,* **25,** 550 (1965).

161. R. R. Rueckert and P. H. Duesberg, *J. Mol. Biol.,* **17,** 490 (1966).

162. R. R. Rueckert, personal communication.

163. Y. Watanabe, L. Prevec, and A. F. Graham, *Proc. Natl. Acad. Sci. U.S.,* **58,** 1040 (1967).

164. A. R. Bellamy and W. K. Joklik, *J. Mol. Biol.,* **29,** 19 (1967).

165. A. R. Taylor, D. G. Sharp, D. Beard, and J. W. Beard, *J. Infect. Diseases,* **72–73,** 31 (1943).

166. P. Cheng, *Virology,* **14,** 124 (1961).

167. R. F. Wachter and E. M. Johnson, *Federation Proc.,* **21,** 461 (1962).

168. F. L. Schaffer, *Cold Spring Harbor Symp. Quant. Biol.,* **27,** 89 (1962).

169. G. L. Ada, A. Abbot, S. G. Anderson, and F. D. Collins, *J. Gen. Microbiol.,* **29,** 165 (1962).

170. E. R. Pfefferkorn and H. S. Hunter, *Virology,* **20,** 446 (1963).

171. E. R. Pfefferkorn and R. L. Clifford, *Virology,* **23,** 217 (1964).

172. R. M. Friedman, H. B. Levy, and W. B. Carter, *Proc. Natl. Acad. Sci. U.S.,* **56,** 440 (1966).

173. N. H. Acheson and I. Tamm, *Virology,* **32,** 128 (1967).

174. W. Schafer, *Bacteriol. Rev.,* **27,** 1 (1963).

175. P. H. Duesberg, *Proc. Natl. Acad. Sci. U.S.,* **59,** 930 (1968).

176. L. W. Mayron, B. Robert, R. J. Winzler, and M. E. Rafelson, *Arch. Biochem.,* **92,** 475 (1961).

177. H. Noll, T. Aoyagi, and J. Orlando, *Virology,* **18,** 154 (1962).

178. M. E. Rafelson, Jr., M. Schneir, and V. W. Wilson, Jr., *Arch. Biochem. Biophys.,* **103,** 424 (1963).

179. J. T. Seto, K. Okuda, and Y. Hokama, *Nature,* **213,** 188 (1967).

180. A. P. Waterson and J. G. Cruickshank, *Nature,* **201,** 640 (1964).

181. R. W. Compans and P. W. Choppin, *Proc. Natl. Acad. Sci. U.S.,* **57,** 949 (1967).

182. R. D. Barry and A. G. Bukrinskaya, *J. Gen. Virol.,* **2,** 71 (1968).

183. L. Prevec and G. F. Whitmore, *Virology,* **20,** 464 (1963).

184. H. M. Chamsy and P. D. Cooper, *Virology,* **20,** 14 (1963).

185. E. Reczko, *Arch. Ges. Virusforschg.,* **10,** 588 (1960).

186. A. F. Howatson and G. F. Whitmore, *Virology,* **16,** 466 (1962).

187. R. W. Simpson and R. E. Hauser, *Virology,* **29,** 654 (1966).

188. R. M. McComb, M. Benyesh-Melnick, and J. P. Brunschwig, *J. Bacteriol.,* **91,** 803 (1966).

189. K. Hummeler, H. Koprowski, and T. J. Wiktor, *J. Virol.,* **1,** 152 (1967).

190. R. J. Huebner, *Proc. Natl. Acad. Sci. U.S.,* **58,** 835 (1967).

191. W. S. Robinson, A. Pitkanen, and H. Rubin, *Proc. Natl. Acad. Sci. U.S.,* **54,** 137 (1965).

192. P. R. Rao and J. W. Beard, *Natl. Cancer Inst. Monograph,* **17,** 673 (1964).

193. M. Johnson and P. Mora, *Virology,* **31,** 230 (1967).

194. H. Bauer and W. Schafer, *Z. Naturforsch.,* **20b,** 815 (1965).

195. D. W. Allen, *Biochim. Biophys. Acta,* **133,** 180 (1967).

196. W. S. Robinson and P. H. Duesberg, in *Molecular Basis of Virology* (H. Fraenkel-Conrat, ed.), Reinhold, New York, 1968, p. 306–332.

VIRUS ARCHITECTURE AS DETERMINED BY X-RAY DIFFRACTION AND ELECTRON MICROSCOPY

CARL F. T. MATTERN

LABORATORY OF VIRAL DISEASES
NATIONAL INSTITUTE OF ALLERGY AND INFECTIOUS DISEASES
NATIONAL INSTITUTES OF HEALTH
BETHESDA, MARYLAND

2-1. Introduction

This review is of necessity limited in scope; it was undertaken with a 4-fold intent. First, to review some of the pertinent background by

which we have seen in the past fourteen years the concepts of rod-
shaped and spherical viruses change to those of helical and icosahedral
viruses. Second, to provide a sufficiently detailed description of icosahedral symmetry to permit virologists who are not oriented toward
ultrastructure to evaluate some of the difficulties in the interpretation
of virus architecture, whether studied by X-ray diffraction or by
electron microscopy. Third, to review some of the major architectural
groups of viruses, irrespective of host. Fourth, to appraise, from the
biologist's viewpoint, our current knowledge of the architecture of
icosahedral viruses and to discuss some of the remaining problems.

As with most reviews, this one is admittedly incomplete, particularly
in regard to the non-English literature. The author apologizes to his
foreign colleagues for his inadequacy in other languages. Other failures
to cite pertinent literature are sins of omission.

In view of a recent and comprehensive review of bacterial viruses by
Bradley (18) many of the classical bacterial viruses are not described
here. As in the case of bacteriophages, insect viruses include many
which differ greatly in morphology and development from the major
virus types discussed in this review. The reader is referred to a book
devoted to this subject by Smith (160).

Since this manuscript was prepared two excellent reviews have
become available. One, by Finch and Holmes (46) is of special interest
to virologists who may wish to seriously study X-ray diffraction and
to those interested in the details of electron microscopic image analysis.
The other, by Horne (75), provides detailed accounts of electron
microscopic techniques and procedures as related to the examination
of isolated viruses.

2-2. X-Ray Diffraction and Virus Structure

No discussion of X-ray diffraction of viruses can be undertaken
without mentioning the work with plant viruses which constitute the
first and most extensively studied viruses. In fact, as will be seen,
poliovirus is the only animal virus in which X-ray crystallography
has yielded information on the structure of the virus particle.

A. Tobacco Mosaic Virus (TMV)

TMV is the first virus to be studied in detail and its structure, which
is now well established, has been largely deduced from X-ray studies

and recently confirmed by electron microscopy (*44*). X-ray studies have been performed on gels of oriented viruses. True crystals of TMV, such as those found naturally (*162, 163*), have not been examined, largely because of their extreme instability.

In view of comprehensive reviews of TMV structure (*24, 25, 97*) only certain highlights of these studies will be briefly outlined.

(1) Initial studies by Bernal and Fankuchen (*12*) and subsequently by Franklin (*55*) indicated that the virus had a basic axial repeating unit of 69 Å with a subrepeating spacing of 23 Å.

(2) Watson (*175*) suggested that the virus consisted of helically arranged subunits of protein with $3n + 1$ (n, integral) identical protein units in three turns of the helix, each turn of the helix advancing 23 Å. Thus there would be a nonintegral number of subunits in each turn and an integral number in three turns of the helix, the basic 69 Å. Further evidence for such a structure was obtained by an analysis of layer line splitting (*58*).

(3) Franklin and Holmes (*57*) determined n to be 16 by employing TMV to which a methyl mercury group had been bound to each protein subunit's single cysteine residue. There are 49 subunits per 69 Å [actually 49.02 ± 0.01 (*58*)].

(4) Caspar (*23*) calculated the radial density distribution of TMV. He found no density out to a radius of 20 Å nor any beyond 90 Å, with four density maxima in between. Thus the particle was a hollow cylinder. The most dense region was at a radius of 40 Å. Concurrently, Franklin (*56*) observed that the absence of this dense peak was essentially the only difference between the radial density distribution of TMV and that of TMV protein which had been reaggregated into TMV-like rods containing no RNA. Therefore it was concluded that the RNA of TMV produced this high-density region at a radius of 40 Å. The basic plan of TMV structure is shown in Fig. 2-1.

Other basic data for the TMV particle and details of their derivation are provided in reviews by Caspar (*24, 25*). This structural information coupled with the determination of the complete amino acid sequence and initial investigations of the nucleotide arrangement in TMV RNA [reviewed by Anderer (*3*)] place this virus in a unique position with respect to our knowledge of its structure.

The knowledge gained from the study of the helical TMV may well have application in the animal virus field since myxoviruses and other viruses, as will be discussed later, appear to have a helical nucleoprotein core component. If such structures could be isolated and

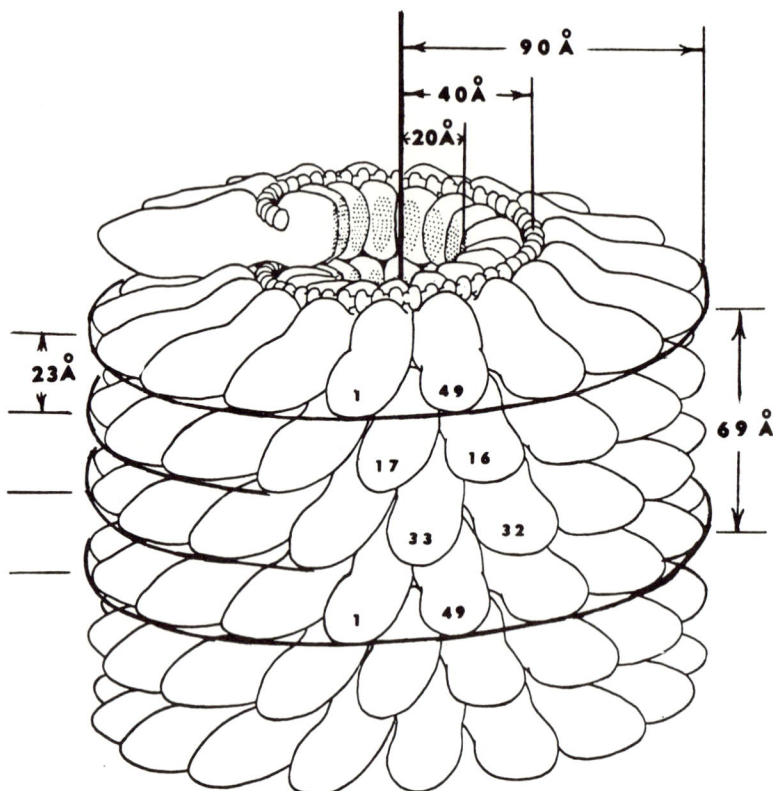

Fig. 2-1. Schematic representation of TMV structure derived largely from X-ray diffraction studies (modified from Ref. 24). TMV protein subunits (M.W. 17,500) are shown arranged to form about 6 helical turns whose pitch is 23 Å. There are 16⅓ protein units per turn and 49 units in the fundamental 69 Å repeat distance. The protein forms a cylinder with a hollow axial core of 40 Å diameter and with the maximum diameter of the protein about 180 Å. The RNA courses between the subunits forming a helix of 80 Å diameter with 49 nucleotides (three per subunit) in each turn. It cannot be determined from X-ray studies whether TMV is a right- or left-handed helix.

oriented, X-ray analysis would seem feasible. Some preliminary X-ray diffraction studies of two other helical plant viruses are discussed in a later section (2-3, D, 2).

B. "Spherical Viruses"

The concept of regular packing of identical viral subunits originated in the study of TMV. This concept was extended to sperical viruses

(*33*) and verified by Caspar (*22*). The understanding of this development requires some knowledge of point-group theory.

1. *Point Groups*

Point groups represent the possible ways in which elements can be
regularly arranged in three dimensions about a point as center of
symmetry. The theory deals only with symmetry elements and has
nothing to do with morphology.

Since the biochemical subunit is protein, and proteins are asymmetric and do not exist in dextro and levo forms, the only possible
symmetry properties applicable to viruses are rotation axes. Mirror
planes and inversion axes and other combinations of symmetries are
impossible without D and L forms of subunits. However, as will be
seen, at the capsomere level of resolution, mirror planes and inversion
axes can be simulated in the electron microscope.

Point groups which could apply to viruses have been enumerated
(*33, 97*). There are only five such point groups. The first has only one
n-fold rotational symmetry axis. The second has a similar n-fold axis
and in addition has n 2-fold axes perpendicular to the primary axis.
The remaining three groups are cubic point groups having **23, 432,** and
532 symmetry. Pertinent information and examples of polyhedra with
these symmetries are presented in Fig. 2-2. The location of asymmetric
units is shown on characteristic faces. Note that these units could have
been drawn on the faces in any of an infinite number of positions; they
only need to conform to the symmetry requirements.

These arrangements are the only possible arrangements for identical, asymmetric protein subunits to be placed in equivalent positions
(identical binding sites being involved) so as to result in a protein
shell. It should also be noted that each morphological subunit (capsomere) could itself be symmetrical, having multiple structure units
(biochemical subunits) and still fill the symmetry requirements. However, in such a case the structure units would not have identical environments, but would be "quasiequivalent" (*27*).

2. *Bushy Stunt Virus (BSV) and 532 Symmetry*

Caspar (*22*) obtained X-ray diffraction pictures of crystals of
bushy stunt virus and was able to deduce from them that the virus
had at least tetrahedral symmetry (**23**) and probably a higher symmetry (**532**).

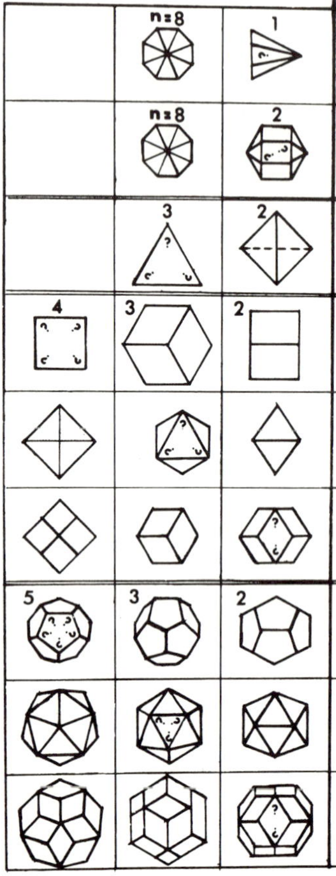

		Name of platonic solid	Crystal. symbol	Rotation axes	Asymmetric units	
					No.	Location
n=8	1	n-fold pyramid	n	1, n-fold	n	1 on each of n faces
n=8	2	bipyramidal n-fold prism	n,2	1, n-fold n, 2-fold	2n	2 on each of n faces
3	2	tetrahedron	23	3, 2-fold 4, 3-fold	12	3 on each of 4 triangular faces
4 / 3	2	cube				4 on each of 6 square faces
		octahedron	432	3, 4-fold 4, 3-fold 6, 2-fold	24	3 on each of 8 triangular faces
		rhombic dodecahedron				2 on each of 12 rhombic faces
5 / 3	2	pentagonal dodecahedron				5 on each of 12 pentagonal faces
		icosahedron	532	6, 5-fold 10, 3-fold 15, 2-fold	60	3 on each of 20 triangular faces
		rhombic triaconta- hedron				2 on each of 30 rhombic faces

Fig. 2-2. Point groups which could represent the symmetry of "spherical" viruses, modified from Crick and Watson (*33*) and Klug and Caspar (*97*). In the case of *n*-fold rotational symmetry, the virus shell could resemble the segments of an orange (*97*). Note that the asymmetric units, shown as question marks, could be projected onto spheres inscribed within each of the platonic solids and maintain the same symmetry axes.

There were two basic concepts involved in arriving at this conclusion. The first is that if a crystal lattice has a given symmetry, in this case **23**, then the unit cell contents must have at least this rotational symmetry, perhaps higher but not a lower symmetry (*74, 108*).

That the unit cell was a body-centered-cubic cell was determined by the X-ray diffraction pictures. The dimensions of the unit cell were calculated and the diameter of the virus was well enough known from electron microscopy to determine that there could only be two viruses per unit cell (one located at each point of the body-centered lattice). The lattice symmetry was determined to be 23 by X-ray patterns produced by "viewing" the lattice down a cubic edge (showing 2-fold symmetry)† and a cubic diagonal (showing 3-fold symmetry). The reciprocal lattice and its symmetry revealed that the two viruses per unit cell were similarly oriented with respect to their axes. Therefore each virus must have at least tetrahedral symmetry (23) and consist of 12 (or a multiple of 12) subunits.

The second concept involved the appearance of spikes of intensity which were not coincident with the crystallographic axes, that is, not coincident with axes related to tetrahedral symmetry. Since the two viruses in each unit cell were oriented identically, the X-ray precession pattern represented a transform of the individual particle. The location of the spots in the X-ray picture along with systematic absences of spots is dependent upon the crystal lattice and its symmetry, but the relative intensity of all spots permissable by the lattice symmetry is dependent upon the summated contribution of every atom within the unit cell diffracting in those directions producing spots. Repeating aggregates of atoms such as subunits can be expected to manifest themselves in the X-ray pattern as reinforcements of intensity of certain spots, particularly those lying along axes of symmetry of the unit cell contents. This has been subsequently shown in the optical transform of a 60 unit structure having 5, 3, 2 symmetry (27). Caspar (22) noted that the spikes in the bushy stunt pattern were related to axes of a particle of higher symmetry than the lattice. There were the 3, 2-fold axes and 4, 3-fold axes of tetrahedral symmetry plus additional 2-, 3-, and 5-fold axes unrelated to tetrahedral symmetry but rather to icosahedral symmetry. The coincidence of such noncrystallographic 3- and 5-fold axes with those of an object with 5, 3, 2 symmetry may be seen in Fig. 2-3. Noncrystallographic

---

† The mirror symmetry shown by the X-ray pattern normal to a 2-fold axis does not imply a similar mirror symmetry of the unit cell. It follows from Friedel's law that the symmetry of the diffraction pattern of a crystal is that of the point-group symmetry plus (usually) a center of symmetry. The center of symmetry plus 2-fold axes results in planes of symmetry.

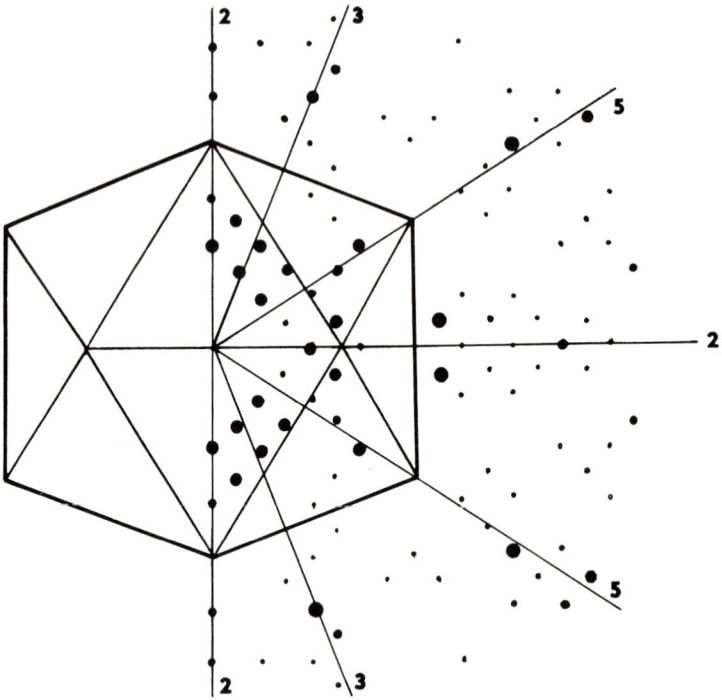

FIG. 2-3. X-ray pattern (from *22*) of a bushy stunt virus crystal "viewed" normal to a cube edge (2-fold crystallographic axis). The pattern represents the basal plane (110) of the reciprocal lattice. An icosahedron is shown in "reciprocal space" showing the coincidence of its 3- and 5-fold (noncrystallographic) axes (arrows) with the high-intensity spikes of the X-ray pattern. Two-fold axes of the icosahedron and the lattice coincide in this view.

2-fold axes were only seen in a view (not reproduced) down a 3-fold axis of the lattice (and virus). Unfortunately the 5-fold axes being unrelated to the lattice axes renders their direct demonstration (viewing on a 5-fold axis of the virus) difficult.

Thus it was deduced that bushy stunt virus had some sort of 532 symmetry. If there were a strict 532 symmetry of the protein shell, then there would have to be $n \cdot 60$ (n, integral) biochemical (asymmetric) subunits on its surface. Biochemical data on molecular weight of the virus and amino acid analysis, available at that time, suggested that there might be many as 300 chemical subunits (*22*). Subsequent estimates (*97*) based on end-group analysis suggested possibly

120 protein subunits. It should be noted that this was, in fact, the first evidence for "icosahedral" symmetry in a "spherical" virus.

3. *Turnip Yellow Mosaic Virus* (*TYMV*)

The structure of this virus is extensively discussed in a recent paper by Klug et al. (*102*). Previous investigations (*98, 101*) showed that the lattice of TYMV crystals was complicated by the fact that half of the particles in the unit cell were oriented with 90° rotation from the other half when viewed on the lattice cube face. This produced in the X-ray precession patterns a double set of spikes again demonstrating 532 symmetry. Top component produced relatively stronger spikes than the virus indicating that the protein was the source of the 532 symmetry. It was again concluded that the virus contained 60 (or $n \cdot 60$) protein subunits.

However, electron microscopic observations by Huxley and Zubay (*88*) and Nixon and Gibbs (*140*) convincingly demonstrated that there were 32 morphological units (capsomeres) on the surface of TYMV arranged with some form of 532 symmetry, namely one capsomere at each of 12 vertices of the icosahedron and one in each of the 20 faces.

A solution to this apparent discrepancy was presented by Klug and Finch (*98*) and Klug and Caspar (*97*) which led to the concept of "quasi-equivalent" positions (*27*). It was shown earlier that only 60 asymmetric subunits could be arranged in identical environments to give 532 symmetry although $n \cdot 60$ asymmetric subunits could satisfy 532 symmetry. Caspar and Klug (*27*) proposed that a shell of $n \cdot 60$ subunits were held together by similar types of binding and that the amount of nonequivalence of environments could be compensated for by a reasonable amount of flexion of these identical bonds. Thus on the basis of quasi-equivalent environments they proposed that the protein shell of TYMV consisted of 20 hexameric capsomeres, one on each icosahedral face, and 12 pentameric capsomers, one at each vertex, thus a total of 32 capsomeres. The 180 protein subunits were visualized as being chemically identical and held together throughout by the same type of binding.

The most recent biochemical data [reviewed by Matthews and Ralph (*126*)] are entirely compatible with the 180 subunit model. Chemical analysis of the protein indicates 189 amino acids and a subunit molecular weight of 20,000 (*64, 164*). The best value for the RNA content is 33.5% as calculated by Kaper and Litjens (*90*). Mitra and Kaesberg (*131*) have estimated the RNA to have a molecular weight of 1.9×10^6.

Since

$$\%\text{RNA} = \frac{\text{M.W. RNA} \times 100}{\text{M.W. RNA} + s \times \text{M.W. protein subunit}}$$

the number of subunits, s, can be calculated from the above data to be 190. Calculation from an assumed virus molecular weight of 5.5×10^6 [estimates of 5.13–5.8×10^6 cited by Matthews and Ralph (*126*)] and the same subunit and RNA molecular weight figures, give a value of 180 subunits. These data undoubtedly are accurate enough to eliminate 120 and 240 as possible subunit numbers, but they do not in themselves establish the value as precisely 180.

Several other points have developed from the most recent studies of Klug et al. (*102*) on virus and top component crystals of TYMV.

(a) The RNA of TYMV also has icosahedral symmetry to a resolution of about 20 Å.

(b) The protein subunits have an effective radius of 145 Å and that of the RNA is about 125 Å. Presumably the RNA protrudes into the hollow capsomeres.

(c) Calculated transforms of models were compared to the X-ray diffraction transforms. The calculated transforms dealt with a level of resolution of only about 30 Å but probably revealed locations of subunits with respect to their rotational positions about 5- and 3-fold virus axes.

The arrangement of 180 protein subunits and 32 knobs of RNA as proposed by Klug et al. (*102*) is presented in Fig. 2-4.

4. *Poliomyelitis Virus*

The symmetry of the poliovirus crystal lattice (*47*) is lower than that of either BSV or TYMV. Since it is at best orthorhombic (to 8 Å resolution) the virus need only have two subunits. However, precession diffraction pictures of a basal reciprocal lattice normal to a 2-fold axis revealed spikes of intensity corresponding to noncrystallographic 5- and 3-fold axes indicating that polio virus also had cubic point-group symmetry 532.

The biochemical data on poliovirus proteins is complicated by the identification of multiple peptide chains and presumably multiple

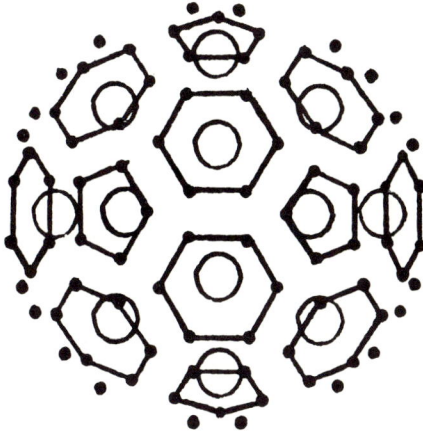

FIG. 2-4. The arrangement of 180 protein subunits (dots) at an effective radius of 145 Å and 32 knobs of RNA (circles) at an effective radius of 125 Å, proposed by Klug et al. (*102*) as the structure of TYMV.

biochemical subunits (*111*). There may be three capsomere proteins and a fourth internal protein. Proteins VP-1 to VP-3 are present in purified virus and artificially produced top component. VP-4 may be an internal protein since it is present in small amount (5%) only in complete virus and is solubilized by conversion of virus into top component. Natural top component contains proteins VP-1 to VP-3 but VP-2 is greatly reduced in amount over whole virus. However, another protein, VP-6, appears in large quantity in natural top component but is present in small amount in whole virus and is easily removed by purification without affecting virus infectivity. It is not seen in artificial top component.

At the present time it is not possible to determine how these proteins fit into the 532 symmetry scheme derived from X-ray diffraction. Knowledge of the molar ratios and molecular weights of the three capsomere proteins may facilitate such a determination.

2-3. The Electron Microscope

Electron microscopy has been the most frequently utilized technique in the study of virus structure, although a number of the more fundamental properties of virus architecture, as was seen, were first determined by X-ray diffraction studies which led to the new concepts of viruses with helical and icosahedral symmetry in place of rod-shaped and spherical viruses.

A. METAL SHADOWING

The first generally employed technique for enhancing the contrast of isolated virus particles (necessary for examination of details) was that of metal shadowing originated by Williams and Wyckoff (*182*). A film of any of a number of metals (*16*) may be applied by evaporation, generally at an angle to the grid surface, resulting in the piling up of metal on the near edge of the virus and the casting of a shadow (absence of metal) on the grid distal to the virus. Unfortunately this technique has not been especially useful in any way other than to elucidate the general morphology and size of viruses. Fine structure of viruses is usually obscured by the metal coating or by a granular texture of the background similar in size to structures sought on the surface of the virus. There are several notable exceptions to these generalities.

Employing direct metal shadowing, Williams (*179*) showed that rabbit papilloma virus had a surface consisting of somewhat ordered knobs.

Steere (*162*) replicated a cleaved surface of a frozen crystal of *turnip yellow* mosaic virus and demonstrated a six-coordinated (six around one) nodular surface structure, the true nature of which was understood only after the development of negative staining techniques. Similarly, Labaw and Wyckoff (*103*) noted a "granular texture" on the surface of *tobacco necrosis* virus in replicated crystals.

The first determination of the icosahedral morphology of a virus by electron microscopy was accomplished by Williams and Smith (*181*). They employed double shadowing from two different angles, and from the shadow shapes they were able to determine the icosahedral form of *Tipula* irridescent virus, a large insect virus. Hall et al. (*62*) clearly resolved the 12 capsomere (morphological subunit) structure of ϕX174, a small bacterial virus. The latter study appeared simultaneously with the first successful application of negative staining in the determination of the icosahedral nature of adenovirus by Horne et al. (*76*).

More recently the *blue green* algal virus was shown to have icosahedral morphology by means of metal shadowing (*157*).

Metal shadowing has only rarely revealed virus substructure and this has only been at the level of the capsomere or morphological sub-unit involving resolution of about 70–90 Å. Actually, a considerable confusion has arisen from the use of metal shadowing in the study of the fine structure of tobacco mosaic virus. Many investigators were unable to demonstrate any surface structure; others, a variety of surface structures incompatible with the 23 Å periodicity unequivo-cally established by X-ray diffraction studies. Some of these difficul-ties may be explained by regular or semiregular perturbations of the virus surface structure upon drying (*24, 119*). Hart (*68, 69*) has been the only investigator who felt he had resolved the 23 Å periodicity by shadowing. However, his evidence is not as convincing as that subse-quently obtained by negative staining (*44*).

For additional information on earlier studies of viruses which utilized metal shadowing, the interested reader may refer to a discus-sion by Williams (*180*).

B. NEGATIVE STAINING, THE ICOSAHEDRON, AND ICOSAHEDRAL VIRUSES

1. *Negative Staining*

The technique of negative staining is extremely simple. Virus is mixed with a solution of neutral salt of phosphotungstic acid to a final concentration of 1 or 2% PTA. Solutions of uranyl acetate or formate have also been employed (*105*). The mixture may be sprayed onto a membrane-coated grid or, more frequently, applied as a droplet with the excess drawn off with a pipet or filter paper. The dry film may be immediately examined. The PTA film is relatively electron dense; the virus relatively electron lucent. The virus and projecting subunits or other protruding structures are often clearly delineated by the stain of contrasting density. The greatest utility of the technique, therefore, is in determining virus surface structure.

The first structural detail of a virus to be revealed by negative staining was accomplished by Huxley (*87*) who confirmed the hollow axial core in tobacco mosaic virus which had previously been deter-mined from X-ray diffraction studies. Horne et al. (*76*) employed neutral phosphotungstate (PTA) (*20*) to visualize the icosahedral shape of an adenovirus whose surface consisted of approximately spherical subunits of about 70 Å center-to-center spacing. The sub-units were regularly arranged on triangular faces, 6 subunits on shared triangle edges. Two axes of 5-fold rotational symmetry were identified on a particle viewed on a 2-fold rotational axis, establishing that the surface of this virus consisted of 252 subunits (now called capsomeres) arranged so as to form an icosahedral shell.

A very great number of viruses have now been examined by this technique, with perhaps the most rewarding results with icosahedral viruses and bacteriophages. However, substantial disagreement has arisen in the interpretation of electron micrographs of some negatively stained viruses, it therefore would seem appropriate to discuss icosahedral symmetry and interpretation in sufficient detail to permit understanding by persons other than cystallographers and electron microscopists.

2. *The Geometric Icosahedron*

The icosahedron is seen normal to 5-, 3-, and 2-fold axes of rotational symmetry in Fig. 2-5. There are 6, 5-fold inversion axes ($\bar{5}$) which pass through opposite vertices (12 vertices); 10, 3-fold inversion axes ($\bar{3}$) which pass through the centers of opposite faces (20 faces); and 15, 2-fold axes with perpendicular mirror planes ($\frac{2}{m}$) pass through midpoints of opposite edges (30 edges). The symmetry class or point group is $\bar{5}\ \bar{3}\ \frac{2}{m}$, but since the $\frac{2}{m}$ axes imply the rotoinversion axes, symmetry class is generally expressed $5\ 3\ \frac{2}{m}$. There is only one view, namely that normal to a 2-fold axis, in which the upper and lower surfaces of the polyhedron are mirror imaged through the plane of the paper which passes through the center of the polyhedron and normal to a 2-fold axis. That is, only as seen on a 2-fold axis do the vertices, lines, and faces of the upper and lower surfaces coincide. This is seen in Fig. 2-5 where the dashed lines, which represent the under surface, do not coincide with lines in the upper surface when viewed on 5- or 3-fold rotational symmetry axes, but are offset by 180°.

Therefore in examining icosahedral viruses by negative staining, it is only on a 2-fold axis that capsomeres on upper and lower surfaces will produce a superimposed image (reinforcement); in all other views contribution to the image by both surfaces will cause some degree of confusion. As will be seen, however, reinforcement of the image seen on a 2-fold axis does not occur with all "icosahedral" viruses.

Another point to be made from Fig. 2-5 concerns the dimensions of the polyhedron. Considering an edge to be of unit length, the face-to-opposite-face distance (F) is 1.539, the edge-to-edge distance (E) is

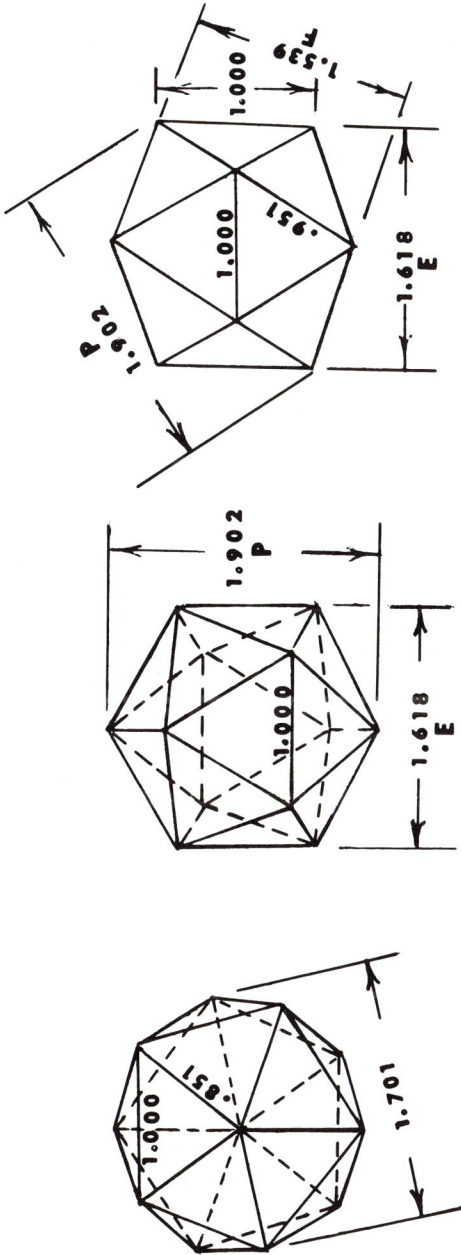

FIG. 2-5. The icosahedron seen normal to 5-, 3-, and 2-fold rotational axes. Dimensions are given relative to an edge length of unity; P, point-to-point; E, edge-to-edge; and F, face-to-face. Note the superposition of edges of upper and lower surfaces only in the projection normal to the 2-fold axis.

1.618, and the point-to-point distance (P) is 1.902. This clearly points
up the difficulty in determining the dimension of such a polyhedral
virus unless its orientation is known and its peripheral portions can
be clearly visualized, conditions which are seldom met even in orderly
arrays of viruses. However, if the centers of 5-fold axes can be
identified and measured, other specific dimensions can be calculated
with a reasonable degree of accuracy.

3. Viruses and Icosahedral Symmetry

The term "icosahedral" is applied to any virus which has 532 point-
group symmetry. Strictly speaking it is incorrect since the icosahedron has $5\,3\,\frac{2}{m}$ symmetry, and only a limited number of viruses even
mimic this symmetry if one considers capsomeres as spherically symmetrical (low resolution). As previously discussed viruses "seen" at
high resolution cannot have mirror symmetry. However, the terminology has crept into general use and has the virtue of brevity.

Icosahedral viruses can be divided into two groups: (1) closed-shell
icosahedra in which vertices and faces are completely covered or
filled with uniformly spaced capsomeres and (2) structures in which
there may be regular holes or openings at vertices, or in lines or faces.

Goldberg (61) has shown how all such closed polyhedra can be
generated by selecting various 5-fold axes from a hexagonal net. His
scheme is modified in Fig. 2-6. In the general case, the number of
capsomeres $C = 10(h^2 + hk + k^2) + 2$, where h and k are integers,
representing coordinates on the inclined (60°) axes h and k. After the
notation of Caspar and Klug (27), $h^2 + hk + k^2 = T$, the triangula-
tion number. Three classes of polyhedra are generated. Along the
h axis, $k = 0$, and the equation simplifies to $C = 10h^2 + 2$ ($T = 1$
series). Thus selecting 5-fold axes from units along the h axis generates
a series of polyhedra containing 12, 42, 92, 162, 252, etc., capsomeres.
These are true icosahedral structures to the extent that capsomered
structures can be called polyhedra. Their faces can be seen in Fig.
2-6B to have close-packed capsomeres aligned along their edges.

A second set of polyhedra is generated by selecting 5-fold axes along
the line which is inclined 30° from the h axis (Fig. 2-6C). Here $h = k$
and the equation for capsomere number simplifies to $30h^2 + 2$ ($T = 3$
series), generating series of deltahedra in which $C = 32, 122, 272, 482$,
etc. The facets of these deltahedra are formed by lines joining neighboring capsomeres.

In this series capsomeres are not close packed along the edges. These

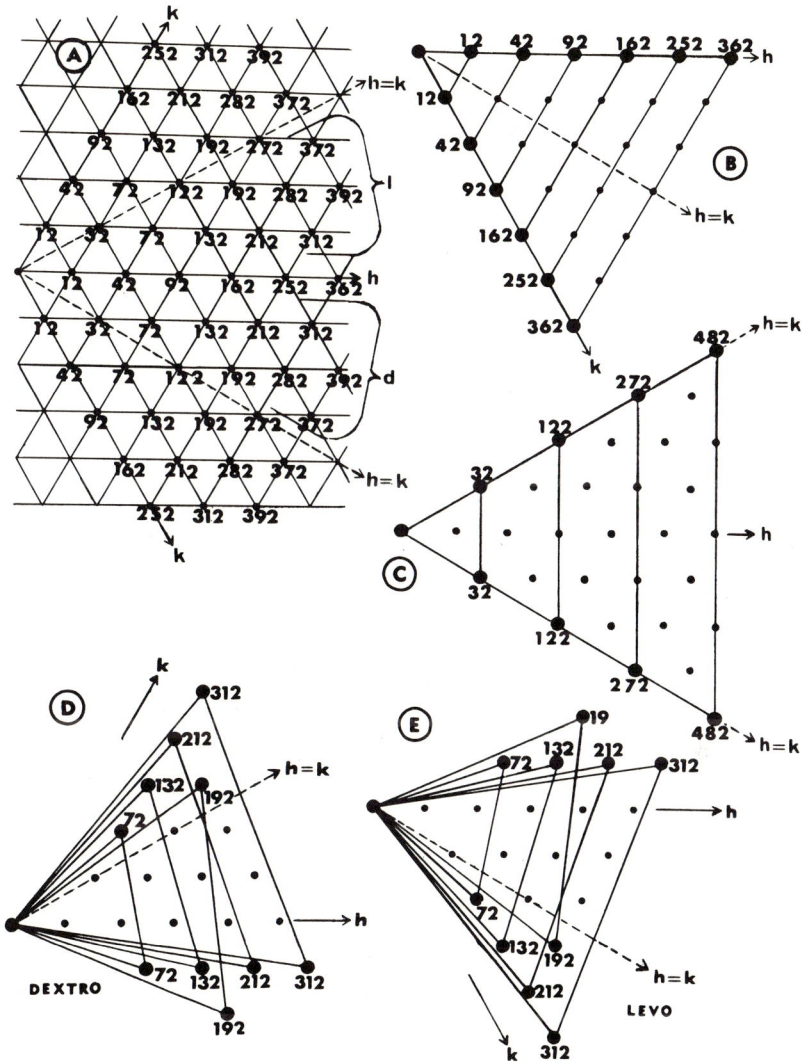

FIG. 2-6. The derivation of all possible deltahedra from a hexagonal net, modified from Goldberg (*61*). In B to E one of the 20 triangular faces of a number of possible deltahedra is shown. The cutting and folding of such net planes into deltahedra has been described in detail by Caspar and Klug (*27*).

two groups of polyhedra also have planes-of-mirror symmetry defined by opposite edges. Thus for every unit on one side of such a plane there is another unit in mirror-image position on the other side. Note the mirror symmetry about line $h = k$ (an edge for $T = 3$ series) and

the h axis (an edge for $T = 1$ series). Because of this mirror symmetry, capsomeres on upper and lower surface of $T = 1$ and $T = 3$
series viruses will coincide when viewed on a 2-fold axis. Thus these
viruses would have icosahedral symmetry $(5\ 3\ \frac{2}{m})$ to the capsomere
level of resolution.

The remaining polyhedra are deltahedra of $T = 7$, 13, 19, 21, etc.,
series. They result from h and k values such that $h, k \neq 0$ and $h \neq k$.
They include values of $C = 72$, 132, 192, 212, etc. They possess no
planes-of-mirror symmetry and have been referred to as the skew
class of deltahedra (27). Therefore there is no view of viruses of this
group which could result in coincidence of images on upper and lower
surfaces. That both surfaces of the negatively stained virus contribute
(albeit unequally) to the image has been clearly demonstrated (48,
49, 96, 100).

In retrospect it is probable that this contributed greatly to the
confusion over capsomere number in the papova (polyoma-like)
viruses. Either 42 or 92 capsomeres should have been relatively easy
to verify by observation on a 2-fold axis as was originally done with
the 252 capsomere, adenovirus (76), and subsequently with a number
of other viruses of the $T = 1$ or $T = 3$ series. The convincing determination that the papova viruses contain 72 capsomeres (26, 95, 99,
100) offered considerable difficulty which is not surprising in view of
the above considerations.

In addition to the lack of mirror symmetry planes these deltahedra
all exist as enantiomorphs; that is, there are dextro and levo forms
which are related to one another by mirror symmetry. These may be
seen in Fig. 2-6D and 2-6E where the 72-unit structure is seen as the
lowest number to exist in the two forms, which are topologically
related structures of the same number C.

Goldberg (61) also observed that there were topologically unrelated
polyhedra with the same unit number. These occur when different
values of h and k generate the same number in $h^2 + hk + k^2$. The
first such duplication occurs when $C = 492$ which results from $h = 7$,
$k = 0$ as well as $h = 5$, $k = 3$. The large number of capsomeres involved may preclude this from becoming of concern in virus morphology.

Pingpong ball models in Fig. 2-7 should be compared with Fig. 2-6
to visualize some of the polyhedra that can be formed by this scheme.

From these considerations a point of primary importance in the
electron microscopic observation of viruses is apparent. Since all

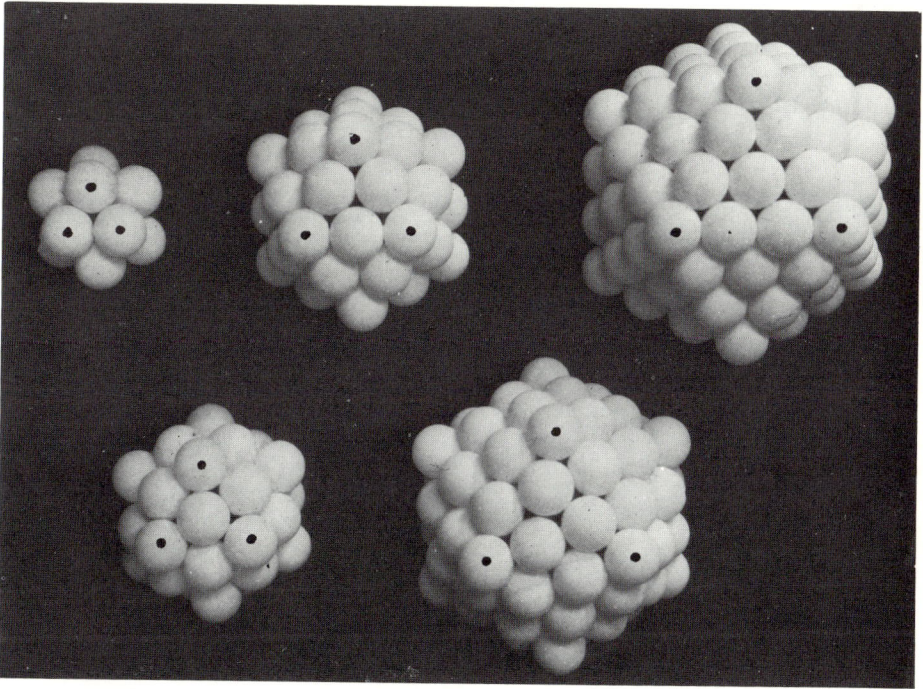

FIG. 2-7. Pingpong ball models of 12, 42, and 92 unit structures (upper row), and 32 and 72d unit models (lower row). The models are seen on 3-fold axes, and 3, 5-coordinated units are indicated by dots. Compare with Fig. 2-6.

classes of closed "icosahedral" structures can be considered as formed by the selection of various points as 5-fold symmetry axes (Fig. 2-6), the positive identification of the capsomere number and of the D or L symmetry is impossible without the unequivocal identification of at least one pair of neighboring 5-fold axes, and the distribution of capsomeres along the edge joining a pair of axes, as has been previously stressed (27). In addition, the assignation of D or L symmetry is absolutely dependent upon knowing whether the capsomeres in question are on the upper or lower surface of the virus (Fig. 2-7). Thus, while it is probably true that the dominant surface is usually that in contact with the grid (26, 48), this may not be true for an individual particle. Hence a reasonable number of particles should be evaluated, and conclusions drawn cautiously.

Possible variations of 532 symmetry are infinite. The faces of polyhedral viruses may not be flat, but may have varying degrees of concavity or convexity; the case of equal radii for all capsomeres results in a "spherical" shell with the required symmetry. The extent of curvature has not been determined for any virus. However, it would appear to be limited or absent in the case of adenovirus (76) and probably also herpes virus (177), whereas TYMV probably has a spherical shell of icosahedral symmetry (102). It is, at present, impossible to determine by electron microscopy the presence and extent of curvature in the smaller viruses with relatively few capsomeres. Also, naturally hydrated viruses may differ in this respect from the dehydrated viruses which are seen in the electron microscope.

In addition to the closed polyhedra, a series of open structures with strict 532 rotational symmetry can be generated from the lattice of Fig. 2-6, by removal of select units leaving holes of equal spacing and size over the surface. Some such arrangements are seen in Fig. 2-8. The numbers 12, 32, 42, etc., refer to corresponding closed-shell structures in Fig. 2-6.

In structure No. 32, the presence of a unit in the center of a face (3-fold axes) requires that this morphologic unit be a triplet (or $n \cdot 3$). Similarly, structures No. 42, 122, 282A, and 362 have units centered on edges (2-fold axes) which must therefore be dimers (or $n \cdot 2$). If the morphological units are identical then all must be appropriate multimers.

Each of these structures is topologically different and from Fig. 2-8, different structures with 12, 32, etc., holes can be visualized. It has been proposed that reovirus consists of 92 regularly spaced holes (89, 170). If so, it could have a shell of 180 morphological subunits as in No. 272, Fig. 2-8, where there could be $n \cdot 180$ structure units [$n = 3$ is suggested as a possibility by Caspar (25); however, n is not restricted by symmetry requirements]. Other possibilities are 270 morphological subunits (540 structure units) as in No. 362, Fig. 2-8, or an arrangement shown by Caspar (25) involving clustering of 540 structure units (his Fig. 15A). Without resolution of the morphological structure units as well as the pattern of holes, it is not possible to distinguish between alternative forms.

A second utility of Fig. 2-8 is realized by considering the net points and circles as loci of structure units (biochemical subunits) rather than as capsomeres (morphological subunits). Thus, a ring of five circles about a hole can be used to represent a capsomere on a 5-fold axis. In this way either No. 72 or No. 92 of Fig. 2-8 can be used as a

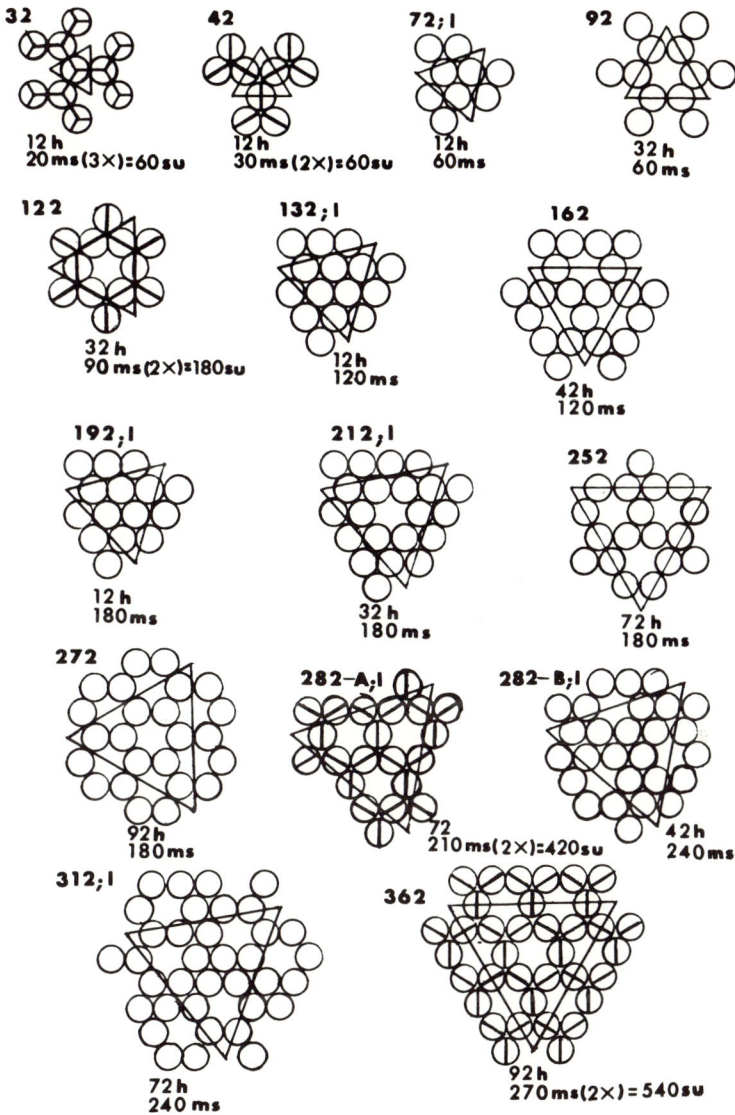

FIG. 2-8. Derivation of open-shell structures from the hexagonal net. h = holes; ms = morphological subunits; su = structure units; (3x), (2x) = morphological subunits required to consist of 3 or 2 structure units (or multiple thereof), respectively.

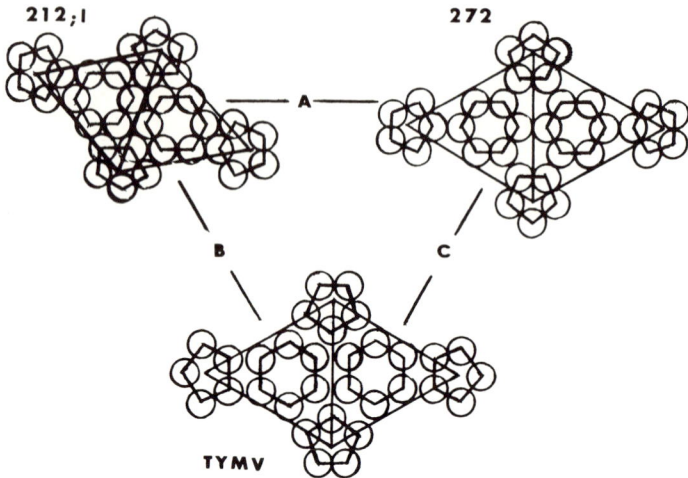

Fig. 2-9. The relationship of the current concept of TYMV structure (*102*) to the hexagonal net. Capsomeres lying on the net must be separated and/or rotated as indicated to conform to the virus model. The model is also a spherical shell rather than a deltahedron. A. rotation, 15° + expansion (explained in text); B. rotation, 15° + expansion; C. rotation, 30°.

model for possible distribution of structure units in ϕX174 (*167*) or in a small polyoma virus-associated particle (*125*) both of which appear to consist of 12 capsomeres and possible 60 structure units. Note that the difference between these two models consists only of a difference in rotation of the structure units about the 5-fold axes coupled with different edge lengths. Finch et al. (*51*) have shown that the surface structure of polyheads of bacteriophage T4 is probably that of a hexagonal net as in No. 272 folded into cylindrical sheath. Similarly, the pattern of No. 282, B, Fig. 2-8, closely approximates the apparent structure of the large tubular structures associated with turnip yellow mosaic virus (*72*).

In a similar fashion the 32 capsomere (180 structure unit) model of turnip yellow mosaic virus can be derived from Nos. 212 or 272 of Fig. 2-9 by the indicated rotations about 3- and 5-fold axes; No. 212 would also require edge expansion by a factor $\sqrt{27/21} = 3/\sqrt{7}$ which follows from the relative edge length, $\sqrt{T} = \sqrt{h^2 + hk + k^2}$. It should be noted that the morphological subunits or capsomeres of TYMV do lie on a hexagonal net but not the structure units as proposed by Klug et al. (*102*).

4. *Viruses with Icosahedral Symmetry*

In the light of the preceding considerations the list of viruses presented in Table 2-1 is comprised of those icosahedral viruses in which the capsomere number seems to be well established.

Most of the viruses listed as having **72** capsomeres were originally thought to consist of **42** capsomeres. This interpretation was questioned independently by Mattern (*120, 122*) and Caspar and Klug (*27*). Mattern et al. (*123*) felt that this group (papovaviruses) probably had **92** capsomeres but that **72** could not be excluded. Klug and Finch (references in Table 2-1) convincingly demonstrated that these viruses

TABLE 2-1

Icosahedral Viruses with Established Capsomere Number

Capsomere number	Virus	Reference
12	φX174	*62*
		167
	Coliphage φR	*91*
32	Turnip yellow mosaic	*140*
		88
	Cowpea chlorotic mottle	*8*
	Broad bean mottle	*50*
	Cucumber mosaic	*52*
42	*a*	
72	Polyoma	*95*
	Human papilloma	*99, 100*
	Rabbit papilloma	*48*
	SV40	*99*
		95
		4
	K virus	*95*
92	*a*	
162	Herpes (probable)	*177*
	Varicella (probable)	*1*
252	Adenoviruses	*76*
	GAL (probable)	*176*
	Infectious canine hepatitis	*37*

a Discussed in text.

did, in fact, consist of **72** capsomeres. The report of wound tumor viruses consisting of **92** capsomeres (*14*) is open to similar question.

Although capsomere numbers have not been confirmed, it would seem likely that herpes zoster virus (*178*), avian infectious laryngotracheitis virus (*34, 174*), pseudorabies virus (*154*), equine abortion (rhinopneumonitis) and equine herpes viruses (*152*), and feline rhinotracehitis virus (*42*) have the same structure, presumably **162** capsomeres, as do herpes and varicella viruses.

Poliovirus and echovirus **24** have been reported to have **32** capsomeres (*127*). However, in view of the presentation of single particles in each case, the general difficulty in resolving capsomeres [also true for Coxsackie viruses (*121*)], and the previously discussed complexity of poliovirus proteins these observations should be confirmed.

Other viruses have been shown to have some sort of icosahedral morphology, but neither their capsomere nor structure unit organization have been definitively resolved. These include blue green algal virus (*157*), tipula irridescent virus (*181*), reovirus (*89, 170*) and possibly some of the larger bacteriophages (*18*), and lymphocystis virus (*187*).

C. IMAGE ANALYSIS TECHNIQUES

1. *Image Enhancement*

Markham et al. (*112*) applied image reinforcement techniques to viruses with rotational symmetry or translational repetition. In the case of n-fold rotational symmetry the image or photographic paper is rotated $360/n$ degrees with each of n exposures. Each exposure is of $1/n$ the duration necessary for optimum contrast without rotation. In this manner the background "noise level" is reduced with a possible increase in resolution up to $\sqrt{n}$.

In the case of translational repetition, the image is successively moved by the translation period and photographed at each step.

A more sophisticated method of performing these operations has also been employed (*112, 113*) by utilizing a stroboscope and synchronizing the flashes with the rotated or translated images.

Two contradictory features of virus structure have been revealed by these methods. First, Markham et al. (*112*) apparently resolved a capsomere of turnip yellow mosaic virus which contained five structure units, yet was surrounded by six nearest neighbors (6-coordinated). A 6-coordinated pentameric capsomere is entirely incompatible with the current concept of TYMV structure as was discussed by Klug et al. (102). Second, Markham et al. (113) appeared to resolve a translational (actually helical) period on TMV of only 17.9 Å† instead of the 23 Å pitch measured by X-ray diffraction. They attributed this difference to extensive shrinkage of the TMV particle on drying for electron microscopy. However, they were able to resolve the 45.4 Å double period of the stacked disk structure on the TMV protein rod, suggesting no appreciable shrinkage. Also, Finch (44) was able to directly resolve the 23 Å period in negatively stained TMV, although his method of calibration was not described.

It would seem that the potential of such methods will not be understood until such discrepancies are clarified.

2. *Optical Diffraction*

Caspar and Klug (27) employed optical diffraction to show that a model consisting of 60 points in the surface of a sphere with icosahedral symmetry would produce spikes of intensity corresponding in symmetry to those seen on the X-ray diffraction pattern of poliovirus (and other virus) crystals.

This type of analysis has been more recently extended by Klug and Berger (96) to include the comparison of optical diffraction by models with optical diffraction by electron microscopic images of negatively stained TMV particles.

Bancroft et al. (8) have also utilized optical diffraction of models and virus images in the analysis of the capsomere structure of cylindrical forms of the protein of cowpea chlorotic mottle virus. They have also similarly analyzed images of tubular structures from preparations of rabbit kidney virus and human wart virus as published by Chambers et al. (29) and Noyes (143), respectively.

3. *Computer-Determined Images and Other Analog Techniques*

Klug and Finch (100) have programmed a computer to calculate and project for photography a series of images which would be expected from 72 capsomere viruses viewed at various angular incre-

† Lauffer and Stevens (104) have concluded on the basis of various biophysical data that this periodicity is an artifact.

ments and with contribution to the image by both upper and lower surfaces of the virus. A variety of different electron microscopic images were compared with the computed images and shown to be entirely compatible with the 72 capsomere model.

This approach has considerable potential for image analyses and should also be useful for more complex images as might result from resolution of structure units.

Caspar (26) has employed a somewhat different analog method which consisted of embedding X-ray transmitting models in an X-ray absorbing substance, and obtaining X-ray pictures of "negatively stained" models. Some other similar analog techniques of restricted usefulness were discussed by Caspar (26).

Broad bean mottle virus (BBMV) was examined by negative staining and compared with computer-determined images (50). Its structure was found very similar to that of turnip yellow mosaic virus, consisting of 32 capsomeres and 180 subunits. However, the capsomeres of BBMV appeared slightly skewed (rotated about their axes) relative to those proposed for TYMV (Fig. 2-4). The icosahedral structure has been confirmed by X-ray diffraction studies (53) which showed the characteristic spikes of intensity corresponding to 2-, 3-, and 5-fold axes of an icosahedral structure. The virus also seems to have a central hole 100–120 Å in diameter.

D. STRUCTURAL DETAILS IN OTHER VIRUSES

1. Myxoviruses

Viruses included within this group are the influenza viruses, fowl plaque virus, Newcastle disease virus, Sendai and other parinfluenza viruses, and measles virus. All of these viruses contain RNA, are highly pleomorphic, and share several other properties in common; their surface is covered with projections, they have a membrane beneath this layer, and they have a helical core structure or nucleocapsid.

Influenza viruses are approximately spherical (although highly elongate forms also exist) and are about 800–1000 Å in diameter (80) their largest dimension sometimes as much as 1800 Å (79). Their outer membrane is studded with projections about 100 Å long and 70–80 Å apart (78, 79, 84). There is also evidence of a 60–100 Å thick membrane beneath the layer of projections. Interactions of influenza virus particles with particles from normal host cells suggest that the first stage of virus–cell interaction consists of disruption of the viral membrane releasing lipids and the inner core component (85). The interior of the particle contains a fine helical strand, presumably a ribonucleic acid–protein complex, which, when removed from the virus particle, has a helical appearance similar to the filamentous beet yellows virus (77). The helix is 90–100 Å in diameter and it has been suggested that it exists as a double helical structure (84) and in the virus particle as a coiled helical structure (6, 79, 85, 132). An artist's conception of the appearance of influenza A2 virus is shown in Fig. 2-10.

Recently Flewett and Apostolov (54) have demonstrated an interesting hexagonal reticular pattern (70 Å along a hexagonal edge) on filamentous forms and a similar network of both hexagons and pentagons on "spherical" forms of influenza virus C. The relationship to the projections and the viral membrane has not yet been established.

Fowl plague virus is similar in size and morphology to influenza viruses and also appears to contain a folded helical nucleocapsid (79, 80).

Tern virus has also been shown to closely resemble influenza viruses (9).

Mumps, Sendai, and Newcastle disease viruses are somewhat larger than influenza virus. Mumps virus particles vary considerably in size, from about 2500 to more than 5000 Å wide (79); Sendai virus from 1600–8000 Å (81); and NDV, about 1500 Å (80). Helical nucleocapsids have been observed after disruption of the three viruses (78). Mumps virus nucleocapsid is flexible and about 170 Å in diameter, with a 50 Å periodicity. Sendai virus nucleocapsid is similar and filaments up to 1 mμ in length were observed, although they seemed to be somewhat more rigid than the mumps filaments (80). Horne and Waterson (78) suggested that the Sendai nucleocapsid may also be a double helical structure. However more recent evidence indicates that mumps, NDV, and Sendai virus nucleocapsids are single helical strands (30). Recently, Hosaka et al. (81), studied the pleomorphism of Sendai virus (HVJ). By sucrose density centrifugation they were able to isolate four overlapping size groups of virus, ranging from 1600 to 8000 Å in diameter. The first group consisted of 1600 to 2500 Å particles whose nucleocapsid lengths were about 1 mμ. Three heavier bands contained larger particles, with some overlapping with the smallest class. Their nucleocapsid lengths were 1 mμ, of multiples of 1 mμ the maximum length encountered being more than 20 mμ. On the basis of these observations they proposed that the smallest particle had a mononucleocapsid of 1 mμ length and that the larger particles

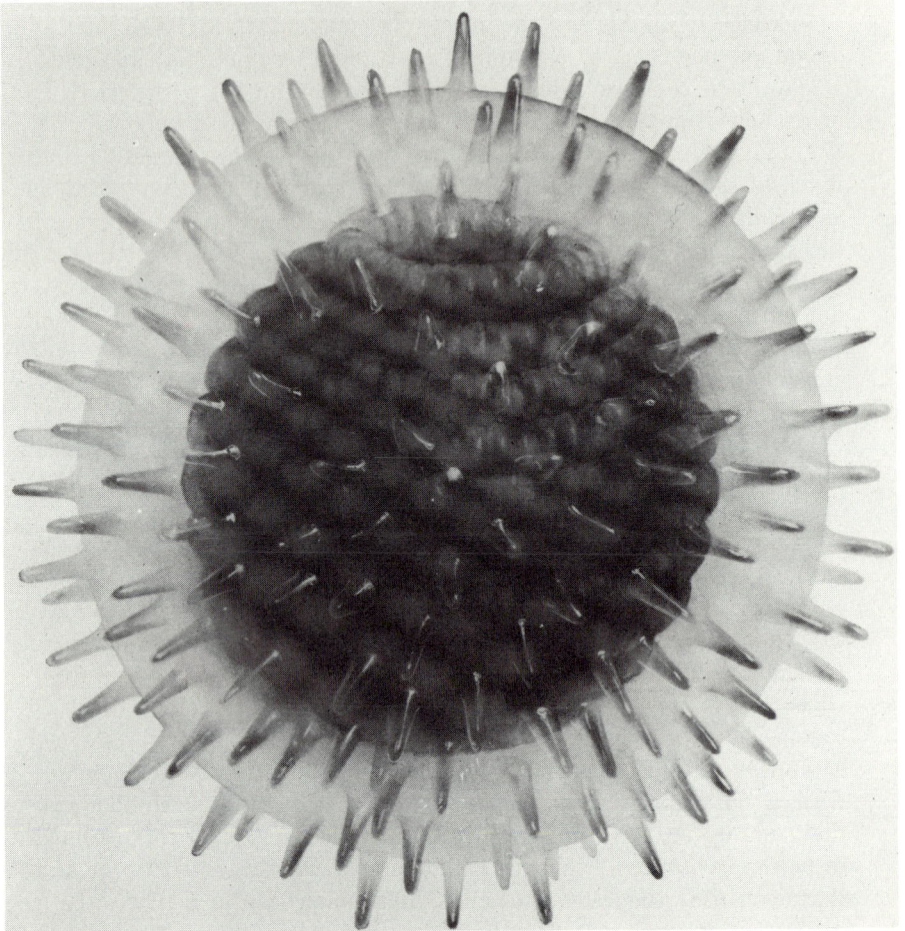

Fig. 2-10. An artist's conception of the structure of influenza virus. (Courtesy of E. I. duPont de Nemours & Co.)

had a variety of polynucleocapsids of up to 20 or more units lengths. Compans and Choppin (*31*) likewise found the unit length of NDV nucleocapsid to be slightly more than 1μ; Kingsbury and Darlington (*92*), 1.3–1.4 μ.

Parainfluenza virus type 2 (SV5 virus) is similar to Sendai virus and Choppin and Stoeckenius (*30*) felt that the nucleocapsid consisted of a single helical strand. Compans and Choppin (*32*) have isolated the helical nucleocapsid of the SV5 virus and determined it to contain 4.1% RNA and to have a mean length of about 1 mμ. Parinfluenza virus 3 (*173*) has also been shown to resemble the mumps, NDV group of viruses. *Yucaipa virus* has been classified as a paramyxovirus virus and has been shown to closely resemble NDV structurally (*40*).

Measles virus was found to have a very similar structure to the above viruses (*171*). The particles were 1200–2500 Å in diameter, possessed typical surface projections and membrane, and disrupted particles revealed helical nucleocapsids 170–180 Å in diameter with a 50–60 Å period.

Canine distemper virus (*35*) and rinderpest virus (*151*) both closely resemble measles virus, and there is some evidence for immunologic cross reaction between these three viruses (*5*).

Respiratory syncytial virus has been shown to resemble the parainfluenza-NDV group of myxoviruses, being 1200–3000 Å in diameter, having surface projections, and possessing a helical nucleocapsid (*172, 184*).

Viruses Superficially Resembling Myxoviruses. Visna virus, which produces a chronic disease in sheep, has also been shown to resemble the influenza viruses in size and morphology: 900–1000 Å in diameter; surface studded with projections about 100 Å in length; pleomorphism; and suggestion of a similar internal structure (*166*).

Avian infectious bronchitis virus (IBV) and related human respiratory viruses, strains 229 E and B814 (*2, 13, 129*), have a size range and pleomorphism similar to the influenza viruses, but their surface projections are of different morphology, longer, often bulbous, and more densely spaced than those of influenza virus, and as yet no helical nucleocapsid has been identified associated with these three viruses. Their morphogenesis (*10*) also distinguishes them from the other myxoviruses and at this time their classification is unsettled.

2. Elongated Viruses

In addition to the tobacco mosaic virus, previously described in detail, a number of rigid, rod-like viruses, baciliform viruses, bullet-shaped viruses flexible filamentous viruses, and filamentous virus-related particles have been described among plant, vertebrate, insect, and bacterial viruses.

a. *Other Helical Viruses*. Dolichos enation mosaic virus, which is serologically related to TMV, has recently been shown by negative staining to be structurally similar to TMV (*128*).

Sugar beet yellows virus has a readily observed helical structure, appears more loosely organized, and more flexible than TMV (*77*). It appears to have a diameter about 100 Å; a helical pitch of about 30 Å; $6\frac{1}{3}$ globular subunits, about 30 Å in diameter, per turn; and an integral number, 19, in three turns (*156*).

Tobacco rattle virus (TRV) was shown by Harrison and Nixon (*66*) and Nixon and Harrison (*141*) to be a tubular virus with a periodic striation of 25 Å, a diameter of 250 Å, and to exist in two classes with respect to length, 750 or 1850 Å average lengths. Markham et al. (cited in Ref. *45*), employing the rotation method of image enhancement, showed that reaggregated disks of TRV apparently had 25-fold rotational symmetry. Finch (*45*) investigated oriented gels of TRV by X-ray diffraction and showed it to have a helical structure with a pitch of 25 Å and an integral number of subunits ($3q + 1$; q, integral) in 3 turns, or nearly 3, of the helix. The value of q is believed to be 25 on the basis of the observations of Markham et al. (cited in Ref. *45*). The X-ray studies showed the interparticle distance to be 340–400 Å in gels where the viruses are oriented with respect to their long axes into a hexagonal pattern. Some of these details have been further supported by electron microscopic studies of Offord (*144*).

Barley stripe mosaic virus and lychnis ringspot virus (*59, 67*) on the other hand, appeared to have a helical structure very similar to TMV on the basis of electron microscopic observations.

Barley stripe, mosaic virus was also examined by X-ray diffraction (*45*) and was shown to have a helical structure of 26 Å pitch and an integral number of subunits ($5q + 1$ or possibly $5q + 4$) in five turns of the helix. It also appeared to have an empty central core of 15–20 Å radius and a maximum diameter of 210–230 Å. A radial density peak at 55 Å may indicate the location of its RNA.

Brandes and Wetter (*19*) have classified a large number of elongate plant viruses on the basis of their fundamental length, their rigidity, diameter, and other properties. Varma et al. (*169*) have examined 13 such viruses by negative staining, including only one virus, sugar beet yellows, previously discussed in this review. The remaining 12 viruses, listed below, all showed helical features with pitches of 33–37 Å, and the number of subunits per turn was estimated for five of these viruses and they were found to range between 10 and 14. The viruses examined by Varma et al. were as follows: white clover mosaic, hydrangea ringspot, potato virus X, potato aucuba mosaic, centrosema mosaic, red clover vein mosaic, carnation latent, ryegrass mosaic, potato virus Y, henbane mosaic, bean yellow mosaic, and clover yellow vein virus.

b. *Bullet-Shaped and Baciliform Viruses.* A substantial number of viruses of vertebrates, insects, and plants have been shown to have a complex cylindrical structure with one end rounded and the other flat, giving the particle a bullet-like appearance, or sometimes both ends are rounded resulting in a bacilliform appearance.

Vesicular stomatitis virus may be considered the prototype of such viruses. Early studies employing negative staining were undertaken by Howatson and Whitmore (*83*); these have been confirmed and extended by Simpson and Hauser (*159*). The virus appears to have a complex cylindrical structure with one end rounded, the other flattened. It presents a variable appearance on negative staining and it is not clear which of the various structures are infectious particles.

(1) The complete particle. The "most complete" particle is about 900 Å in diameter and 1750 Å in length, and it reveals only surface structure. The outer layer appears as distinct surface projections about 100 Å thick. The interior of the virus is electron lucent and "structureless" due to failure of stain to penetrate the particle.

(2) Frequently particles appear devoid of surface projections, as "structureless" particles 700 Å in diameter.

(3) The above particles very often demonstrate at the flattened end a stain-filled core of extreme variability. The core may extend nearly to the rounded end or it may be very short; it may be partially or completely filled with stain and it may vary in diameter from scarcely 100 to about 340 Å. This core material which is variably extruded is likely the source of the relatively unstructured "tails" frequently seen with this and related viruses, and also the helical structures 180 Å in diameter seen by Simpson and Hauser (*159*). The core is presumably a ribonucleoprotein complex.

(4) Particles with an outer membrane occasionally appear to have stain-filled cores of about 340 Å diameter and often demonstrate the classical transverse striations with a 45–50 Å periodicity and about 500 Å O.D. This internal structure is now known to represent a large helix of 45 Å pitch.

(5) Still other particles appear to have lost their outer membrane as well as their core contents and these appear to consist solely of a helix 500 Å in diameter and with a pitch of about 45 Å.

TABLE 2-2

Bullet-Shaped or Bacilliform Viruses Resembling VSV in Structure

Virus	Reference
Cocal	*41*
Kern Canyon	*135*
Egtved	*186*
Sigma	*11*
Rabies	*7, 36, 86, 117, 118, 150*
Flanders	*134*
Wheat striate mosaic	*106, 107*
Lettuce necrotic yellows	*28, 65, 145, 183*
Barley stripe mosaic	*59, 67*
Lychnis ringspot	*59*
Gomphrena	*93*
Potato yellow dwarf	*109, 110*
Maize mosaic	*70, 71*
Plantain	*73*
Aglais urticae nuclear polyhedrosis[a]	*63*
Granulosis viruses[a]	*160, 161*

[a] These insect viruses bear a superficial resemblance to the VSV-like viruses.

On the basis of these observations and consideration of the investigations of a number of VSV-like viruses listed in Table 2-2, a schematic diagram of the structure of VSV and some related viruses is presented in Fig. 2-11. It should be emphasized that there is considerable uncertainty over many of the structural features, particularly concerning the organization of the core region.

c. *Elongated Viruses of Nonhelical or Unknown Structure*

(1) *Lucerne mosaic virus.* Lucerne mosaic virus (*60*) is an unusual virus in that apparently it exists in three or more classes of length distributed about 360, 480, and 580 Å in width. Small forms 200–300 Å in length were found in old preparations and there was doubt as to their infectivity. The authors' interpretation of the surface structure as open hexagonal rings forming a cylindrical shell (see Fig. 8 No. 272, this paper) should be confirmed.

(2) *Filamentous bacterial viruses containing single-stranded DNA.* At least four filamentous bacteriophages which infect only male strains of *E. coli* have been recently described; fl phage (*185*), fd

phage (*115*), M13 phage (*158*), and ZJ/2 phage (*17*). These viruses
are reported as 50–78 Å in diameter and to have a unit length of
8000–8500 Å. Some structural detail was also seen in ZJ/2 phage (*17*)
as the occasional appearance of cross striations of 20–30 Å period and
the appearance of a fine strand of negative stain in the core of the
virus. Marvin and Schaller (*116*) presented evidence that fd DNA
existed as a single-stranded ring. The electron microscopic and X-ray
diffraction studies of Marvin (*114*) revealed other structural features.
Electron micrographs showed the particle to consist of two strands
loosely woven about one another. His X-ray diffraction pictures on
oriented fibers and gels showed an interparticle spacing of 55.4–65 Å,

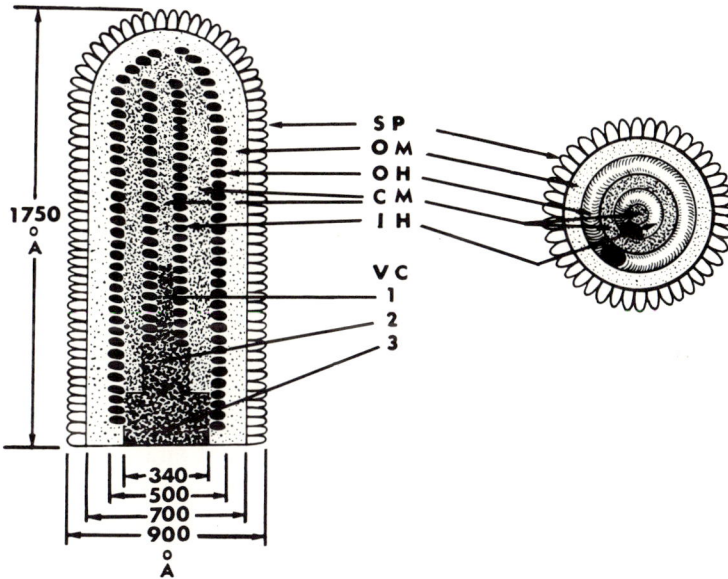

FIG. 2-11. A schematic diagram of the possible structure of vesicular stomatitis
virus and related viruses. The dimensions are approximate and the structure of
the core is largely speculative. Some of the isolated helices of Simpson and Hauser
have more than twice the 35 helical turns of 50 Å pitch which could be accom-
modated as shown in a virus of this length. This suggests a possible continuity
between the two helices. SP, surface projections; OM, outer membrane; OH,
outer helix; CM, core matrix; IH, inner helix; VC, variable core extrusion;
1, inner core matrix (70 Å O.D.); 2, inner helix + 1 (180 Å O.D.); 3, outer core
matrix + 1 and 2.

depending on degree of hydration, and an axial repeat of 32.2 Å. Strong reflections of 10 Å spacing on layer lines 0 and 1 suggested a substantial degree of alpha helical structure of the protein, with the helix axes being slightly off of parallel to the virus axis. The data of Day (*38*) using optical rotatory dispersion are compatible with 90% of the protein being alpha helical. It is not yet clear whether these viruses are constructed as closed strings (more like rubber bands) enclosed in a continuous coating of protein, the two sides of the flattened circle being loosely wound about each other (*114*) or whether only the DNA is loosely coiled and contained within a single tubule of protein (*18*). Phages which appear similar from electron microscopic observations have been isolated from *Pseudomonas*, *Pf1* and *Pf2*, (*130, 165*) and from *Vibrio parahaemolyticus* (*139*). Another phage, V6, appears from preliminary studies to be filamentous.

(3) *Filamentous forms of icosahedral viruses*. The filamentous forms of icosahedral viruses are considered in the Discussion.

3. *The Poxviruses*. A number of this group of DNA viruses have very similar structure. Vaccinia and orf (contagious pustular dermatitis) viruses have probably been examined in the electron microscope in the greatest detail. Others include molluscum contagiosum virus (*153, 178*); bovine papular stomatitis virus (*21, 138*); cowpox and rabbitpox viruses (*43*), the latter possibly being a strain of vaccinia virus; and variola major, alastrim, ectromelia, monkeypox, fowlpox, myxoma, fibroma, and paravaccinia viruses (*155*).

They are among the largest viruses [vaccinia, 230 × 340 mμ; orf, 157 × 263 mμ (*136*)], being somewhat flattened and rectangular or elliptical, and possessing some characteristic features. Again, the appearance of these viruses, when negatively stained, appears to depend upon the degree of penetration of the contrasting stain (*142, 148*).

The surfaces of orf and bovine papular stomatitis viruses, in particular, present an unusual picture (*136–138, 147*). A "basket weave" appearance consisting of superficial "tubules" or filaments is seen when both surfaces of the virus contribute to the particle's image. With metal shadowing (*137, 147*), only the upper surface is visualized and is seen to consist of a series of widely spaced, parallel ridges diagonally crossing the virus surface. Negative staining reveals them to circumscribe the particle diagonally and Nagington et al. (*137*) have drawn models which may account for this appearance. Similar, but less well-defined structures, have been seen on the surface of vaccinia virus

(*80, 142*). A drawing depicting this surface structure is presented in
Fig. 2-12A.

The internal structure of vaccinia virus has been investigated by
Peters and Müller (*146*), who observed the same details employing
both negative staining and thin sectioning. Beneath the outer membrane, which presumably contains the "tubules," there is a rather
thick inner membrane in which some periodic structure has been
resolved (*148* cited in *80, 142*). The inner membrane appears biconcave
in shape, its narrowest part coinciding with two central bulges on the
virus surface apparently caused either by material lying between the
two membranes or by a thickened membrane. Beneath the inner membrane Peters and Müller observed a curious triplet structure, consisting of what appear to be three hollow filaments about 500 Å thick and
which may in fact represent one continuous filament in the form of a
flattened S. Their interpretation that the large filaments may consist
of two interwoven smaller filaments does not seem adequately supported by their electron micrographs. Most of these structural features
are shown in Fig. 2-12B. Very similar internal structures have been
shown in orf virus (*147*) and bovine papular stomatitis virus (*21*).

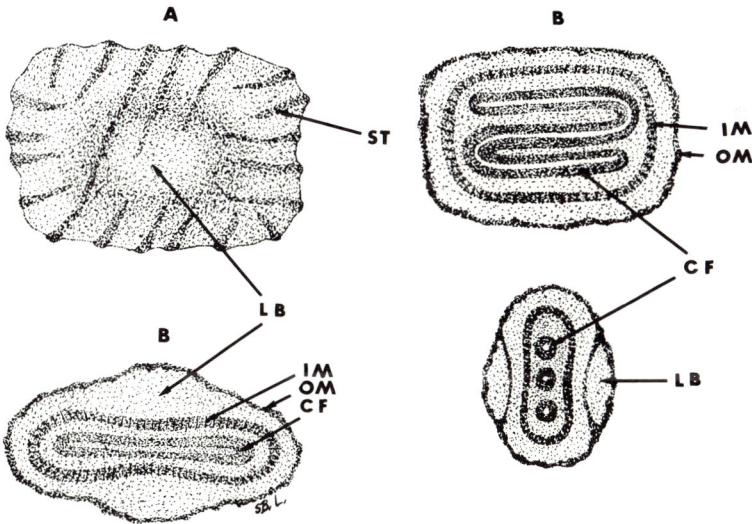

FIG. 2-12. A composite drawing of the surface (A) and internal structure (B)
of vaccinia virus based upon electron micrographs discussed in text. ST, superficial
"tubules"; LB, lateral body; IM, inner membrane; OM, outer membrane; CF,
core filaments (possibly S shaped as shown).

4. *"Relatively Unstructured" Viruses*

A large number of viruses can be arbitrarily placed under this heading, but probably not because they do not have a highly organized architecture, but rather that such has not yet been well established.

Among them are several viruses of the avian sarcoma–leucosis–murine leukemia complex. Although they have been extensively examined in sectioned materials, relatively little is known beyond their multilayered structure, their typical budding from the cell membrane, and the appearance of a "tail" or more properly a membranous appendage on some viruses. Probably the most extensive investigation of a virus of this group was that of de-Thé and O'Connor (*39*) who studied the structure of the Rauscher leukemia virus. They confirmed the multiple layering in negatively stained, partially disrupted "immature" viruses. Two of the rings had a periodic structure. However, in "mature" viruses, little or no structural detail was observed.

Lactic dehydrogenase (LDH) virus, first seen by Bladen and Notkins (*15*), has as yet not revealed significant structural details, beyond its double-membraned structure.

The arboviruses are divided into three major groups on immunological and other grounds and a large, motley group of left-over viruses (*5*). Many of the arboviruses which have been examined in the electron microscope appear rather pleomorphic and externally structureless. However, among the Group A arboviruses, Klimenko et al. (*94*) have observed a helical nucleocapsid-like structure associated with Venezuelan equine encephalomyelitis virus. Others, such as Colorado tick fever virus and possibly Chenuda virus have recently been shown to closely resemble reovirus both in structure and morphogenesis (*133*). It seems probable that in the future a number of arboviruses will be found to belong to one or another of the well-defined morphological groups, and a large number will remain as one or more arbovirus types, whose architectural details are as of now unknown.

2-4. Discussion

Comments on the current concept of the structure of viruses and capsomeres.

(1) Hollow pentagonal and hexagonal prismatic capsomeres *in situ*. Early models of a variety of viruses were constructed of capsomeres shaped like hollow prisms. The concept of hollow capsomeres arose

from their appearance in negative staining where capsomeres of certain icosahedral viruses frequently contained stain in their central region. The concept of an elongated prism arose largely from the appearance of capsomeres on the lateral surfaces of negatively stained viruses. However, this region of the intact virus is especially hazardous for deducing capsomere structure due to the superposition of multiple capsomere images.

To conform to the X-ray diffraction interpretation of strict 532 symmetry these prisms are now visualized as consisting of either five or six biochemical subunits, the former located at icosahedral vertices, the latter in all other positions. It should be remembered that the only viruses for which 532 symmetry requires such capsomere substructure are those few which have been investigated by X-ray diffraction. It is assumed by analogy that other icosahedral viruses have similar capsomere substructure, which is quite possible. However, a number of viruses could satisfy 532 symmetry if they were constructed entirely of pentameric capsomeres. Consider the group of 72 capsomere viruses. There are no capsomeres on either 2-fold or 3-fold axes and therefore, considering only symmetry, they could be constructed entirely of pentamers, which are required on the vertices. Other similar possibilities can be readily seen in Fig. 2-8, and these include 252 capsomere structures such as the adenoviruses.

The most recent model of TYMV (*102*) presents the two classes of capsomere as hollow aggregates of either five or six roughly elipsoidal protein subunits. Thus there are 180 biochemical subunits and 32 capsomeres as previously described. Klug et al. (*102*) feel that this model is confirmed by electron microscopic evidence presented in a companion paper (*49*). However, two aspects of this confirmation should be noted. First, on their Plates 5 and 6, the images of "hexamer" capsomeres are usually confused by the appearance of one or more regions of electron lucency near their centers. In others the capsomere image is not resolvable into biochemical subunits. Without the X-ray model it is doubtful that these images would have been interpreted in the manner so done. Second, in Plates 8 and 9, they note a railway appearance in two-dimensional arrays of viruses. Whether this effect is produced by intra- and intercapsomere strain or by the delineation of biochemical subunits which are not visible in the images as reproduced is not clear. Thus, while the X-ray model of TYMV may be entirely correct, these pieces of electron microscopic evidence offered as confirmation are open to question.

Recognizing this difficulty, Finch and Klug (*50*) examined RNA-free TYMV top component and observed well-defined, stain-filled holes in these capsomeres. They compared broad bean mottle virus (BBMV) with TYMV and correlated the RNA content 22% for BBMV versus 33.5% for TYMV [found by Kaper and Litjens (*90*)] with the presence or absence of negative staining of the hole of the capsomere. Klug et al. (*102*) had previously presented evidence that the RNA of TYMV extended substantially into the hollow capsomeres. This provided a plausible explanation for the failure of stain to penetrate the capsomeres of intact TYMV, whereas it did penetrate capsomeres of BBMV and TYMV top component. One possibility that was not considered is that the capsomeres of the intact particle have more tightly bound subunits than those of top component particles, due to the presence of RNA. Evidence for variation in packing of TMV subunits in the presence and in the absence of RNA have been described in detail (*24, 25*). The difference in appearance of TYMV and BBMV capsomeres could represent major structural differences in the two protein subunits which presumably accounts for similar capsomere differences in other viruses, for example, adenoviruses and herpesviruses (*76*). Also, the difference in RNA content between TYMV and BBMV could instead, explain the fact that intact BBMV often contains a substantial central core of negative stain, possibly representing the 100–120 Å hole deduced from X-ray diffraction (*53*), whereas this appearance was not noted on TYMV (*50*).

The most convincing evidence for hexameric capsomeres of TYMV is that of Hitchborn and Hills (*72, 73*). They have studied the structure of large protein tubules isolated from cells infected with a necrotic strain of TYMV. The tubules were found to be mostly protein which was serologically related to TYMV protein. Negatively stained tubes often showed an ill-defined surface structure, but with linear integration, with or without secondary linear or rotational integration, hexameric "rosettes" appeared to be clearly resolved. However, some caution in interpretation seems in order, in view of the two anomalous observations previously discussed (Sec. 2-3, C) which appeared with this type of image enhancement and which have not as yet been resolved. Of some concern also is the inordinate size of the negatively stained holes after integration. Hitchborn and Hills did present one micrograph to which these criticisms do not apply. A sheet consisting of a single layer of "rosettes" (compared to two or more layers in the tubes) clearly showed hexagonally oriented "rosettes," many of which strongly suggest a hexameric substructure. The only question here seems to be whether the "rosettes" are structurally identical to capsomeres on the intact virus.

The only evidence which directly confronts the current model of TYMV is that of Markham et al. (*112*). Their demonstration of a pentameric capsomere surrounded by six neighboring capsomeres, if true, would in itself invalidate the model. Although Finch and Klug (*49*) have offered plausible reasons to suggest that this observation may be an artifact, it should not be entirely dismissed without more convincing confirmation of the TYMV model.

(2) Isolated capsomeres. In a number of electron micrographic studies free capsomeres of disrupted viruses have appeared. Their appearance is either that of an approximately spherical particle with a central staining region or no recognizable details can be discerned. It is difficult to imagine that hollow prismatic structures, even with an axial ratio nearly unity, would not occasionally be seen in side view to resemble a hollow cylinder. I am unaware of any report demonstrating isolated capsomeres to be hollow cylinders. A recent study of purified "hexon" antigen of type 2 adenovirus (*149*) shows these capsomeres to have multiple knobs (many more than six), a central region containing stain, and no suggestion of a cylindrical structure.

(3) Filamentous and icosahedral forms of the same viruses. Filamentous forms associated with icosahedral viruses have been reported in a number of the papova group of viruses. It has been recently proposed that in the case of one strain of polyoma virus that filamentous forms play a role in the development of this icosahedral virus (*124, 125*). Subsequent observations (unpublished) combining uranyl acetate staining and metal shadowing have shown that some of the isolated filaments contain DNA as well as the coat of capsomeres, as was suggested by the appearance of filaments in sections of infected nuclei. Budding of filaments into icosahedral forms or terminal spherical enclosure was also suggested by negative staining of a number of "papova" viruses (*29, 82, 143*). The problem generated by these observations is shown by the following two alternative possibilities: *i*. If filaments consist exclusively of hexameric capsomeres (they are all 6-coordinated), budding of a filament into an icosahedral form containing 60 hexamers and 12 pentamers obviously requires the replacement of select hexamers by pentamers. This could occur by conversion of hexamers to pentamers, or by substitution by pentamers,

preformed, or otherwise. Conversion seems unlikely since capsomeres appear to be relatively stable structures; the other possibility requires not only removal of certain capsomeres but also the ready availability of the appropriate replacement. *ii*. If filaments consist partially or entirely of pentamers, this would require pentamers to be 6-coordinated which would be especially awkward with mixed capsomeres. It is possible to visualize filaments containing only pentamers. Such filaments would have an even number of longitudinal (or, if helical, somewhat off longitudinal) rows of capsomeres, capsomeres in a given row oriented alike, and capsomeres in adjacent rows rotated 180°. These could only form capsids consisting exclusively of pentamers. Thus, if filamentous forms convert to icosahedral forms, a considerable strain is placed on the concept of a general hexamer–pentamer structure applying to these viruses.

The extremely interesting system of assembly of capsomeres of disassociated cowpea chlorotic mottle virus into a number of forms including filaments, ellipsoids, and spheres (*8*) may lead to any explanation for the above difficulty.

(4) Icosahedral viruses and strict 532 symmetry. It would seem premature to assume that all icosahedral viruses have strict 532 symmetry. The interesting observations of Valentine and Pereira (*168*), on an adenovirus which shows thin protein fibers protruding from vertex capsomeres and ending in a small globule, may well be an example of an icosahedral virus not having strict 532 symmetry of its protein shell. Such symmetry would require that the globule and fine fiber also have 5-fold symmetry which at this time seems doubtful.

The terms hexon and penton antigens appear in established usage, but it should be understood that there is no evidence, in the case of most icosahedral viruses, that these capsomeres consist of six and five biochemical subunits, respectively (or multiples thereof). The terms, for the present, should be used in the sense of coordination; the hexon capsomere having six neighboring capsomeres, the penton capsomere five.

Future investigations undoubtedly will elucidate the structure of capsomeres of many viruses and it is to be hoped that some generalized structural principles will be verified.

(5) The internal structure of viruses. The available information on the internal structure of a number of viruses has been previously discussed. In general knowledge of the packing of RNA and DNA in most viruses is rather limited, particularly for the icosahedral viruses.

It is likely that, as is apparently the case with turnip yellow mosaic virus, the nucleic acid will be shown to have some sort of icosahedral symmetry, at least to the extent that it conforms to the symmetry of an outer protein layer with which it is in contact.

The observation of polyoma virus shells and filaments of three different sizes and what was interpreted as an inner shell of capsomeres *in situ,* led to the proposal that polyoma virus may consist of four concentric layers: an outer **72** capsomere shell, a **32** capsomere shell immediately beneath it, a shell of DNA, and possibly an innermost **12** capsomere shell (*125*). However, Anderer et al. (*4*) observed what they felt were four different shells (72, 42, 32, and 12 capsomeres) in SV40 preparations and concluded that they could not represent internal protein shells since four such shells could not fit into an SV40 virus particle. This difference in interpretation has as yet not been resolved. However, it is likely that a number of icosahedral viruses larger than **30** mμ have a more complex structure than simply an outer shell of capsomeres and an interior consisting only of a folded mass of nucleic acid.

REFERENCES

1. J. D. Almeida, A. F. Howatson, and M. G. Williams, *Virology,* **16,** 353–355 (1962).

2. J. D. Almeida and D. A. J. Tyrrell, *J. Gen. Virol.,* **1,** 175–178 (1967).

3. F. A. Anderer, *Advan. Protein Chem.,* **18,** 1–35 (1963).

4. F. A. Anderer, H. D. Schlumberger, M. A. Koch, H. Frank, and H. J. Eggers, *Virology,* **32,** 511–523 (1967).

5. C. H. Andrewes, *Viruses of Vertebrates,* Williams and Wilkins, Baltimore, 1964, pp. 136–137.

6. K. Apostolov and T. H. Flewett, *Virology,* **26,** 506–508 (1965).

7. P. Atanasiu, P. Lépine, J. Sisman, C. Dauguet, and M. Wetten, *Compt. Rend.,* **256,** 3219–3221 (1963).

8. J. B. Bancroft, G. J. Hills, and R. Markham, *Virology,* **31,** 354–379 (1967).

9. W. B. Becker, *Virology,* **20,** 318–327 (1963).

10. W. B. Becker, K. McIntosh, J. H. Dees, and R. M. Chanock, *J. Virol.,* **1,** 1019–1027 (1967).

11. A. Berkaloff, J. C. Bregliano, A. Ohanessian, *Compt. Rend.,* **260,** 5956–5959 (1965).

12. J. D. Bernal and I. Fankuchen, *J. Gen. Physiol.,* **25,** 111–165 (1941).

13. D. M. Berry, J. G. Cruickshank, H. P. Chu, and R. J. H. Wells, *Virology,* **23,** 403–407 (1964).

14. R. F. Bils and C. E. Hall, *Virology,* **17,** 123–130 (1962).

15. H. A. Bladen, Jr., and A. L. Notkins, *Virology,* **21,** 269–271 (1963).

16. D. E. Bradley, "Replica and Shadowing Techniques," in *Techniques for Electron Microscopy* (D. Kay, ed.), Blackwell Scientific Publications, Oxford, 1961, pp. 82–137 (Charles C Thomas, Publisher, Springfield, Ill., 1961).

17. D. E. Bradley, *J. Gen Microbiol.,* **35**, 471–482 (1964).

18. D. E. Bradley, *Bacteriol. Rev.,* **31**, 230–314 (1967).

19. J. Brandes and C. Wetter, *Virology,* **8**, 99–115 (1959).

20. S. Brenner and R. W. Horne, *Biochem. Biophys. Acta.,* **34**, 103–110 (1959).

21. D. Büttner, H. Giese, G. Müller, and D. Peters, *Arch. Ges. Virusforsch.,* **14**, 657–673 (1964).

22. D. L. D. Caspar, *Nature,* **177**, 475–476 (1956).

23. D. L. D. Caspar, *Nature,* **177**, 928 (1956).

24. D. L. D. Caspar, *Advan. Protein Chem.,* **18**, 37–121 (1963).

25. D. L. D. Caspar, "Design Principles in Virus Particle Construction," in *Viral and Rickettsial Infections of Man,* 4th ed. (F. L. Horsfall, Jr., and I. Tamm, eds.), Lippincott, Philadelphia, 1965, p. 51–93.

26. D. L. D. Caspar, *J. Mol. Biol.,* **15**, 365–371 (1966).

27. D. L. D. Caspar and A. Klug, *Cold Spring Harbor Symp. Quant. Biol.* **27**, 1–24 (1962).

28. T. C. Chambers, N. C. Crowley, and R. I. B. Francki, *Virology,* **27**, 320–328 (1965).

29. V. C. Chambers, S. Hsia, and Y. Ito, *Virology,* **29**, 32–43 (1966).

30. P. W. Choppin and W. Stoeckenius, *Virology,* **23**, 195–202 (1964).

31. R. W. Compans and P. W. Choppin, *Virology,* **33**, 344–346 (1967).

32. R. W. Compans and P. W. Choppin, *Proc. Soc. Natl. Acad. Sci. U.S.,* **57**, 949–956 (1967).

33. F. H. C. Crick and J. D. Watson, *Nature,* **177**, 473–475 (1956).

34. J. G. Cruickshank, D. M. Berry, and B. Hay, *Virology,* **20**, 376–378 (1963).

35. J. G. Cruickshank, A. P. Waterson, A. D. Kanarek, and D. M. Berry, *Res. Vet. Sci.,* **3**, 485–486 (1962).

36. M. C. Davies, M. E. Englert, G. R. Sharpless, and V. J. Cabasso, *Virology,* **21**, 642–651 (1963).

37. M. C. Davies, M. E. Englert, M. R. Stebbins, and V. J. Cabasso, *Virology,* **15**, 87–88 (1961).

38. L. A. Day, *J. Mol. Biol.,* **15**, 395–398 (1966).

39. G. de-Thé and T. E. O'Connor, *Virology,* **28**, 713–728 (1966).

40. Z. Dinter, S. Hermodsson, and L. Hermodsson, *Virology,* **22**, 297–304 (1964).

41. J. Ditchfield and J. D. Almeida, *Virology,* **24**, 232–235 (1964).

42. J. Ditchfield and I. Grinyer, *Virology,* **26**, 504–506 (1965).

43. F. Fenner, *Virology,* 5, 502–529 (1958).

44. J. T. Finch, *J. Mol. Biol.,* **8**, 872–874 (1964).

45. J. T. Finch, *J. Mol. Biol.,* **12**, 612–619 (1965).

46. J. T. Finch and K. C. Holmes, "Structural Studies of Viruses," in *Methods in Virology,* Vol. III (K. Maramorosch and H. Koprowski, eds.), Academic Press, New York, 1967, pp. 351–474.

47. J. T. Finch and A. Klug, *Nature,* **183**, 1709–1714 (1959).

48. J. T. Finch and A. Klug, *J. Mol. Biol.,* **13**, 1–12 (1965).

49. J. T. Finch and A. Klug, *J. Mol. Biol.,* **15**, 344–364 (1966).

50. J. T. Finch and A. Klug, *J. Mol. Biol.,* **24**, 289–302 (1967).

51. J. T. Finch, A. Klug, and A. O. W. Stretton, *J. Mol. Biol.,* **10,** 570–575 (1964).

52. J. T. Finch, A. Klug, and M. H. V. vanRegenmortel, *J. Mol. Biol.,* **24,** 303–305 (1967).

53. J. T. Finch, R. Leberman, and J. E. Berger, *J. Mol. Biol.,* **27,** 17–24 (1967).

54. T. H. Flewett and K. Apostolov, *J. Gen. Virol.,* **1,** 297–304 (1967).

55. R. E. Franklin, *Nature,* **175,** 379–383 (1955).

56. R. E. Franklin, *Nature,* **177,** 928–930 (1956).

57. R. E. Franklin and K. C. Holmes, *Acta. Cryst.,* **11,** 213–220 (1958).

58. R. E. Franklin and A. Klug, *Acta Cryst.,* **8,** 777–780 (1955).

59. A. J. Gibbs, B. Kassanis, H. L. Nixon, and R. D. Woods, *Virology,* **20,** 194–198 (1963).

60. A. J. Gibbs, H. L. Nixon, and R. D. Woods, *Virology,* **19,** 441–449 (1963).

61. M. Goldberg, *Tohoku Math. J.,* **43,** 104–108 (1937).

62. C. E. Hall, E. C. MacLean, and I. Tessman, *J. Mol. Biol.,* **1,** 192–194 (1959).

63. K. A. Harrap and B. E. Juniper, *Virology,* **29,** 175–178 (1966).

64. J. I. Harris and J. Hindley, *J. Mol. Biol.,* **13,** 894–913 (1965).

65. B. D. Harrison and N. C. Crowley, *Virology,* **26,** 297–310 (1965).

66. B. D. Harrison and H. L. Nixon, *J. Gen. Microbiol.,* **21,** 569–581 (1959).

67. B. D. Harrison, H. L. Nixon, and R. D. Woods, *Virology,* **26,** 284–289 (1965).

68. R. G. Hart, *J. Mol. Biol.,* **3,** 701–702 (1961).

69. R. G. Hart, *Virology,* **20,** 636–638 (1963).

70. F. Herold, G. H. Bergold, and J. Weibel, *Virology,* **12,** 335–347 (1960).

71. F. Herold and K. Munz, *J. Gen. Virol.,* **1,** 227–234 (1967).

72. J. H. Hitchborn and G. J. Hills, *Virology,* **35,** 50–70 (1968).

73. J. H. Hitchborn, G. J. Hills, and R. Hull, *Virology,* **28,** 768–772, (1966).

74. D. C. Hodgkin, *Cold Spring Harbor Symp. Quant. Biol.* **14,** 65–78 (1950).

75. R. W. Horne, "Electron Microscopy of Isolated Virus Particles and Their Components," in *Methods in Virology,* Vol. III (K. Maramorosch and H. Koprowski, eds.), Academic Press, New York, 1967, pp. 521–574.

76. R. W. Horne, S. Brenner, A. P. Waterson, and P. Wildy, *J. Mol. Biol.,* **1,** 84–86 (1959).

77. R. W. Horne, G. E. Russell, and A. R. Trim, *J. Mol. Biol.,* **1,** 234–236 (1959).

78. R. W. Horne and A. P. Waterson, *J. Mol. Biol.,* **2,** 75–77 (1960).

79. R. W. Horne, A. P. Waterson, P. Wildy, and A. E. Farnham, *Virology,* **11,** 79–98 (1960).

80. R. W. Horne and P. Wildy, *Advan. Virus Res.,* **10,** 101–170 (1963).

81. Y. Hosaka, H. Kitano, and S. Ikeguchi, *Virology,* **29,** 205–221 (1966).

82. A. F. Howatson and J. D. Almeida, *Biophys. Biochem. Cytol.,* **7,** 753–760 (1960).

83. A. F. Howatson and G. F. Whitmore, *Virology,* **16,** 466–478 (1962).

84. L. Hoyle, R. W. Horne, and A. P. Waterson, *Virology,* **13,** 448–459 (1961).

85. L. Hoyle, R. W. Horne, and A. P. Waterson, *Virology,* **17,** 533–542 (1962).

86. K. Hummeler, H. Koprowski, and T. J. Wiktor, *J. Virol.,* **1,** 152–170 (1967).

87. H. E. Huxley, Some observations on the structure of tobacco mosaic virus; Proc., Stockholm Conference Electron Microscopy, 1956, pp. 260–261 (1957).

88. H. E. Huxley and G. Zubay, *J. Mol. Biol.,* **2,** 189–196 (1960).

89. L. E. Jordan and H. D. Mayor, *Virology,* **17,** 597–599 (1962).

90. J. M. Kaper and E. C. Litjens, *Biochemistry,* **5,** 1612–1617 (1966).

91. D. Kay and D. E. Bradley, *J. Gen. Microbiol.*, **27**, 195–200 (1962).

92. D. W. Kingsbury and R. W. Darlington, *J. Virol.*, **2**, 248–255, (1968).

93. E. W. Kitajima and A. S. Costa, *Virology*, **29**, 523–539 (1966).

94. S. M. Klimenko, F. I. Yershov, Y. P. Gofman, A. P. Nabatnikov, and V. M. Zhdanov, *Virology*, **27**, 125–128 (1965).

95. A. Klug, *J. Mol. Biol.*, **11**, 424–431 (1965).

96. A. Klug and J. E. Berger, *J. Mol. Biol.*, **10**, 565–569 (1964).

97. A. Klug and D. L. D. Caspar, *Advan. Virus Res.*, **7**, 225–325 (1960).

98. A. Klug and J. T. Finch, *J. Mol. Biol.*, **2**, 201–215 (1960).

99. A. Klug and J. T. Finch, *J. Mol. Biol.*, **11**, 403–423 (1965).

100. A. Klug and J. T. Finch, *J. Mol. Biol.*, **31**, 1–12 (1968).

101. A. Klug, J. T. Finch, and R. E. Franklin, *Biochem. Biophys. Acta.*, **25**, 242–252 (1957).

102. A. Klug, W. Longley, and R. Leberman, *J. Mol. Biol.*, **15**, 315–343 (1966).

103. L. W. Labaw and R. W. G. Wyckoff, *J. Ultrastruct. Res.*, **2**, 8–15 (1958).

104. M. A. Lauffer and C. L. Stevens, *Advan. Virus Res.*, **13**, 1–63 (1968).

105. R. Leberman, *J. Mol. Biol.*, **13**, 606 (1965).

106. P. E. Lee, *Virology*, **33**, 84–94 (1967).

107. P. E. Lee, *Virology*, **34**, 583–589 (1968).

108. B. W. Low, *Proteins*, **1A**, 235–391 (1953).

109. R. MacLeod, *Virology*, **34**, 771–777 (1968).

110. R. MacLeod, L. M. Black, and F. H. Moyer, *Virology*, **29**, 540–552 (1966).

111. J. V. Maizel, Jr., B. A. Phillips, and D. F. Summers, *Virology*, **32**, 692–699 (1967).

112. R. Markham, S. Frey, and G. J. Hills, *Virology*, **20**, 88–102 (1963).

113. R. Markham, J. H. Hitchborn, G. J. Hills, and S. Frey, *Virology*, **22**, 342–359 (1964).

114. D. A. Marvin, *J. Mol. Biol.*, **15**, 8–17 (1966).

115. D. A. Marvin and H. Hoffmann-Berling, *Z. Naturforsch*, **18b**, 884–893 (1963).

116. D. A. Marvin and H. Schaller, *J. Mol. Biol.*, **15**, 1–7 (1966).

117. S. Matsumoto, *Virology*, **17**, 198–202 (1962).

118. S. Matsumoto, *J. Cell Biol.*, **19**, 565–591 (1963).

119. C. F. T. Mattern, *Virology*, **17**, 76–83 (1962).

120. C. F. T. Mattern, *Science*, **137**, 612–613 (1962).

121. C. F. T. Mattern, *Virology*, **17**, 520–532 (1962).

122. C. F. T. Mattern, *J. Royal Microscop. Soc.*, **83**, 179–181 (1964).

123. C. F. T. Mattern, A. C. Allison, and W. P. Rowe, *Virology*, **20**, 413–419 (1963).

124. C. F. T. Mattern, K. K. Takemoto, and W. A. Daniel, *Virology*, **30**, 242–256 (1966).

125. C. F. T. Mattern, K. K. Takemoto, and A. M. DeLeva, *Virology*, **32**, 378–392 (1967).

126. R. E. F. Matthews and R. K. Ralph, *Advan. Virus Res.*, **12**, 273–328 (1966).

127. H. D. Mayor, *Virology*, **22**, 156–160 (1964).

128. D. McCarthy and R. D. Woods, *J. Gen. Virol.*, **2**, 9–12 (1968).

129. K. McIntosh, J. H. Dees, W. B. Becker, A. Z. Kapikian, and R. M. Chanock, *Proc. Natl. Acad. Sci. U.S.*, **57**, 933–940 (1967).

130. Y. Minamishima, K. Takeya, Y. Ohnishi, and K. Amako, *J. Virol.*, **2**, 208-213 (1968).

131. S. Mitra and P. Kaesberg, *J. Mol. Biol.*, **14**, 588–571 (1965).

132. D. H. Moore, M. C. Davies, S. Levine, and M. E. Englert, *Virology*, **17**, 470–479 (1962).

133. F. A. Murphy, P. H. Coleman, A. K. Harrison, and W. G. Gary, Jr., *Virology*, **35**, 28–40, (1968).

134. F. A. Murphy, P. H. Coleman, and S. G. Whitfield, *Virology*, **30**, 314–317 (1966).

135. F. A. Murphy and B. N. Fields, *Virology*, **33**, 625–637 (1967).

136. J. Nagington and R. W. Horne, *Virology*, **16**, 248–260 (1962).

137. J. Nagington, A. A. Newton, and R. W. Horne, *Virology*, **23**, 461–472 (1964).

138. J. Nagington, W. Plowright, and R. W. Horne, *Virology*, **17**, 361–364 (1962).

139. H. Nakanishi, Y. Iida, K. Maeshima, T. Teramoto, Y. Hosaka, and M. Ozaki, *Biken's J.*, **9**, 149–157 (1966).

140. H. L. Nixon and A. J. Gibbs, *J. Mol. Biol.*, **2**, 197–200 (1960).

141. H. L. Nixon and B. D. Harrison, *J. Gen. Microbiol.*, **21**, 582–590 (1959).

142. W. F. Noyes, *Virology*, **18**, 511–516 (1962).

143. W. F. Noyes, *Virology*, **23**, 65–72 (1964).

144. R. E. Offord, *J. Mol. Biol.*, **17**, 370–375 (1966).

145. G. T. O'Loughlin and T. C. Chambers, *Virology*, **33**, 262–271 (1967).

146. D. Peters and G. Müller, *Virology*, **21**, 266–269 (1963).

147. D. Peters, G. Müller, and D. Büttner, *Virology*, **23**, 609–611 (1964).

148. D. Peters, G. Müller, and R. Geister, *Arch. Ges. Virusforsch.*, **12**, 437–440 (1962).

149. U. Pettersson, L. Philipson, and S. Höglund, *Virology*, **33**, 575–590 (1967).

150. L. Pinteric, P. Fenje, and J. D. Almeida, *Virology*, **20**, 208–211 (1963).

151. W. Plowright, J. G. Cruickshank, and A. P. Waterson, *Virology*, **17**, 118–122 (1962).

152. G. Plummer and A. P. Waterson, *Virology*, **19**, 412–416 (1963).

153. R. Postlethwaite and J. A. Watt, *J. Gen. Virol.*, **1**, 269–280 (1967).

154. M. Reissig and A. S. Kaplan, *Virology*, **16**, 1–8 (1962).

155. A. J. Rhodes and C. E. vanRooyen, *Textbook of Virology*, Williams and Wilkins, Baltimore, 1962.

156. G. E. Russell and J. Bell, *Virology*, **21**, 283–284 (1963).

157. R. S. Safferman and M. Morris, *Science*, **140**, 679–680 (1963).

158. W. O. Salivar, H. Tzagoloff, and D. Pratt, *Virology*, **24**, 359–371 (1964).

159. R. W. Simpson and R. E. Hauser, *Virology*, **29**, 654–667 (1966).

160. K. M. Smith, *Insect Virology* Academic Press, New York, 1967.

161. K. M. Smith, Z. M. Trontl, and R. H. Frist, *Virology*, **24**, 508–513 (1964).

162. R. L. Steere, *J. Biochem. Biophys. Cytol.*, **3**, 45–60 (1957).

163. R. L. Steere and R. C. Williams, *Am. J. Bot.* **40**, 81–84 (1953).

164. R. H. Symons, M. W. Rees, M. N. Short, and R. Markham, *J. Mol. Biol.*, **6**, 1–15 (1963).

165. K. Takeya and K. Amako, *Virology*, **28**, 163–164 (1966).

166. H. Thormar and J. G. Cruickshank, *Virology*, **25**, 145–148 (1965).

167. W. J. Tromans and R. W. Horne, *Virology*, **15**, 1–7 (1961).

168. R. C. Valentine and H. G. Pereira, *J. Mol. Biol.*, **13**, 13–20 (1965).

169. A. Varma, A. J. Gibbs, and R. D. Woods, *J. Gen. Virol.*, **2**, 107–114 (1968).

170. C. Vasquez and P. Tournier, *Virology*, **17**, 503–510 (1962).

171. A. P. Waterson, J. G. Cruickshank, G. D. Laurence, and A. D. Kanarek, *Virology*, **15**, 379–382 (1961).

172. A. P. Waterson and D. Hobson, *Brit. Med. J.*, **1962–2**, 1166–1167.

173. A. P. Waterson, K. E. Jensen, D. A. J. Tyrrell, and R. W. Horne, *Virology*, **14**, 374–378 (1961).

174. A. M. Watrach, L. E. Hanson, and M. A. Watrach, *Virology*, **21**, 601–608 (1963).

175. J. D. Watson, *Biochim. Biophys. Acta*, **13**, 10–19 (1954).

176. D. H. Watson, I. A. Macpherson, and M. C. Davies, *Virology*, **19**, 418–419 (1963).

177. P. Wildy, W. C. Russell, and R. W. Horne, *Virology*, **12**, 204–222 (1960).

178. M. G. Williams, J. D. Almeida, and A. F. Howatson, *Arch Dermatol.*, **86**, 290–297 (1962).

179. R. C. Williams, *Cold Spring Harbor Symp. Quant. Biol.*, **18**, 185–195 (1953).

180. R. C. Williams, *Advan. Virus Res.*, **2**, 183–239 (1954).

181. R. C. Williams and K. M. Smith, *Biochim. Biophys. Acta.*, **28**, 464–469 (1958).

182. R. C. Williams and R. W. G. Wyckoff, *Proc. Soc. Exptl. Biol. Med.*, **58**, 265–270 (1945).

183. B. S. Wolanski, R. I. B. Francki, and T. C. Chambers, *Virology*, **33**, 287–296 (1967).

184. L. Ya. Zakstelskaya, J. D. Almeida, and C. M. P. Bradstreet, *Acta Virol.*, **11**, 420–423 (1967).

185. N. D. Zinder, R. C. Valentine, M. Roger, and W. Stoeckenius, *Virology*, **20**, 638–640 (1963).

186. L. O. Zwillenberg, M. H. Jensen, and H. H. L. Zwillenberg, *Arch. Ges. Virusforsch.*, **17**, 1–19 (1965).

187. L. O. Zwillenberg and K. Wolf, *J. Virol.*, **2**, 393–399 (1968).

3

THE REPLICATION OF PICORNAVIRUSES

DAVID BALTIMORE†

THE SALK INSTITUTE FOR BIOLOGICAL STUDIES
SAN DIEGO, CALIFORNIA

† Present Address: Department of Biology, Massachusetts Institute of Technology, Cambridge, Massachusetts.

3-1. Introduction

Extensive investigation of the replication of picornaviruses over the past ten years has produced a general understanding of the various elements of the virus growth cycle. The picture which has emerged is of surprisingly autonomous agents which are able to both inhibit host-cell macromolecular syntheses and induce the necessary processes for their own reproduction, all with a very limited amount of genetic material. In this chapter, numerous aspects of the growth cycle of picornaviruses will be reviewed. The main emphasis will be on those parts of the replication process with which the present author has had the most experience, and much previously unpublished data from the author's laboratory is included. Interspersed are speculations about the mechanisms of certain phenomena; these are generally an attempt to reinterpret old experiments in the light of more recent results and, in most cases, will require direct experimentation to provide clear-cut tests of the hypotheses.

For the purpose of this review, "picornaviruses" will be defined as "poliovirus and other animal viruses which are indistinguishable from poliovirus by gross morphology." This includes the Coxsackie-viruses, echoviruses, and Columbia-SK viruses (EMC, Mengo, ME, etc.), all of which have densities and sedimentation coefficients which are virtually identical with poliovirus (1). Furthermore, all of these viruses which have been studied have genetic material consisting of a single molecule of RNA which has about 6000 nucleotides (1) and no differences in their intracellular mode of replication have been discovered. Since the only distinguishing characteristics of these viruses are not considered in this review (antigenicity, host range, disease production), no distinction will be made between them, and it will be assumed that if studies of two different picornaviruses yield opposite results, the fault does not lie in the virus.

Two other groups of viruses are generally included among the picornaviruses, the rhinoviruses and foot-and-mouth disease viruses.

Rhinoviruses have received very little morphological or biochemical study, but at least by two criteria, sensitivity to low pH and the length of the RNA (*1a*), they are different from poliovirus. Foot-and-mouth disease viruses are more sensitive to low pH, smaller, and denser than poliovirus (*2*). In general, therefore, these viruses will not be discussed in this review.

A specific review concerned with biochemical aspects of picornavirus multiplication has not appeared since Darnell and Eagle's review in 1960 (*3*). Biological and medical questions have been discussed more recently (*1*) and genetic and morphological problems are dealt with in other chapters of the present volume.

3-2. Methodology

Since this chapter describes experiments done in many laboratories, it is impossible to describe all of the techniques which have been used. However, much of the work has been carried out by a limited number of groups all employing basically the same methods, and the rationale for some of the biochemical procedures has never been clearly stated. Thus, this section will be limited to a brief discussion of the methods of analysis of poliovirus growth which have been developed mainly by Darnell, Penman, Girard, Maizel, me, and our colleagues. It will also include discussions of the uptake of RNA and protein precursors by mammalian cells.

A. CYTOPLASMIC EXTRACTS

Since picornavirus multiplication occurs in the cytoplasm of the infected cell, it is possible to discard the nuclei during most biochemical procedures. This is desirable because many such procedures require treating extracts with detergents which cause the formation of an intractable gel when nuclei are present.

To prepare cytoplasmic extracts from cells growing in suspension (*4*), the cells are allowed to swell for 5–10 min in a hypotonic buffer (*5*) containing enough magnesium to prevent disruption of polyribosomes (10^{-3}–$10^{-4} M$) (*6*). They are then broken with either an all-glass (Dounce) homogenizer with a tight-fitting pestle or a stainless-steel homogenizer with a clearance of 0.002 in. (*7*) which gives more reproducible results. For HeLa cells, a few strokes with the pestle liberates at most a few per cent of the DNA from the nuclei. The homogenized cells are then cleared of nuclei by centrifugation to produce the cytoplasmic extract.

Since many structures of interest are at least partially attached to cytoplasmic membranes, it is usually necessary to dissolve the membranes prior to further analysis. The usual procedure is to treat extracts with sodium desoxycholate (DOC)—an anionic detergent derived from bile salts. A final concentration of 0.5–1% suffices to make the cloudy cytoplasmic extract almost clear, indicating solubilization of most of the membranous material. Usually the addition of DOC is followed within less than a minute by the addition of a nonionic detergent, generally BRIJ-58, to 0.5–1% (6). This causes the formation of mixed micelles of the two detergents and effectively sequesters the DOC. It is of value because prolonged exposure of extracts to DOC (especially when very little protein is present) can cause deleterious effects such as breakdown of polyribosomes (8). Further, DOC will cause precipitation of the 40S subribosomal particle unless it is effectively neutralized with a nonionic detergent (6), and DOC itself has a tendency to precipitate—especially in solutions of high ionic strength—which is counteracted by mixed micelle formation. Nonionic detergents by themselves cannot completely solubilize membranous materials, although they have been used for selective release of components from membranes.

The nuclei produced by this method are not free of cytoplasm, but can be purified using a mixture of DOC and the nonionic detergent Tween 40 (7). While DOC alone would disrupt the nuclei and lead to the formation of a gel, the mixture of detergents is much more gentle.

B. PREPARATION OF RNA

The total RNA from cells is most easily prepared by phenol extraction at 60°C in the presence of sodium dodecyl sulfate (SDS) (9). Addition of chloroform to the phenol causes more complete extraction (7) and lowering the temperature to 55°C avoids aggregation of RNA's (10).

For analysis of RNA, a much simpler method is generally applicable. Cytoplasmic extracts can be treated directly with SDS which separates all RNA protein complexes into their component parts (11). [The only documented exceptions to this are certain viruses, including poliovirus (12)—a viral trait which is often of great experimental value.] If an SDS-treated cytoplasmic extract is analyzed on a suc-

rose gradient, the RNA sediments just as it would after phenol deproteinization (*13*) except that interactions which often occur during
phenol extraction are not present, leading to very sharp bands (*8*).
The method is not only very simple but is more efficient in terms of
the yield of RNA and leads to fewer artifacts.

C. Zonal Centrifugation through Sucrose Gradient

This very potent tool for analysis of virus-induced processes was
first described by Britten and Roberts in 1960 (*14*) and has been widely
adopted. Automatic scanning methods (*4, 15*) have greatly increased
the resolution of the technique. Its value to the study of picornaviruses
comes from the fact that although the virus-specified material in
an infected cell represents a very small percentage of the total mass
of the cell, it is possible to selectively tag virus-specific molecules
with radioactive atoms and then separate the molecules, or structures
containing them, on sucrose gradients. The distribution of radioactivity
at various levels of the gradient then allows determination of the
kinds and amounts of the various labeled species. This can be done
with detergent-treated cytoplasmic extracts to find structures like
virus-specific polyribosomes, virus particles, and the replication complex. With SDS-treated extracts, the various species of virus-specific
RNA can be identified.

D. Acrylamide Gel Electrophoresis

The use of electrophoresis for the separation of proteins has a long
history, but it has generally been applied to purified materials. To
analyze all of the proteins in a cytoplasmic extract or to analyze proteins which are insoluble in most solutions has only become possible
recently through Summers et al.'s (*16*) modification of acrylamide gel
electrophoresis. The critical factor is again a detergent, SDS, which
binds avidly to proteins and keeps them separate and in solution. Thus,
treatment of cytoplasm or of viral proteins with SDS plus mercaptoethanol, urea, and acetic acid separates all known proteins into individual, soluble polypeptide chains which can be fractionated on acrylamide gels containing SDS. These gels fractionate proteins almost
exclusively on the basis of size (*17, 18*).

The usefulness of this technique in the study of viral proteins
comes from the fact that picornaviruses inhibit host-cell protein syn-

thesis so that all of the protein made after a certain time of infection is virus specific. Therefore, if radioactive amino acids are added to infected cells at the appropriate time, all of the labeled protein is virus specific and determination of the distribution of radioactivity in the gel allows identification and quantification of the viral proteins.

The technique of acrylamide gel electrophoresis has recently been extended to the fractionation of RNA (19). It gives much finer resolution than sucrose gradients and promises to become the method of choice for the analysis of RNA.

E. THE UPTAKE OF URIDINE INTO MAMMALIAN CELLS

Many of the experiments discussed in this chapter involve labeling of viral RNA with uridine. For this reason, a short summary of unpublished data (8) on the accumulation of uridine in mammalian cells will be given here.

Uridine is actively accumulated in mammalian cells. If a low concentration of radioactive uridine is added to the medium of growing (infected or uninfected) cells, they will take up a large percentage of it and all of the intracellular material will be in the form of nucleotides (UMP, UDP, UTP). For instance, if 0.04 μmole of radioactive uridine is added to 10^7 HeLa cells in 2 ml of medium, the cells will take up 80% or more of it even if RNA synthesis has been inhibited with actinomycin. Since the intracellular uridine is phosphorylated and cannot escape from the cells, addition of nonradioactive uridine will not lower the amount of radioactivity in the cellular pool of nucleotides.

Furthermore, it appears that the cell itself has a larger internal pool of uridine nucleotides than the maximal amount of uridine it is able to take up from the medium. Therefore, even if the cells have taken in only a trace of uridine from the medium, the nucleotide pool cannot be extensively diluted by adding a large excess of unlabeled uridine. This is shown by the data in Fig. 3-1. In this experiment, infected cells (5×10^6/ml) were exposed to 1.8 mμmole/ml of ^{3}H-uridine and at 30 sec an aliquot was "chased" by the addition of a 100-fold excess of unlabeled uridine. The incorporation of radioactivity into RNA was followed in the two samples. The chase did not stop uridine incorporation into RNA but caused its rate to become linear; the linear rate extrapolated to approximately the time of addition of the unlabeled uridine. This shows that the unlabeled uridine is not

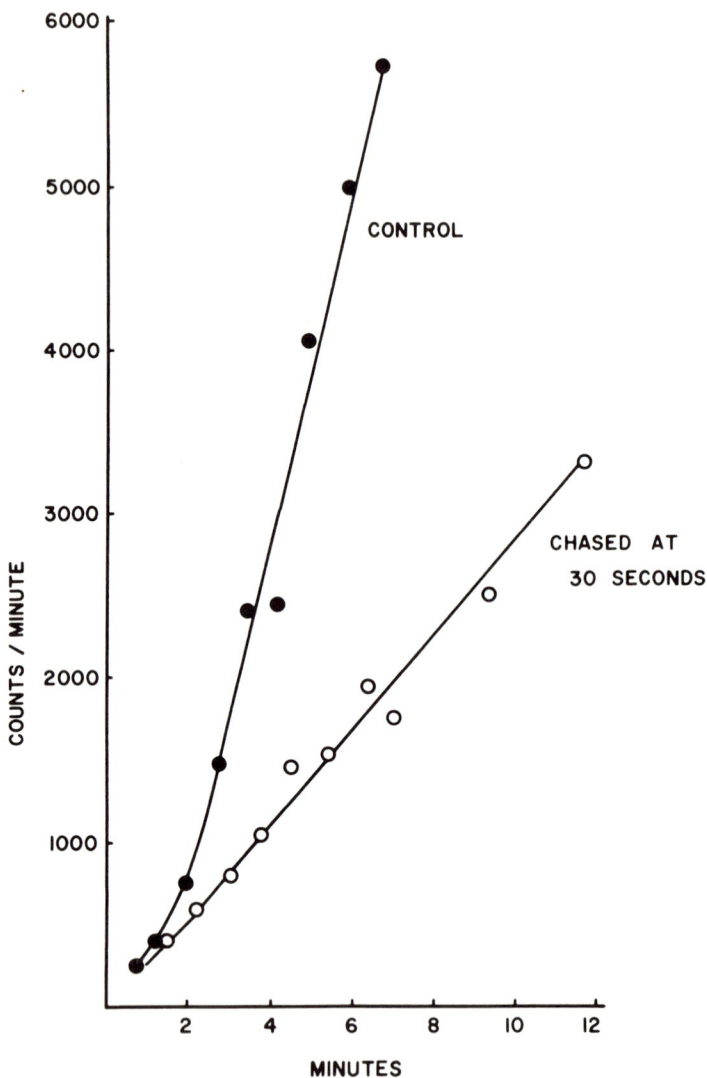

Fig. 3-1. Linear RNA synthesis following a 30-sec exposure to radioactive uridine. ³H-uridine was added to a culture of poliovirus-infected HeLa cells and after 30 sec a 100-fold excess of unlabeled uridine was added to one-half of the culture. At various times, samples from the two cultures were precipitated with cold 5%/trichloracetic acid. The precipitates were collected by centrifugation, redissolved in 0.3 N NaOH, and reprecipitated. The acid-precipitable material was collected on filters and counted. (Unpublished data.)

able to extensively dilute the internal pool, even though the amount of unlabeled uridine which was added was low enough so that almost all of it would be taken up by the cells.

F. The Uptake of Amino Acids

In contradistinction to uridine, at least some amino acids are not taken from the medium by HeLa cells in large amounts and can easily be chased. In fact, if 1.6 mμmoles of radioactive leucine are added to cells, almost all of the intracellular radioactivity is acid precipitable —very little can be detected in a free pool (*20*). Addition of a large excess of unlabeled leucine therefore leads to an immediate cessation of incorporation of radioactivity into protein.

3-3. Over-all Aspects of Virus Development

This section will provide the framework for further discussion. After describing the structure of the virion and its components, the replication of the virus will be discussed from three points of view: the kinetics of appearance of progeny, the dependence of virus synthesis on host-cell functions, and the subcellular localization of the replicating virus.

A. Size of Picornaviruses, Their RNA, and Their Protein

It is generally agreed that all picornaviruses have about the same size by electron microscopy (about 27 mμ), the same sedimentation rate in sucrose gradients, and the same density in cesium chloride (*1, 21*). However, the absolute value of the particle weight of the virion is unclear. Two recent papers report particle weights based on sedimentation and diffusion constants—one finds a value of 5×10^6 daltons (*22*); the other a value of $8.3 \pm 0.7 \times 10^6$ daltons (*23*). The reason for the difference is discrepancies in the diffusion constants and partial specific volumes which were used in the calculations (see Ref. *1*, p. 97).

In contrast, these same papers report molecular weights for the isolated RNA of the virions as 2×10^6 daltons (*22*) and $1.7 \pm 0.2 \times 10^6$ daltons (*23*). Considering that these numbers were arrived at by different methods and using different salt conditions, the value of about 2×10^6 daltons of RNA per isolated molecule can be accepted. There must be one such RNA molecule per virion because a virion contains about 1–3×10^6 daltons of RNA [25–30% of the virion is RNA (Ref. *1*, p. 101) and 6–8×10^6 daltons is the molecular weight of the virion].

Early work on the molecular weight of the minimal protein subunit of the capsid of poliovirus produced values of 27,000 daltons in two reports (*22*, *24*), 26,000 in a third (*25*), and 42,000 in a fourth (*26*). Recently, electrophoretic analysis has revealed that the capsid consists of an approximately equal molar ratio of four different proteins (see Sec. 3-5, B). The molecular weights of these proteins have been estimated as 35,000; 28,000; 24,000; and about 8000 daltons (*20*, *26a*).

An X-ray diffraction study of poliovirus crystals indicated that there are 60 identical subunits in the capsid (*27*). If each of the 4 polypeptides is present in 60 copies in each virion, there would be 5.7×10^6 daltons of protein per particle. Add to this the 2×10^6 daltons of RNA and a particle weight of 7.7×10^6 daltons emerges. This is within the range of published values and is close to the particle weight calculated by Schwerdt (6.8×10^6 daltons) using the dry weight density and the diameter of the particle (*28*).

From this discussion, it can be said that, although the data are not in complete agreement, an idealized picture of a picornavirus is a particle with a diameter of 27 mμ, consisting of one RNA molecule of 2×10^6 daltons and 240 polypeptide chains with a total weight of 5.7×10^6 daltons. The particle is therefore 26% nucleic acid and has a molecular weight of 7.7×10^6 daltons.

An interesting aspect of the properties of picornavirus RNA's is that they sediment ahead of the larger ribosomal RNA in 0.1 M salt, but behind it in lower salt concentrations (less than 0.01 M salt) (*29*, *30*). This differential effect of salt concentration on the structures of viral and ribosomal RNA may be a reflection of the different role these molecules perform. In any case, the experiments show that viral RNA is single stranded because double-stranded nucleic acids have about the same sedimentation rate in high or low salt.

The single strandedness of viral RNA is also indicated by the inequality of adenylate and uridylate in the RNA (29% vs. 24%) (*31*, *32*). This inequality further shows that viral RNA is not an equal mixture of two single-stranded species of RNA related to each other by the Watson–Crick pairing rules (*33*). Attempts to form double-stranded RNA from viral RNA preparations by annealing have been totally unsuccessful (*34*, *35*), indicating the RNA in virions has at most 1% complementary RNA. The implications of the existence of a unique strand to the question of RNA replication are discussed in Sec. 3-7.

B. The Growth Cycle

The timing of viral replication varies considerably, depending on numerous factors such as the specific virus, the host cell, the nutritional state of the cell, and especially the multiplicity of infection (*36*). The general finding is that for 2–4 hr after infection no new virus is detectable, and at least some of the infectivity of the innoculum disappears. The duration of this eclipse period is 2.5–3 hr for poliovirus in monkey kidney cells, but 3.5–4 hr for Coxsackie A9 virus growing in the same cells (*37*). The length of this phase can be reduced by increasing the multiplicity of infection (*36*). After this eclipse period there is an exponential period of virus production lasting 2–4 hr. The timing of the release of the virus from the cell is quite variable and seems to be under the genetic control of the virus (see Sec. 3-9, B).

During the latent period, not only is no virus produced, but there is very little viral RNA or protein synthesis. Darnell et al. (*38*) and Darnell and Levintow (*39*) first showed that the synthesis of poliovirus RNA and protein precedes virion appearance by only a short time. Thus, there is no large pool of either component made prior to maturation. (A detailed discussion of the timing of RNA and protein synthesis is given in Section 3-6, A.)

C. Independence of Virus Growth as Revealed by Inhibitors

For many drugs, at least one mechanism of inhibition of cellular growth is known [e.g., actinomycin inhibits DNA-dependent RNA synthesis (*40, 41*), puromycin inhibits protein synthesis (*42*)], but one can never be sure that the drug does not also cause other effects. Thus, when such a drug *does* inhibit virus growth it may be difficult to determine the reason, but when it *does not* inhibit viral reproduction, the result is much easier to interpret. Two basic facts about picornavirus reproduction have been established using this principle. The well-documented fact that actinomycin, under most conditions, does not inhibit viral growth (*43–45*) shows that DNA-dependent RNA synthesis plays no essential role in viral development. The lack of inhibition by aminopterin (*46*), 5-bromouracil (*46*), and 5-fluorodeoxyuridine (*47*) is proof that DNA replication is not involved in viral multiplication.

The picture which emerges from the use of inhibitors is that picorna-virus synthesis is largely autonomous of host-cell nucleic acid metab-olism. However, the simplicity of this conclusion is marred by recurrent reports that under some conditions picornaviruses are sensitive to inhibition by actinomycin.

Three laboratories have reported sensitivity of poliovirus to actino-mycin (48–50). Some strains of virus are extremely sensitive (yield reduced to 1% or less of normal) while others are virtually unaffected (49). The virus is most sensitive to actinomycin during the early phase of the growth cycle; one group reports that the virus becomes resistant to the drug after only 30 min of infection (50). Cooper (48) has shown that treatment of cells with insulin for 24 hr prior to in-fection abolishes the sensitivity of the virus to actinomycin. The in-sulin treatment does not seem to affect the metabolism of the cells, nor does it enhance viral yield in the absence of actinomycin. The mechanism of the insulin effect is obscure.

D. Localization of the Synthesis of Viral Components in the Cytoplasm

It is now universally agreed that viral RNA is manufactured in the cytoplasm of the infected cell, but for quite a while it was thought that viral RNA might be made in the nucleus (51–54). Franklin and Rosner (55) and Hausen (56) showed conclusively, using autoradiog-raphy, that viral RNA is made in the cytoplasm. Cells making vir-tually only viral RNA were pulse-labeled with tritiated uridine for as short a time as 30 sec, and all of the radioactivity was found in the cytoplasm. More recent studies have confirmed these results, using various criteria such as the subcellular localization of the RNA-synthesizing enzymes (57–59). Evidence from cell fractionation, while satisfying, can never be as convincing as the original autoradiographic analyses because the moment cells are broken, the unanswerable ques-tion arises of whether nuclear materials have leaked into "cyto-plasmic" fractions.

It is generally accepted that viral protein synthesis is carried out in the cytoplasm. The best evidence for this conclusion is the isolation from the cytoplasm of infected cells of virus-specific polyribosomes containing nascent proteins antigenically related to poliovirus (60). Unfortunately, no autoradiographic study of viral protein synthesis has appeared.

All of the evidence in this section leads to the concept that the nucleus has no role in viral development. The final proof of this would be the synthesis of virus in enucleated cells. Such an experiment has been reported. Crocker and Pfendt (61) infected cytoplasmic fragments of a human amnion cell line with poliovirus and found both viral antigen and virus-stimulated RNA synthesis in the fragments. Marcus and Freiman (61a) have also shown virus-stimulated RNA synthesis in infected anucleate fragments of HeLa giant cells.

E. ROLE OF THE CELLS

Up to this point, the independence of poliovirus from its host cell has been stressed, but it is obvious that the cell contributes an enormous amount to virus synthesis. The ribosomes and enzymes for manufacture of viral proteins come from the cell, the precursors and energy for synthesis of viral macromolecules are cellular and the cell provides a milieu in which virus synthesis can occur.

One striking effect of the state of cells has been noticed. Marcus and Robbins (62) have reported that cells held in metaphase will not synthesize poliovirus RNA. Later work by Salb and Marcus (63) showed that the ribosomes from cells in metaphase are affected by a trypsin-sensitive substance so that they synthesize protein, *in vitro*, at a much reduced rate. This is true even when polyuridylic acid is provided as a messenger RNA. The authors suggest that the low rate of protein synthesis may be responsible for the inability of the cell to make viral RNA (no RNA-synthesizing machinery can be made without protein synthesis). High input multiplicities of virus can overcome the inhibition of RNA synthesis in some of the metaphase cells.

3-4. Events Occurring before Progeny Virions Appear

It is convenient to divide the picornavirus growth cycle into two time periods, one from the initiation of infection to the first appearance of progeny virions (eclipse) and a second from the end of eclipse to the end of infection. During eclipse a number of events occur which prepare the cell for the massive synthesis of virus which will ensue. The extent to which the virus is able to divert host-cell syntheses toward viral replication can be appreciated if it is noted that the rate of viral RNA synthesis can reach and even overshoot the rate of synthesis in uninfected cells and the rate of virus-specific protein synthesis can be 40% of normal (*44*). To prepare for this, during the eclipse phase, the virus begins the inhibition of both host RNA and protein synthesis and thus frees almost all of the cell's ribosomes to participate in the synthesis of viral proteins. These inhibitions are caused by proteins synthesized after infection and probably under the direction of the viral genome.

The eclipse phase is not analogous to the "early" phase of DNA bacteriophage replication because during that period there is no synthesis of bacteriophage DNA. During picornavirus eclipse, as will be discussed in detail in Sec. 3-7, B, there is synthesis of viral RNA and, in fact, it has been impossible to demonstrate any time interval before replication begins. Picornavirus eclipse is best viewed as a priming period, a time when host functions are being eliminated and viral functions are accelerating.

A. Inhibition of Host-Cell Protein Synthesis

Very soon after infection of cells with a picornavirus the rate of protein synthesis begins to decline (*44, 64–66*). The kinetics of this decline vary from one system to another but in general the rate of synthesis falls to 25–50% of normal by the time virus-induced protein synthesis becomes evident. Under special conditions (synchronized infection, very high multiplicities of virus), inhibition can be complete within an hour (*67*). The inhibitor continues to function throughout infection, so the rate of host protein synthesis continues to fall while viral proteins are being made.

The first question about this inhibition is whether it is caused by a protein contained in the virion or by a protein synthesized in the infected cell under the direction of the viral genome. The second hypothesis is clearly correct. The earliest evidence for this was that fluorophenylalanine (FPA) is able to prevent the inhibition (*66*). Later results have shown that light inactivation of proflavin-sensitized virus (*68*) or ultraviolet light inactivation of virus (*67*) prevents it from carrying out the inhibition. The inhibition will only take place if prior virus-induced protein synthesis has occurred (*67*). At high multiplicities of infection, inhibition occurs in cells treated with guanidine (*68, 69*) where no viral RNA replication can be detected. At lower multiplicities RNA replication is required, so the inhibitor is made both from input and progeny genomes.

The most interesting question about the inhibitor of protein synthesis is how it is able to differentiate viral protein synthesis from host-cell protein synthesis. No answer to this question is as yet possible, but there is further evidence on the mechanism of the inhibitor. The first fact is that viral infection causes a breakdown of host-cell polyribosomes, liberating free ribosomes (4). Although, at one time, it was believed that the breakdown is not accompanied by a progressive diminution of the size of the remaining polyribosomes (70, 71), it now appears that the polyribosomes do get progressively smaller as the extent of inhibition increases (71a). However, the host messenger RNA left in polyribosomes after partial inhibition of protein synthesis is of normal size (71), and the polyribosomes which are left are synthesizing protein at the same rate as in uninfected cells (70). None of the available evidence strongly supports any specific model for how the virus is able to selectively inhibit host-cell protein synthesis. However, the decreasing size of polyribosomes with undegraded messenger RNA appears to eliminate specific nucleases, or other agents which might inactivate messenger RNA's, as causes of the inhibition. One possible model is a virus-induced elimination of a factor needed for initiation of peptide chains on host-cell messenger RNA's (71a).

B. INHIBITION OF HOST-CELL RNA SYNTHESIS

Soon after infection of cells with a picornavirus, the rate of cellular RNA synthesis begins to decline (44, 61, 72, 73). In mengovirus-infected L-cells this decline is very marked, RNA synthesis falling to less than 10% of normal within 2 hr after infection. In other systems the inhibition is less marked and RNA synthesis goes smoothly from host to virus specific with a slow decline in over-all synthesis as infection proceeds (74). The reason for the different results found with different systems seems to reside in the cells. Mengovirus causes a rapid decline in RNA synthesis in L-cells but not in HeLa cells (75); one line of rat hepatoma cells is very sensitive to the inhibitor while a nutritional mutant of these cells is very insensitive (76).

Studies on cells in which inhibition of RNA synthesis is slow have indicated that functional host messenger RNA and 45S ribosomal precursor RNA continue to be made (77). However, the processing of the ribosomal precursor RNA is aberrant in a way which is not mimicked by a high concentration of cycloheximide (77). In spite of this, the aberrant processing may still be a secondary effect of the virus-induced inhibition of protein synthesis because the rapid and complete inhibition produced by cycloheximide is not comparable to the gradual virus-induced inhibition.

The cessation of nuclear RNA synthesis after infection is paralleled by a loss in function of the DNA-dependent RNA synthetic system measured *in vitro* (41). That is true whether inhibition is gradual or abrupt (41, 65), suggesting that both kinds of inhibition may occur by the same mechanism. The reason for the reduced rate of synthesis does not seem to reside in the DNA, since the DNA, is not broken down and its melting temperature is unchanged (41). Further, the DNA isolated from infected cells as well as the deoxyribonucleoprotein from infected cells acts as template for the *E. coli* DNA-dependent RNA polymerase as efficiently as the same materials isolated from uninfected cells (65). Thus, it appears that the inhibition is due to a direct effect on the polymerase.

It has been reported that the cytoplasm of mengovirus-infected cells contains a trypsin-sensitive factor which will inhibit DNA-dependent RNA synthesis in nuclei from uninfected cells (79). Unfortunately, this report has never been followed up and is contradicted by the inability of Holland (78) to find such an effect with lysates of poliovirus-infected HeLa cells. It is clear that the effect is not simply due to ribonuclease (41), but its mechanism remains obscure.

The inhibition of cellular RNA synthesis is clearly due to a virus-induced protein. This was shown by the ability to prevent inhibition with FPA, puromycin, and ultraviolet light irradiation of the virus (44, 66). The effect is not simply a concomitant of the inhibition of protein synthesis just as the inhibition of protein synthesis is not a side-effect of the inhibition of RNA synthesis (44).

Thus, soon after infection, picornaviruses cause the production of two new proteins (or one bifunctional protein) which inhibit cellular protein and RNA synthesis. The effect on RNA synthesis seems to be of relatively minor importance to the growth of the virus since in certain cells the inhibition is very gradual. The critical necessity for the inhibition of protein synthesis seems reasonable in that the release of single ribosomes for use by the virus is directly necessary for synthesis of viral proteins, but there is a report of a cell line which will grow normal amounts of virus but is not affected by the inhibitor (76). Inhibition of RNA synthesis will reduce the amount of cell messenger RNA which will compete with viral RNA for ribosomes, but this can clearly be overcome by the specificity of the protein inhibitor and so is useful but not necessary.

C. OTHER EVENTS

DNA synthesis is not drastically inhibited by picornavirus infection until long after RNA and protein synthesis have been severely affected (44). This suggests that the effect on DNA synthesis is a secondary result of one of the other inhibitions. Cell division, however, comes to a very abrupt halt within less than an hour after infection (80). In fact, it appears that within minutes after addition of virus, premitotic cells are prevented from entering prophase although cells in prophase at the time of infection go on to complete division. This effect can apparently be mimicked by puromycin and is probably a secondary effect of the inhibition of protein synthesis (80).

3-5. The Manufacture of Virus-Specific Proteins

The only known difference between viral protein synthesis and host protein synthesis is that the virus-induced inhibitor can distinguish one from the other. Otherwise, just as those of uninfected cells, viral proteins are made on a polyribosome and are released from the polyribosome very soon after their completion (4). However, the virus-specific polyribosomes are distinguishable from the cell's polyribosomes on the basis of their size and the kind of messenger RNA which they contain.

A. MECHANISM OF VIRAL PROTEIN SYNTHESIS; THE VIRAL POLYRIBOSOME

Polyribsomes can be identified and quantitated either by their content of ribosomes (which absorb at 260 mμ) or by their content of nascent polypeptide chains (which can be labeled by a short pulse of radioactive amino acids). Both criteria show that HeLa cell polyribosomes, which have an average sedimentation rate of about 200S, begin to disintegrate immediately after poliovirus infection and are replaced later by polyribosomes sedimenting at an average rate of about 350S (4). By 4 hr after infection, these large polyribosomes may contain 50% or more of the total ribosomes of the cell (8).

Virus-specific protein synthesis occurs on these new, large polyribosomes, and some of their nascent polypeptides have the antigenic specificity of capsid protein (60). The messenger RNA in them has the size (*81, 83*) and base composition (*82*) of viral RNA. Whether or not it is infectious has not been directly investigated, but there is infectious RNA in the cells which is not in viral particles (*84, 85*) and it might be from polyribosomes because most of the intracellular viral RNA is in polyribosomes. The infectious RNA isolated from virions will act as a messenger *in vitro* (see Sec. 3-5, D). On the basis of this evidence, it will be assumed throughout this chapter that there is no physical difference between RNA isolated from virions and the messenger RNA of viral polyribosomes.

Using infected cells in which the ribosomal RNA was prelabeled with ^{3}H and the virus-specific RNA was labeled with ^{14}C, Summers et al. (*83*) were able to measure accurately the average number of ribosomes on a 380S polyribosome. They found that there are 35 ribosomes per polyribosome which compares reasonably well with an earlier estimate from electron microscopy of 40 or more (*86*). Late in infection (after 4 hr of a synchronous infection), the polyribosomes begin to decrease in size while the messenger RNA remains intact (*83*). By 4.5 hr, the polyribosomes are sedimenting at 200S and have about 20 ribosomes per unit. The fact that a unit of 20 ribosomes can sediment at the same rate as a unit of 5–6 ribosomes from the reticulocyte (*87*) emphasizes the role of secondary structure and probably of ribosome spacing along the message in determining the sedimentation rate of a polyribosome. The later, smaller polyribosomes are making protein at a slower rate than the larger ones (*83*), possibly because by this time after infection the cell's metabolism is severely damaged. However, unless ribosomes are prematurely dissociating from messenger RNA before translation is completed, the slow translation rate cannot account for the small size of the polyribosomes, and therefore, the number of new ribosomes attaching to the polyribosomes must be reduced. This could either be due to a limitation in the number of available ribosomes or to a direct effect on attachment.

It is not known if all of the components used to synthesize protein are identical in uninfected and infected cells. The same ribosomes that make virus protein, must make host proteins because 50% or more of the infected cell's ribosomes may be in virus-specific polyribosomes (*8*), yet virtually no new synthesis of ribosomes (*77*) or breakdown of old ribosomes (*44*) occurs after infection. But no systematic search for new transfer RNA's in infected cells has been made and the possibility exists that, for instance, a modified transfer RNA for chain initiation might be made. Unfortunately, the chain initiator for mammalian cells has not been identified; the known chain initiator for bacterial cells, N-formylmethionine, cannot be found in either infected or uninfected HeLa cells (70).

B. The Viral Proteins and Control of Their Synthesis

The adaptation of acrylamide gel electrophoresis to the analysis of proteins from both virions and unfractionated cytoplasmic extracts (16) has greatly advanced the study of the proteins made under the control of the viral genome. Since the virus inhibits host-cell protein synthesis, conditions can be found where radioactive amino acids are incorporated solely into virus-specific proteins. Electropherograms of the separated poliovirus protein from labeled virions and the cytoplasm of infected cells are shown in Fig. 3-2.

The poliovirion contains four distinct proteins (16) (Fig. 3-2A). Their molecular weights as estimated from their electrophoretic migration rate are 35,000 (VP-1), 28,000 (VP-2), 24,000 (VP-3), and about 8000 (VP-4) (20, 26a). When poliovirions are labeled with radioactive isoleucine, leucine, and valine, the relative amounts of label in each of the four proteins corresponds approximately to their molecular weights, indicating that they occur in about equal molar ratios in the virion. Treatment of poliovirus at pH 10.5 and 40°C causes the RNA to be released from the virion, leaving an intact spherical particle (89); VP-4 is also released from the particle during this treatment (90).

The pattern of intracellular proteins varies with the time of exposure to radioactive amino acids for reasons which will be discussed below. In Fig. 3-2B, the infected cells had been labeled for 8 min followed by a 5-min chase with unlabeled amino acids (which is necessary to allow completion of nascent chains). The gel pattern shows eight distinct major peaks and a series of minor peaks. In all, Summers et al. (16) estimate that there are 14 reproducible peaks which are numbered as indicated (NCVP = noncapsid viral protein; the reason for changing the name of NCVP-6 to VP-0 will be discussed in Sec. 3-6, D). NCVP-1 is the largest viral protein and VP-4 is the smallest.

It has recently become apparent that proteolytic cleavages play a large role in the formation of the various viral proteins. Three laboratories have shown that after a short exposure of infected cells to radioactive amino acids only the larger viral polypeptides are labeled and that after continued incubation in the presence of unlabeled amino

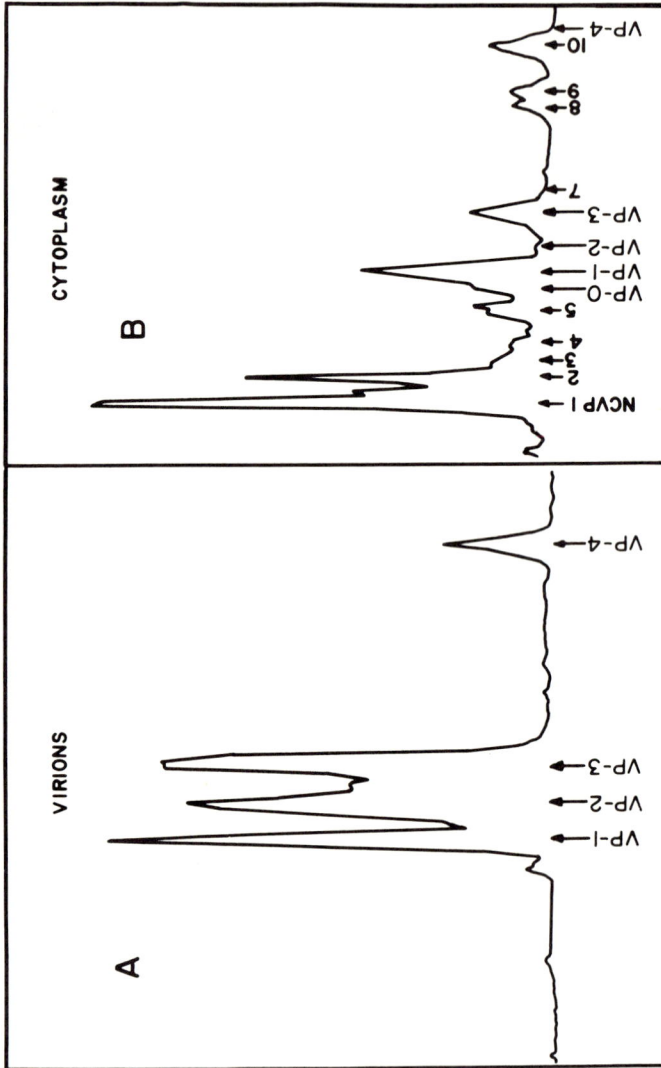

FIG. 3-2. Acrylamide gel electrophoresis of (A) the proteins from poliovirus virions and (B) all of the virus-specific proteins made in the infected cells. Direction of migration is left to right. Redrawn from Refs. *16* and *70*, respectively, by courtesy of the National Academy of Sciences and Academic Press.

acids, the radioactivity is found in shorter polypeptides (*90a, 90b, 90c*). These experiments imply that most of the viral proteins, and especially the capsid proteins, are not formed as primary gene products but are the result of secondary cleavages.

As detailed in Sec. 3-8, B, fluorophenylalanine (FPA) is a potent inhibitor of viral development. This phenylalanine analog is incorporated into viral proteins, thus altering their structure and function. One of its effects is to prevent the proteolysis of the large viral proteins (*90c*), presumably because FPA-containing protein is not a substrate for the proteolytic enzyme. In the presence of FPA plus a number of other amino acid analogs (ethionine for methionine, azetidine-2-carboxylic acid for proline, and canavanine for arginine) a new polypeptide, larger than NCVP-1, is a major product (*90c*) and a second protein, yet larger, is also found (*20d*). This protein has a molecular weight exceeding 200,000 and probably represents all of the genetic potential of the poliovirus genome. It would therefore appear that the genome is translated into a single polypeptide. During the synthesis of this polypeptide, it is usually cleaved before the ribosome has traversed the whole RNA molecule and therefore the largest polypeptides can only be demonstrated when cleavage is inhibited. Thus far, no information is available about the enzymes which are responsible for the cleavage.

This model has an important implication for the kind of control mechanism which might be operative during the growth cycle of picornaviruses. If a single polypeptide is the primary gene product then there could probably not be any control over the ratio of the various viral proteins except by selective degradation and all proteins would have to be made at all times during the cycle. This latter prediction has been verified: From the earliest time that virus-specific proteins can be measured until the end of the growth cycle, the ratio of proteins made during a given length of pulse is approximately constant (*16, 83*). In this regard, it should be noted that in bacteria infected with an RNA bacteriophage, temporal control does exist and its mechanism has been partially elucidated (*91, 92, 93*). Synthesis of the bacteriophage replicase falls to a very low level relative to the coat protein as infection proceeds. Such a control implies that the activity of the bacteriophage replicase is stable; in picornavirus-infected cells the replicase is unstable (Sec. 3-8, A) and is constantly being replaced by new enzyme.

C. *In Vitro* Synthesis of Viral Proteins

Extracts of poliovirus-infected cells can incorporate amino acids into protein *in vitro* (*82, 94*). Much of the incorporation takes place in a large particle fraction and is greatly reduced if the infected cells are treated with guanidine (see Sec. 3-8, C) (*94*). Polyribosomes are the site of the incorporation and some of the labeled nascent chains react with an antibody prepared against capsid protein (*82*). Combination of viral RNA with HeLa cell ribosomes leads to a 2–3-fold stimulation of amino acid incorporation in the polyribosome fraction but none of this is antigenically specific (*82*).

In contrast, when poliovirus RNA is mixed with an *E. coli* extract, prepared according to the method of Matthaei and Nirenberg (*95*), there is more than a 10-fold stimulation of amino acid incorporation, and some of the *in vitro*-labeled protein is antigenically similar to poliovirus capsid protein (*96*). Later work has shown that the protein synthesized in an *E. coli* system under the direction of poliovirus RNA contains peptides which are indistinguishable from peptides which come from virus-specific proteins (*97*). The *in vitro* product contains N-formylmethionine and is heterogeneous in size (*97*).

3-6. Virion Formation

The structure of picornaviruses is discussed elsewhere in this volume. A major unsolved morphologic problem is the relation between the RNA and protein of the particle—are they interpenetrating or is the capsid a true shell enclosing the RNA? The evidence from studies on turnip yellow mosaic virus, which is similar in over-all morphology to poliovirus, is that the RNA is not simply enclosed within a protein shell but is closely interwoven with the protein (*98, 99*). If this is true for poliovirus, then the general conception of how a virion is formed—condensation of the protein around a core of RNA—may not be correct. It is conceivable that the particle is made by condensation of the RNA with a preformed protein shell. This latter view has been strengthened by recent evidence which is presented following a discussion of the kinetics of virus production and the protein composition of virions.

A. KINETICS OF THE SYNTHESIS OF VIRIONS

Virion production first becomes evident in poliovirus-infected HeLa cells at about 2.5 hr after infection and then follows an exponential course. The exponential growth rate is paradoxical in light of the finding that from 3 hr onward the production of viral RNA follows linear kinetics (36). The explanation lies in the fact that virions are formed from recently made RNA (not from a large pool; see Sec. 3-6, B) and the percentage of such RNA found in virus particles increases continuously from 3 hr on. Therefore, the exponential kinetics of virion synthesis result from a larger and larger percentage of RNA being encapsidated as infection proceeds. This presumably results from the increasing number of viral polyribosomes which produce a continually expanding pool of capsid protein which is able to associate with the new RNA. There may be a competition between capsid protein and ribosomes for each new molecule of RNA. If a ribosome wins, the molecule of RNA is destined to serve a messenger function; otherwise, the RNA is encapsidated.

B. VIRUS PRECURSOR POOLS

If poliovirus-infected cells are briefly pulsed with radioactive amino acids followed by a chase with unlabeled amino acids, a minimum of 20 min is required before any labeled protein can be detected in virus particles (20, 81). The maximal labeling of virus requires about an hour of chase following a 10-min exposure to radioactive amino acids (20). These observations suggest that a pool of capsid protein exists in the cell. The size of the pool is hard to establish but it must be large because of the 20-min lag before labeled virus appears.

Another type of experiment also shows the existence of a sizeable pool of virus protein. If cells infected for 3.75 hr are treated with cycloheximide to completely prevent protein synthesis, the cells continue to make RNA for some time (see Sec. 3-8, A). A normal percentage of the RNA made during the next 25 min is encapsidated even though no protein synthesis is occurring (36). This shows the existence of a pool of protein at least sufficient to encapsidate the RNA made during 15 min of the linear phase of RNA synthesis (not 25 min, because of the reduction in RNA synthesis caused by the drug). The state of the protein in this pool will be discussed in Sec. 3-6, C.

When poliovirus-infected cells are exposed to radioactive uridine,

there is only a short lag before RNA begins to appear in virus particles—at most 5 min and probably less (36). Because uridine cannot
be chased in HeLa cells, it is impossible to determine how long a period
is required for maximal encapsidation of the RNA made during a
given time; however, by pulsing cells with uridine and then treating
with guanidine, which very rapidly stops further uridine incorporation
into virus particles (see Sec. 3-8, C), it is possible to follow the movement into virions of the RNA made prior to guanidine addition. An
experiment of this sort showed that the half-time for movement of
newly made RNA into virions is less than 15 min and whatever pool
of RNA exists must therefore be small (101). The situation for RNA
is therefore very different from that for protein.

A pool of virus precursor RNA or protein, as defined by these experiments, consists of two types of molecules: (a) those which are
ready to move on to their final state but have not yet done so and (b)
those which must undergo some change before their final state can be
reached. In none of these experiments have these two been distinguished. In fact, with protein, the long lag before any incorporation
occurs suggests that there are a series of intermediary steps which
must be carried out before a molecule can be integrated into a virion.
The studies on proteolytic cleavage of large viral proteins, in fact, have
elucidated some of these steps (see Sec. 3-5, B; others are indicated
in Sec. 3-6, C). Therefore, the use of the term "pool" here must be
understood to include all intervening stages between the initial determination of the sequence of amino acids in the polypeptide and the
incorporation of the protein into a virion. For RNA, the rapid appearance of labeled virus suggests that if any intermediate stages occur
they have a very transient existence.

These results have a bearing on another question. Viral RNA can
either become part of a virion or a polyribosome, but is polyribosomal
RNA a precursor of virion RNA? The rapidity with which labeled
RNA appears in virions suggests that no polyribosomal stage need
occur before RNA is encapsidated. Also, the small pool of RNA, at a
time when there is known to be a large amount of RNA in polyribosomes, indicates that polyribosomal RNA is never a precursor of
virions. Therefore, the pathways to virions or polyribosomes are
probably separate—the fate of a viral RNA molecule is determined
shortly after its synthesis.

The most striking observation in support of the conclusion that
polyribosomal RNA is not the precursor of viral RNA is that the

same percentage of newly made viral RNA is encapsidated whether or not protein synthesis is abolished by cycloheximide (*36*). This drug is known to greatly reduce or abolish the movement of ribosomes along the messenger RNA in a polyribosome. Thus, if there were an obligate polyribsomal stage before RNA becomes part of a virion, inhibition of encapsidation by cycloheximide would be predicted. The absence of any inhibition argues strongly that there is no obligate polyribosomal stage for a molecule of RNA before virion formation.

Yet another approach leads to the same conclusion (*102*). In this experiment, infection was allowed to proceed in the presence of a concentration of FPA which allows viral RNA and protein synthesis to occur in an almost normal fashion, but totally inhibits virion formation (see Sec. 3-8, B). The FPA was then removed and the cells incubated further in the absence or presence of guanidine. Guanidine stops any newly made RNA from entering particles (see Sec. 3-8, C) and so provides a very sensitive test of whether pre-existing RNA can be incorporated into virions. It was found that after removal of FPA, if guanidine was absent virion formation occurred, but in the presence of guanidine no virions were produced. Thus, the pool of RNA in polyribosomes, which is built up in the presence of FPA, will not serve as precursor of virions after removal of FPA.

C. MORPHOGENESIS OF THE VIRION; ROLE OF THE TOP COMPONENT

Associated with many virus infections is the production of a particle called "top component" (*103*). This is a particle the size of a virion which contains no nucleic acid (*104*), sediments slower than virions during velocity sedimentation (*103*), and bands lighter than virions after equilibrium sedimentation (*104, 105*). It has been assumed that such particles, which accompany many picornavirus infections (*105*), arise by aggregation of viral protein in its normal configuration but without nucleic acid. Closely associated with this assumption is the idea that virions are formed by aggregation of viral protein around the RNA. It was thus surprising to find that the protein composition of poliovirus "top component" is different from that of virus (*90*). Specifically, there are two virion proteins lacking in the top component (VP-2 and VP-4) and one extra one which is not found in virions (VP-0, originally called NCVP-6). This finding suggests that top component may not be merely a by-product of infection.

Direct investigation of the mechanism of virion synthesis has provided evidence that top component plays a critical role in virion formation (*100*). These experiments depend on the use of guanidine. As detailed in Sec. 3-8, C, RNA synthesis is only slowly inhibited after the addition of the drug but none of the RNA made in its presence can be incorporated into virions. Protein synthesis continues after the addition of the inhibitor, and the newly made protein will aggregate to form top component (*100*). Since no virions are being formed, this experiment rules out the hypothesis that top component is a breakdown product of virions.

In another type of experiment, radioactive top component is allowed to accumulate in the presence of guanidine, and then the guanidine is removed. Concurrent with the resumption of viral RNA synthesis, radioactive top component disappears and an equal amount of radioactivity appears in virions (*100*). In a parallel culture in which guanidine was not reversed, top component did not disappear. This result indicates that top component is an intermediate in the morphogenesis of poliovirus. The kinetics of appearance of radioactivity in top component and virons in cells not treated with guanidine are also consistent with the idea that top component is a precursor of virions (*100*).

It is possible that the conversion of top component to virions after guanidine removal is not direct. Top component might break down to smaller units which would then be reorganized into virions. The argument against this is that top component, under *in vitro* conditions and in the cell, is very stable; however, such an argument is not proof and some doubt about the conclusion must remain.

Further evidence on the conversion of top component to virions comes from studies on the radioactive proteins from the top component synthesized in the presence of guanidine and from the virions derived from that top component. It was found that top component made in the presence of guanidine has the same three protein species as are found in normal top component and all four proteins in the virions formed after guanidine removal are labeled in the usual relative amounts (*100*). Since all of the radioactivity in the proteins was incorporated in the presence of guanidine, the VP-2 and VP-4 components of the virions which formed must have been synthesized in the presence of guanidine. But no labeled VP-2 appears in the total cytoplasm if guanidine is not removed and, therefore, VP-2 and probably also VP-4 are not primary gene products. They are only formed when virions are made, are never found outside of virions, and must be derived from some other protein which is made in the presence of guanidine. The existence of VP-0 in top component but not virions suggests that this is the protein which gives rise to VP-2 and VP-4. The hypothesis is strengthened by the observation that the tryptic peptides of the proteins from top component and virions are indistinguishable (105a) and the recent evidence that most of the peptides of VP-2 are found in VP-0 (20).

Therefore, the present model of the final stage of virion morphogenesis is a union of top component with viral RNA, with a concomitant cleavage of VP-0 to form VP-2 and VP-4. The cleavage might be carried out by one of the top component proteins or by a soluble protein. It is unlikely that the RNA is inserted into the central cavity of a protein shell because this is topologically difficult to imagine and requires that there be a hole in top component, which would ruin the symmetry of the structure. Also, there are probably 60 (and certainly a large number of) molecules of VP-0 in top component and the hypothesis requires cleavage of all of them (no VP-0 is demonstrable in purified virions). Therefore, virion production may occur by the interaction of viral RNA with proteins in top component followed by a cleavage of VP-0 which changes the structure of the particle so that the RNA becomes surrounded by protein. The evidence that VP-4 is not closely associated with the other proteins of the virion (90) suggests that, as part of VP-0, it may bind to the RNA and then be severed from the rest of VP-0. Because top component appears to be the protein precursor of the virion, it has been given the name *procapsid* (100).

In the scheme of virion morphogenesis, it seems probable that top component forms by aggregation of subunits. Such subunits, having the proteins of top component and sedimenting at 10S, have been identified in extracts of poliovirus-infected HeLa cells (105a, 105b). Incubation of these 10S particles with extracts of infected cells at 37°C converts them to material which cosediments with top component, whereas neither extracts of infected cells incubated at 0° nor extracts of uninfected cells produce this conversion.

Indirect evidence suggests that there may be carbohydrate moieties attached to the capsid proteins (106).

D. POLIOVIRUS ANTIGENS

A complicated picture of the poliovirus antigens emerged a number of years ago (103, 107–109) and has, unfortunately, not been correlated with the more recent structural work on viral proteins.

The basic immunological finding is that virions and top component have different antigens (107). The most convincing demonstration is that a monotypic antibody against one or the other component will specifically aggregate only the homologous class of particles in a mixture (110). The virion antigen is called D (or N) and the top component antigen is known as C (or H). Heating virions at 56°C releases the RNA from the particles, causing them to become noninfectious (111) and at the same time produces transition from D to C antigenicity. The finding that the particles produced by heating virions have the same antigenicity as top component was one reason for believing that the top component was simply a virion containing no RNA. This view is no longer tenable because natural top component and heated virions are composed of electrophoretically different proteins (90).

Although no adequate explanation has yet been proposed, the following considerations make the difference between the antigenicity of virions and top component less mysterious. The D antigen is associated with virions in which RNA and protein are closely associated. Thus, the D antigen itself might well consist of both RNA and protein or of protein constrained by its combination with RNA. The C antigen might then be a part of the capsid which is hidden from the antibody in the very tight structure·of the virion but which becomes accessible when the RNA is removed from the particle. The impenetrability of the virion structure is indicated by the inability of photosensitizing dyes to combine with the RNA of the virion (112, 113), and by the inability of phosphotungstate to penetrate the virion (114). The heated virion and the natural top component are penetrable by phosphotungstate (115). The possibility that C antigen may be present in the virion, but not accessible, could be tested by carrying out antibody–antigen reactions under the conditions which Wallis and Melnick showed would allow dyes to photosensitize the virus (pH 8.0, 0.1 M phosphate buffer, in the absence of organic small molecules) (116). In fact, it would be of interest to know what conditions are able to loosen the structure of the virion and what the effect of such treatments would be on the properties of the virion.

A third poliovirus antigen, the S (soluble) antigen, can be uncovered by degradation of virions with 6.35 M guanidine to produce protein with a sedimentation coefficient of 2S (117). In the infected cell, there is no S antigen sedimenting at 2S; however, there is a 5S particle which has S antigenicity. Since radioactivity will chase from

S antigen to D and C antigens, the 5S particle must be a precursor of virions. The 5S particle is probably the initial aggregate of the capsid proteins which later further aggregates to form top component. The relation of this S antigen to the 10S subunit (*105a, 105b*) (see Sec. 3-6, C) is not clear.

E. PHENOTYPIC MIXING AND GENOMIC MASKING

Phenotypic mixing is defined (*118*) as the occurrence of two *serologically* distinct types of protein in the capsid of a single virion. It occurs when two types of viruses are multiplying in the same cell and their capsid proteins form a common pool. The genome in any one particle can still only specify a single serotype even though the capsid is mixed. "Genomic masking" is an extension of this phenomenon where a given genotype is found in a capsid which has only the serotype specified by another genotype.

Both phenomena have been shown to occur in the picornaviruses. Hybrid capsids of pairs of the three poliovirus serotypes (*118, 119*) and even of poliovirus and Coxsackie B-1 (*120*) have been reported. Genomic masking occurs at high frequency when a cell infected with one type of virus is then superinfected with a different virus (*119*).

3-7. Viral RNA Synthesis

Certain aspects of the synthesis of picornavirus RNA have already been discussed. The RNA is made in the cytoplasm of the infected cell, and its synthesis is not affected by actinomycin. The accumulation of RNA occurs in two phases, an early period when it follows an exponential course and a later, linear period. The RNA which is made can serve as the genetic material of the virion or as messenger RNA in polyribosomes.

The central problem in the study of the mechanism of RNA synthesis arises from the fact that viral RNA is a unique, single-stranded molecule (Sec. 3-3, A), but the only model for the duplication of a nucleic acid, the Watson–Crick model for DNA replication, requires two types of molecules each of which is complementary to the other. Two possible resolutions of this difficulty are (1) that a different mechanism for duplication is at work, one that does not involve two complementary strands of RNA, or (2) that the DNA model or a close relative is operative. There is now strong evidence for, and general agreement on, the second alternative. The crucial fact was the demonstration that a double-stranded RNA can be found in the infected cell. Following that, evidence rapidly accumulated showing that double-stranded regions of RNA are involved in replication. This evidence derived from analysis of both bacteria infected with RNA bacteriophages and mammalian cells infected with picornaviruses. Since the work on RNA bacteriophages is described elsewhere (120a), the following discussion will be based almost entirely on results with picornaviruses.

A. Viral RNA Molecules Found in Infected Cells

Three virus-specific RNA's are found in infected cells. One is viral RNA (74), the second is double-stranded RNA (121), and the third is the replicative intermediate RNA (122). Before considering the interrelations between these various species, the properties of each will be discussed.

1. Viral RNA

Some of the properties of the RNA which can be isolated from virions were discussed in Sec. 3-3, A, and it was concluded that the RNA is an unique single strand of about 6000 nucleotides. A second unique molecule can be theoretically specified, one related to the viral RNA by the Watson–Crick pairing rules (adenine and uracil being interchanged, guanine and cytosine being interchanged and the 3' to 5' polarity being reversed) (33). We will call this RNA *complementary RNA*.

The bulk of the RNA made in actinomycin-treated, infected cells has the same properties as viral RNA (74). Its sedimentation rate is indistinguishable from viral RNA, its base ratios are identical, and it will not form double-stranded molecules with viral RNA (8). This is a crucial fact in any attempt to understand viral RNA synthesis. It strongly suggests that synthesis is asymmetric—little or no free single-stranded complementary RNA is ever produced, only viral RNA. There is, of course, another possibility—equal production of two strands followed by selective degradation. This has been ruled out by the observation that no free complementary RNA is detectable when newly made RNA or RNA still attached to its template is investigated (8).

2. Double-Stranded RNA

The initial description of double-stranded RNA in picornavirus-infected cells was made by Montagnier and Sanders (*121*) who noticed that the infected cell contains a small amount of infectious RNA which resists digestion with ribonuclease and sediments at about 20S. The possibility that this was double stranded was suggested by the observation that double-stranded RNA is ribonuclease resistant (*123*). The putative double-stranded RNA was shown by Montagnier and Sanders to have a lower buoyant density than single-stranded RNA in cesium sulfate, to be soluble in 1.5 M NaCl in which single-stranded RNA is insoluble (but double-stranded DNA is soluble), and to have a fairly sharp melting point.

Extensive investigation has confirmed and amplified the results of Montagnier and Sanders on the sharp melting point, sedimentation rate, buoyant density, and infectivity of the RNA (*124–127*). Furthermore, using poliovirus, it has been possible to apply the criterion of base ratios to the structure of the RNA because of the lack of equality of adenylate and uridylate in the viral RNA (*31, 128*). From Table 3-1, it is clear that the base ratios do correspond to a double-stranded

TABLE 3-1
Base Ratios of Poliovirus RNA's

Nucleotide	Viral RNA	Double-stranded RNA[a]	Complementary RNA[a]
AMP	28.9	26.6(26.5)	24.1(24.1)
UMP	24.1	26.6(26.5)	28.5(28.9)
GMP	22.9	23.1(23.5)	24.3(24.1)
CMP	24.1	23.8(23.5)	23.1(22.9)

[a] Values in parentheses are theoretical ratios derived from the viral RNA base ratios assuming Watson–Crick base pairing. Reprinted from Ref. *128* by courtesy of Academic Press.

molecule. Furthermore, if ^{32}P-labeled double-stranded RNA is melted and reannealed in the presence of a large excess of unlabeled viral RNA, 50% of the label becomes ribonuclease resistant and the base ratios of this material correspond very closely to that predicted for complementary RNA (Table 3-1).

The last piece of evidence for the structure of this RNA was given by Katz and Penman (*129*), who obtained whole single-stranded molecules from the double strand by denaturation, using dimethyl sulfoxide. This completes the proof that there is a molecule of RNA consisting of one viral strand and one complementary strand in cells infected with picornaviruses.

After cells have been exposed to radioactive uridine for 15 min, double-stranded RNA is detectable as a peak of ribonuclease-resistant RNA at about 18S in a sucrose gradient (*128*). Simple deproteinization of cytoplasmic extracts with SDS is sufficient to produce this free double-stranded RNA (*128*); therefore, it appears to be a material which is native to the infected cell and not a degradation product of some more complex RNA. It represents more than 10% of the total virus-specific RNA in the cell at late times of infection (6 hr), but much less at earlier times (*122*). This increased percentage probably arises from the degradation of single-stranded RNA rather than from an enormous hyperproduction of double-stranded RNA (*8*).

The ease of extraction of double-stranded RNA and the ability to anneal it with single-standed RNA suggests that studies of the base-sequence homology between the RNA's from various types of picornaviruses could be carried out. One report of this kind has appeared in which (*35*) the lack of controls makes the results uncertain. The conclusion was that very little homology exists between type 1 poliovirus and type 2, type 3, Coxsackie B-1, and Coxsackie B-5. A second, more complete, paper indicates that there is only 25–33% homology among the various poliovirus types and 74% homology between two type 1 poliovirus strains (*129a*). Further work along this line might clarify the evolutionary relationships between the different viruses.

The biological properties of double-stranded poliovirus RNA have been extensively studied. Using optimum conditions of assay, Bishop and Koch (*127*) found that the specific infectivity of double-stranded RNA is higher than single-stranded RNA by a factor of 30. While this difference may reflect the ribonuclease insensitivity of the double-stranded RNA or the ease with which this RNA can penetrate a cell, the fact remains that the double-stranded RNA can initiate infection with a surprisingly high efficiency. The infectivity of double-stranded RNA is less sensitive to ultraviolet light inactivation than is single-stranded RNA (*130*) and actinomycin treatment of cells decreases their sensitivity to infection with double-stranded RNA, while increasing their sensitivity to single-stranded RNA (*131*). These results emphasize the difference between the interaction of infection by double- and single-stranded RNA but provide little information about the nature of the difference.

In principle, it is hard to imagine how double-stranded RNA can initiate an infection. Presumably the first process which an infecting RNA molecule must carry out is synthesis of the viral replicase. Double-stranded RNA will not act as messenger *in vitro* because, for one thing, it will not bind to ribosomes. So it appears that either the duplex molecule is replicated by a host cell enzyme—an unlikely prospect—or it is separated into its component strands in order to initiate replicase formation.

3. *Replicative Intermediate RNA*

This species of RNA has never been isolated in large amounts from picornavirus-infected cells but labeled material has been studied. The RNA bacteriophage replicative intermediate has been purified and its physical properties extensively studied (*120a*).

The replicative intermediate is a class of molecules with a heterogeneous distribution in a sucrose gradient extending from 20S to 70S (*132*). Two properties of this RNA are critical to understanding its structure: It is at least partially ribonuclease resistant and yet is precipitated by salt concentrations in which double-stranded RNA is soluble (*122*). The conclusion from this is that the replicative intermediate is partially single stranded and partially double stranded. Replicative intermediate sediments faster than double-stranded RNA, but after ribonuclease treatment the resistant portion sediments at the same rate as double-stranded RNA (*122*). Thus, the replicative intermediate contains a full-length double-stranded RNA combined with single strands.

These conclusions were reached on the basis of experiments in which the replicative intermediate was not separated from the other RNA species found in the infected cell (*122*). The discovery that beaded 2% agarose gels can be used to fractionate RNA's (*133*) has allowed isolation of the replicative intermediate. Agarose gels act as molecular sieves and the pore sizes are such that the coiled single-stranded viral RNA penetrates the gel while the rigid double-stranded RNA is excluded. Thus, a mixture of the two species elutes in two peaks: a front peak containing the double-stranded RNA, and a later, broader peak containing the single-stranded RNA (*132*). The existence of a double-stranded moiety in the replicative intermediate suggests that it would also elute in the front peak and this prediction has been verified (*132*). The resolution on an agarose column is good enough so that if double-stranded RNA is first eliminated by salt precipitation, 95% pure replicative intermediate can be isolated from a preparation of labeled RNA by one passage through the column (*132*).

B. The Time Course of Viral RNA Synthesis

The combined use of actinomycin and sucrose gradients has revealed that the first 10–20% of viral RNA made in the infected cell accumulates exponentially (*36*). By 1 hr after infection, 0.1% of the final yield of RNA is found and the rate of synthesis doubles approximately once every 15 min for the following 2 hr until about 20% of the final yield has been synthesized. During the next hour of the growth cycle, there is a constant rate of synthesis and 80% of the RNA is made. The timing of the switchover from exponential to linear synthesis is dependent on the input multiplicity (*36*). When the exponential portion of the accumulation curve is extrapolated back to the number of particles which initiated the infection, it intersects the time axis at about 30 min, suggesting that RNA synthesis is already in progress during the second half-hour of infection. Since the processes of absorption, penetration, and uncoating require much of this time, it is doubtful whether any significant time passes between the liberation of the infecting RNA into the cytoplasm of the cell and the beginning of viral RNA synthesis (*36*). Thus the idea, brought over from the study of DNA bacteriophages, of an "early phase" as a period before viral nucleic acid synthesis begins is not relevant to the study of picornaviruses.

The biochemical evidence showing synthesis of RNA during eclipse is corroborated by evidence showing that infectious centers become progressively more resistant to ultraviolet light inactivation starting before 1 hr after infection (*134*). This stabilization to inactivation is prevented by puromycin (*135, 136*) and by guanidine (*136*), agents which inhibit viral RNA synthesis (see Secs. 3-8, A and C). It is therefore likely that the early stabilization to ultraviolet light inactivation is due to early RNA replication. Fenwick and Pelling (*134*) have rejected this interpretation because they found that a virus which does not produce progeny at 40°C is still stabilized to ultraviolet light inactivation at this temperature. In the light of the results with guanidine and the direct biochemical evidence, it is more likely that a virus which makes no progeny at 40°C does make RNA during the earliest stages of infection. This would be in accord with the finding that such a virus can carry out the events of the latent period in a normal fashion (*180*). In fact, unpublished results (*8*) indicate that the effect of supraoptimal temperatures may only be to stop RNA synthesis earlier than usual rather than to cause a total blockage of RNA production (see Sec. 3-8, D, for further discussion of the effect of temperature).

C. THE TIME OF SYNTHESIS OF A MOLECULE OF VIRAL RNA

In principle, it should be very easy to determine the amount of time which is required for the manufacture of a molecule of viral RNA (the "synthetic time"). At any time there exists in the cell a pool of nascent chains. It must be assumed that this pool contains an equal number of molecules of all sizes (this is, that the growth rate along the chain is constant). When cells are exposed to radioactive uridine, assuming that the uridine equilibrates with the cellular pool of uridine nucleotides instantaneously, in one synthetic time all of the nascent chains which were in the cell at the time of addition of uridine will be finished and a new crop of nascent chains will have been made. Therefore, after one synthetic time has passed, 50% of the radioactivity is in nascent chains and 50% in finished chains (see Fig. 2, of Ref. 77).

When experiments were performed to determine the synthetic time, two difficulties arose. The first is that uridine does not equilibrate instantaneously; in fact, it requires a long time (see Sec. 3-2, E). Second, it is difficult to distinguish nascent chains from completed chains because sucrose gradients do not clearly separate the replicative intermediates from finished molecules. (These experiments were performed before agarose was available.) By making a correction for the first difficulty and by assuming that the second is a minor problem, it was possible to estimate the synthetic time as about one minute (*77*). However, another method was clearly desirable.

The analysis of uridine uptake by cells which is given in Sec. 3-2, E and Fig. 3-1 showed that by a rapid pulse-chase sequence, a linear incorporation rate for uridine is achieved, starting at 30 sec after the addition of the radioactive material. This fact has allowed a more convincing analysis of the synthetic time for a molecule of viral RNA (*8*). At about 3 hr after infection, cells were exposed for 30 sec to ^{3}H-uridine followed by a 100-fold excess of unlabeled uridine. At 30-sec intervals aliquots were pipetted into a mixture of SDS and phenol. RNA was extracted at 60°C and analyzed on sucrose gradients, the patterns of which are shown in Fig. 3-3A. The RNA made during

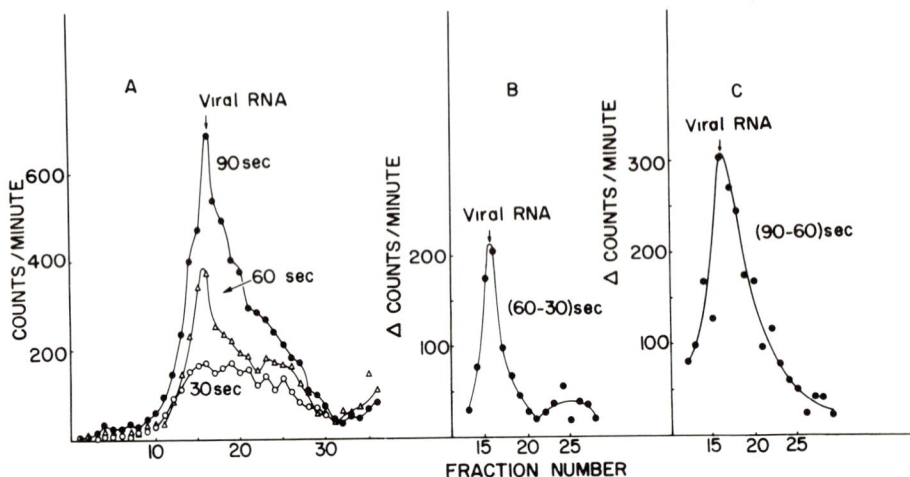

Fig. 3-3. The synthetic time for a molecule of poliovirus RNA. (A) Sucrose-gradient patterns of RNA from poliovirus-infected cells labeled for 30 sec with ^{3}H-uridine followed by 30, 60, and 90 sec in the presence of a 100-fold excess of unlabeled uridine. The samples were prepared by pouring the culture onto a mixture of phenol and sodium dodecyl sulfate followed by phenol extraction (procedures described in Ref. *122*). (B) The data for the 30-sec sample in part (A) subtracted from the data for the 60-sec sample. (C) The 60-sec sample subtracted from the 90-sec sample. (Unpublished data.)

the second and third 30-sec intervals after the chase were determined by subtraction and are shown in Figs. 3-3B and 3-3C. During the first 30 sec, a heterogeneous class of molecules is labeled, most of it sedimenting more slowly than 35S viral RNA (this is a mixture of nascent chains and replicative intermediate, see Sec. 3-7, D and Ref. *132*). During the second 30-sec period, 35S RNA predominated and during the third period only the 35S peak is evident. Thus, it required about 45 sec for the nascent chains and the replicative intermediate to be completely labeled and this is therefore the synthetic time for a molecule of viral RNA.

D. Structure and Role of the Replicative Intermediate

The name "replicative intermediate RNA" implies knowledge of the function of this species of virus-specific RNA. Although the molecule was named before its function or structure was clearly defined, the name turns out to be quite appropriate. Three lines of evidence show that it is an intermediate in viral RNA synthesis. The first involves analysis of short pulse labels, the second comes from agarose chromatography and sucrose-gradient analysis of heated samples, and the third results from analysis of RNA synthesis *in vitro*.

A priori, no matter how RNA molecules are synthesized, there must exist some number of partially completed (nascent) molecules in the infected cell and, if cells are exposed to radioactive uridine for a short enough time, the nascent molecules should represent a large proportion of the labeled RNA. Since the synthetic time for a molecule is 45 sec, and since the nonlinearity of uridine uptake during the first few minutes after its addition effectively shortens the period of exposure, when cells are exposed to radioactive uridine for up to 3 min, the label is mostly contained in nascent RNA molecules. These nascent molecules appear as RNA sedimenting more slowly than viral RNA and elute from agarose after viral RNA.

Contrary to expectation, none of the RNA prepared from cells which were exposed to uridine for 2.75 min is free nascent chains but 67% of the RNA is in replicative intermediate (*132*). The fraction of the RNA which is replicative intermediate falls rapidly as the labeling time is extended; by 15 min replicative intermediate is a very minor component and 90% of the labeled RNA is now viral RNA (*122*). The absence of free nascent chains and the very high proportion of radioactivity in replicative intermediate in pulse-labeled RNA suggest that the nascent chains are part of the replicative intermediate.

The existence of nascent chains in the replicative intermediate can be directly demonstrated by heating pulse-labeled RNA samples for 5 min at 60–70°C (20–30°C below the melting temperature of double-stranded RNA) (*132*). The effect of this subcritical heating can be demonstrated by either sucrose-gradient sedimentation or agarose chromatography: (1) the percentage of the pulse-labeled RNA which is in the replicative intermediate drops markedly; (2) the sedimentation rate of the remaining replicative intermediate is greatly reduced; (3) much of the RNA which is lost from the replicative intermediate appears as molecules sedimenting more slowly than viral RNA and eluting from agarose after viral RNA; (4) there is no cleavage of viral RNA molecules during the heating so no covalent bond breakage has occurred. The conclusion from this experiment is that nascent chains of viral RNA occur in the replicative intermediate and many of them are released by subcritical heating (*132*).

In a recent study, Girard (*132a*) has shown that the replicative intermediate is the precursor of finished viral RNA chains in an *in vitro* poliovirus replicase reaction. After a very short incubation period with radioactive ribonucleoside triphosphates, the majority of the incorporated label is found in the replicative intermediate. Upon addition of excess unlabeled triphosphate, the radioactivity is lost from the replicative intermediate and accumulates in viral RNA molecules. This experiment also showed that the replicative intermediate is a precursor of completely double-stranded RNA molecules.

The existence of a double-stranded RNA segment in the replicative intermediate was shown in Sec. 3-7, A, 3. If this region is isolated from pulse-labeled RNA samples by treatment with ribonuclease, almost all of the radioactivity is in viral RNA, whereas from long-labeled samples it is 50% in viral and 50% in complementary RNA (*132*). At all times, the only detectable complementary RNA in the replicative intermediate is in the double-stranded region. The existence of complementary RNA and its relatively slow rate of labeling suggests that it is a template for viral RNA synthesis. The removal of the nascent chains by heat in the absence of covalent bond cleavage further suggests that they are attached to the complementary RNA by hydrogen bonds. This is consistent with its role as a template. That subcritical heating is sufficient to remove the nascent chains has been tentatively attributed to a competitive annealing reaction (*132*).

Therefore, the replicative intermediate contains nascent chains and the template for their synthesis—what then is its structure? Two likely models for its structure are shown in Fig. 3-4. For either model, the number of nascent chains in the average replicative intermediate can be calculated from the per cent of ribonuclease resistance in totally labeled replicative intermediate. From the fact that there is one double-stranded RNA in the replicative intermediate (containing 2 molecule-equivalents of viral RNA), and N growing chains which must be assumed to have an average length of one-half molecule-equivalent, the per cent of ribonuclease-resistant RNA can be expressed as

$$\% \text{ ribonuclease resistance} = 100 \left(\frac{2}{2 + N/2} \right) \qquad (1)$$

Since the totally labeled replicative intermediate is 38% ribonuclease resistant (*132*), from Eq. (1) the average number of growing chains per replicative intermediate (N) is 6.5. The replicative intermediate which remains after subcritical heating is 79% (*132*) ribonuclease

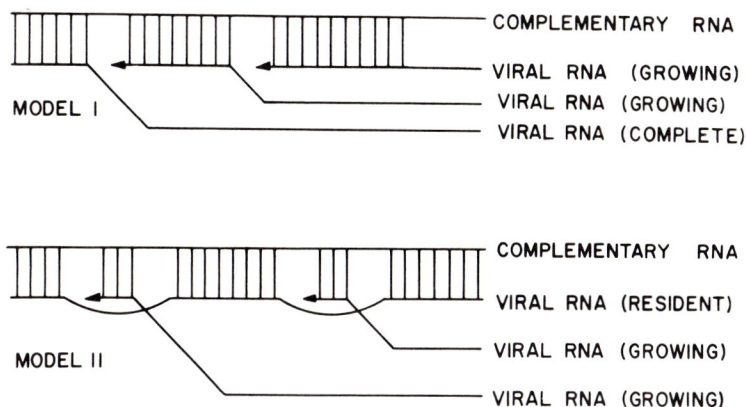

FIG. 3-4. Models of the replicative intermediate RNA. Two possible structures for the replicative intermediate. Both structures can be generated from a double-stranded RNA. Structure I (the semiconservative model) arises by assuming that each new nascent chain displaces the resident viral strand in the duplex. In Structure II (the conservative model) the duplex remains largely intact except for the region in which nascent viral chains are attached by hydrogen bonds at their growing ends. The arrows show the direction of growth of the nascent chains.

resistant, corresponding to 1.1 nascent chains (see Ref. *132* for a more complete discussion).

The facts about the replicative intermediate established above are compatible with model I in Fig. 3-4, but not with model II. The arguments which show this are complicated and will be presented in several steps.

(1) The outline of the argument is as follows: The percentage of pulse-labeled RNA that is ribonuclease resistant in the replicative intermediate is 37%, and most of this is in viral (rather than complementary) RNA. If it is assumed that all of this radioactivity is in the growing end of nascent chains then *a long region of the growing chain must be hydrogen bonded to the complementary RNA*. This would be consistent with model I where the viral strand of the duplex is constantly being displaced. It would also be formally consistent with model II but requires that the growing chains have a long stretch associated with complementary RNA.

(2) To calculate the length of the hydrogen-bonded region of the nascent chains, it is first necessary to look more closely at the value of 37% for the ribonuclease resistance of the pulse-labeled replicative intermediate. This number is somewhat misleading because the rate

of uridine incorporation is constantly increasing during the pulse period (see Sec. 3-2, E) and the growing end of a nascent molecule will therefore be synthesized from uridine of a higher specific activity than that which was used for beginning part of the chain. The doubling time for the rate of uridine incorporation into RNA is about one minute (*8, 77*), that is, approximately the same time as is required for the synthesis of a molecule of RNA. This means that, at most, the growing end of the nascent molecule has twice the specific activity of the initiating end and therefore has no more than 1.5 times the average specific activity expected of the whole nascent chain. The maximum effect that the slow uptake of uridine could have on the percentage of ribonuclease resistance would then be to change it from 37% to 25% and this lower figure will be used for further calculations.

The actual percentage of newly made RNA which is resistant to ribonuclease digestion can be calculated from other considerations if model I is assumed to be correct. During a pulse label of a few minutes, little radioactivity appears in complementary RNA so there is only one molecule-equivalent of labeled RNA in a double-stranded configuration and the per cent of ribonuclease resistance is expressed by

$$\% \text{ ribonuclease resistance} = 100 \left(\frac{1}{1 + N/2} \right) \tag{2}$$

When this equation is solved for N, using 37% ribonuclease resistance, a value of 3.4 chains is found. But from the data on totally labeled replicative intermediate, N is known to be 6.5 so the value of 37% ribonuclease resistance must be too high. Substituting $N = 6.5$ gives 24% ribonuclease resistance which agrees with the 25% estimated from the increasing rate of uridine incorporation.

(3) If 25% of the newly made RNA in the replicative intermediate is double stranded, model II becomes very unlikely because each nascent chain must have a long stretch of bases hydrogen-bonded to the template. The length of this stretch can be calculated from the fact that the replicative intermediate contains an average of 6.5 growing chains, each of which is one-half molecule-equivalent in length. The viral RNA in the double-stranded portion of the replicative intermediate must then consist of $(0.5 \times 0.25 \times 6.5 =)$ 81% growing chains and only 19% resident strand, and the average growing strand would have 25% of its length hydrogen bonded. If the actual percentage of ribonuclease resistance is nearer to 37% than 25%, model

II becomes totally impossible because there would have to be *more than* an equivalent of one strand of viral RNA in the double-stranded portion of the replicative intermediate [$(0.37 \times 0.5 \times 6.5) = 1.2$]. In any case, this analysis shows that model II is only possible if the growing chains are attached by at least 25% of their length to the complementary RNA.

(4) The above argument is based on the assumption that all of the radioactivity incorporated during a pulse label is in the nascent chains. Experiments have shown that there is very little label in complementary RNA. Therefore, the assumption is certainly correct for model I because by this model all of the viral strands are nascent except for the single complete chain which is constantly replaced. In model II, however, there may be some radioactivity in the "resident" strand. This would negate the calculation in the prior paragraph because it was assumed that all of the ribonuclease resistance in a pulse-labeled replicative intermediate is in nascent chains. Fortunately, another fact allows this assumption to be dropped: About 40% of the over-all ribonuclease-resistance in a pulse-labeled sample is lost upon heating.

To show quantitatively how this fact relates to model II, a new set of parameters is necessary. Thus, let

$1 =$ length of a molecule of viral RNA
$N =$ number of growing chains per replicative intermediate
$X =$ fraction of a molecule of viral RNA by which the average nascent chain is attached to the complementary RNA
$S_c =$ average specific radioactivity of the nascent strands
$S_r =$ average specific radioactivity of the resident strand
$A = (1 - XN)S_r =$ radioactivity in the hydrogen-bonded region of the resident strand
$B = XNS_r =$ radioactivity in the nonhydrogen-bonded region of the resident strand
$C = XNS_c =$ radioactivity in the hydrogen-bonded region of the nascent strands.

Making the assumption that the nonbonded regions of the resident strand are ribonuclease sensitive, then for the pulse-labeled material

$$\% \text{ ribonuclease resistance} = 100 \left(\frac{A + C}{\frac{1}{2}NS_c + S_r} \right) \qquad (3)$$

where the numerator is the radioactivity in the ribonuclease-resistant RNA and the denominator is the total radioactivity in the replicative intermediate.

A second equation can be derived which reflects the loss of ribonuclease-resistant radioactivity on heating. It assumes that the subcritical heating causes the loss of all nascent chains and that their double-stranded regions are replaced by resident strands. Thus

$$\left(\frac{A + B}{A + C}\right) = 0.6 \tag{4}$$

since only 60% of the ribonuclease-resistant RNA is retained after heating. Equations (3) and (4) can be solved for X and for (S_c/S_r) and the result is that $X = 1/N$. This result implies that for model II to be correct, the complete complementary strand would have to be covered by growing chains. But this is exactly model I. Because of the possible errors in the data used in these calculations, the only conclusion which can be made is that if model II is correct, a large percentage of the growing chains must be hydrogen bonded to the template.

(5) In summary, the data available at present are consistent with model I and possibly consistent with a version of model II where the region of nascent chain which is bound to the complementary RNA is very long. In fact, if model II is correct, at the most 20% of the resident strand could be hydrogen bonded to the complementary strand. Hayashi (137) has shown that a structure analogous to model II is involved in DNA-dependent RNA synthesis and the region of hydrogen bonding is less than 80 nucleotides or about 1% of the length of a poliovirus RNA molecule.

Up to this point, the analysis of the structure of the replicative intermediate has been purely from the point of view of structure. But the replicative intermediate is actually a dynamic structure *in vivo*, and when the possible models are considered from this point of view, model II becomes very difficult to accept. There is no difficulty imagining how a model I structure would function; the nascent chain has a replicase molecule at the growing point which is able to separate the double strands ahead of itself in order to allow replication. A model II structure presents the difficulty that the growing point of a nascent chain must be hydrogen bonded to the complementary RNA, but some time after a given region of RNA has been synthesized, that region must be displaced from the complementary RNA by the resident strand. If the region of hydrogen bonding is short, this is easy to understand because the replicase molecule can simply assure that the resident strand re-attaches to the complementary RNA. However, if the region of hydrogen bonding of the nascent strand is long, as it must be from the above analyses, then it becomes very difficult to imagine why the resident strand should ever displace the nascent chain, and the model does not make dynamic sense. Therefore, model II is provisionally rejected as a structure for the replicative intermediate.

Thus, it is concluded that a structure like model I in Fig. 3-4, containing 6.5 growing chains, one complete viral RNA strand and one complete complementary RNA strand, is very probably an intermediate in viral RNA synthesis. The double-stranded portion of the molecule arises from Watson–Crick base pair complementarity and is presumably a reflection of the mechanism by which the sequence of nucleotides in the growing molecule is specified. Viral RNA synthesis is then very much like DNA synthesis or DNA-dependent RNA synthesis in that base pairing is the method by which sequence is specified, but the exact architecture of the complex between nascent chains and template appears to be quite unique to RNA viruses. The question of how the replicase is able to choose one strand of the duplex rather than the other—to carry out an asymmetric synthesis —is unanswered.

E. The Viral Replicase (Viral RNA Polymerase)

The enzyme responsible for picornavirus RNA synthesis, when it was first reported, was called the viral RNA polymerase (138). The discoverers of the RNA bacteriophage polymerase called it either the "synthetase" (139) or the "replicase" (140). This latter name seems to be the most widely accepted and we will adopt it here.

Mengovirus or poliovirus replicase is easily demonstrable in cytoplasmic extracts of infected cells (138, 141, 142), whereas no similar activity is found in uninfected cells. The replicase activity is not stimulated by adding viral RNA to the crude extract and so incorporation is due to replicase-template complexes. The synthesis of the enzyme in the infected cell occurs with approximately the same kinetics as viral RNA synthesis in vivo. No purification of any picornavirus replicase has been reported, although the activity is demonstrable in many systems (143–146). The activity of the enzyme is not affected by actinomycin which shows that DNA is not involved in its activity and easily allows its differentiation from the nuclear DNA-dependent RNA polymerase.

The ribonucleotides incorporated by the enzyme *in vitro* are found in high molecular weight RNA (*142*), some of which is ribonuclease-sensitive viral RNA, some ribonuclease-resistant double-stranded RNA, and some replicative intermediate (*8*). Thus, at least qualitatively, the *in vitro* action of the polymerase is identical to the *in vivo* manufacture of viral RNA (*147*, *148*).

Two major questions are as yet unsolved. One is whether the replicase is synthesized under the direction of the viral genome or is a modified version of a host-cell enzyme. Only the clear demonstration of conditional lethal replicase mutants will provide convincing evidence of a viral origin, but the finding that only actively multiplying foot-and-mouth disease virus elicits the formation of antibodies against its replicase (*149*) suggests that at least part of that enzyme is synthesized under the direction of the viral genome. The second question is how many replicases there are. No evidence for more than one enzyme has been adduced, but the mere fact that there are two types of reactions in the infected cell (synthesis of double strands and their use in making single strands) raises the possibility that two proteins are involved. Genetic experiments with the RNA bacteriophage have clearly shown that the two activities are independently mutable (*150*) but this does not prove the existence of two different proteins. For simplicity in further discussions, it will be assumed that only one replicase exists.

F. The Replication Complex

The replicase and the replicative intermediate must be united during the process of RNA synthesis. The unit formed by these two entities, called the *replication complex,* can be found by sucrose gradient analysis of cytoplasmic extracts from poliovirus-infected cells pulse labeled with uridine (*151*). The replication complex sediments heterogeneously with a modal rate of about 250S. The high sedimentation rate suggests the association of ribosomes with the complex but two facts mitigate against this: neither EDTA treatment of extracts nor puromycin treatment of cells—treatments which release ribosomes from messenger RNA—lower its sedimentation rate. Moreover, using cells with labeled ribosomes, no evidence could be found for ribosome attachment to the replication complex after EDTA treatment. The replicative intermediate can be isolated from the complex, and the complex contains all of the active replicase in the cell (*151*). Thus, the replication complex is the site of viral RNA synthesis.

The fast sedimentation rate of the complex may be explicable by its content of replicative intermediate and of replicase. If the replicase is as large as the bacterial DNA-dependent RNA polymerase (about 9×10^5 daltons) (152) and there are 6.5 replicase molecules in the average replication complex plus a replicative intermediate of $[(2 + 6.5/2)(2 \times 10^6)] = 10.5 \times 10^6$ daltons, then the average replication complex has a molecular weight of about 17×10^6 daltons. In comparison, a reticulocyte polyribosome containing 5 ribosomes sediments at about the rate of the replication complex and has a molecular weight of about 21×10^6 daltons. So, it is possible that the replication complex contains only replicase and replicative intermediate along with other proteins which would probably bind to any part of the replicative intermediate which was free of replicase (153).

G. ROLE OF DOUBLE-STRANDED RNA

Most of the true double-stranded RNA in the cell seems to be a by-product of infection. This is suggested by the fact that it is not found in the replication complex (8) and does not appear in pulse-labeled samples but predominates after long periods of labeling (122). Furthermore, double-stranded RNA is a minority of the ribonuclease-resistant RNA during the exponential phase of RNA synthesis; only towards the end of the growth cycle does it become the predominant species (122). The most likely source for the double-stranded RNA is breakdown of the replicative intermediate. From the kinetics of synthesis of double-stranded RNA, the replicative intermediate population must then be rapidly turning over all of the time. An experiment to test this idea was performed by determining whether the complementary RNA in the replicative intermediate is continually replaced. Annealing studies on the ribonuclease-resistant portion of the replicative intermediate showed that this is the case; the complementary strand is continually replaced with a half-life of about 10 min (132). This confirms the idea that the formation of double-stranded RNA results from the loss of replicative intermediates.

H. QUANTITATIVE ASPECTS OF VIRAL RNA SYNTHESIS

1. Final Yield of Viral RNA

The total amount of viral RNA which is made in one infected cell is independent of the multiplicity of infection (36); the actual number of molecules which is synthesized can be determined from the total yield of virions and the percentage of the RNA which is encapsidated. The yield of virions varies according to the cell type and the physiological state of the cells; the value of 1.5×10^5 poliovirions/cell which has been reported for suspended HeLa cells is used here (*154, 155*). The per cent of the viral RNA synthesized in a cell which is ultimately encapsidated is difficult to define. The values of 65% (*74*) and 25% (*45*) encapsidation which have been reported were based on very different criteria. The higher estimate was based on the total amount of 35S RNA remaining in the cell at 7 hr after infection; it is clearly too high because at least 25% of the RNA present at 4 hr is degraded by 6 hr (*74*) and it should be corrected to about 50%. The value of 25% encapsidation was based on total incorporation, not 35S RNA and therefore is too low. For the present calculations we will assume a value of 50% and conclude that 2×10^5 molecules of viral RNA are made per cell.

2. *The Exponential and Linear Phases of RNA Synthesis*

When the time course of RNA synthesis is investigated at a number of multiplicities of infection, two phases are always evident, an early exponential phase and a later linear phase (see Sec. 3-7, B). As the multiplicity is decreased from 30 to 1 particle per cell, the curves of RNA accumulation during the linear phase are parallel but are displaced to later times (Fig. 3-5). The final linear rate is independent of multiplicity, but the lower the multiplicity, the longer the time before it is reached. The duration of the linear phase is always approximately 1 hr, after which the rate of RNA synthesis declines. The final yield of RNA is independent of the initiating multiplicity. These results (*36*) imply that the following series of events ensues after addition of virus to cells:

Each virus particle initiates infection independently of other particles entering the same cell and each RNA strand begins to multiply exponentially, producing new templates and thus continually increasing the rate of synthesis. When 4×10^4 molecules of RNA have been produced in the whole cell, the rate of RNA synthesis becomes constant and there follows a period of approximately 1 hr of linear accumulation of viral RNA succeeded by a period of quite rapid diminution of the rate. This 1-hour period occurs later as the multiplicity of the initial infection decreases, because it requires more time

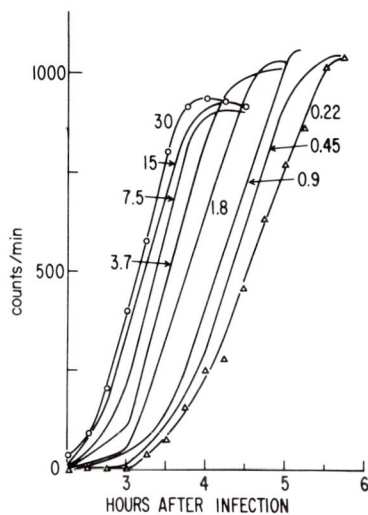

FIG. 3-5. The time course of accumulation of poliovirus RNA in cells infected at different multiplicities. Reprinted from Ref. *36*, p. 183, by courtesy of Academic Press.

to reach the final synthetic rate when fewer articles initiate the infection.

Two points in this scheme require explanation: the switch from exponential to linear synthesis and the cessation of synthesis after the hour of linear synthesis. Neither phenomenon has been adequately explained. The final cessation of RNA synthesis is accompanied by a declining rate of protein synthesis (*83*) which suggests that the cell has been so severely damaged that either energy production ceases and/or that there is leakage of small molecules from the cell. The switch from exponential to linear synthesis is more interesting but only a formal analysis of a possible mechanism for its occurrence can be given at present.

The analysis starts from the fact that the replicative intermediates are constantly turning over with a half-life of about 10 min (*132*). Let us assume that the rate by which they are lost is zero order, that is, does not depend on the concentration of replicative intermediates. The synthesis of new replicative intermediates is, unfortunately, not at all understood but if a simple process of synthesis of complementary RNA on viral RNA molecules is assumed, then the formation of replicative intermediates will depend on the concentration of these viral RNA molecules. If this concentration were to be limited to a constant value at any time, the number of new replicative intermediates being formed would become constant and the zero-order loss of replicative intermediates would cause their concentration to remain constant. This would set the rate of viral RNA synthesis at a constant value, which is the observed fact.

This argument focuses attention on a poorly understood aspect of viral replication: how complementary RNA is made. The crucial aspect is the template for synthesis of complementary strands. It must be a molecule of viral RNA, but what molecule is it? It could be a molecule which is part of a polyribosome and it could be a molecule which is free of ribosomes. Whatever the pool of templates for complementary RNA, the argument says that the mechanism for the switch to linear synthesis lies in the fact that the pool is no longer expanding. It may be a coincidence, but the switch occurs at just about the time when the cell is beginning to manufacture virions from the newly made RNA (*36*) and it could be that this limits the size of the pool.

It is interesting to note that the switch point for the rate of RNA synthesis represents a critical time in the evolution of the infectious cycle. Up to this time the centers of RNA synthesis have apparently been acting independently; from this point on, the centers interact and the cell is homogeneous.

3. *The Multiplication of RNA Templates*

The basis for the exponentially increasing rate of RNA synthesis during the first hours of a picornavirus infection can be easily seen in terms of an increasing number of templates during this time. Since the template has already been identified as part of the replicative intermediate, this explanation predicts that the concentration of replicative intermediate should increase during the first hours of infection. This increase should follow an exponential time course for the following reasons: (1) accumulation of viral RNA follows an exponential course; (2) the *rate* of RNA synthesis is obtained by differentiating this exponential accumulation function and is therefore itself exponentially increasing; (3) since the rate of RNA synthesis depends on the *concentration of templates*, that too must increase exponentially. This prediction of an exponential increase of replicative intermediates during the first hours of infection has been verified by direct experimentation (*122*).

Before the role of the replicative intermediate as template had been clarified, an important formal question was debated: Is there multiplication of templates or are all the templates either contained in the input virions or specified by the initially infecting RNA molecules? An argument was presented which shows the implications of assuming no multiplication of templates (36). The observation that, even at a multiplicity of infection one, 80% of the final yield of RNA is made in 1 hr means that the synthetic rate in the cell is about 3000 molecules/min. Since it requires about a minute to make a molecule of RNA, if there were only one template in the cell it would have 3000 growing chains on it and therefore 3000 replicase molecules. This seems impossible, and, of course, more recent evidence that there are 6.5 nascent chains per template directly shows that there is multiplication of templates during the growth cycle.

3-8. The Action of Inhibitors on Viral Growth

A. PUROMYCIN AND CYCLOHEXIMIDE

These two drugs are powerful inhibitors of protein synthesis which work by very different mechanisms. Puromycin is an analog of aminoacyl transfer RNA which acts by being coupled to the growing peptide chain, thereby releasing it from the ribosome (156) and causing breakdown of polyribosomes (157). Cycloheximide, also known as actidione, acts by stopping the elongation of nascent protein molecules (possibly by directly inhibiting peptide bond formation) and freezes the ribosome in place on the messenger RNA (158). Cycloheximide is not known to have any other effects; puromycin may not be completely specific (159). Because the drugs are chemically and functionally so different it is likely that any effect caused by both of them is due to their inhibition of protein synthesis and not to any side effect. The effects discussed below are caused by either drug and, in fact, are also caused by high concentrations of FPA.

Inhibition of protein synthesis at any time after infection has a drastic effect on viral RNA synthesis (160, 161). If RNA synthesis has not yet begun, it does not begin; if it has already begun, it rapidly declines to a very low level after a lag of approximately 5 min. This is true whether infectious RNA production or virus-specific uridine incorporation is measured. Even the earliest detectable viral RNA synthesis (before 2 hr after infection) is prevented by the drugs (*36*).

An explanation of this inhibition of viral RNA synthesis was suggested by the observation that replicase activity, measured *in vitro*, declined markedly after cells were treated with puromycin (*162*). It was postulated that the replicase is unstable in the infected cell and must be continually replaced by *de novo* protein synthesis; so, inhibition of protein synthesis would lead to inhibition of RNA synthesis (*163*). Support for this hypothesis comes from recent experiments showing that after addition of cycloheximide, there is a decay of the replication complex (*164*), without any effects on encapsidation (*36*) or on the relative amounts of the various virus-specific RNA's (*8*). This means that RNA is leaving the complex normally but production of new RNA is being inhibited, which supports the hypothesis that the replicase is unstable.

The original explanation for the inhibition of RNA synthesis by puromycin was that viral RNA and protein synthesis (*165*) are directly coupled but this was ruled out by the finding that viral RNA synthesis *in vitro* is insensitive to puromycin (*166*). The further possibility that a lack of capsid protein causes difficulty in the release of RNA chains from the template is unlikely in view of the large pool of capsid protein (*36*), the absence of inhibition of RNA synthesis by levels of FPA which prevents the formation of virions (see Sec. 3-8, B), the fact that puromycin inhibits RNA production at a time when virion synthesis is not yet detectable (*36*), and the continued movement of RNA from the replication complex in the presence of actidione (*164*).

B. Fluorophenylalanine (FPA)

This drug has only a slight effect on the over-all rate of protein synthesis but it causes the formation of fraudulent proteins because it is incorporated in the place of phenylalanine (*161*). The absolute level of FPA is not what determines its effects but rather the ratio of phenylalanine to FPA. For this reason, the terms "high" and "low" concentration will be used here because the absolute levels of FPA which will produce a given effect vary from one laboratory to another, depending on the specific medium used. Phenylalanine is an ubiquitous constituent of proteins, and since there are several proteins involved in the viral replication cycle, it might be expected that FPA would have multiple effects which should become evident at different concentrations of the drug, depending on the relative amounts of phenylalanine in different proteins and its importance to their function. This prediction is borne out by the following observations.

At high concentrations, FPA acts like puromycin or cycloheximide in that it prevents viral RNA synthesis and stops on-going RNA synthesis (*161, 165*). At lower concentrations, however, its selectivity becomes evident. RNA synthesis as measured by radioactivity or infectious RNA production is much more resistant to the drug than infectious virus formation, and conditions can be achieved where infectivity is reduced to 1% of normal while RNA synthesis is 70% of normal (*160, 161, 165*).

In FPA-treated cells which are not making infectious particles but are synthesizing infectious RNA, much of the RNA does not sediment with virus particles and both the sedimentable and nonsedimentable RNA is largely ribonuclease sensitive (*160, 167*). D and C antigens are synthesized (*165*) and both the antigens and infectious RNA are synthesized slightly later in treated than in untreated cells (*165*). Determination of the amounts of D and C antigens using isotopically labeled cell extracts showed that D antigen was inhibited to a somewhat greater extent than C (*104*), and that S antigen was affected much more slowly (*168*). Actually, the synthesis of S antigen and of infectious RNA was inhibited coordinately.

As in the untreated cell, D antigen in the FPA-treated cell is particulate and is associated with viral RNA. The D particles, however, have a lower infectivity per physical particle than those found in untreated cells (*104*). These findings are supported by experiments which showed that the inhibition of formation of RNA-containing particles resistant to SDS was slower than the inhibition of infectivity (*101*). Aberrant particles with disordered capsids have been reported in FPA-treated infected cells, but the lack of quantitative data makes it difficult to assess their frequency and RNA content (*167*).

In summary, as the concentration of FPA in an infected culture increases, each of a number of parameters has its own level of sensitivity. The most sensitive parameter is the infectivity of particles. In low concentrations of FPA there are particles made which appear normal by criteria such as sedimentation rate, SDS-resistance, and antigenicity but which are not infectious. Unfortunately, it has never been shown that the RNA in these particles is normal, and the report of formation of sedimentable particles containing ribonuclease-sensitive infectious RNA (*167*) illustrates the importance of this question.

The viral function which is least sensitive to FPA is the activity of the replicase, although at high concentrations of FPA, it too can be completely abolished.

The addition of phenylalanine to FPA-treated cells prevents additional damaged proteins from being made, but neither the RNA nor the protein made in presence of the drug is rescued by the reversal. Only protein and RNA made after drug reversal contribute to the formation of virions (160), a result which is consistent with the idea that only newly made RNA can be incorporated into virions (see Sec. 3-6, B).

C. HBB and Guanidine

The growth of a number, but not all, of the picornaviruses is inhibited by either 2-(α-hydroxybenzyl)-benzimidazole (HBB) or guanidine hydrochloride (used at about 2 mM). Earlier data on these inhibitors and on viral mutants whose growth is resistant to or dependent on these drugs have been reviewed (169) and are discussed in this book by Baron and Levy; they will not be discussed here. The only point which will be considered is the mechanism by which guanidine inhibits poliovirus replication. The mechanism of action of guanidine is much better understood than that of HBB, and it is quite possible that they act by different mechanisms.

The basic facts to any understanding of the action of guanidine are: (1) The drug has no known effect on host-cell macromolecular syntheses. Cells grow normally in its presence (169).

(2) The drug is able to inhibit on-going viral RNA synthesis but does not inhibit the activity of the replicase *in vitro* (141). Even the earliest detectable synthesis of viral RNA (before 2 hr of infection) (36) is prevented by guanidine. If the compound is added during the linear phase of viral RNA synthesis, RNA production decreases to 10% of normal in about 20 min and remains at approximately this level for at least an hour (164).

(3) Inhibition of RNA synthesis is evident soon after addition of the drug, whereas protein synthesis is affected only slowly. One-half hour after the addition of guanidine, the rate of protein synthesis is at least 50% of the rate which pertained at the time the drug was added (8, 100). Also, the spectrum of proteins which are made in its presence is normal (100).

(4) When guanidine is added during the linear phase of RNA

synthesis, incorporation of newly made RNA into virions halts much more abruptly than does RNA synthesis (164).

These results suggest that the primary effect of the drug may not be to inhibit RNA synthesis but rather to inhibit the processing of RNA after it is synthesized (164). This hypothesis is strengthened by the observation that the RNA which is made in the presence of guanidine is found in a structure which sediments more rapidly than either viral polyribosomes or replication complex, called the *guanidon* (164). The guanidon is the site of RNA synthesis in the presence of guanidine but the RNA is apparently unable to leave the structure and all the viral RNA remains associated with it. The size of the guanidon is unchanged by puromycin, indicating that functioning ribosomes are not responsible for its large size. Also, there is little or no protein synthesis in the guanidon as measured by an amino acid pulse label. The density of the guanidon in CsCl also suggests that ribosomes are not responsible for its large size (170). The sedimentation rate of the guanidon is greatly reduced by addition of EDTA and most of its attached viral RNA chains are released.

Even though some of the guanidon sediments in the region of the polyribosomes, it is clear from buoyant density determinations that no RNA moves from the guanidon to polyribosomes (170). It therefore appears that the primary action of guanidine is to cause newly made RNA to accumulate in the replication complex, thereby creating the entity called the guanidon.

At one time, it was suggested that guanidine acts by inhibiting synthesis of functional replicase (141). First of all, this hypothesis is not in accord with several of the facts mentioned above. Secondly, a comparison of inhibition by guanidine and by cycloheximide shows that the inhibitor of protein synthesis acts by a different mechanism than does guanidine (164). Cycloheximide shows a lag (presumably reflecting a pool of replicase molecules) whereas guanidine shows none. Furthermore, cycloheximide is able to inhibit RNA synthesis more completely than is guanidine.

If the proposed hypothesis of guanidine action is correct, then the inhibition of viral RNA synthesis must be a secondary consequence of the aberrant processing of RNA, and the kinds of RNA which continue to be made should be of interest. Figure 3-6 shows the sucrose-gradient analysis of the RNA made in the presence of guanidine. The drug was present for 20 min before the addition of radioactive uridine, so these patterns represent RNA made at 10% of the normal rate. If uridine is added earlier, sucrose-gradient pat-

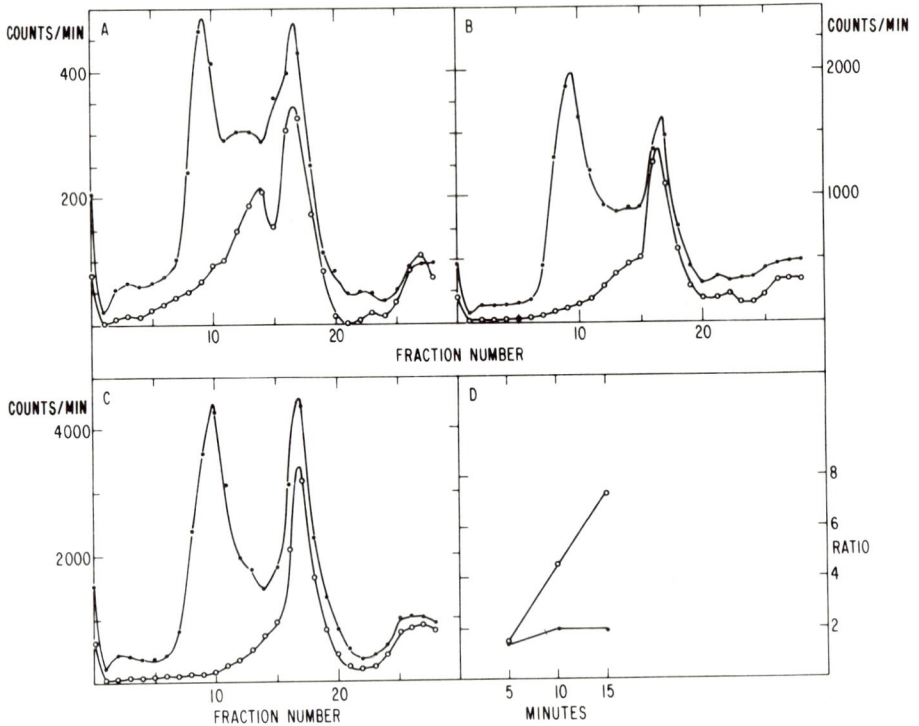

Fig. 3-6. Sucrose-gradient profiles of the RNA made in poliovirus-infected HeLa cells treated with guanidine. At 3 hr after infection, a 30-ml culture of infected cells at 4×10^6 cells/ml was treated with 2 mM guanidine. After 20 min, 1 mCi of ³H-uridine (20 Ci/mmole) was added, and at various times samples were removed and the large particle fraction (P-20) from the cytoplasm was prepared and treated with SDS to release the RNA (procedure in Ref. *132*). (A), (B), and (C) represent 5, 10, and 15 min of labeling with ³H-uridine. The lower curves in each of these panels represent the ribonuclease-resistant portion of each fraction of the sucrose gradients (*132*). The left peak in the upper curves of these patterns represents molecules sedimenting at the rate of viral RNA. The right peak is double-stranded RNA. (D) The total ribonuclease-resistant radioactivity in replicative intermediate (●-●-●) and double-stranded RNA (○-○-○) in each of the samples. This shows that the replicative intermediate behaves kinetically like an intermediate in RNA synthesis while the double-stranded RNA behaves like a product. (Unpublished data.)

terns intermediate between these and normal ones are seen. It is evident that whether the pulse length is 5, 10, or 15 min there is single-stranded viral RNA (35S), double-stranded ribonuclease-resistant RNA (20S), and heterogeneous ribonuclease-resistant RNA (the replicative intermediate) synthesized. The striking feature of these curves is the abnormally high percentage of the RNA, even in a 5-min pulse, which is double stranded. (This RNA is soluble in 2 M LiCl, so it is totally double stranded.) When the duration of the pulse is increased, the amounts of both the single- and double-stranded RNA's increase and the amount of replicative intermediate RNA shows a relative decrease. This is in accord with the earlier discussions which concluded that the replicative intermediate is a precursor of both single- and double-stranded RNA. Thus, the normal species of RNA are made in the presence of guanidine, but the amounts of RNA being made are greatly reduced. This reduction is much more pronounced in viral RNA than in double-stranded RNA; in fact, experiments have shown that after 20 min in guanidine, the rate of double-stranded RNA synthesis is at least 50% of the normal rate (8). So the effect of guanidine is to greatly reduce the amount of single-stranded RNA being made, an effect which seems to be secondary and results from the inability of viral RNA to leave its site of synthesis. What role this type of feedback regulation may play in the untreated, infected cell is not known.

A few further observations are in accord with these ideas on the mechanism of guanidine inhibition. The first is evident from Fig. 3-6, which shows that the bulk of the replicative intermediate is quite small in the presence of guanidine (less than 35S) while in untreated cells the replicative intermediate is much larger (132). The small size of the replicative intermediate is in accord with the idea that the synthesis of viral RNA molecules is greatly reduced so that there are fewer nascent chains. Also, annealing studies have shown that the RNA which is made in the presence of guanidine is not complementary RNA (8). The further observation that cycloheximide can inhibit the synthesis which continues in the presence of guanidine suggests that replicase turnover continues in the presence of guanidine (8).

As noted above, double-stranded RNA synthesis proceeds in guanidine-treated cells at nearly its normal rate. This means that the turnover of replicative intermediate must be continuing. In fact, one way to interpret these results is that guanidine inhibition results in decreased synthetic activity by each replicative intermediate. Rather than making about 9 chains/min (6.5 chains at 45 sec per chain), the replicative intermediate is making fewer than one. As discussed in Sec. 3-7, D, a replicative intermediate having approximately one nascent chain does sediment at less than 35S. The pattern of incorporation after a 5-min label (Fig. 3-6) suggests that the rate of synthesis of a given molecule is almost normal. In contrast, if the rate were one-tenth of normal there should be very little 35S RNA made after 5 min and the replicative intermediate would be the same size as normal.

The picture which emerges from these studies is that the target site for guanidine inhibition is involved in a function about which there is very little knowledge. This function is the release of newly made RNA molecules from the replication complex. There seems to be a common process for the release of molecules which are destined to become either part of virions or part of polyribosomes. The function is specified by the viral genome since guanidine-resistant virus mutants can be isolated. It may be carried out by the replicase itself, or alternatively, by some other viral protein. This view of guanidine action is not accepted by Caliguiri and Tamm (170a, 170b). They agree that the number of new viral RNA molecules being initiated in the presence of guanidine is greatly reduced in comparison with an untreated sample but they believe that this is the primary effect of the drug.

The existence of guanidine-resistant and -dependent mutants has made possible a series of experiments which shows that the guanidine-sensitive protein is able to function on heterologous RNA. The experiments involve simultaneous infection with guanidine-sensitive and guanidine-dependent virus and show that in the absence of guanidine the dependent virus can still grow and, in the presence of the drug, the sensitive virus still multiplies (171–173). The rescue of dependent virus by sensitive virus is more efficient than the inverse situation (173). It is clear that the rescued virus actually multiplies and that the apparent multiplication is not due merely to encapsidation of input RNA (174). Generally, RNA of the rescued virus is coated by protein with the specificity of the helper virus (119, 120).

Experiments on the rescue phenomenon have also been performed by superinfecting cells with dependent virus 2 hr after addition of sensitive virus (175). To overcome interference (see Sec. 3-9, A), the growth of the sensitive virus was transiently inhibited with guanidine at the time the dependent virus was added. Once the dependent virus became established, guanidine was removed and it was observed that the dependent virus multiplied. Its growth curve was the same as that of the sensitive virus, which means that its latent period was an hour shorter than it would have been in the presence of guanidine.

These experiments were all interpreted in the original publications as showing that a given replicase could multiply a heterologous genome. But, if the interpretation of guanidine action presented here is correct, the experiments require a different explanation. Specifically, the protein which carries out the "release" function must be able to act on heterologous genome. This may or may not be the replicase. But the experiment on the growth of a superinfecting guanidine-dependent virus in the absence of guanidine (175) raises another problem—why is the latent period of the superinfecting virus reduced? The answer may be that it is a matter of definition—the latent period is the time before mature virions are formed, not before new RNA is made. The guanidine-dependent virus which emerges earlier than expected is mostly coated by the protein of the sensitive virus and so it is probably an example of premature encapsidation rather than premature viral RNA synthesis. The discussion of interference in Sec. 3-9, A supports this interpretation.

The final aspect of guanidine inhibition to be considered concerns compounds which antagonize the inhibitory action of the drug: the antiguanidines (176–178). A large number of compounds have such activity; most of them contain a methyl group, an amino group, or both. Such compounds as choline, methionine, and trimethylamine are active. The possibility that they act by methylating the guanidine to an inactive derivative has been rejected because no methylated guanidine could be found. In fact, guanidine does not seem to be metabolized by the cell under any conditions (177). Methyl groups are not an obligate part of an antiguanidine; analogs containing ethyl moieties are active, as are such compounds as norleucine, while betaine is inactive, showing that methyl groups are not sufficient to make a compound an antiguanidine.

The mechanism by which these chemicals can antagonize guanidine inhibition is not known. Any explanation must take into account one striking fact: The suppression of guanidine inhibition is not competitive and only occurs when a minimal inhibitory dose of guanidine is used (177). For instance, 3×10^{-4} M guanidine totally inhibits polio-virus growth in a medium composed of Earle's saline plus embryo extract, but addition of 3.3×10^{-3} M choline or 4×10^{-4} M methionine results in the formation of 20–30% of the normal yield. However, 6×10^{-4} M guanidine is not antagonized at all by even 8-fold higher concentrations of choline and methionine (177). The antiguanidines also suppress the growth-promoting effect of suboptimal concentrations of guanidine on the guanidine-dependent strain of poliovirus (*179*).

D. THE EFFECT OF TEMPERATURE

One of the most effective ways to stop or depress poliovirus growth is to increase the temperature by a relatively small amount. Lwoff (*177*), for instance, has shown that a 0.1°C increase in temperature can produce a 2-fold decrease in the yield of poliovirus. The exact temperature at which this exquisite sensitivity is manifested is a genetic trait of the specific virus strain and can be manipulated by selection of appropriate mutants (*180*). The effect can be demonstrated at temperatures which have little or no effect on the growth of uninfected cells.

The effect of infraoptimal and supraoptimal temperatures on the growth of poliovirus have been extensively investigated. Two review articles by Lwoff (*177, 180*) summarize this work and discuss the interrelated effects of temperature, heavy water, and urea. Most interpretations of these results have been in terms of effects of temperature on the RNA synthetic apparatus. However, a series of observations on mutants which have different temperature sensitivities suggests an alternative explanation. It has been shown (see Ref. *180*, p. 169–170) that various mutants sensitive to different temperatures are distinguishable antigenically, as well as in their affinities for cells and resins and in their different sensitivities to degradation by urea. Thus, virus strains whose growth is inhibited at varying temperatures have virions with different physical properties. It is likely, therefore, that the thermosensitive event in the growth of these viruses is virion formation. In Sec. 3-6, C it was proposed that the formation of top component occurs by spontaneous aggregation of proteins, and that subsequently RNA associates with this particle to form the virion. The aggregation of capsid proteins must involve the simultaneous formation of a large number of equivalent noncovalent bonds, and such a process would be expected to have a temperature threshold above which it would fail to occur. Therefore, it is likely that top component aggregation would be thermosensitive. Such a hypothesis is easily tested.

The effect of temperature on viral growth is not entirely on capsid aggregation; there is also a decreased synthesis of viral RNA as the temperature is raised (*8, 180*). It is not clear whether the effects on the formation of virions and on RNA synthesis are interdependent or independent. Unpublished results (*8*) indicate that high temperature affects RNA production by reducing the length of the linear phase of RNA synthesis. For instance, RNA production may follow the same time course at 37° and 39°C up to the middle of the linear phase, but at 39°C the cells will then stop synthesizing RNA while at 37°C they will continue. At 40°C only the exponential phase of RNA synthesis will occur (*180*). It may be that the process which ordinarily limits the linear phase to 1 hr occurs faster at a higher temperature. In accord with this idea is the observation that the replicase isolated from cells infected with a thermosensitive strain of poliovirus does not display thermosensitivity during *in vitro* incubation at a restrictive temperature (*180a*).

In summary, there are two effects of supraoptimal temperature on the growth of poliovirus: a premature cessation of RNA synthesis and an inhibition of virion formation. Since no one study of both effects has appeared, it is impossible to know if they are related. It might be that the lack of correctly assembled top component or the formation of incorrectly assembled protein aggregates is responsible for the cessation of RNA synthesis. It is also possible that there are a number of genetically separable temperature-sensitive functions. Further experimentation is clearly required to settle these questions.

The thermosensitive events discussed above are ones which occur in virus strains which have been isolated from nature or submitted to very heavy selective pressure. Cooper (*181, 182*) has isolated thermosensitive poliovirus mutants using 5-fluorouracil (FU) mutagenesis on a strain which ordinarily grows at a relatively high temperature. The FU-induced mutations are presumably single-step events. Cooper has mutants which are in a number of apparently different viral functions, and some of them complement the growth of others (*182*). The various mutants produce differentiable effects (*183a*) but it is difficult to correlate all of the criteria used in these studies with the known events occurring during the growth cycle. Recently Cooper has mapped his mutants by recombination (*183b*).

E. 5-FLUOROURACIL (FU)

This compound, if added to the medium of infected cells, will be incorporated in the place of uracil residues in poliovirus RNA (*184*).

At 10^{-3} M it reduces the yield of infectious progeny to 20% of normal but the virions which are made have a normal specific infectivity despite the replacement of 35% of their uracil by FU. A large percentage of these virions are abnormally temperature sensitive in their growth and they give rise to temperature-sensitive mutants (181). Replicase induced by the progeny of FU virus has a component which is more resistant to heating than wild-type replicase and is more sensitive to inhibition by manganese (185). This is probably due to the mutagenic effect of FU and represents the best evidence that at least part of the replicase activity is specified by the viral genome.

3-9. Miscellaneous Topics

A. INTERFERENCE

This word is used to cover a multitude of effects. The only form of interference which will be discussed here is the intracellular interference occurring between picornaviruses which is not mediated by interferon or interferon-like substances (186, 187).

Interference can be readily demonstrated between different enteroviruses, between different serotypes of one virus, and even between mutants of a given virus strain (186–188). A typical experiment involves infecting cells with one virus (called the interfering virus) and then superinfecting the cell 2 hr later with a distinguishable virus (called the challenge or superinfecting virus). The yield of the challenge virus will be greatly reduced compared to the yield which would be produced in the absence of the interfering virus. The time which must elapse between the addition of interfering virus and the addition of challenge virus is dependent on the relative multiplicities of the two viruses—with multiplicities of 10 for each virus, a separation of 60 min is sufficient to produce complete interference. When the interfering virus is in great excess, interference may be quite evident after 30 min and complete in 40 min (186).

Interference occurs during the intracellular stages of the viral growth cycle; adsorption and penetration of the challenge virus are normal (186). Furthermore, interference is reversible; if the interfering virus is guanidine sensitive and the challenge virus is guanidine resistant, then addition of guanidine with the challenge virus (thus stopping the replication of the interfering virus) reverses the interference and the challenge virus grows normally (187). In fact, if the multiplication of the interfering virus is merely inhibited for an hour, then the challenge virus is able to grow to a certain extent (*175*).

The facts about picornavirus interference seem to suggest that in the infected cell some virus-induced function produces an immunity to superinfection. However, no activity which is known to exist in the virus-infected cell would explain interference, and its reversibility is difficult to understand on this basis. An alternative hypothesis is suggested by certain facts presented in earlier sections. This simple hypothesis invokes no new virus-induced process and explains all of the experimental observations.

The hypothesis derives from a consideration of the biphasic nature of viral RNA synthesis (see Sec. 3-7, B). A given template for RNA synthesis (a molecule of complementary RNA in a replicative intermediate) can only produce a fixed amount of RNA per unit time (about 9 molecules/min). So for any input virus to contribute significantly to the yield of progeny (which is about 2×10^5 molecules/cell) it must specify the production of a large number of new templates for RNA synthesis. The turnover of replicative intermediates further accentuates this requirement for new templates. For a virus to contribute significantly to the yield of progeny, it must execute the exponential phase of RNA synthesis, which is the time when templates multiply. The length of the exponential phase can be shortened by initiating infection with more virions, but it must occur at any reasonable multiplicity. The linear phase which follows the exponential is independent of multiplicity and has a fixed duration of about an hour after which viral RNA synthesis stops. The final cessation of viral RNA synthesis is probably a result of cell damage (leakage of precursors, reduced energy production) and should be general: *When the RNA synthesis induced by one virus is completed, the synthesis of all viral RNA in the cell will stop.*

This analysis provides an obvious explanation for picornavirus interference. *Interference is a reflection of the fact that the virus which is in the highest multiplicity determines the time at which all viral RNA synthesis ceases.* Any viruses whose replication is still in the exponential phase will not have produced enough RNA and certainly not enough protein to contribute significantly to the virus yield.

This interpretation of interference says, in essence, that if two distinguishable types of virus are present in different amounts in a single cell, the majority species will stop all viral replication in the cell before the minority species can finish its cycle of multiplication. So, if an hour elapses between the addition of equal multiplicities of the challenge and the interfering viruses, the challenge virus would never reach its linear phase and would contribute only about 10% to the final yield of RNA. If 2 hr elapsed, the challenge virus would still be in its exponential phase and would only contribute about 0.3% to the yield.

To test this hypothesis requires experiments which are not complicated by phenotypic mixing or genomic masking (Sec. 3-6, F), because the large excess of protein from the interfering virus will encapsidate any challenge virus RNA. The most extensive study of interference in the literature unfortunately is probably misleading because the two viruses were distinguished by antigenic criteria where phenotypic mixing was not rigorously excluded (186). The best way to avoid this difficulty would be to differentiate the challenge virus from the interfering virus by a criterion which reflects an intracellular process and which is thus independent of the capsid, such as guanidine resistance or thermoresistance. No systematic study of such a situation has appeared, but the observations which have been reported are consistent with the present hypothesis on the mechanism of interference (135, 175, 187).

B. RELEASE OF VIRUS AND CELL DAMAGE

It seems very probable that picornaviruses are released from infected cells by actual rupture of the cell membrane (189). The study of single cells has shown that release occurs in a burst and not by a protracted leakage (190). Since two closely related viruses (Mengovirus and EMC) differ markedly in the time of cell lysis (191), and Mengovirus mutants can be isolated which induce cell lysis at different times after infection (190), the kinetics of the processes leading to lysis seem to be under viral control.

Amako and Dales (192) have presented evidence which suggests that lysis is the culmination of a series of events including leakage of enzymes from lysosomes (193) and proliferation of cytoplasmic membranes (194). These events are evident much before lysis occurs and the stimulation of choline uptake which is evident by 3 hr after infection (195) may be a reflection of membrane hyperplasia. The site of action of the viral protein responsible for these cytopathic effects has not been discovered. The protein is only formed in sufficient amounts to cause cytopathic effects after the time when new virions appear

(*196*); the coat protein of the virus might even be responsible. The inhibitions of cellular macromolecular synthesis which occur soon after infection do not seem to be responsible for the cytopathology (*197*).

C. THE INCOMING VIRUS

It has been impossible so far to follow the RNA of input particles as it initiates infection. Such experiments require the use of labeled virus and, unfortunately, so few particles of a preparation actually initiate an infection that the uncoated, partially coated, and degraded particles obscure the interesting materials such as viral polyribosomes (*198*). Joklik and Darnell (*155*) were able to show that many particles attach and then are eluted in a noninfectious form. The RNA of others is broken down to acid insoluble material and some RNA is released in a high-molecular-weight form. Tobey (*199, 200*) claimed to have investigated the fate of entering RNA, but he did not show that the labeled RNA which he was investigating was not still part of virions. Recently, Levy and Carter (*200a*) have apparently successfully demonstrated the association of the RNA of incoming particles with polyribosomes, using magnesium precipitation to reduce the amount of undegraded virions in cell extracts.

The question of whether any nucleic acid of the input particles is reincorporated into progeny particles has been investigated by Goodheart (*201*). Using a naturally occurring density variant of EMC virus (*202*), he showed that virtually none of the input RNA was encapsidated anew during the growth cycle. Whether this result, which has also been found with the RNA bacteriophage (*203, 204*), is due to an ability of some element of the replication mechanism to distinguish input RNA from progeny RNA or whether it is due to other causes has not been settled.

In order to initiate an infection, it seems likely that the input RNA must serve two functions: to make viral proteins (specifically the replicase) and to act as a template for the synthesis of complementary RNA by the replicase. If, however, there is a cellular protein which can carry out the functions of the replicase (the manufacture of both viral and complementary RNA), the first requirement might not exist. No such cellular protein is known, but one reason for suggesting its existence is the high infectivity of double-stranded RNA (see Sec. 3-7, A, 2). Unless a double-stranded RNA is split into its component chains, it cannot act as a messenger RNA, and one possibility is that it acts directly as a template for RNA synthesis using a cellular enzyme to catalyze the synthesis of new viral RNA molecules.

If the RNA must serve as a messenger and then as a template for complementary RNA synthesis, a new problem arises. In the process of protein synthesis, RNA is translated in the $5' \rightarrow 3'$ direction (205). Present evidence from DNA-dependent RNA synthesis indicates that DNA is transcribed in the $3' \rightarrow 5'$ direction (206), which suggests that an RNA molecule serving as a template for RNA synthesis will be copied in the $3' \rightarrow 5'$ direction. Thus, ribosomes proceed down an RNA molecule in the opposite direction from a replicase which is using that RNA as template. Unless the replicase can displace ribosomes, an RNA molecule cannot serve as a template until it is free of ribosomes. Therefore, the initial viral RNA molecule will have to act as a messenger, then be cleared of ribosomes and finally act as a template for complementary RNA synthesis. Alternatively, complementary RNA might be made in the $3' \rightarrow 5'$ direction, or the replicase might displace ribosomes.

D. Role of Membranous Structures in Virus Replication

The idea that possibly all processes associated with viral replication are carried by membrane-bound particulates has been put forward on the basis of biochemical studies of disrupted cells (81). The evidence for such a situation, however, is not compelling. The basic finding is that most viral polyribosomes are not free in a cytoplasmic extract but are attached to a large DOC-sensitive structure (81). Furthermore, both the replicase and pulse-labeled RNA sediment quite readily (30, 81). However, when this question is investigated by sucrose-gradient analysis of DOC-treated and -untreated extracts, the results are equivocal: About half of the polyribosomes are apparently free and half must be freed with DOC (8, 58). Similarly, some of the replication complexes appear to be free and some attached to a larger structure (8). In electron micrographs of infected cells, some of the polyribosomes seem to be free and some appear to be membrane bound (194). Therefore, it is difficult to know if the apparent membrane attachment has any functional significance (58).

It was also suggested that virion formation may occur within membrane-bound structures (81). On this point the electron microscopy is clear—newly made particles are found free in the cytoplasmic matrix (*194*). Furthermore, analysis by sucrose-gradient sedimentation of extracts indicates that little or no virus is in large structures (*8, 58*). In general, electron micrographs show no evidence of virus-specific membranous structures except for a massive synthesis of small, membranous bodies in the centrosphere (*191, 194*), completely separated from viral polyribosomes. There is also the appearance of "viroplasm," which consists of dense aggregates of undefined but apparently virus-specific materials (*194*); these are found free in the cytoplasmic matrix. The earliest recognizable progeny virions show no special spatial relationship to either polyribosomes or viroplasm (*194*). From the electron microscopic and biochemical evidence there is no reason to believe that viral RNA synthesis, viral protein synthesis, or virion formation are in any way spatially related. It is quite likely that each process occurs separately and that the various components get together by diffusion through the cytoplasmic matrix.

3-10. General View of Replication

This section will be an attempt to integrate many of the findings discussed above into a general picture of the process of picornavirus replication.

A. The Number of Virus-Specific Proteins Involved in the Growth Cycle

Table 3-2 shows all of the proteins which are known to be involved in the multiplication of poliovirus. There may be other proteins; a second replicase may exist, and there might be as-yet-undiscovered processes. On the other hand, any one of the proteins might be multifunctional. The only proteins which have been correlated with the proteins which are identifiable by acrylamide gel electrophoresis are those found in the procapsid and virion.

TABLE 3-2

Virus-Specific Proteins Involved in Poliovirus Replication

1. Inhibitor of host protein synthesis (polyribosome breakdown agent)
2. Inhibitor of host RNA synthesis
3. Replicase (viral RNA polymerase)
4. Lysis factor
5. Caspid proteins

B. Temporal Aspects of the Growth Cycle

A central thesis of this essay has been the lack of temporal control of virus-induced processes. This means that viral (or host) genes are not selectively "turned on" or "turned off" at specific times nor are activities of the gene products regulated; at all times during the growth cycle the same set of viral gene products are being made and all products function throughout the cycle. Of course, this thesis may be an oversimplification, but no data available at present compel the postulation of controlling elements active during virus growth.

The apparent existence of unique times in the virus growth cycle does not contradict the lack of control. For instance, the fact that virion production, no matter how it is measured, is not detectable for the first 2 hr after infection can easily be explained by chemical kinetics; a capsid is composed of many protein chains and aggregation will depend very strongly on the concentration of capsid protein. A process requiring 100 or more chains to come together, even if they need not collide simultaneously but can be aggregated in a stepwise manner, will have a threshold concentration below which it will not measurably occur. To be consistent with the data, the threshold concentration for top component aggregation must be reached at about 2.5 hr after infection.

Other apparently unique times in the cycle have been explained already. This is true of the switch from exponential to linear RNA synthesis (see Sec. 3-7, H, 2), the final cessation of viral RNA synthesis (Sec. 3-7, H, 2), and cell lysis (Sec. 3-9, B). The processes occurring before virion maturation begins are also not instances of temporal control. The inhibitions of cellular RNA and protein synthesis, which occur before virion formation, are not completed during that period but continue during the later stages of infection.

The best evidence for the lack of temporal control is the fact that no differences in the relative amounts of the various viral proteins can be detected between 85 min and 5 hr after a synchronized infection (16, 83). This means that whether the cell is in the exponential phase of RNA synthesis, the linear phase, or the declining phase, the protein spectrum is grossly similar (there could be small differences, or large differences in minor protein species, which would have been missed in the analysis).

Although it would appear that there is no way to distinguish between periods of the growth cycle on the basis of the proteins which are made, it is still of heuristic value to make such distinctions. Traditionally, the time before progeny virions appear is known as the "eclipse" or "early" phase. Because this period is operationally defined, it is a valid partition of the growth cycle. The end of this phase corresponds roughly to the time at which RNA accumulation changes from exponential to linear, and so it is possible to equate the exponential phase with eclipse.

C. THE HISTORY OF AN RNA MOLECULE

In Sec. 3-7 the process of RNA synthesis and handling was discussed in detail. Integrating that material with other evidence presented in this chapter leads to the picture of viral RNA synthesis shown in Fig. 3-7. In this scheme the replication complex has a central role.

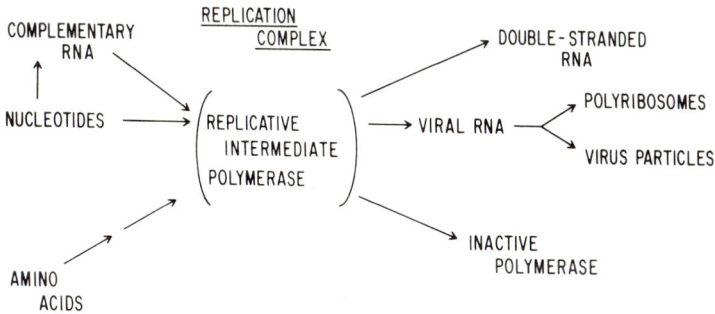

FIG. 3-7. Scheme of the synthesis and processing of poliovirus RNA.

Both of its known components, the replicative intermediate and the replicase (or viral RNA polymerase), are replaced continually, releasing double-stranded RNA and inactive replicase. Once a molecule of RNA is synthesized, it is released from the replication complex, and it is this step which is inhibited by guanidine. After release of the molecule it may become part of either a polyribosome or a virion; whichever choice is made seems to depend on a competition, probably between ribosomes and top component. As infection proceeds, a continually increasing percentage of the RNA becomes part of virions.

3-11. Major Unsolved Problems

A number of questions have been raised in the course of this assay for which there was either no answer or only a very speculative one.

These are accumulated in this last section in order to underscore the need for further investigation of the growth cycle of picornaviruses.

A. ARE THERE UNDISCOVERED VIRAL FUNCTIONS?

This question is best investigated by isolating enough conditional lethal mutants (e.g., temperature sensitive) so that all of the obligate functions can be uncovered (*181*). Then all of the functions should be correlated with specific proteins identifiable in acrylamide gel patterns.

B. WHAT IS THE MECHANISM OF THE INHIBITIONS OF HOST-CELL FUNCTION?

Very little is known about how the viral inhibitors work, and there is only one short report of any demonstrable activity of the inhibitors *in vitro*. Although these inhibitions are very general—they affect total RNA and protein synthesis of the host cell rather than being selective for a class of genes or a class of cell messages—it would be of great importance to know how they occur. Such processes as inhibition of RNA and protein synthesis during mitosis might be clarified and the selectivity of the inhibitor of protein synthesis for host (as opposed to viral) messages could be elucidated.

C. WHAT ARE THE DETAILS OF THE MECHANISM OF VIRAL RNA SYNTHESIS?

There are a number of open questions: How many replicases are there? What is the detailed structure of the replicative intermediate? Why is the replicative intermediate unstable? What releases finished molecules from the replication complex, and thus what protein is the target for guanidine? How is the replicase able to recognize one strand of the duplex to carry out an asymmetric synthesis of RNA? Most of these questions will have to be answered by studying the action of the replicase(s) *in vitro* which will require isolation and characterization of the protein(s). Since this has already been accomplished for the RNA bacteriophage replicase (*140*), it is likely that answers will be forthcoming from that system, but until a picornavirus replicase is studied there will always be doubt whether the systems are identical.

D. How Does the Input RNA Initiate the Infection?

This problem was discussed in some detail in Sec. 3-9, C. It is one of the most important unanswered questions.

E. Why Does Increased Temperature Inhibit Virus Replication?

The question was discussed at length in Sec. 3-8, D. The basic difficulty is in defining the number of thermosensitive processes.

F. What Are the Steps in Virion Maturation?

This was discussed in Sec. 3-6, C and a number of separate questions were raised. Are there modifications of the gene products which are precursors of capsid protein which allow them to aggregate? Is the aggregation into top component a completely spontaneous process? What is the organization of the RNA and protein of the particle and how does it come about? Are there any morphogenetic proteins—proteins which catalyze an aspect of virion formation but are not parts of the top component or virion?

G. What Is the Evolutionary Status of RNA Viruses and Is RNA Self-Duplication Restricted to This Group of Parasites?

This final question has recently been discussed by Luria and Darnell (*207*). It is part of the general question of the origin of viruses about which there is no evidence. The only present hope for an answer lies in finding virus-like processes in normal cells, and the most likely candidate is RNA self-duplication. Only one example of such a system has been reported—the metagon which seems to be self-replicating in *Didinium* (*208*)—but no systematic search has been made.

ABBREVIATIONS

DOC   sodium desoxycholate
SDS   sodium dodecyl sulfate
FPA   fluorophenylalanine
FU   fluorouracil

ACKNOWLEDGMENTS

I would like to express my gratitude to Alice Huang, Michael Jacobson, and Andre Lwoff for their critical reading of the manuscript and their many helpful suggestions.

Supported by Grant No. 97592 from United States Public Health Service.

REFERENCES

1. *Viral and Rickettsial Infections of Man*, 4th ed. (F. L. Horsfall, Jr., and I. Tamm, eds.), Lippincott, Philadelphia, 1965, Chaps. 5, 17–22.
1a. S. McGregor and H. D. Mayor, *J. Virol.*, **2**, 149 (1968).
2. H. L. Bachrach, R. Trautman, and S. E. Breese, Jr., *Am. J. Vet. Res.*, **25**, 333 (1964).
3. J. E. Darnell, Jr., and H. Eagle in *Advan. Virus Res.*, **7**, 1–26 (1960).
4. S. Penman, K. Scherrer, Y. Becker, and J. E. Darnell, *Proc. Natl. Acad. Sci. U.S.*, **49**, 654 (1963).
5. J. R. Warner, P. Knopf, and A. Rich, *Proc. Natl. Acad. Sci. U.S.*, **49**, 122 (1963).
6. M. Girard, H. Latham, S. Penman, and J. E. Darnell, *J. Mol. Biol.*, **11**, 187 (1965).
7. S. Penman, *J. Mol. Biol.*, **17**, 117 (1966).
8. D. Baltimore, unpublished results (1963–1967).
9. K. Scherrer and J. E. Darnell, *Biochem. Biophys. Res. Commun.*, **7**, 486 (1962).
10. E. K. Wagner, L. Katz, and S. Penman, *Biochem. Biophys. Res. Commun.*, **28**, 152 (1967).
11. W. Gilbert, *J. Mol. Biol.*, **6**, 389 (1963).
12. B. Mandel, *Virology*, **17**, 288 (1962).
13. H. Latham and J. E. Darnell, *J. Mol. Biol.*, **14**, 1 (1965).
14. R. J. Britten and R. B. Roberts, *Science*, **131**, 32 (1960).
15. F. O. Wettstein, T. Staehelin, and H. Noll, *Nature*, **197**, 430 (1963).
16. D. F. Summers, J. V. Maizel, Jr., and J. E. Darnell, Jr., *Proc. Natl. Acad. Sci. U.S.*, **54**, 505 (1965).
17. E. Viñuela, M. Salas, and S. Ochoa, *Proc. Natl. Acad. Sci. U.S.*, **57**, 729 (1967).
18. A. Shapiro, E. Viñuela, and J. Maizel, *Biochem. Biophys. Res. Commun.*, **28**, 815 (1967).
19. D. H. L. Bishop, J. R. Claybrook, and S. Spiegelman, *J. Mol. Biol.*, **26**, 373 (1967).
20. M. Jacobson and D. Baltimore, unpublished data (1966–1968).
21. F. L. Schaffer and L. H. Frommhagen, *Virology*, **25**, 662 (1965).
22. V. F. A. Anderer and H. Restle, *Z. Naturforsch.*, **19b**, 1026 (1964).
23. D. G. Scraba, C. M. Kay, and J. S. Colter, *J. Mol. Biol.*, **26**, 67 (1967).
24. J. V. Maizel, Jr., *Biochem. Biophys. Res. Commun.*, **13**, 483 (1963).
25. R. R. Rueckert, *Virology*, **26**, 345 (1965).

26. A. Boeye, *Virology*, **25**, 550 (1965).

26a. J. V. Maizel, Jr., and D. F. Summers, *Virology*, **36**, 48 (1968).

27. J. T. Finch and A. Klug, *Nature*, **183**, 1709 (1959).

28. C. E. Schwerdt, in *Cellular Biology, Nucleic Acids and Viruses*, Vol. 5 (O. V. St. Whitelock, ed.), N.Y. Academy of Sciences, New York, 1957, pp. 157–172.

29. L. Montagnier and F. K. Sanders, *Nature*, **197**, 1177 (1963).

30. D. Baltimore, *Proc. Natl. Acad. Sci. U.S.*, **51**, 450 (1964).

31. J. M. Bishop, D. F. Summers, and L. Levintow, *Proc. Natl. Acad. Sci. U.S.*, **54**, 1273 (1965).

32. D. Baltimore, *J. Mol. Biol.*, **18**, 421 (1966).

33. J. D. Watson and F. M. C. Crick, *Nature*, **171**, 964 (1953).

34. D. Baltimore, Y. Becker, and J. E. Darnell, *Science*, **143**, 1034 (1964).

35. Y. Watanabe, *Biochim. Biophys. Acta*, **95**, 515 (1965).

36. D. Baltimore, M. Girard, and J. E. Darnell, *Virology*, **29**, 179 (1966).

37. H. J. Eggers, N. Ikegami, and I. Tamm, *Ann. N.Y. Acad. Sci.*, **130**, 267 (1965).

38. J. E. Darnell, Jr., L. Levintow, M. M. Thoren, and J. L. Hooper, *Virology*, **13**, 271 (1961).

39. J. E. Darnell, Jr., and L. Levintow, *J. Biol. Chem.*, **235**, 74 (1960).

40. J. Hurwitz and J. J. Furth, *Proc. Natl. Acad. Sci. U.S.*, **48**, 1222 (1962).

41. D. Baltimore and R. M. Franklin, *Proc. Natl. Acad. Sci. U.S.*, **48**, 1383 (1962a).

42. M. B. Yarmolinsky and G. L. de la Haba, *Proc. Natl. Acad. Sci. U.S.*, **45**, 1721 (1959).

43. E. Reich, R. M. Franklin, A. J. Shatkin, and E. L. Tatum, *Proc. Natl. Acad. Sci. U.S.*, **48**, 1238 (1962).

44. R. M. Franklin and D. Baltimore, *Cold Spring Harbor Symp. Quant. Biol.*, **27**, 175 (1962).

45. A. J. Shatkin, *Biochim, Biophys. Acta*, **61**, 310 (1962).

46. E. H. Simon, *Virology*, **13**, 105 (1961).

47. N. P. Salzman, *Biochim. Biophys. Acta*, **31**, 158 (1959).

48. P. D. Cooper, *Virology*, **28**, 663 (1966).

49. F. L. Schaffer and M. Gordon, *J. Bacteriol.*, **91**, 2309 (1966).

50. C. Grado, S. Fischer, and G. Contreras, *Virology*, **27**, 623 (1965).

51. E. M. Martin and T. S. Work, *Biochem. J.*, **81**, 514 (1961).

52. E. M. Martin and T. S. Work, *Biochem. J.*, **83**, 574 (1962).

53. A. J. D. Bellett and A. T. H. Burness, *J. Gen. Microbiol.*, **30**, 131 (1963).

54. H. B. Levy, *Virology*, **15**, 173 (1961).

55. R. Franklin and J. Rosner, *Biochim. Biophys. Acta*, **55**, 240 (1962).

56. H. Hausen, *Z. Naturforsch.*, **17b**, 158 (1962).

57. J. J. Holland and D. W. Bassett, *Virology*, **23**, 164 (1964).

58. L. Dalgarno and E. M. Martin, *Virology*, **26**, 450 (1965).

59. R. Eason and R. M. S. Smellie, *J. Biol. Chem.*, **240**, 2580 (1965).

60. M. D. Scharff, A. J. Shatkin, and L. Levintow, *Proc. Natl. Acad. Sci. U.S.*, **50**, 686 (1963).

61. T. T. Crocker and E. Pfendt, *Science*, **145**, 401 (1964).

61a. P. I. Marcus and M. E. Freiman, in *Methods in Cell Physiology,* Vol. **II** (D. M. Prescott, ed.), Academic Press, New York, 1966, p. 93.

62. P. I. Marcus and E. Robbins, *Proc. Natl. Acad. Sci. U.S.,* **50,** 1156 (1963).

63. J. M. Salb and P. I. Marcus, *Proc. Natl. Acad. Sci. U.S.,* **54,** 1353 (1965).

64. P. Hausen and D. W. Verwoerd, *Virology,* **21,** 617 (1963).

65. J. J. Holland and J. A. Peterson, *J. Mol. Biol.,* **8,** 556 (1964).

66. D. Baltimore and R. M. Franklin, *Biochim. Biophys. Acta,* **76,** 425 (1963).

67. S. Penman and D. Summers, *Virology,* **27,** 614 (1965).

68. J. J. Holland, *J. Mol. Biol.,* **8,** 574 (1964).

69. R. Bablanian, H. J. Eggers, and I. Tamm, *Virology,* **26,** 100 (1965).

70. D. F. Summers and J. V. Maizel, Jr., *Virology,* **31,** 550 (1967).

71. M. Willems and S. Penman, *Virology,* **30,** 355 (1966).

71a. R. Liebowitz and S. Penman, personal communication (1968).

72. J. J. Holland, *Proc. Natl. Acad. Sci. U.S.,* **49,** 23 (1963).

73. M. L. Fenwick, *Virology,* **19,** 241 (1963).

74. E. F. Zimmerman, M. Heeter, and J. E. Darnell, *Virology,* **19,** 400 (1963).

75. W. McCormick and S. Penman, *Virology,* **31,** 135 (1967).

76. P. G. W. Plagemann and H. E. Swim, *J. Bacteriol.,* **91,** 2317 (1966).

77. J. E. Darnell, M. Girard, D. Baltimore, D. F. Summers, and J. V. Maizel, in *Molecular Biology of Viruses* (J. S. Colter and W. Paranchych, eds.), Academic Press, New York, 1967, p. 375.

78. J. J. Holland, *Biochem. Biophys. Res. Commun.,* **9,** 556 (1962).

79. I. G. Balandin and R. M. Franklin, *Biochem. Biophys. Res. Commun.,* **15,** 27 (1964).

80. R. A. Tobey, D. F. Petersen, and E. C. Anderson, *Virology,* **27,** 17 (1965).

81. S. Penman, Y. Becker, and J. E. Darnell, *J. Mol. Biol.,* **8,** 541 (1964).

82. D. F. Summers and L. Levintow, *Virology,* **27,** 44 (1965).

83. D. F. Summers, J. V. Maizel, Jr., and J. E. Darnell, Jr., *Virology,* **31,** 427 (1967).

84. T. Tolbert and R. Engler, *Nature,* **200,** 498 (1963).

85. D. Gelfand and D. Baltimore, unpublished results (1967).

86. A. Rich, S. Penman, Y. Becker, J. Darnell, and C. Hall, *Science,* **142,** 1658 (1963).

87. J. R. Warner, A. Rich, and C. Hall, *Science,* **138,** 1399 (1962).

88. R. R. Rueckert and P. H. Duesberg, *J. Mol. Biol.,* **17,** 490 (1966).

89. A. V. Elsen and A. Boeye, *Virology,* **28,** 481 (1966).

90. J. V. Maizel, B. A. Phillips, and D. F. Summers, *Virology,* **32,** 692 (1967).

90a. D. F. Summers and J. V. Maizel, Jr., *Proc. Natl. Acad. Sci. U.S.,* **59,** 966 (1968).

90b. J. J. Holland and E. D. Kiehn, *Proc. Natl. Acad. Sci. U.S.,* **60,** 1015 (1968).

90c. M. F. Jacobson and D. Baltimore, *Proc. Natl. Acad. Sci. U.S.,* **61,** 77 (1968).

91. E. Viñuela, I. Algranati, and S. Ochoa, *European J. Biochem.,* **1,** 3 (1967).

92. T. Sugiyama and D. Nakada, *Federation Proc.,* **26,** Abst. 1984 (1967).

93. K. Eggen and D. Nathans, *Federation Proc.,* **26,** Abst. 1070 (1967).

94. D. Baltimore, H. J. Eggers, and I. Tamm, *Biochim. Biophys. Acta* **76,** 644 (1963).

95. J. H. Matthaei and M. W. Nirenberg, *Proc. Natl. Acad. Sci. U.S.,* **47,** 1580 (1961).

96. J. Warner, M. J. Madden, and J. E. Darnell, *Virology*, **19**, 393 (1963).
97. D. Rekosh, M. Jacobson, H. Lodish, and D. Baltimore, unpublished results (1968).
98. A. Klug, W. Longley, and R. Leberman, *J. Mol. Biol.*, **15**, 315 (1966).
99. J. G. Finch and K. Klug, *J. Mol. Biol.*, **15**, 344 (1966).
100. M. Jacobson and D. Baltimore, *J. Mol. Biol.*, **33**, 369 (1968).
101. M. Girard and D. Baltimore, unpublished results (1966).
102. S. Halperen, H. J. Eggers, and I. Tamm, *Virology*, **24**, 36 (1964).
103. M. M. Mayer, H. J. Rapp, B. Roizman, S. W. Klein, K. M. Cowan, D. Lukens, C. E. Schwerdt, F. L. Schaffer, and J. Carney, *J. Exptl. Med.*, **78**, 435 (1957).
104. M. D. Scharff and L. Levintow, *Virology*, **19**, 491 (1963).
105. S. Halperen, H. J. Eggers, and I. Tamm, *Virology*, **23**, 81 (1964).
105a. Y. Watanabe, K. Watanabe, S. Kataggiri, and Y. Hinuma, *J. Biochem.*, **57**, 733 (1965).
105b. B. A. Phillips, D. F. Summers, and J. V. Maizel, Jr., *Virology*, **35**, 216 (1968).
106. J. R. Tillotson and A. M. Lerner, *Proc. Natl. Acad. Sci. U.S.*, **56**, 1143 (1966).
107. B. Roizman, W. Hopken, and M. M. Mayer, *J. Exptl. Med.*, **78**, 386 (1957).
108. B. Roizman, M. M. Mayer, and H. J. Rapp, *J. Exptl. Med.*, **81**, 419 (1958).
109. K. Hummeler and J. J. Tumilowicz, *J. Exptl. Med.*, **84**, 631 (1960).
110. K. Hummeler, T. F. Anderson and R. A. Brown, *Virology*, **16**, 84 (1962).
111. G. L. LeBouvier, *Lancet*, **1955-II**, 1013.
112. D. Crowther and J. L. Melnick, *Virology*, **14**, 11 (1961).
113. F. L. Schaffer, *Virology*, **18**, 412 (1962).
114. R. W. Horne and J. Nagington, *J. Mol. Biol.*, **1**, 333 (1959).
115. N. Dimmock, *Virology*, **31**, 715 (1967).
116. C. Wallis and J. L. Melnick, *Virology*, **21**, 332 (1963).
117. M. D. Scharff, J. V. Maizel, Jr., and L. Levintow, *Proc. Natl. Acad. Sci. U.S.*, **51**, 329 (1964).
118. N. Ledinko and G. K. Hirst, *Virology*, **14**, 207 (1961).
119. E. Wecker and G. Lederhilger, *Proc. Natl. Acad. Sci. U.S.*, **52**, 705 (1964).
120. J. J. Holland and C. E Cords, *Proc. Natl. Acad. Sci. U.S.*, **51**, 1082 (1964).
120a. H. F. Lodish, *Progress in Biophysics and Molecular Biology*, **18**, 285 (1968).
121. L. Montagnier and F. K. Sanders, *Nature*, **199**, 664 (1963).
122. D. Baltimore and M. Girard, *Proc. Natl. Acad. Sci. U.S.*, **56**, 741 (1966).
123. E. P. Geiduschek, J. W. Moohr, and S. B. Weiss, *Proc. Natl. Acad. Sci. U.S.*, **48**, 1078 (1962).
124. D. Baltimore, Y. Becker, and J. E. Darnell, *Science*, **143**, 1034 (1964).
125. P. Hausen, *Virology*, **25**, 523 (1965).
126. M. Pons, *Virology*, **24**, 467 (1964).
127. J. M. Bishop and G. Koch, *J. Biol. Chem.*, **242**, 1736 (1967).
128. D. Baltimore, *J. Mol. Biol.*, **18**, 421 (1966).
129. L. Katz and S. Penman, *Biochem. Biophys. Res. Commun.*, **23**, 557 (1966).
129a. N. A. Young, B. H. Hoyer, and M. A. Martin, *Proc. Natl. Acad. Sci. U.S.*, **61**, 548 (1968).
130. J. M. Bishop, N. Quintrell, and G. Koch, *J. Mol. Biol.*, **24**, 125 (1967).
131. G. Koch, N. Quintrell, and J. M. Bishop, *Virology*, **31**, 388 (1967).
132. D. Baltimore, *J. Mol. Biol.*, **32**, 359 (1968).

132a. M. Girard, *J. Virol.,* in press (1969).

133. B. Oberg and L. Philipson, *Arch. Biochem. Biophys.,* **119,** 504 (1967).

134. M. L. Fenwick and D. Pelling, *Virology,* **20,** 137 (1963).

135. D. Tershak, *Virology,* **23,** 1 (1964).

136. H. J. Eggers, N. Ikegami, and I. Tamm, *Virology,* **25,** 475 (1965).

137. M. Hayashi, *Proc. Natl. Acad. Sci., U.S.,* **54,** 1736 (1965).

138. D. Baltimore and R. M. Franklin, *J. Biol. Chem.,* **28,** 3395 (1963).

139. C. Weissmann, L. Simon, and S. Ochoa, *Proc. Natl. Acad. Sci. U.S.,* **49,** 407 (1963).

140. I. Haruna, K. Nozu, Y. Ohtaka, and S. Spiegelman, *Proc. Natl. Acad. Sci. U.S.,* **50,** 905 (1963).

141. D. Baltimore, H. J. Eggers, R. M. Franklin, and I. Tamm., *Proc. Natl. Acad. Sci. U.S.,* **49,** 843 (1963).

142. D. Baltimore, *Proc. Natl. Acad. Sci. U.S.,* **51,** 450 (1964).

143. E. Horton, S. L. Liu, L. Dalgarno, E. M. Martin, and T. S. Work, *Nature,* **204,** 247 (1964).

144. P. G. W. Plagemann and H. E. Swim, *J. Bacteriol.,* **91,** 2327 (1966).

145. E. Horton, S. L. Liu, E. M. Martin, and T. S. Work, *J. Mol. Biol.,* **15,** 62 (1966).

146. P. G. W. Plagemann and H. E. Swim, *Bacteriol. Rev.,* **30,** 288 (1966).

147. L. Dalgarno, E. M. Martin, S. L. Liu, and T. S. Work, *J. Mol. Biol.,* **15,** 77 (1966).

148. P. G. W Plagemann and H E. Swim, *Bacteriol. Rev.,* **30,** 288 (1966).

149. J. Polatnick, R. B. Arlinghaus, J. H. Graves, and K. M. Cowan, *Virology,* **31,** 609 (1967).

150. H. F. Lodish and N. D. Zinder, *Science,* **152,** 372 (1966).

151. M. Girard, D. Baltimore, and J. E. Darnell, *J. Mol. Biol.,* **24,** 59 (1967).

152. J. P. Richardson, *Proc. Natl. Acad. Sci. U.S.,* **55,** 1616 (1966).

153. M. Girard and D. Baltimore, *Proc. Natl. Acad Sci. U.S.,* **56,** 999 (1966).

154. L. Levintow and J. E. Darnell, Jr., *J. Biol. Chem.,* **235,** 70 (1960).

155. W. K. Joklik and J. E. Darnell, Jr., *Virology,* **13,** 439 (1961).

156. D. Nathans, *Proc. Natl. Acad. Sci. U.S.,* **51,** 585 (1964).

157. H. Latham and J. E. Darnell, *J. Mol. Biol.,* **14,** 13 (1965).

158. L. Felicetti, B. Colombo, and C. Baglioni, *Biochim. Biophys. Acta,* **119,** 120 (1966).

159. M. M. Appleman and R. G. Kemp, *Biochem. Biophys. Res. Commun.,* **24,** 564 (1966).

160. L. Levintow, M. M. Thoren, J. E. Darnell, Jr., and J. L. Hooper, *Virology,* **16,** 220 (1962).

161. D. Baltimore and R. M. Franklin, *Biochim. Biophys. Acta,* **76,** 431 (1963).

162. H. J. Eggers, D. Baltimore, and I. Tamm, *Virology,* **21,** 281 (1963).

163. D. Baltimore and R. M. Franklin, *Cold Spring Harbor Symp. Quant. Biol.,* **28,** 105 (1963).

164. D. Baltimore, *Medical and Applied Virology, Proceedings of the Second International Symposium* (M. Sanders and E. H. Lennette, eds.), Green, St. Louis, 1968, p. 340.

165. E. Wecker, K. Hummeler, and O. Goetz, *Virology,* **17,** 110 (1962).

166. D. Baltimore and R. M. Franklin, *Biochem. Biophys. Res. Commun.,* **9,** 388 (1962).

167. K. Hummeler and E. Wecker, *Virology,* **24,** 456 (1964).

168. M. D. Scharff, D. F. Summers, and L. Levintow, *Ann. N.Y. Acad. Sci.,* **130,** 282 (1965).

169. I. Tamm and H. J. Eggers, *Science,* **142,** 24 (1963).

170. A. S. Huang and D. Baltimore, unpublished results (1967).

170a. L. A. Caliguiri and I. Tamm, *Virology,* **35,** 408 (1968).

170b. L. A. Caliguiri and I. Tamm, *Virology,* **36,** 223 (1968).

171. V. I. Agol and G. A. Shirman, *Biochem. Biophys. Res. Commun.,* **17,** 28, (1964).

172. C. E. Cords and J. J. Holland, *Proc. Natl. Acad. Sci. U.S.,* **51,** 1080 (1964).

173. N. Ikegami, H. J. Eggers, and I. Tamm, *Proc. Natl. Acad. Sci. U.S.,* **52,** 1419 (1964).

174. D. R. Tershak, *Biochem. Biophys. Res. Commun.,* **27,** 189 (1967).

175. E. Wecker and G. Lederhilger, *Proc. Natl. Acad. Sci. U.S.,* **52,** 246 (1964).

176. L. Philipson, S. Bengtsson, and Z. Dinter, *Virology,* **29,** 317 (1966).

177. A. Lwoff, *Biochem. J.,* **96,** 289 (1965).

178. B. Loddo, G. L. Gessa, M. L. Schivo, A. Spanedda, G. Brotzu, and W. Ferrari, *Virology,* **28,** 707 (1966).

179. A. Lwoff and M. Lwoff, *Compt. Rend.,* **260,** 4116 (1965).

180. A. Lwoff, *Cold Spring Harbor Symp. Quant. Biol.,* **27,** 159 (1962).

180a. H. Priess and H. J. Eggers, *Nature,* **220,** 1047 (1968).

181. P. Cooper, *Virology,* **22,** 186 (1964).

182. P. Cooper, *Virology,* **25,** 431 (1965).

183. P. D. Cooper, R. T. Johnson, and D. J. Garwes, *Virology,* **30,** 638 (1966).

183a. B. B. Wentworth, D. McCahon, and P. D. Cooper, *J. Gen. Virol.,* **2,** 297 (1968).

183b. P. D. Cooper, *Virology,* **35,** 584 (1968).

184. W. Munyon and N. P. Salzman, *Virology,* **18,** 95 (1962).

185. D. R. Tershak, *J. Mol. Biol.,* **21,** 43 (1966).

186. N. Ledinko, *Virology,* **20,** 29 (1963).

187. C. E. Cords and J. J. Holland, *Virology,* **22,** 226 (1964).

188. G. D. Hsiung, *Arch. Ges. Virusforsch.,* **11,** 343 (1961).

189. K. Amako and S. Dales, *Virology,* **32,** 184 (1967).

190. A. Lwoff, R. Dulbecco, M. Vogt, and M. Lwoff, *Virology,* **1,** 128 (1955).

191. S. Dales and R. M. Franklin, *J. Cell Biol.,* **14,** 281 (1962).

192. K. Amako and S. Dales, *Virology,* **32,** 201(1967).

193. J. F. Flanagan, *J. Bacteriol.,* **91,** 789 (1966).

194. S. Dales, H. J. Eggers, I. Tamm, and G. E. Palade, *Virology,* **26,** 379 (1965).

195. S. Penman, *Virology,* **25,** 148 (1965).

196. R. Bablanian, H. J. Eggers, and I. Tamm, *Virology,* **26,** 114 (1965).

197. R. Bablanian, H. J. Eggers, and I. Tamm, *Virology,* **26,** 100 (1965).

198. S. Penman and D. Baltimore, unpublished results (1965).

199. R. A. Tobey, *Virology,* **23,** 10 (1964).

200. R. A. Tobey, *Virology,* **23,** (1964).

200a. H. B. Levy and W. A. Carter, *J. Mol. Biol.,* **31,** 561 (1968).

201. C. R. Goodheart, *J. Mol. Biol.,* **23,** 183 (1967).

202. C. R. Goodheart, *Virology,* **26,** 466 (1965).

203. R. H. Doi and S. Spiegelman, *Proc. Natl. Acad. Sci. U.S.,* **49,** 353 (1963).

204. J. E. Davis and R. I. Sinsheimer, *J. Mol. Biol.,* **6,** 203 (1963).

205. H. Kossel, A. R. Morgan, and H. G. Khorana, *J. Mol. Biol.,* **26,** 449 (1967).

206. U. Maitra and J. Hurwitz, *Proc. Natl. Acad. Sci. U.S.,* **54,** 815 (1965).

207. S. E. Luria and J. E. Darnell, Jr., *General Virology,* 2nd ed., Wiley, New York, 1967, p. 99.

208. I. Gibson and T. M. Sonneborn, *Proc. Natl. Acad. Sci. U.S.,* **52,** 869 (1964).

THE GENETIC ANALYSIS
OF POLIOVIRUS

PETER D. COOPER†

DEPARTMENTS OF CELL BIOLOGY AND OF MICROBIOLOGY AND IMMUNOLOGY
ALBERT EINSTEIN COLLEGE OF MEDICINE
NEW YORK, NEW YORK
AND
DEPARTMENT OF MICROBIOLOGY
JOHN CURTIN SCHOOL OF MEDICAL RESEARCH
AUSTRALIAN NATIONAL UNIVERSITY
CANBERRA, AUSTRALIA

† This chapter was written while the author was a Visiting Professor at the Albert Einstein College of Medicine and on study leave from the Australian National University.

4-1. Introduction

The virus of poliomyelitis originally became a model for intensive study because it was the specific agent of an emotionally provocative epidemic disease. This early impetus was maintained because poliovirus soon showed many now familiar technical virtues: it grew rapidly and reproducibly in culture, could easily and accurately be assayed, was physically and chemically stable, and could readily be purified to the point of crystallization. It remained a primate pathogen, leaving the hope that in time one might define the specific viral functions involved in a well-characterized human disease.

More recently, two new and valuable virtues have become apparent. First, infection by poliovirus *in vitro* and in presence of actinomycin can almost totally suppress cellular protein and RNA synthesis, so that newly made viral protein and RNA species can be resolved even though they are not incorporated in the virion. Secondly, poliovirus undergoes limited genetic recombination that makes it possible to obtain a reproducible genetic map. This combination of properties is almost unique among the plant, animal, and bacterial riboviruses studied so far.

So many unusual attributes have led to a large body of information on the growth of poliovirus. Perhaps an even greater inducement to its study has been its apparent simplicity: the virion is small, containing only a single strand of RNA and a few species of protein. The RNA itself seems also to be small and simple, containing sufficient information only for about ten average-sized proteins. There is thus a reasonable expectation that the functions of this virus can be completely described. Other agents, for example reovirus, a somewhat simple virus that nevertheless contains several species of RNA (1), or the large poxviruses, or viruses with complex cellular interrelations (e.g., the leucosis viruses) will certainly pose much more formidable problems, although the information content of the results will be correspondingly larger. It will be seen, however, that the small size of the poliovirus genome is somewhat deceptive, and that the poliovirus system is not as simple as it seemed at first.

This chapter attempts to set out the technical facilities available for the total genetic analysis of poliovirus, together with a review of the present status of such analysis in relation to the viral growth process. The main emphasis is on the use of temperature-sensitive mutants.

4-2. The Goals of Genetic Analysis

A. The Total Genetic Map

Genetic analysis, broadly speaking, is an attempt to describe the structure and function of the genetic material by means of its behavior in an altered (mutant) condition. The function of the genetic material is 2-fold, namely to replicate itself, and to express the genetic information it contains in order, ultimately, to expedite the replicative process.

Thorough reviews of the "genetics" of animal viruses (i.e., certain characters and types of interaction that are pertinent to genetic analysis) have been given by Fenner and Sambrook (2) and Fenner (3), and recent aspects of their genetic analysis have been briefly discussed by Cooper (4).

The ultimate goal of genetic analysis may be said to be the obtaining of a total genetic map. This can be conceived of as a diagram of the entire genome, comprising all genes, their number, sequence, extent, origins and terminations, and methods of replication, transcription, and translation, in relation to the identity, structure, function, and interaction of each of their gene products. As has been pointed out (4), some approach to this rather monumental goal is now possible for animal viruses. Such a full description would provide a definitive catalog of function of the virus, the complete molecular basis for all the ramified effects of an infective agent growing through populations of multicellular hosts. In theory, the total genetic map provides both a systematic way of accounting for all properties of a virus, and the method for defining the mutants needed to study altered functions.

B. The Proteins Specified by Poliovirus

The number of genes carried by the poliovirus genome, assuming all to be concerned with specifying proteins, can be approximately estimated from the familiar assumptions of a triplet nonoverlapping genetic code and about 200 amino acids for the average gene product. This gives an expectation of about 5 genes per 3000 nucleotides for single-stranded RNA.

The molecular weight of poliovirus RNA is generally taken to be about 2×10^6 daltons (5), i.e., to represent about 6000 nucleotides or some 10 genes. However, the confidence limits in this estimate of size may be $\pm 50\%$ (see below), and this uncertainty is important at the resolution now being considered. Only two direct estimates of the molecular weight of poliovirus RNA appear to be available. One mention, without supporting data, states the S value of extracted type 1 poliovirus RNA to be 20S, which in comparison with TMV RNA of 22S and 1.94×10^6 daltons presumably gives a molecular weight for poliovirus RNA of about 1.7×10^6 (5). The other, measuring viscosity as well as sedimentation coefficient (6), gives a molecular weight of 2×10^6 daltons for type 2 poliovirus RNA; there seems to be no a priori reason for assuming types 1 and 2 RNA to be precisely identical in size. Changes in salt concentration, however, have such a large effect on the sedimentation coefficient of poliovirus RNA (5, 7) that it is very difficult to assess the value of this parameter in determining molecular weight in this case.

Hydrodynamic estimates of the weight of the intact particle gave values for the type 1 poliovirion of 6.8×10^6 (8) or 5.5×10^6 (6) daltons, although Schaffer and Schwerdt (9) consider that the latter value is better interpreted as 6.2×10^6 daltons. Analyses of the proportion by weight of the RNA (9, 10) gave 22% (by absorption at 250 mμ), 24% (by the orcinol reaction), or 29% (from N:P ratios); the latter value will be high if free phosphate or phosphoprotein is present. These data provide values for RNA molecular weights that range between 1.21 and 1.97, with a mean about 1.5×10^6 daltons. Thus the "accepted value" of 2×10^6 is higher than all but one of the actual measurements; a value of a little under 1.7×10^6 for type 1 poliovirus RNA would seem more in accord with the data unless one particular measurement is preferred for some reason, and although the N:P ratio appears to be the preferred datum there does not seem to be any direct demonstration that all the P resides in RNA. The length of ribonucleoprotein fibers obtained by heat treatment of poliovirus and measured in the electron microscope (10a) is the same as that to be expected from 2.0×10^6 daltons of RNA; the authors state that removal of the protein does not affect the fiber length. Quite recently, work stimulated by these uncertainties (82) has shown that the mobilities and sedimentation constants of tobacco mosaic virus and poliovirus RNA's in acrylamide gel electrophoresis and in formalin–sucrose gradient centrifugation (with and without salt) are in fact very similar, poliovirus RNA behaving, if anything, as slightly the larger. Some problems of configuration remain, but these findings strongly support the "accepted value" of $2.0–2.1 \times 10^6$ daltons for poliovirus RNA.

RNA with an average molecular weight of 346 daltons/nucleotide [derived from the base composition given by Schaffer, Moore, and Schwerdt (11)] and totaling 2.08×10^6 daltons contains 6000 nucleotides, representing an information content of 2000 amino acids; the actual value probably lies between 1800 and 2100 amino acids. A more precise chemical determination would appear to be called for.

A valuable analytical tool for the purposes of a total map is that of gel electrophoresis, coupled by Maizel (12, 13) and Summers et al. (14) with an isotope labeling and gel fractionation procedure that permits a high resolution of viral proteins in the presence of a large excess of cellular proteins.

These methods have unequivocally shown the presence of at least four major peptides in the poliovirion (Fig. 4-1). Figure 4-1 also shows that the virion proteins (particularly VP-1) of strains ts^+ and Mahoney are not precisely identical, despite the fact that ts^+ is a descendent of Mahoney. The fact that a peptide's position in the gel pattern is closely related to molecular weight (15) has permitted the definition of the molecular weight of the Mahoney virion proteins (designated VP) as 35,000 (VP-1), 28,000 (VP-2), 24,000 (VP-3), and 5500 (VP-4) daltons (16).

Work similar to that illustrated in Fig. 4-1 has shown that the VP-1 component of ts^+ (VP-1 B) has a molecular weight of about 31,000, and that the VP-2 components of both Mahoney and ts^+ are actually duplex, with molecular weights of about 26,500 and 29,000. Not only are the VP-1 components of ts^+ and Mahoney different, but their proportion in the virion also differs reproducibly, as does the proportion of VP-2. The VP-1 component of ts-2 differs in size from that of its parent ts^+.

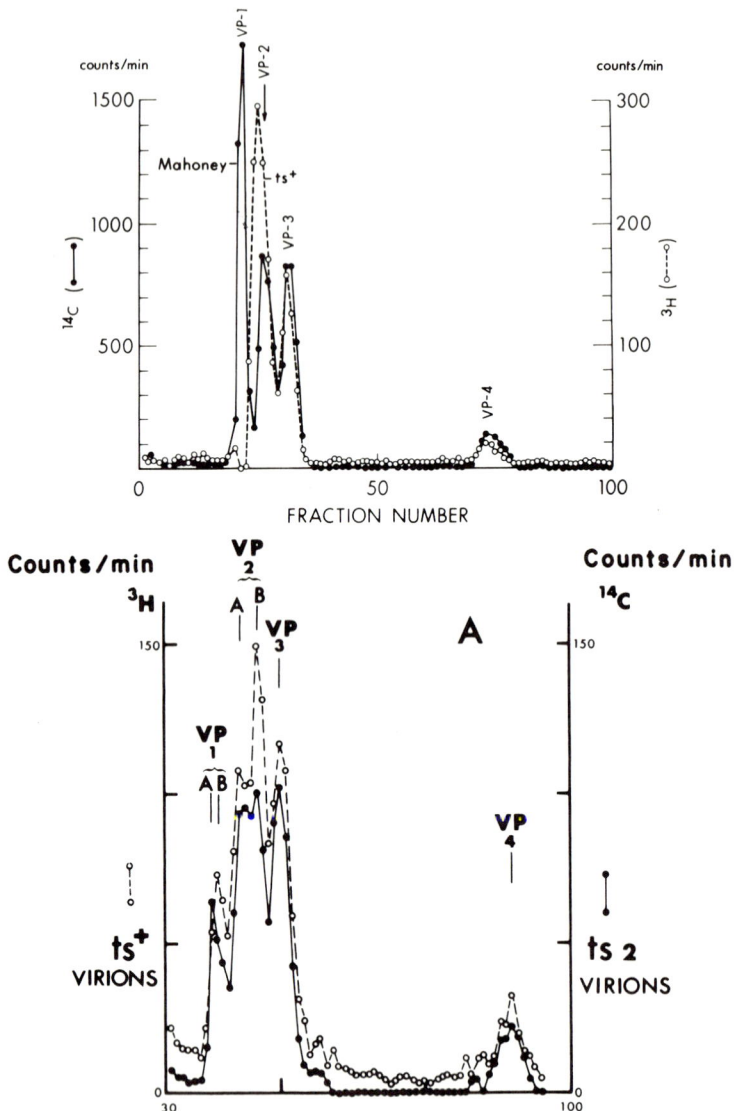

FIG. 4-1. The structural proteins of poliovirus type 1 (strains Mahoney, *ts*-2, and *ts*⁺) as compared by co-electrophoresis of labeled purified virions on an acrylamide gel (*14a*).

The total molecular weight of the 5 components in each virion is 120,000 (Mahoney and *ts*-2) and 116,000 (*ts*⁺). If it is accepted that the 2-3-5 symmetry of the crystallized icosahedral virion implies that the capsomers on the surface must occur in multiples of 60 (*19*), then an equimolar content of these proteins would involve a total weight of protein per particle of 7.2 or 6.96 $\times$ 10^6 daltons. This value is significantly in excess of the protein content of 4.2–4.8 $\times$ 10^6 daltons calculated from the observed particle weight of 6.2–6.8 $\times$ 10^6. Thus the components are probably not present in equimolar amounts, a conclusion implied by the virion structure of 20 hexamers and 12 pentamers suggested by Mayor (*19a*). However, it is difficult to reconcile even such a structure with the multiplicity of virion components actually found, and it seems likely at present that certain of these components have peptide sequences in common and are to a considerable extent mutually replaceable. A way in which this could arise is discussed below. Analyses of purified virus indicate that the average molecular weight of amino acids in Mahoney poliovirions is 118.0 (*17*) or 119.9 (*18*), so that the five virion proteins contain a total of about 1000 amino acids, equivalent to just half of the genetic information of poliovirus. The likelihood of some duplication of virion peptide sequences suggests that, in fact, rather less than half of the genome specifies virion protein.

However, some ten other proteins appear in the cytoplasm of infected cells [revealed by the same electrophoretic method after allowing the virus to suppress cellular protein synthesis in presence of guanidine (*14*)]; certain of them are resolved in Fig. 4-2, in which the difference between *ts*$^+$ and Mahoney in VP-1 is also apparent. These proteins appear to be virus coded but their total molecular

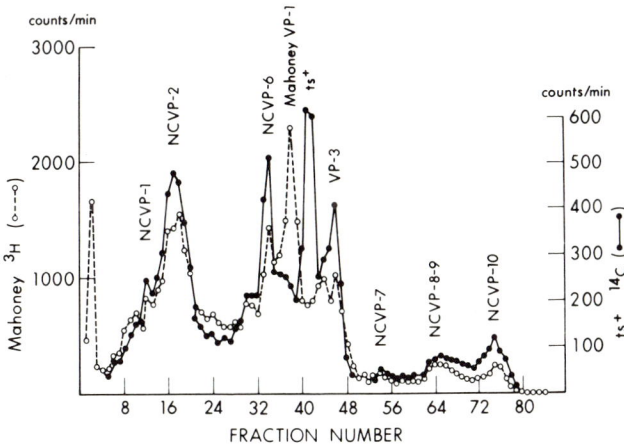

Fig. 4-2. Structural and nonstructural proteins of poliovirus type 1 (strains Mahoney and *ts*$^+$) as compared by co-electrophoresis of labeled infected-cell cytoplasms on an acrylamide gel (*14a*).

weight is greater than that to be expected from the information content of the viral RNA. It is unlikely that they represent de-repressed cellular proteins, or artifacts produced by aggregation before electrophoresis. The gel pattern of the nonstructural viral protein (designated NCVP) is reproducible, and their molecular weights are (in daltons): NCVP-*1*, 79,000; NCVP-*2*, 71,000; NCVP-*3*, 67,000; NCVP-*4*, 58,000; NCVP-*5*, 50,000; NCVP-*6*, 41,000; NCVP-*7*, 17,000; NCVP-*8*, 11,000; NCVP-*9*, 9500; NCVP-*10*, 6500 (*16*). Their total molecular weight is 410,000 daltons, which when combined with the 120,000 of the structural proteins is equivalent to about 4460 amino acids, or well over twice the information content to be expected. The simplest, indeed the only apparent, explanation for these findings is that some of the viral information is expressed more than once in these proteins, i.e., that certain peptides exist in more than one covalently linked form. In support of this, Summers and Maizel (*20*) have now shown that the largest protein (NCVP-1) is unstable in pulse-chase experiments, suggesting that it is a precursor for some other viral protein(s). This most interesting observation of apparent co-translation of viral proteins implies a novel method of genetic punctuation, and raises the question of what advantage, if any, such a complex departure from the "normal" has for virus growth. A clear need is to establish what are the primary translation products and their relation to the genetic map, what are the end-product peptides and what are the pathways of interconversion.

Jacobson and Baltimore (*20a*) and Holland and Kiehn (*20b*) draw the same conclusions as Summers and Maizel. In addition, Jacobson and Baltimore make the pregnant suggestion, based on several independent data, that mammalian cells may lack punctuation mechanisms for polycistronic messages. Certainly a system of random post-translational cleavage of a polypeptide precursor (presumably by a very specific peptidase) must produce many intermediates. These could readily account for the many poliovirus proteins found, as follows:

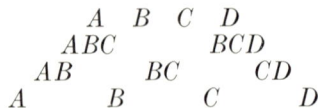

$$A \quad B \quad C \quad D$$
$$ABC \qquad\quad BCD$$
$$AB \qquad BC \qquad CD$$
$$A \qquad B \qquad C \qquad D$$

The diagram shows that a linear precursor $ABCD$, with three potential cleavage sites yielding the four final products A, B, C, and D by a series of random cuts, will involve nine different species of product. In general, n cleavage sites will yield

$$\frac{(n+1)(n+2)}{2} - 1$$

products, so that the 14 electrophoretic peptides of poliovirus would be exactly accounted for by only 4 cleavage sites. However, the balance sheet of genetic information and molecular weights of the gene products makes it likely at present that more cleavage sites exist, and that certain products are not yet resolved in the gel pattern. Incomplete cleavage of certain of the intermediates might account for the peptides with overlapping sequences that one expects to find in the virion.

C. The Functions of Poliovirus Proteins

Despite the intensity with which poliovirus has been studied, the primary functions of its nonstructural proteins remain almost entirely obscure. Even the precise roles of the six structural proteins, i.e., in maintaining virion structure as capsid or internal (core) protein, or proteins responsible for adsorption to cells or reaction with neutralizing antibody, are unknown. Nor is it known whether virion proteins have other parts to play in intracellular growth, although this possibility has often been considered (21, 22); they undoubtedly play a part in penetration, but apart from possessing some specificity for receptors on primate cell surfaces their function in initiating infection remains undefined, even in general terms. The existence of one nonstructural protein [NCVP-6, (23)] in the empty capsid-like top component of zonal centrifugation tests suggests that NCVP-6 may play a part in maturation or assembly, but this role is not yet demonstrated. It is equally likely that this is a sterile side-product, resulting from an incompletely cleaved virion protein that blocks incorporation of RNA.

Many complex changes in the infected cell, for example, decrease in polysome content (14, 24) and prevention of cellular synthesis of DNA, RNA, and protein (22, 25–32), follow quickly on infection by the virus, but no specific viral product has yet been implicated in any of them. It is not known which of these or how many other effects [e.g., the early nuclear changes (33, 34) and the late changes in membrane permeability leading to cell lysis (34a)] are primary or secondary effects of gene functions, and what role if any these effects play in virus growth. Under some circumstances, cellular RNA synthesis is stimulated by the virus [references in (35)]; choline incorporation is also stimulated (36). The sensitivity of poliovirus growth to actinomycin in insulin-starved cells (37) and the strain dependence of this phenomenon (38) remain a puzzle. It is to be expected that some portion of the viral genome is concerned with countering inter-

feron action, although this has not been shown, yet it cannot be assumed that all the genome is concerned with protein synthesis, although experiments examining alternative possibilities are lacking
also. Despite the existence of large polysomes involving viral mRNA
(*24*), the mechanism of their formation and function in circumstances where cellular polysomes are in decline is not known. The
factors that determine the fate of each progeny RNA molecule, i.e.,
whether it is to become a messenger, a template for new complementary ("minus" strand) RNA, or the RNA of a virus particle, are
quite unspecified.

Perhaps the function that can most directly be studied is the *in
vitro* RNA-incorporating activity of a fraction derived from poliovirus-infected cells (*7*). This fraction can catalyze incorporation of
uridine into both single- and double-stranded viral RNA. However,
although a variety of complexes of viral "plus" and "minus" RNA
strands and protein are found *in vivo* (*39–41*), the *in vitro* syntheses
appear often to be limited, and the identity, number, and nature of
polymerase and other functions involved are obscure.

Thus, although poliovirus RNA codes for many proteins and many
changes occur in infected cells, the molecular relations between the
two remain almost totally undefined.

4-3. The Methods of Genetic Analysis

A. The Types of Experiment

Three types of experiments can be made, namely tests for recombination, complementation, and physiological function. In recombination tests, mixed infection by mutants with markers in different loci
produces some recombinants, i.e., progeny carrying certain characters
of both parents. The further apart in the genome are the markers, the
higher should be the proportion of recombinants, and by crossing many
mutants a linear or circular picture (genetic map) of the sequences
and relative spacing of the markers can be drawn. In complementation tests, mixed infection by mutants with functional defects in different genes allows the mutants to multiply. Each mutant produces the
gene product lacking for the other, and growth is accordingly symbi-
otic or complementary. In the ideal situation for mapping purposes,
mutants with defects in the same gene will not complement each
other, and mutants can be classified into noncomplementing groups
that should correspond with each defective gene or cistron. In physiological tests, mutants are classified according to their functional dif-

ferences from the parent virus, and can, even without sufficient mutants to permit a classification, give a good idea of the various functions involved. The results of these various tests with poliovirus mutants will be returned to below; it should be noted that in practice the results are rarely as simple as has been suggested above.

For genetic analysis, small alterations (mutations) are made in the genetic material of the parental virus and their effects observed in the three genetic tests; a large number of mutants is needed. The alterations may be substitutions, insertions, or deletions of one or more nucleotides, but only one part of the genome must be affected at one time. This demands the comparison of pure clones of single-step mutant progeny with unaltered parent (wild type); adaptation of strains to new conditions and choice of natural strains with different properties (procedures often followed with animal viruses) give intolerable ambiguities because many genetic changes may have occurred. Even the chance isolation of a double mutant will cause much toil and trouble until it is recognized. Thus an important part of this work is to identify the number and general nature of the mutations involved.

B. The Markers Available for Poliovirus

A mutation is recognized by a change in a particular character (marker) of a virus. Probably more markers have been developed for poliovirus than for any other animal virus, initially to follow genetic changes in liberated vaccine strains. These markers are listed in Table 4-1.

Most of these characters have not been rigorously identified with viral functions, but the majority, especially of the nonselective markers, would appear to involve virion proteins. Several markers that seem to be concerned with virion proteins involved in adsorption to cell receptors (d, m, e, r, s, and probably MS and others) are likely to reflect different properties of the same or closely linked mutational events; others (e.g., certain ts, Δ, cy and g^+, as indicated below) may also be similarly related to one another. Probably only the ts character is "universal" in that it can be expected to occur in most or all genes, but many (e.g., cy, Δ, g, m) appear to exist in multiple alleles, that is, mutation at more than one locus can change these particular characters. Because of the complexity of the system, in particular the covariation discussed below and the likelihood of double mutations, the relationships between these markers are almost impossible to interpret until they can be related to specific gene products, preferably

TABLE 4-1

The Genetic Markers Available for Poliovirus

Character	Symbols in literature	References	Symbols used in this chapter
Selective markers (i.e., can be used to score progeny in the assay plate)			
Sensitivity/resistance to agar inhibitor at acid pH	d/d^+	96–98a	d/d^+
Sensitivity/resistance to agar inhibitor at neutral pH	m/m^+	55, 98a, 99	
Sensitivity/resistance to dextran sulfate inhibitor	m/m^+; ds, ds^s/ds^r	49, 100–105	m/m^+
Enhancement by/independence of dextran sulfate	$r?$; ds^-/ds^+	99a, 100–102, 105	r/r^+
Small plaque size	s	47, 48, 96, 106	
Inability/ability to multiply at high temperature	T/T^+; rct-40$^-$/rct-40$^+$; ts/ts^+ (originally hd/hd^+)	47, 50, 107	ts/ts^+
Ability to multiply at low temperature		84, 108	
Inability/ability to grow in MS cells	MS^-/MS^+	109	MS/MS^+
Cystine inhibition of multiplication	cy^i, cy^t	51a, 85, 110	cy^i
Cystine dependence/independence	cy^+/cy^i	85, 110	cy/cy^+
Tryptophan dependence/independence		111	
Adenine resistance/sensitivity		112	
Guanidine sensitivity/resistance	g^+/g	60, 115	g^+/g
Guanidine dependence	g^d	60, 113, 114	
HBB resistance/sensitivity	hbb^r/hbb^s	104, 115	
Sensitivity/resistance to normal bovine serum inhibitor	i^+/i; bo^+/bo	59, 60, 101, 116	bo^+/bo
Sensitivity/resistance to normal horse serum inhibitor	ho^+/ho	59, 60, 101, 117	ho^+/ho
Nonselective markers (i.e., need to be scored on isolated clones)			
Sensitivity/resistance to heat inactivation of virion	t; t^s/t^r; H^+/H, $-/\Delta^r$	60, 87, 118, 118a, 119, 120	Δ/Δ^+
Sensitivity/resistance to heating in AlCl₃	A^-/A^+	121	
Fast/slow elution from Al(OH)₃ gel		122	
Stronger/weaker adsorption to DEAE cellulose	E; e^-/e^+	123–125	e/c^+

TABLE 4-1 (*Continued*)

Character	Symbols in literature	References	Symbols used in this chapter
Stronger adsorption to calcium phosphate		*126*	
Intratypic antigenic variants		*127*	
Sensitivity/resistance to 7.2 *M* urea inactivation		*90, 91*	
Neurovirulence in monkeys	N, nv/nv^+	*47, 51a*	nv/nv^+
Morphological changes in CNS of monkeys	M	*47*	
Intracerebral neurovirulence in mice	mM	*47*	
Prevention of host-cell leucine ^{14}C incorporation		Table 4-2	pli/pli^+
Prevention of host-cell leucine ^{14}C incorporation in actinomycin		Table 4-2	$plia/plia^+$
Prevention of host-cell lactic acid synthesis		Table 4-2	pls/pls^+
Prevention of ^{32}P incorporation into host-cell 28S ribosomal RNA	ppi_1/ppi_1^+	*35*	ppi_1/ppi_1^+
Prevention of ^{32}P incorporation into host-cell 16S ribosomal RNA	ppi_2/ppi_2^+	*35*	ppi_2/ppi_2^+
Prevention of host-cell thymidine incorporation	pti/pti^+	*35*	pti/pti^+
Damage to cellular chromatin		Table 4-2	chr/chr^+
Induction of cell staining by trypan blue		Table 4-2	tb/tb^+
Prevention of cellular reduction of tetrazolium		Table 4-2	tet/tet^+
Prevention of ^{32}P incorporation into host 4-10S RNA	ppi_3/ppi_3^+	*35*	ppi_3/ppi_3^+
Production of antigen by fluorescent antibody test	a/a^+	*35*	a/a^+
Maturation efficiency of infectious RNA		*78*	
Maturation efficiency of serum blocking antigen		*78*	

via the genetic map. Unfortunately, many of the strains referred to in Table 4-1 have been obtained by serial adaptation of parental stocks to various altered growth conditions, and therefore are not likely to fulfill the desirable criterion of containing only a single mutation. The only character extensively examined for singularity is the *ts* character (see below).

In general, it is desirable to increase the mutation rate by a mutagen, not only to make it easier to find mutants but also to ensure that the isolates are not "sisters" or preexisting double mutants, and to know the general nature of the mutational changes. Rather few mutagens have been used with poliovirus, most of the clonal isolates having arisen spontaneously. The $d \to d^+$ conversion was enhanced by proflavine (*42, 42a*); this mutagen is usually presumed to act on DNA by intercalation leading to a frame-shift mutation (*43*), but in the case of poliovirus it also has a photodynamic effect on the RNA like that of neutral red (*44, 45*). Perhaps in this case the mutagenic effect is light-mediated and of a different mechanism. The most frequently used mutagen is nitrous acid (*46–49*), leading to many types of mutant including *ts* (*49a*). However, perhaps the most potent mutagen is 5-fluorouracil (*50*), which at high doses converts the majority of the progeny to *ts*, and probably other, mutants. Hydroxylamine was relatively ineffective in making poliovirus *ts* mutants (*49a*); ultraviolet light does not appear to be a mutagen for single-stranded riboviruses.

Most of the mutations are likely to be of the missense type, in which one amino acid is substituted for another in the gene product. No animal virus mutants of the nonsense (e.g., "amber") or frame-shift types have yet been identified, nor has any genetic deletion been shown, although many defective strains of virus are known and several nonpermissive host cells have been described (*3*).

C. Problems in Handling Poliovirus Mutants

Several difficulties have arisen in handling poliovirus mutants and most of these problems either do not seem to have a counterpart or are less severe in the larger DNA bacteriophages. The first difficulty (*51*) was of pleiotropism or covariation, in which an alteration in one of seven characters (cy^i, *cy*, Δ, *d*, *ts*, *MS*, *nv*) would often but not always accompany an alteration in another. Two characters (*ho* and *bo*), however, were always independent. Probably all of these characters were associated with virion proteins. It is likely that some of the covariation was the result of double mutations, some the result of a dual expression of the same mutation (identity of genetic determination of two different characters), and some due to suppression, in which a change in one part of a gene alters the whole configuration of the gene product and thus suppresses the activity of another part of the gene.

Some clear examples of covariation between a number of structural protein characters (certain *ts*, Δ, *cy* and *g*⁺ markers) are given below in Fig. 4-6. In these cases, the original *ts* mutation was shown to be single and the covariation to be due to that mutation by a second change in the covariant characters when the *ts* mutation reverted to *ts*⁺. Recombination in poliovirus was not demonstrated until non-covariant characters (*ho*, *bo*, or *g*) were used. Another form of pleiotropism (asymmetric covariation) was met with *ts* mutants (see below).

A second problem, for *ts* mutants the most troublesome, was that of reversion. The reversion frequencies of the most stable *ts* mutants of poliovirus were about 10^{-4} per particle per duplication (*52*); probably the most frequently isolated type of *ts* mutant was even less stable (10^{-1} or 10^{-2} per particle per duplication). It was necessary to retain all but the most unstable mutants, since with stringent selection criteria one might finish up solely with double mutants. Because the genome is small, the chance of effective mutations is probably also small, and so mutants that are more leaky or unstable than those of large bacteriophages, for example, may have to be accepted. The *d* to *d*⁺ reversion frequency was also 0.5–1×10^{-4} per particle per duplication (*42*), as were the frequencies of formation of plaque size mutants of EMC virus (*53*) and of *ts*⁺ revertants of the ribophage fCan1 (*54*). The frequencies of mutation from the *m* (minute plaque) to *m*⁺ for a type 2 poliovirus were 3×10^{-7} per particle per duplication (*55*), for type 1 poliovirus *ts* to *ts*⁺ were 2–25×10^{-6}, and for type 1 guanidine-sensitive to -resistant were 7.5×10^{-8} to 3.8×10^{-6}, per particle per duplication (*56*), but whether or not these low frequencies represent single-step mutations is not known.

Numerous re-clonings did not produce revertant-free stocks of poliovirus *ts* mutants, but gave stocks that varied up to 100-fold in revertant content. This probably reflects the fact that, by chance, the first reversion occurred at different times in different clones. It is highly likely with such unstable viruses that no clone is pure, and even possible that no two RNA strands in a population are precisely alike; many of the mutations will be lethal, probably accounting for a proportion of the "noninfective" particles usually found in poliovirus preparations. The *ts*⁺ content of poliovirus *ts* mutant stocks was minimized by subculturing with the largest *ts*⁺-free mutant inoculum (50–

500 PFU), by selecting those cultures that by chance contained least *ts*⁺, and by limiting the total stock made to 1–5 × 10⁹ PFU (*52*). Many of the *ts*⁺ revertants, even those derived from single mutants, did not show a fully wild phenotype, indicating the presence of a suppressor effect.

Other problems include "leakiness," or incomplete penetrance, whereby some wild-type function appears under restrictive conditions despite the mutant state. Because of the need to avoid double mutants, and also because a completely defective gene may nevertheless allow some virus to be made at low efficiency, a fairly high proportion of leaky *ts* poliovirus mutants was retained (*35*). It is not known whether all infected cells produce a small leak yield or a few cells produce a full yield. The possibility of leak has always to be taken into account before concluding that a mutant is not different from the wild type in a particular character. Single mutants of a type that can be predicted to be nonleaky (nonsense, frame-shift, or deletion) have not yet been demonstrated for animal viruses, although many animal-virus characters do have complete penetrance; it is clearly desirable to find out what kind and what number of mutations are involved in such characters.

4-4. Results of Recombination Tests: The Genetic Map of Poliovirus

A preliminary suggestion of genetic interactions in poliovirus was given in a report of multiplicity reactivation of UV-inactivated virus (*57*). However, although several alternative explanations were shown not to cause the small increases in infectivity found in multiple infection, no mention was made of the possibility of aggregation of inoculum particles. It has often been observed by the writer that the infectivity of poliovirus preparations may be increased by contact with cells (e.g., Fig. 5 in Ref. *58*), and this effect may be suppressed with increase in titer by pretreatment of the inoculum with pH 2.5 glycine buffer. It seems that the validity of the observation of multiplicity reactivation in poliovirus should be re-examined in the light of possible effects of aggregation.

The first report of recombination in poliovirus was given by Hirst in 1962 (*59*). The markers used were the noncovariant *ho* and *bo*, the strains were stable multistep mutants, and the double *ho bo* mutants in the progeny were 15 times more frequent than in the self-crosses. The recombination frequency was 0.37%. Ledinko (*60*) repeated these experiments using *ho* and a stable character obtained by development of several strains resistant to guanidine (g). The ho and g strains showed a consistent recombination frequency of 0.4% ho g double mutants that was 15–20 times the spontaneous frequencies in the self-crosses (3×10^{-6} for ho, and 6×10^{-4} for g). One strain (ho) was developed to be resistant to heat inactivation *in vitro* (ho Δ^+), and crosses between ho Δ^+ and g still gave ho g recombination frequencies of 0.3–0.4%, with almost all recombinants now Δ^+. This greatly strengthened the evidence that the excess of double mutants was not caused by an increase in spontaneous reversion, and so, like older genetic systems, this excess could then safely be presumed to involve genetic recombination. This test was incidentally a form of three-factor cross, giving a presumptive sequence in the genetic map of g-ho-Δ^+, or at least of close linkage of ho and Δ^+. It is likely that ho and Δ^+ are located in virion protein genes, as is g (see below).

Genetic recombination using Ledinko's method was found between *ts* mutants of poliovirus (*61*), but attempts to obtain a self-consistant genetic map by such procedures were not at that time successful. The reasons for this were found to be 4-fold (*52*). First, the virus was usually aggregated, giving variable adsorption and recombination characteristics; this was overcome by disaggregation with pH 2.5 glycine buffer. Second, the unstable character of the mutants mentioned in the previous section meant that even repeatedly cloned preparations had high and variable revertant contents; this difficulty was bypassed by selecting randomly obtained "clean" stocks of very small size from the most stable mutants. Third, criteria for strict control of assay conditions and selection for "nonleaky" mutants enabled ts^+ recombinants to be assayed reproducibly while losing <20% of ts^+ plaques; mutant plaque formation was kept below 10^{-4}, so that it was possible to detect 0.01% recombinants in some crosses. Fourth, and most important, 60% of the 15 mutants that were sufficiently stable and nonleaky to be used in these tests were identified as double mutants. It was essential for these to be found and rejected; their non-singularity was shown by mapping and physiological tests. It is the writer's expectation from these results that any *ts* ribovirus mutant that fails to show significant revertant or "leak" plaques under the restrictive condition is almost certain to be a double mutant. Until this fact is known such a mutant is a great source of confusion.

When these difficulties were overcome, it was possible to obtain an additive linear genetic map of poliovirus *ts* mutants that comprised one linkage group (*52*) (Fig. 4-3). The mean recombinant frequencies were reproducible and characteristic of each pair of mutants (Fig.

FREQUENCY OF ts⁺ RECOMBINANTS (%)

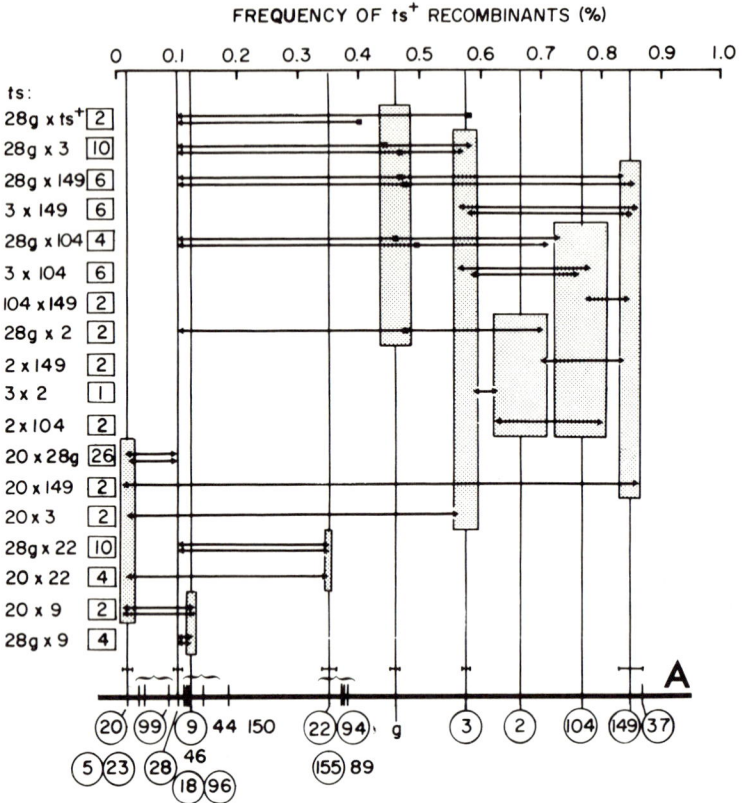

FIG. 4-3. Genetic map of poliovirus *ts* mutants (Ref. *52*). The solid squares show the frequency of *g* recombinants, and the stippled boxes show the discrepancies in the additivity.

4-4), being between 5.2 and 31.4 times the total background reversion rates and ranging from 0.02% to 0.85% of the total progeny.

A major factor contributing to this map was the obtaining of a *ts* mutant (*ts*-28) in a guanidine-resistant form without altering its *ts* defect. This enabled many sequences to be determined by a type of three-factor cross, and several mutants to be mapped that would otherwise have been too leaky or too unstable. It also allowed this particular guanidine-resistance locus (*g*) to be accurately sited at about the middle of the map (Fig. 4-3); *ts*-28*g* crosses with *ts⁺* and four *ts* mutants gave identical map locations for *g*. The physiological significance of this map is returned to in subsequent sections of this chapter.

Bengtsson (*62*) has used the comparatively high recombination between *g* and *ho* strains to show the relative locations of certain Δ and *m* markers in three-factor crosses, which markers were all found to lie close to *ho*. Like the *ts* mutants, Bengtsson's strains were developed from the type 1 vaccine strain LSc (which is $g^+ \cdot ho^+ \cdot \Delta \cdot m \cdot ts$); multi-step *g* and *ho* strains were first separately selected, from each of which m^+ and Δ^+ variants were independently adapted. His recombination frequencies were variable (0.024–0.4% *g ho* double mutants) and so cannot yet be firmly related to the *ts* map, but one experiment gave frequencies as high as that found for *g ho* recombinants by Ledinko (*60*). Recombination between ts^+ revertants was also studied, but the results suggested only that LSc contains several *ts* defects. The same conclusion was drawn from crosses between LSc and several *ts* mutants (*49a*); partially *ts* recombinants only were obtained, a finding that failed to allow the several *ts* defects of strain LSc to be mapped.

Hirst (*62a*) has also used three-factor crosses to show that the sequence in the map is *g-ho-bo,* and the markers used were approximately equidistant.

It is clearly desirable to establish the relation between the various poliovirus markers. A possible relation between the three maps is given in Fig. 4-5, which shows that the three sets of data are mutually

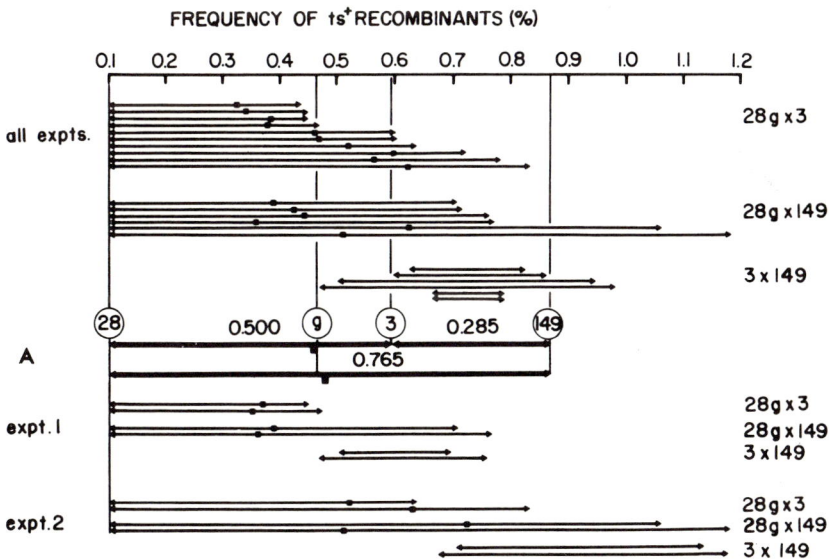

FIG. 4-4. Frequency of *ts*+ recombinants in the crosses *ts*-28g × *ts*-3, *ts*-28g × *ts*-149, and *ts*-3 × *ts*-149 (Ref. *52*).

FREQUENCY OF $\underline{ts}^+$ RECOMBINANTS (%)

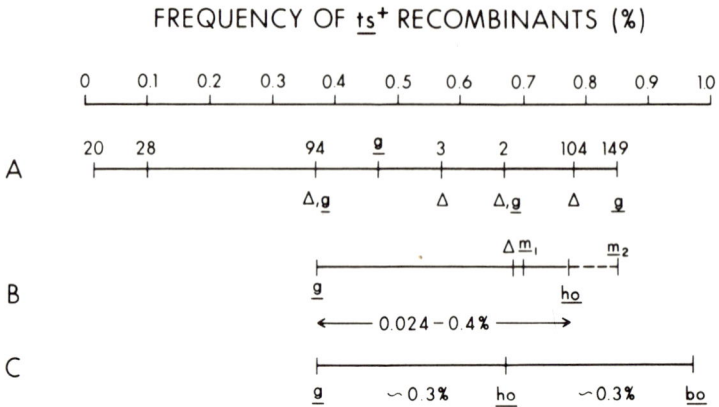

Fig. 4-5. A possible relation between the *ts* genetic map of poliovirus (A), and the maps developed by Bengtsson (B) and Hirst (C).

compatible, but unfortunately the correct polarities, relative spacings, and exact correlation can only be guessed at the moment. It is noted in the figure that several *ts* mutants also represent Δ and *g* markers. One difficulty will be that many characters are likely to exist in multiple alleles, necessitating the comparison of several mutants carrying each marker in the same system.

Ledinko (*60*) presented evidence suggesting that the proportion of poliovirus recombinants may increase almost 2-fold during growth. This was reexamined under conditions where a relatively small increase could be shown to be statistically significant (*52*). It was found that the proportion of recombinants did increase significantly ($p < 10^{-4}$), but only by 50% during the period 3.1–7.0 hr after infection. The large majority of RNA replicative events occurred between these times, indicating that the mating events are not obligatorily linked with replicative events. An increase in recombination during replication does not necessarily imply that some progeny RNA can take part in recombination, but only that there is still a significant chance of recombination between possibly preexisting mating forms in the second half of the cycle; progeny strands could not as a rule be available to participate in mating. In fact, the marked nonrandomness with respect to time suggests that multiple rounds of mating do not occur to any great extent, and also that single-stranded RNA is not the mating form. Whatever recombination mechanism is proposed, one must suppose that some type of double-stranded or complementary structure must be involved in recombination in order to provide the recognition sites to maintain continuity of the genetic message. It is to be wondered which genome (host or virus) specifies the enzymes presumably involved in RNA recombination; small amounts of double-stranded RNA have been reported in uninfected animal cells [(Montagnier, Vigier, and Goldé, quoted in reference (63)].

Although doubly defective *ts* recombinants were not looked for, the self-consistent nature of the *ts* map suggests that recombination was reciprocal and that both parental strands were involved. The maximum frequency found was 1.7% of presumptive total recombinants. The amount of recombination to be expected from about 6000 nucleotide pairs of bacteriophage λ (64), and for bacteriophage T_4D (65) for a single round of mating, was also 0.5–3%; Tessman (66) found a maximum of 0.6% presumptive total recombinants for the small single-stranded deoxyribophage S13. These similarities between phage and poliovirus recombination rates seem remarkable when account is taken of the large qualitative and quantitative differences between the two types of cell-virus system; they suggest a close resemblance between the fundamental mechanisms involved.

The question arises as to what proportion of the total genome of poliovirus is represented by the map of Fig. 4-3. The amount of recombination obtained (0.85% ts^+ recombinants) suggests, in comparison with the other systems mentioned in the previous paragraph, that it is a significant proportion; the amount of genetic information represented by the mutants, as discussed below, would suggest that it is about half. However, only about four or five functions have been tentatively identified, and the distribution of the mutants in the map of Fig. 4-3 is interestingly nonrandom. It is possible that map distance is not proportional to nucleotide number in this system, and that the highly polar structure represented by the replicative intermediate of riboviruses might encourage a higher rate of recombination at one end than at the other, leading to an asymmetrically condensed form of map. The clearest answer to this is likely to lie in the identification of different map regions with peptides of known molecular weight.

4-5. Results of Complementation Tests with Poliovirus Mutants

Mixed infection of various poliovirus strains has given quite effective complementation provided that the restrictive condition is guanidine sensitivity or dependence (67–69). However, this restriction

really provides asymmetric rescue rather than reciprocal complementation, since one of the strains is always free to grow; if the
guanidine-induced defect lies in capsid proteins, as is suggested below,
then it is possible that the restricted strain is rescued by a process
like phenotypic mixing.

Phenotypic mixing between poliovirus type 1 and type 2 strains
is actually highly efficient (70), as is the recombination between *ts*
mutants (mentioned above) when account is taken of the small size
of the genome, and so it does not seem likely that the replicating units
are segregated from each other to any marked extent. Nevertheless,
the degree of complementation found between a large number of
poliovirus *ts* mutants was very small (49a). In this case, both mutants
were under restrictive conditions, that is, neither could grow. One such
complementing pair (*ts*-5 and *ts*-19) was studied in some detail (71).
It was shown that complementation reproducibly occurred at 39.5°C
up to 14 times the background leak rate but only to some 0.1% of *ts*⁺
yields, and only one strain appeared in the yield. Variation of experimental conditions did not significantly improve the complementation. Strains *ts*-5 and *ts*-19 have subsequently been shown to be double
mutants (52), but the same low efficiency was found with many other
pairs of single mutants with different phenotypes. This contrasts with
complementation efficiencies of up to 5% of wild-type yields with
Sindbis and Semliki Forest viruses (72, 73), and 20% or more with
poxviruses (74). Such low values in the poliovirus system were insufficient to distinguish between inter- and intragenic complementation. In addition, some mutants could be shown to have interfering
properties at the restrictive temperature (75, 76), which render dubi-
ous the interpretation that the few pairs that did not show complementation belong to the same cistron.

At present, therefore, the complementation test is of no value for
demonstrating cistron groups in the poliovirus system, for reasons
that are currently inexplicable. It is possible that the unusual "co-
translation" of several gene products, implied by the work of Summers, Maizel, and Darnell (14, 20) discussed above, might drastically
affect complementation efficiencies between certain cistrons.

4-6. Results of Temperature-Shift Experiments with Poliovirus Mutants

If the defective function of a *ts* mutant is important early in the
growth cycle (for example, in a stage of penetration or in some

activity needed to "prime" the cell for replication of viral nucleic acid), then a shift from permissive to restrictive conditions after this function has been expressed would no longer inhibit growth of the mutant. Similar considerations would apply to a shift from restrictive to permissive conditions, and such changes made at intervals during a one-step cycle can be used to determine the stage of growth affected by the *ts* defect.

Temperature-shift experiments with poliovirus *ts* mutants *(35)* showed that all but one of the mutants (*ts*-23) were defective in 'late' functions. This was the case whether the experiments were of the temperature step-up or step-down type. Thus the mutants were defective in functions that were still needed for new virus formation at a time when virus maturation had already begun. The gene products responsible for these functions must be presumed either to be heat labile or always to be freshly made at the time of use; an RNA polymerase or coat protein would be suitable candidates. However, a brief pulse of restrictive or permissive temperatures had relatively little effect at any time of the cycle for *ts*-23 or any of the others. It is therefore presumed that no particular time is critical for any gene function; i.e., all genes represented are in action continuously during the growth cycle and that division into "early" and "late" phases in terms of type of function is not valid, at least for the mutants so far available. This concept is also suggested by the finding that replication of RNA is detected as soon as 1 hr after infection *(77)*.

The "early" mutant *ts*-23 (which happens also to be a double mutant) presumably specifies a *ts* gene product that is heat stable once formed and is only *ts* if synthesized at 39.5°C. It is probably defective in the same function as *ts*-20, -5, -81, and -99, which it resembles in phenotype (see below) and map position (Fig. 4-3).

4-7. Results of Physiological Tests with Poliovirus Mutants

Some 33 out of 155 *ts* poliovirus isolates were sufficiently stable and nonleaky to examine by a variety of biochemical tests. The main comparison was between wild type (*ts*+) and mutant at restrictive temperature (39.6°C), since the important factor was the change induced by the *ts* mutation. Controls of mutant and *ts*+ at permissive temperature (37°C) were also included to confirm that the mutant behavior was significantly temperature sensitive and that *ts*+ behavior was not. It can be expected that some mutant gene products

TABLE 4-2
Characters of Poliovirus *ts* Mutants[a]

	20	99	5	23	81	28	18	46	96	63	9	44	150	19	94	22	89	155	147	151	123	3	2	104	149	37	*ts*+
Map position	0.015	0.02–0.1			<0.1 0.10		0.10–0.12				0.12–0.14			0.37					0.3–0.4		>0.3	0.57	0.67	0.78	0.85	0.86	
Double or single mutant	S?	D	D	D	S?	S	D	S?	S?	S?	D	S?	S?	D	S?	D	S?	S?	S?	S?	B	S	S	S	S?	D	
Phenotype (physiological group)	A	A	A	A	A	B	B	B	B	B	B	B	.	C	C	B	C	C	.	.	B	D	D	B	.	C	
pli (prevention of leucine ¹⁴C incorporation)	+	±	±	±	±	±	±	+	+	+	+	+	+	+	+	+	±	+	±	±	±	±	−	+	+	+	+
plia (prevention of leucine ¹⁴C incorporation in presence of actinomycin)	+	±	±	±	±	+	±	+	.	+	+	.	+	+	+	+	±	+	±	+	+	.	.	±	+	+	+
pls (prevention of lactic acid synthesis)	+	+	+	+	+	.	.	.	.	+	+	.	.	+	+	+	+	+	.	.	+	.	±	+	.	.	+
ppri, prri (prevention of ³²P phosphate incorporation into ribosomal RNA)	+	+	+	.	.+	+	+	+	+	+	+	+	.	+	+	+	+	+	.	.	+	+	+	+	.	+	+
pti (prevention of thymidine incorporation)	+	+	+	−	−	+	+	+	+	+	+	+	+	+	+	+	+	+	.	.	+	+	+	+	+	+	+
chr (damage to cellular chromatin)	−	−	−	−	−	.	. +	+	+	+	+	.	.	+	+	+	.	.	.	.	+	+ .	+ +	+	.	+ .	+
tb (induction of cell-staining by trypan blue)	−	−	−	−	−	. +	+	+	.	+	+	.	.	+	+	+	+	.	.	.	.	.	+	.	.	.	+
tet (prevention of cellular reduction of tetrazolium)	±	±	±	.	±	+	+	+	+	+	+	+	.	.	.	+	.	.	.	.	+	+	.	+	.	+	+

Genetic marker / property	Values across mutants
ppi₃ (prevention of ³²P phosphate incorporation into 4S–10S RNA)	+ + − − + − − + − − − − − − − − − − −
a (production of antigen by fluorescent antibody test)	+ − − − − − − − − − − − − − − − − − −
Serum-blocking antigen production (%)	100 0.9 — 27.6 2.5 — 6.1 — 11.5 2.8 7.6 1.8 3.0 3.0 2.9 3.4 0.6 — 5.6 3.5 0.6 6.3
Infectious RNA production (%)	100 3.1 <1 88 3.3 — 44 6.8 14 60 22 7.4 4.0 10 5.2 0.35 0.22 3.4 2.9 3.0 3.0 0.09
Interference with ts⁺ at 39.5°C	+ · − + + − + + − + + + + + + + + + + + + +
iRNA maturation efficiency	+ − − − − · − − − − + + · · + + + + · + + +
SBA maturation efficiency	+ · − − − · + − − − + + · · − · · + + + + ·
Δ45°C (stability at 45°C)	+ + + − − − − + + − + + + + + + + + − + − +
Δ39.5°C (stability at 39.5°C)	+ · − · − − · · · ± + · · · · · · · · · · ·
cy (cystine dependence)	+ + + − − − − − − + + + + + + + + + + + + +
g/120 (sensitive to 120 μg/ml guanidine)	+ · − + · · · − · − + + · · · · · · · · · ·
Defect in temperature shift experiment (early or late)	· L L L · L L L L L · L · L L E L L · · · ·
Production of double-stranded RNA	+ · + + · · · · + · · · · · · + · · + · · −

ᵃ + = wild type; − = mutant; ± = values between those designated + and −; · = not determined.

will also perform rather inefficiently at 37°C, and such small defects in mutant behavior at 37°C were occasionally found.

The results of these tests are given in Table 4-2, and the remainder of this chapter is concerned with them. Mutants known to contain double *ts* defects are also considered in case any serious anomaly is revealed.

A. PRODUCTION OF INFECTIOUS RNA AND SERUM-BLOCKING ANTIGEN

Tests for infectious RNA synthesis at 39.6°C showed that no mutant produced fully wild-type yields, i.e., all were somewhat *ts* (*78*). However, some produced large yields (up to 88% of *ts*⁺), some produced no detectable new RNA (<0.5%), and some gave intermediate yields (2–10%). Table 4-2 shows that large producers are to the right in the genetic map, and, with some exceptions, mutants producing little RNA are to the left. The exceptions (*ts*-19, -22, -89, -104, and -149) also have defects in virion protein, and are discussed below in Sec. 4-10.

Thus there was no clear separation into RNA⁺ and RNA⁻ mutants. The reasons for this are probably 2-fold. First, some mutants (e.g., *ts*-150) are somewhat leaky, producing rather large amounts of virus and RNA at 39.6°C. This may be a matter of chance selection, or it may result from the type of defect involved since it is possible that a completely defective gene may yet allow some RNA synthesis at low efficiency. As mentioned before, relatively leaky mutants were not excluded from the mutant series because of this possibility. Secondly, other mutants (e.g., *ts*-89) will interfere with *ts*⁺ at 39.6°C, and, as elaborated below, this effect is likely to reduce their own yield of RNA. Thus information on the absolute amount of RNA produced may by itself be misleading (see next section).

Tests for production of serum-blocking antigen (*78*) gave results similar to the RNA tests, except that the largest amounts produced came only to 27% of *ts*⁺ values. Once again the absolute amount produced was not very informative (see next section).

B. MATURATION EFFICIENCIES AND OTHER CRITERIA OF ASSEMBLY DEFECTS

Because some mutants may be more leaky than others, and some may show autointerference, the absolute amount of RNA made at

39.6°C is not a reliable indication of type of defect. More valuable is the efficiency with which this RNA is converted to virus: clearly a mutant that produces much RNA and little virus has some defect in maturation or assembly, in contrast to a mutant that converts its yield of RNA efficiently to infective virus. The efficiencies with which the mutant infective RNA formed at restrictive temperature were converted to infective virus (i.e., the ratio of virus:RNA PFU) were compared with those of ts^+ (78), and the results are summarized in Table 4-2. In this case, a fairly clear distinction between wild and defective phenotypes could be made. A maturation efficiency of 0.44–1.73 ($ts^+ = 1$) was regarded as not significantly different from 1 and is scored as wild type ('+'), while a maturation efficiency of $\leqslant 0.14$ is scored as defective ('−'). Those mutants producing negligible new RNA could not be scored. Of the mutants to the left of ts-19, five matured their RNA efficiently and two did not. The two exceptions (ts-99 and -9) have been shown by mapping experiments (52) to be double mutants, with a minor ts mutation to the right. Mutant ts-19 and eight mutants to the right of it did not mature their RNA efficiently, and accordingly possessed some defect in maturation; most of these were probably single mutants. An exception is the single mutant ts-104, which has no defect in maturation efficiency; however, it shows other evidence of a virion protein defect (see below), and such a defect need not necessarily show up as a low efficiency of maturation. Thus in general the right-hand region is concerned with infective virion assembly, and the left-hand region is apparently not.

A similar conclusion is drawn from tests for serum-blocking antigen. Four mutants to the left of ts-19 (all single mutants) converted their small yield of antigen efficiently to infective virus, while the two that did not (ts-5 and -9) are double mutants with minor ts defects to the right. Mutant ts-19 and three mutants to the right (two of them single mutants) were defective; in this case ts-155 and -104 were exceptional.

The presence of defects in the virion itself could in some mutants be shown by the effect of heat (15 min in buffer at 45°C) on mutant virus grown at 37°C (79). This treatment had little effect on ts^+ infectivity, and the behavior of mutants to the left of ts-19 was indistinguishable from ts^+ (the exceptions were the double mutants ts-99 and -18 which have minor defects to the right). However, mutants to the right of ts-19 were (except for ts-149, -89, and -37, and ts-19 itself) 10 to 100 times less stable than ts^+, and therefore differ in structural protein from the wild type. Once again, the exceptional mutants may be ignored, since a *ts* virion protein defect need not necessarily show up as a defect in thermostability. In all cases tested, the mutants unstable at 45°C were also unstable at 39.5°C, and so thermolability of the virion itself is sufficient to account for their *ts* defect. The possibility that the thermolability at 45°C was unrelated to the *ts* defect was ruled out by showing that most *ts*⁺ revertant clones isolated from the thermolabile mutants had also reverted to thermostability.

Thus all of the single mutants mapping to the right of *ts*-19 are shown by one or more of the three tests for maturation efficiency or thermolability to possess defects in assembly or in virion proteins. Accordingly, the whole of the right-hand segment of the map is concerned with specifying structural or other assembly protein. The most that can be said for the single mutants to the left of *ts*-19 is that they have all failed to show such defects, and therefore are likely (but not proven) to have defects in other functions. The double mutants present no anomaly, and support these conclusions without exception.

C. Defects in Synthesis of RNA

The previous section shows that the single mutants to the left of *ts*-19 (*ts*-20, -81, -28, -46, -96, -44, -150) have all failed to exhibit a sign of defects in assembly proteins. Nevertheless, the temperature-shift experiments indicate that their defective function is "late," i.e., needed for formation of new progeny at a time when progeny is already being made. As they fail to make much infectious RNA at 39.5°C, it is therefore likely that this function is directly concerned with RNA synthesis.

These mutants form the two physiological groups A and B (*35*, and below). The species of viral RNA made by the two representative mutants *ts*-20 and -28 were accordingly examined (*80*). Step-up temperature-shift experiments, which involve a 2-hr incubation at 37°C before transfer to 39.5°C, were used to ensure the presence of adequate "plus" and "minus" strands as templates before label was added. Both mutants produced the normal pattern of RNase sensitive 35S (viral) and RNase resistant 16-22S (replicative intermediate) RNA at 37°C (*39, 40, 81*). At 39.5°C *ts*-20 produced practically no new RNA of either 35S or 16-22S, but *ts*-28 produced a normal yield of RNase-resistant 16-22S material and small amounts of RNase-sensitive RNA of widely varying S value, including some apparently degraded acid-insoluble material of small molecular weight; ts-81 resembled ts-28.

Thus, ts-20 behaves as though defective in a function needed to make or maintain the double-stranded form, whereas ts-28 behaves as though defective in a function needed to make or maintain intact progeny $35S$ viral RNA. It is interesting that the defects of ts-20 and -28 are so distinct; either defect suggests that more than one virus-coded activity is involved in viral RNA synthesis. However, there is no direct evidence yet that either of these functions are RNA polymerases; the *in vitro* RNA incorporating system (7) of ts-20 appears to be thermostable (82), although it does differ from that of ts^+ in its temperature optimum.

Several single mutants with defects in assembly proteins also appear to have major defects in RNA synthesis (e.g., ts-104). They are further discussed below in Sec. 4-10.

D. Characters Showing No Significant Defect

A variety of viral genetic characters was examined for as many mutants as possible in an attempt to relate primary gene functions to the observable effects of poliovirus on the infected cell. These characters, some of which have been reported previously (35), fall into the four groups A–D described in that paper.

Table 4-2 shows that a number of characters failed to reveal a ts defect with any mutant, despite the testing of some 33 mutants, a number theoretically sufficient to place at least one mutation in each of the presumed 10 genes of poliovirus. Principal among these characters was the prevention of synthesis of cellular 28s and 16s ribosomal RNA (ppi_1 and ppi_2 characters), which effect is initiated soon after infection (30, 31).

Other characters that failed to show significant defects were the prevention of cellular protein synthesis [measured as prevention of leucine incorporation, or pli character (83)], the prevention of lactic acid synthesis that occurs 1–2 hr after infection [pls character (83)], and the destruction of cellular polysomes (24) measured as the increased decay of leucine incorporation in presence of actinomycin [$plia$ character (83)]. These characters are summarized in Table 4-2 (the apparent defect in pli for ts-2 is likely to be spurious, influenced by the large amount of viral protein made by this mutant).

The nature of all these characters is such that all could conceivably stem from a single gene function of the virus, or could be initiated by structural protein(s) introduced by the inoculum virions.

E. CHARACTERS COVARIANT IN PHYSIOLOGICAL GROUP A MUTANTS

Physiological group A was defined by a defect in prevention of cellular DNA synthesis [pti^+ character (35)]. Five such mutants were found, all restricted to the extreme left of the genetic map. Several other characters (see Table 4-2) were found to be covariant with pti (83): the chr character, or ability to change the structure of host-cell chromatin, apparent 2–4 hr after infection; the tb character, or ability to induce cellular staining with trypan blue, apparent 12–16 hr after infection; the tet character, or ability to prevent cellular reduction of 2-iodophenyl-3-nitrophenyl-5-phenyl tetrazolium chloride, apparent 4–8 hr after infection. The tet character appeared to polygenic, and group A mutants were only partially defective. In addition, group A mutants were defective in the characters ppi_3^+ and a^+; this is discussed below in Sec. 4-8 on asymmetric covariation.

F. GROUP B MUTANTS

Physiological group B was defined by an apparent defect in the prevention of cellular-fraction-3 RNA production [ppi_3^+ character, (35)] in the absence of a defect in pti^+. Eleven such mutants were found, most of them mapping between group A mutants and ts-19; three (ts-22, -123, and -104) occurred in the right-hand half of the map, however. The nature of the "fraction 3" RNA whose production is permitted at 39.5°C is not known; it may be a breakdown product of cellular messenger RNA or of viral RNA. In addition, group B mutants were defective in the a^+ character (see Sec. 4-8).

G. GROUP C MUTANTS

Physiological group C was defined by a defect in antigen synthesis [a^+ character, (35)] in the absence of defects in the pti^+ or ppi_3^+ characters. Five mutants were found, four of them (ts-19, -94, -89, -155) located in the middle of the map, and all were identified as defective in virion proteins.

H. GROUP D MUTANTS

Physiological group D was defined by the absence of a defect in production of antigen [a^+ character, (35)]. Two mutants (ts-2 and -3) were found, located together to the right of the map. They are both defective in virion protein, and for ts-2 the a^+ character is reflected in a high production of serum-blocking antigen. These mutants produce very few intact virions or empty capsids at 39.5° (79); since one similar mutant (ts-89) is thermostable but still fails to make detectable particles at 39.5°, it is possible that their defect is effective before assembly.

4-8. Asymmetric Covariation

The upper half of Table 4-2 shows that many of the characters discussed above (namely pli^+, $plia^+$, pls^+, ppi_2^+, pti^+, chr^+, tb^+, tet^+, ppi_3^+, a^+, production of serum-blocking antigen and production of infectious RNA) fall into a consistent pattern. That is (with a few exceptions), mutants to the left of the map were defective in many of these characters, and mutants to the right were defective in rather few; for example, all mutants defective in pti^+ were also defective in chr^+, tb^+, tet^+, ppi_3^+ and a^+, but mutants defective in ppi_3^+ were not necessarily defective in pti^+. In other words, the defects were covariant in a fashion that was asymmetric.

At first it was thought that mutants of group A, for example, had a primary lesion in a function whose ultimate effect was to prevent DNA synthesis, and that the other characters covariant in group A were similarly related to this primary function. Unfortunately, the likelihood that ts-20 and -28 were defective in functions actually associated with RNA synthesis, and the finding that group B mutants are in both halves of the map, makes this simple explanation seem unlikely. It remains possible that there is some polar defect in translation mechanisms. However, a more likely supposition is that this covariation simply reflects the amount of gene products available, since the physiological grouping seems to follow the amount of RNA made; this is especially noticeable for mutants like ts-104. In this case, ts-104 is presumed to be a simply because it makes too little RNA to function effectively as a messenger specifying the gene product needed to express the a^+ character. This illustrates the fact that single-stranded riboviruses probably form highly interrelated systems, in that the ultimate product of a function (viral RNA) is also the messenger required to increase the amount of that function, and may also be its substrate if the function is a polymerase. Thus any defective function may well be covariant with many other functions because of this type of interrelation.

This conclusion is somewhat disappointing, since it means that most of the many viral characters that have been explored are once again orphans and not identified with any region of the map. However, it has the merit of explaining the correlation of the "physiological map" with the genetic map, if the additional assumption is made, for example, that the ppi_1^+ character requires practically no messenger function of progeny RNA, the pti^+ character needs only a little, the ppi_3^+ character rather more, and the a^+ character the maximum of messenger function. The probable reasons for low RNA production by some virion protein mutants are discussed in Sec. 4-10.

It is curious that not one of the ten or so effects of poliovirus on host cells described above is directly represented by a ts mutant; perhaps such functions are effected by a very small proportion of the genome, or the gene products with identified primary functions, such as assembly and RNA-synthesis proteins, also have secondary functions whose defect is obscured by asymmetric covariation.

4-9. Cystine Dependence

There are several early reports of poliovirus strains that are either independent of, dependent on, or inhibited by cystine (*84, 85*). The cystine concentrations are generally higher than those required to provide building blocks for viral protein synthesis. This phenomenon is largely unexplained, but the work of Pohjanpelto (*86, 87*) may provide some indications. She showed that L (but not D) cystine stabilized the virion *in vitro;* the function of the cystine may be either to protect specific active sites or to maintain a particular configuration of the protein by an allosteric effect. In support of the latter possibility is the finding (Table 4-2) that all mutants mapping between and including ts-94 and ts-2 were cystine dependent at 37°C (*79*). Their plaque formation in agar medium containing normal cystine levels (25–40 μg/ml) was normal and indistinguishable from ts^+, but in 2 μg/ml or less their plating efficiency at 37°C was 10^{-2} to 10^{-4} and plaques were pin-point in size. Strain ts^+, and the mutants scored "$+$" in Table 4-2, grew normally in low cystine media. The ts defect of cystine-dependent mutants was also much ameliorated by cystine, and the thermolability *in vitro* at 39.5 or 45°C markedly reduced; most ts^+ revertants from these mutants were no longer cystine dependent, and most cystine-independent revertants had also reverted to ts^+.

Thus the virion protein(s) specified by the central region of the map are likely to be particularly sensitive to configurational factors, and it seems significant that this region corresponds almost exactly with the region determining guanidine sensitivity (see next section).

4-10. Guanidine Sensitivity and Interference-without-Multiplication

As mentioned above, the provision of a genetic map for poliovirus ts mutants has enabled the guanidine resistance (g) locus of one strain (ts-28g) to be mapped rather accurately. Figure 4-6 presents the map of the single mutants only, together with some salient characters of the mutants abstracted from Table 4-2. It shows that this particular g locus lies in the middle of the region concerned with specifying virion proteins, away from the regions to the left of ts-150 apparently concerned with RNA synthesis. This was unexpected, in view of the known fact that guanidine prevents viral RNA synthesis (*32, 88*). Guanidine also prevents the appearance of RNA polymerase activity (*89*), but this deficiency may simply reflect the lack of substrate or of messenger to specify the enzyme (in both cases viral RNA), rather than a direct effect of guanidine on enzyme synthesis; guanidine has no effect on the enzyme *in vitro*.

In view of this surprising result, independent evidence was sought that guanidine resistance was indeed determined by a virion protein (*76*). It was shown that six out of eight apparently single ts virion-protein mutants were even more sensitive to guanidine than the ts^+ parent (Fig. 4-7). These mutants also straddle the g locus on the genetic map (Fig. 4-6). It is most unlikely that, by chance, all six mutants contain a second mutation conferring guanidine sensitivity, and this remote possibility was ruled out by showing that all ts^+ revertants from these mutants had also reverted toward the ts^+ guanidine phenotype, i.e., had become more guanidine resistant. In addition, a strain of poliovirus selected from strain VS by urea treatment to have stronger capsid bonding without changing its ts defect

Frequency of ts⁺ Recombinants (%)

| 0 | 0.1 | 0.2 | 0.3 | 0.4 | 0.5 | 0.6 | 0.7 | 0.8 | 0.9 |

ts⁺	Single mutants	20	81	28	46	96	44	150	94	89	155	3	2	104	149
	Group	A	A	B	B	B	B	·	C	C	C	D	D	B	·
	ppi₁	+	+	+	+	+	+	·	+	+	+	+	+	+	·
	pti	−	−	+	+	+	+	+	+	+	+	+	+	+	+
	ppi₃	+	−	−	−	−	−	·	−	−	−	+	+.	−	·
	a	+	−	−	−	−	−	−	−	−	+	+	+	−	·
100	i RNA	.09	3.4	2.9	3.0	7.4	22	60	6.8	44	34	88	3.3	3.1
	Interference	+	·	+	+	+	+	+	−	+	+	+	+	+	−
	Maturation efficiency	+	+	+	+	·	+	+	−	−	−	−	−	+	−
	Δ 45°C	+	+	+	+	+	+	+	+	+	−	−	−	−	+
	cy	+	+	+	+	+	+	+	−	−	−	−	−	+	+
	g/120	+	+	+	+	+	+	+	−	−	−	+	−	+	−

g

RNA Synthesis   Virion Protein

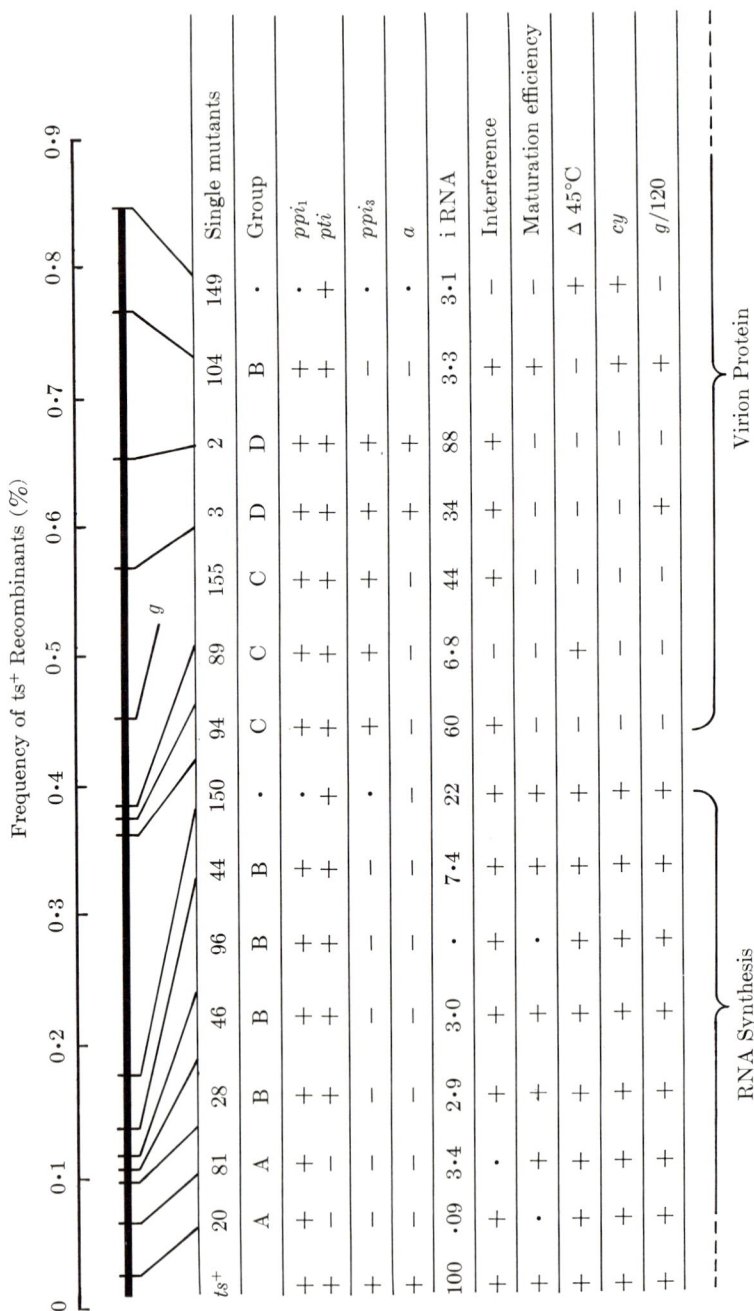

Fig. 4-6. Correlation between position on the genetic map and physiological character of single poliovirus mutants (abstracted from Table 4-2).

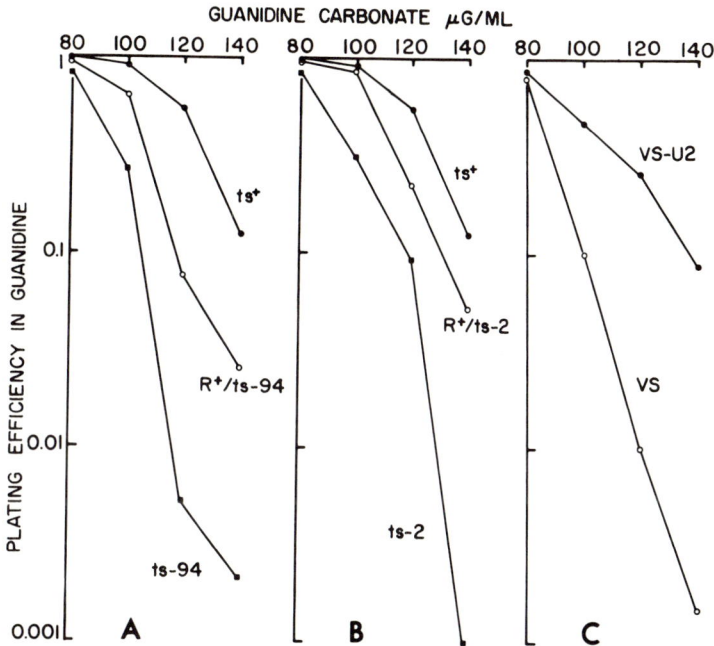

FIG. 4-7. Plaque formation in various concentrations of guanidine of several poliovirus type 1 strains (see text). Strains R+/*ts*-94 and R+/*ts*-2 are *ts*+ revertants picked from 39.5°C assays of *ts*-94 and *ts*-2, respectively (Ref. *76*).

[strain VS.U2, (*90, 91*)] was also increased in guanidine resistance (Fig. 4-7), while all seven *ts* strains apparently mutant in non virion proteins were unchanged in guanidine resistance.

Thus a single mutational change in a virion protein can lead to a change in guanidine resistance. It must therefore be concluded that virion protein is a determinant of guanidine resistance, in confirmation of the map position of the *g* locus. As a mutation in this protein alone can change the resistance, it is also an essential determinant; it is, of course, not necessarily the only determinant.

However, these experiments do not show whether or not the virion protein determining resistance merely suppresses a primary site of sensitivity in another gene, for example, a polymerase. This would imply a situation of genetic dominance of the resistant (*g*) strains, i.e., in mixed infection with *g*+ strains a *g* strain should be unaffected by guanidine. In negation of this possibility, it was shown (*75, 76*) that the growth of *g* strains in presence of guanidine was inhibited

by the presence of g⁺ strains, even though the g⁺ strains were themselves also inhibited. Thus it is likely that the virion protein gene is
the only one mainly affected in a direct fashion by guanidine, a conclusion supported by the finding that the g locus behaves as a single
character in recombination. In support of this implication, it has been
found (*32, 88, 92*) that guanidine added 2–3 hr after infection was still
highly effective, i.e., like the late ts mutants a function was perturbed
that was needed for further multiplication at a time when maturation
was under way. As the inhibition by guanidine in mixed as well as
single infection also extended to RNA synthesis, it again follows that
the affected virion protein is in some way a determinant of RNA synthesis. However, the outstanding implication of the dominance of
guanidine sensitivity is that a novel virus protein with interfering
capabilities is produced.

A similar situation exists with certain ts mutations in virion protein
loci adjacent to the g locus (Table 4-2). These mutants (ts-19, -22,
-89, -123, and -149) were capable of interfering with the growth of
ts^+ at 39.5°C but not at 37°C (*75, 76*). The involvement of interferon
was ruled out. Again, the interference extended to RNA synthesis of
the nonrestricted strain, and all these mutants were also inhibited in
their own RNA synthesis under the restricted conditons, despite the
fact that their primary defect was in virion protein. Once again it
follows that virion protein is a determinant of RNA synthesis; these
mutants are only inhibited in production of progeny 35s RNA under
the restrictive conditions and produce normal yields of double-
stranded RNA (*80*).

In the case of the ts interference with ts^+ it can be presumed that
the primary defect is a misfolding of virion proteins at 39.6°C, since
the mutants grow well and do not interfere at 37°C. The resemblance
between the effects of their ts defect and the defect induced by guanidine, and their adjacency to the g locus in the map, make it very
likely that the primary defect induced by guanidine is similar. Guanidine probably forms a few weak bonds at certain loci during
folding of the critical virion protein, thus slightly deranging a structure that is particularly sensitive to such perturbations. In support
of this concept, it was found (*76*) that the urea sensitivity of poliovirus strains was significantly altered by growth in subinhibitory levels
of guanidine (10–30 μg/ml) suggesting that guanidine in very low doses
may interact with virion protein during synthesis or assembly. Single
mutations in many places in this same virion protein are able to re-

modify its structure so that the presence of guanidine or of restrictive temperatures are not harmful. It seems likely that guanidine dependence, or the modification of guanidine sensitivity by amino acids (*92, 93*), results from similar interactions.

The unusual sensitivity of this region of the map to small modifications of structure by guanidine and amino acids, perhaps including cystine (see above), is of considerable interest; it may be relevant to the correct scission of the protein intermediates (*20*), itself an unusual phenomenon.

The question of how defects in virion proteins, induced either by *ts* lesions or by guanidine, can affect viral RNA synthesis is clearly of importance to the growth process of the virus. It implies that virion protein synthesis somehow controls RNA synthesis, perhaps to keep the two in step. Many hypotheses are possible. One model is supplied by Sugiyama and Nakada (*94*), who showed that MS2 ribophage capsid protein represses the synthesis of noncapsid proteins, which presumably include the polymerase; such a repression would be expected to reduce RNA synthesis. Another is suggested by Martin (*95*), in which a viral (conceivably structural) protein may "lead off" progeny RNA from a replicating complex; failure to do so may simply block the template. It should be noted that in either model the effector protein may or may not be the determinant protein (i.e., the one containing the primary lesion) since faulty assembly caused by the determinant protein could increase or decrease the amount of a free effector virion protein. In addition, a single defect in a precursor protein (*20*) may cause faulty scission, and hence affect the function of several products at one blow; the same is true of products that function as multimers (e.g., capsid proteins). Perhaps the simplest model of all is that the determinant virion protein is itself also a polymerase, but the high RNA production by *ts* mutants surrounding the *g* locus in the genetic map suggests that this is not the case.

However, a more likely explanation for the action of guanidine follows from the following facts. (a) As discussed above in Sec. 4-2, poliovirus proteins appear to be produced by post-translational cleavage of a large precursor polypeptide. (b) Guanidine (*95a*) and the *ts* defect of several interfering mutants, notably *ts*-89 (*14a*), both hinder the efficiency of this cleavage, whereas the defects of noninterfering mutants do not. (c). The viral gene specifying the enzyme blocked by both guanidine (*95b, 95c, 80*) and the interfering *ts* defect (*80*) appears to control the second step in RNA synthesis, namely the production of progeny plus strands, rather than the first step of double-stranded RNA synthesis. (d) This gene is adjacent to the coat-protein loci of g and the interfering mutants in the genetic map (Fig. 4-6). One might therefore expect that cleavage of the affected enzyme from the adjacent temperature- and guanidine-sensitive virion protein will be slowed under restrictive conditions (which are also the conditions where interference is manifest) by the latter's unsuitable configuration. It is accordingly most attractive to suggest that the incompletely separated enzyme possesses the polymerase's specificity for the site on the RNA involved in the initiation of plus-strand replication but without its full polymerase activity. In such a model, the polymerase–virion protein dimer would be the novel interfering substance supposed to be the effector both of the *ts* interference and of guanidine inhibition, and would block the second step in replication of all viral RNA templates in the cell by competing with the normal enzyme for the RNA initiation site.

4-11. Conclusions

The genetic analysis of poliovirus, so far as it has been taken, appears to provide a reasonably self-consistent picture. The incomplete genetic map is small and simple, with virion protein genes at one end and RNA synthesis genes at the other, and a small gap in the middle. Two points, however, have become clear. The first is that, despite progress with the map and the enumeration of gene products, most of this work is still preliminary and not much new information is yet available about the actual growth process, namely the detailed function of the various gene products. The second is that considerable complexities are now emerging, revealing the system to be far from the simple one originally envisaged. As with small high-performance radios and European cars, it may be that compactness is achieved by sophisticated internal relationships rather than by simplicity.

ACKNOWLEDGMENTS

I am very grateful to Dr. H. Eagle and Dr. E. Hehre for their support and hospitality, and to Dr. D. Summers and Dr. J. Maizel for access to their unpublished material, for many delightful discussions, and for the opportunity of collaboration in their laboratories.

REFERENCES

1. A. R. Bellamy, L. Shapiro, J. T. August, and W. K. Joklik, *J. Mol. Biol.,* **29,** 1 (1967).

2. F. Fenner and J. F. Sambrook, *Ann. Rev. Microbiol.,* **18,** 47 (1964).

3. F. Fenner, *Biology of Animal Viruses,* Academic Press, New York, 1968.

4. P. D. Cooper, *Brit. Med. Bull.,* **23,** 155 (1967).

5. F. L. Schaffer, *Cold Spring Harbor Symp. Quant. Biol.,* **27,** 89 (1962).

6. F. A. Anderer and H. Restle, *Z. Naturforsch.,* **19b,** 1026 (1964).

7. D. Baltimore, *Proc. Natl. Acad. Sci. U.S.,* **51,** 450 (1964).

8. C. E. Schwerdt, in *Cellular Biology, Nucleic Acids and Viruses,* N.Y. Academy of Science, New York, 1957, p. 159.

9. F. L. Schaffer and C. E. Schwerdt, in *Viral and Rickettsial Infections of Man,* 4th ed. (F. L. Horsfall and I. Tamm, eds.), Lippincott, Pennsylvania, 1965, p. 194.

10. C. E. Schwerdt and F. L. Schaffer, *Ann. N.Y. Acad. Sci.,* **61,** 740 (1955).

10a. S. McGregor and H. D. Mayor, *J. Virol.,* **2,** 149 (1968).

11. F. L. Schaffer, H. F. Moore, and C. E. Schwerdt, *Virology,* **10,** 530 (1960).

12. J. V. Maizel, *Biochem. Biophys. Res. Commun.,* **13,** 483 (1963).

13. J. V. Maizel, *Science,* **151,** 988 (1966).

14. D. F. Summers, J. V. Maizel, and J. E. Darnell, *Proc. Natl. Acad. Sci. U.S.,* **54,** 505 (1965).

14a. P. D. Cooper, D. F. Summers, and J. V. Maizel, in preparation, 1968.

15. A. L. Shapiro, E. Vinuela, and J. V. Maizel, *Biochem. Biophys. Res. Commun.,* **28,** 815 (1967).

16. J. V. Maizel and D. F. Summers, *Virology,* **36,** 48 (1968).

17. L. Levintow, and J. E. Darnell, Jr., *J. Biol. Chem.,* **235,** 70 (1960).

18. W. Munyon and N. P. Salzman, *Virology,* **18,** 95 (1962).

19. J. T. Finch and A. Klug, *Nature,* **183,** 1709 (1959).

19a. H. D. Mayor, *Virology,* **22,** 156 (1964).

20. D. F. Summers and J. V. Maizel, *Proc. Natl. Acad. Sci. U.S.,* **59,** 966 (1968).

20a. M. F. Jacobson and D. Baltimore, *Proc. Natl. Acad. Sci. U.S.,* **61,** 77 (1968).

20b. J. J. Holland and E. D. Kiehn, *Proc. Natl. Acad. Sci. U.S.,* **60,** 1015 (1968).

21. A. Lwoff, *Cold Spring Harbor Symp. Quant. Biol.,* **27,** 159 (1962).

22. R. Bablanian, H. J. Eggers, and I. Tamm, *Virology,* **26,** 100 (1965).

23. J. V. Maizel, B. A. Phillips, and D. F. Summers, *Virology,* **32,** 692 (1967).

24. S. Penman, Y. Becker, and J. E. Darnell, *J. Mol. Biol.,* **8,** 541 (1964).

25. H. Goldfine, R. Koppelman, and E. A. Evans, *J. Biol. Chem.,* **232,** 577 (1958).

26. N. P. Salzman, R. Z. Lockart, and E. D. Sebring, *Virology,* **9,** 244 (1959).

27. H. B. Levy, *Virology,* **15,** 173 (1961).

28. J. J. Holland, *Proc. Natl. Acad. Sci. U.S.,* **48,** 2044 (1962).

29. J. J. Holland and J. A. Peterson, *J. Mol. Biol.,* **8,** 556 (1964).

30. M. L. Fenwick, *Virology,* **19,** 241 (1963).

31. E. F. Zimmerman, M. Heeter, and J. E. Darnell, *Virology,* **19,** 400 (1963).

32. J. J. Holland, *Proc. Natl. Acad. Sci. U.S.,* **49,** 23 (1963).

33. M. Riessig, D. W. Howes, and J. L. Melnick, *J. Exptl. Med.,* **104,** 289 (1956).

34. H. D. Mayor, *Texas Repts. Biol. Med.,* **19,** 106 (1961).

34a. D. J. Garwes and P. D. Cooper, unpublished.

35. P. D. Cooper, R. T. Johnson, and D. J. Garwes, *Virology,* **30,** 638 (1966).

36. S. Penman, *Virology,* **25,** 148 (1965).

37. P. D. Cooper, *Virology,* **28,** 663 (1966).

38. F. L. Schaffer and M. R. Gordon, *J. Bacteriol.,* **91,** 2309 (1966).

39. J. M. Bishop, D. F. Summers, and L. Levintow, *Proc. Natl. Acad. Sci. U.S.,* **54,** 1273 (1965).

40. D. Baltimore, *J. Mol. Biol.,* **18,** 421 (1966).

41. J. E. Darnell, M. Girard, D. Baltimore, D. F. Summers, and J. V. Maizel, in *The Molecular Biology of Viruses* (J. S. Colter and W. Paranchych, eds.), Academic Press, New York, 1967.

42. R. Dulbecco and M. Vogt, *Virology,* **5,** 220 (1958).

42a. R. Dulbecco and M. Vogt, *Virology,* **5,** 236 (1958).

43. F. H. C. Crick, L. Barnett, S. Brenner and R. J. Watts-Tobin, *Nature,* **192,** 1227 (1961).

44. F. L. Schaffer, *Virology,* **18,** 412 (1962).

45. J. N. Wilson and P. D. Cooper, *Virology,* **26,** 1 (1965).

46. A. Boeyé, *Virology,* **9,** 691 (1959).

47. Yu. Z. Ghendon, *Acta Virol.,* **7,** 16 (1963).

48. R. I. Carp and H. Koprowski, *Virology,* **17,** 99 (1962).

49. R. Klein, D. Sergiescu, and M. Teodorescu, *Virology,* **30,** 145 (1966).

49a. P. D. Cooper, unpublished.

50. P. D. Cooper, *Virology,* **22,** 186 (1964).

51. R. Dulbecco, in *Poliomyelitis: Papers and Discussion, 5th International Poliomyelitis Conference,* Copenhagen, Lippincott, Pennsylvania, 1961, p. 21.

51a. R. Dulbecco, in *Second International Conference on Live Poliovirus Vaccines,* Pan American Sanitary Bureau, Washington, D. C., 1961, p. 47.

52. P. D. Cooper, *Virology,* **35,** 584 (1968).

53. D. C. Breeze and H. Subak-Sharpe, *J. Gen. Virol.,* **1,** 81 (1967).

54. C. I. Davern, *Australian J. Biol. Sci.,* **17,** 726 (1964).

55. N. Takemori and S. Nomura, *Virology,* **12,** 171 (1960).

56. R. I. Carp, *Virology,* **21,** 373 (1963).

57. J. W. Drake, *Virology,* **6,** 244 (1958).

58. M. L. Fenwick and P. D. Cooper, *Virology,* **18,** 212 (1962).

59. G. K. Hirst, *Cold Spring Harbor Symp. Quant. Biol.,* **27,** 303 (1962).

60. N. Ledinko, *Virology,* **20,** 107 (1963).

61. P. D. Cooper, unpublished observations, 1963.

62. S. Bengtsson, *Acta Pathol. Microbiol. Scand.* **73,** 592 (1968).

62a. G. K. Hirst, private communication.

63. L. Montagnier, in *The Molecular Biology of Viruses* (L. V. Crawford and M. G. P. Stoker, eds.); *Symp. Soc. Gen. Microbiol.,* Cambridge Univ. Press, Cambridge, 1968, p. 125.

64. A. D. Kaiser, *J. Mol. Biol.,* **4,** 275 (1962).

65. F. W. Stahl, R. S. Edgar, and J. Steinberg, *Genetics,* **50,** 539 (1964).

66. E. S. Tessman, *Virology,* **25,** 303 (1965).
67. E. Wecker and G. Lederhilger, *Proc. Natl. Acad. Sci. U.S.,* **52,** 246 (1964).
68. V. I. Agol and G. A. Shirman, *Biochem Biophys. Res. Commun.,* **17,** 28 (1964).
69. C. E. Cords and J. J. Holland, *Proc. Natl. Acad. Sci. U.S.,* **51,** 1080 (1964).
70. N. Ledinko and G. K. Hirst, *Virology,* **14,** 207 (1961).
71. P. D. Cooper, *Virology,* **25,** 431 (1965).
72. B. W. Burge and E. R. Pfefferkorn, *Virology,* **30,** 214 (1966).
73. J. F. Sambrook, Ph.D. thesis, Australian National University, Canberra, 1966.
74. B. L. Padgett and J. K. N. Tomkins, *Virology,* **36,** 161 (1968).
75. P. Pohjanpelto and P. D. Cooper, *Virology,* **25,** 350 (1965).
76. P. D. Cooper, B. B. Wentworth, and D. McCahon in preparation 1968.
77. D. Baltimore, M. Girard, and J. E. Darnell, *Virology,* **29,** 179 (1966).
78. B. B. Wentworth, D. McCahon, and P. D. Cooper, *J. Gen. Virol.,* **2,** 297 (1968).
79. D. McCahon and P. D. Cooper, in preparation, 1968.
80. P. D. Cooper, D. Stanček, and D. F. Summers, in preparation, 1968.
81. D. Baltimore and M. Girard, *Proc. Natl. Acad. Sci. U.S.,* **56,** 741 (1966).
82. G. Tannock and P. D. Cooper, unpublished results, 1968.
83. D. Garwes and P. D. Cooper, unpublished results, 1965.
84. G. R. Dubes and M. Chapin, *Science,* **124,** 586 (1956).
85. W. D. McBride, *Virology,* **18,** 118 (1962).
86. P. Pohjanpelto, *Virology,* **6,** 472 (1958).
87. P. Pohjanpelto, *Virology,* **15,** 231 (1961).
88. H. J. Eggers, E. Reich, and I. Tamm, *Proc. Natl. Acad. Sci. U.S.,* **50,** 183 (1963).
89. D. Baltimore, H. J. Eggers, R. M. Franklin, and I. Tamm, *Proc. Natl. Acad. Sci. U.S.,* **49,** 843 (1963).
90. P. D. Cooper, *Virology,* **16,** 485 (1962).
91. P. D. Cooper, *Virology,* **21,** 322 (1963).
92. A. Lwoff, *Biochem. J.,* **96,** 289 (1965).
93. L. Philipson, S. Bengtsson, and Z. Dinter, *Virology,* **29,** 317 (1966).
94. T. Sugiyama and D. Nakada, *Proc. Natl. Acad. Sci. U.S.,* **57,** 1744 (1967).
95. E. M. Martin in *Genetic Elements* (D. Shugar, ed.), Academic Press, New York, 1966, p. 117.
95a. M. F. Jacobson and D. Baltimore, *J. Mol. Biol.,* **33,** 369 (1968).
95b. L. A. Caliguiri and I. Tamm, *Virology,* **35,** 408 (1968).
95c. L. A. Caliguiri and I. Tamm, *Virology,* **36,** 223 (1968).
96. M. Vogt, R. Dulbecco, and H. A. Wenner, *Virology,* **4,** 141 (1957).
96a. G. D. Hsiung and J. L. Melnick, *J. Immunol.,* **80,** 282 (1958).
97. V. I. Agol and M. Ya. Chumakova, *Virology,* **17,** 221 (1962).
98. V. I. Agol and M. Ya. Chumakova, *Acta Virol.,* **7,** 97 (1963).
98a. S. Bengtsson and L. Philipson, *Virology,* **20,** 176 (1963).
99. S. Nomura, and N. Takemori, *Virology,* **12,** 154 (1960).
99a. R. Dulbecco and M. Vogt, *Ann. N.Y. Acad. Sci.,* **61,** 790 (1955).
100. K. K. Takemoto and H. Liebhaber, *Virology,* **17,** 499 (1962).
101. J. S. Pagano, *Ann. N. Y. Acad. Sci.,* **130,** 398 (1965).

102. S. Bengtsson, L. Philipson, H. Persson, and T. C. Laurent, *Virology*, **24**, 617 (1964).

103. S. Bengtsson, *Proc. Soc. Exptl. Biol. Med.*, **118**, 47 (1965).

104. D. Sergiescu, F. Horodinceanu, and R. Crainic, *Nature,* **215**, 313 (1967).

105. K. K. Takemoto and R. L. Kirschstein, *J. Immunol.*, **92**, 329 (1964).

106. R. Dulbecco, in *Poliomyelitis: Papers and Discussion,* International Poliomyelitis Congress, Lippincott, Pennsylvania, 1958, p. 366.

107. R. I. Carp, *Virology*, **18**, 151 (1962).

108. N. Groman, A. Lwoff, and M. Lwoff, *Ann. Inst. Pasteur*, **98**, 357 (1960).

109. Y. Kanda and J. L. Melnick, *J. Exptl. Med.*, **109**, 9 (1959).

110. G. R. Dubes and M. Chapin, *J. Gen. Microbiol.*, **18**, 320 (1958).

111. M. Chapin and G. R. Dubes, *J. Infect. Diseases*, **110**, 210 (1962).

112. M. Chapin and G. R. Dubes, *J. Infect. Diseases,* **112**, 247 (1963).

113. B. Loddo, A. Ferrari, A. Spaneddi, and G. Brotzu, *Experientia,* **18**, 518 (1962).

114. M. Nakano, S. Iwami, and I. Tagaya, *Virology*, **21**, 264 (1963).

115. I. Tamm and H. J. Eggers, *Virology,* **18**, 439 (1962).

116. N. Takemori, S. Nomura, M. Nakano, Y. Morioka, M. Hemui, and M. Kitaoka, *Virology*, **5**, 30 (1958).

117. K. K. Takemoto and K. Habel, *Virology*, **9**, 228 (1959).

118. R. Dulbecco, *Ciba Foundation Symposium: The Nature of Viruses* (G. E. Wolstenholme and E. C. Millar, eds.), Churchill, London, 1957, p. 127.

118a. J. S. Youngner, *J. Immunol.*, **78**, 282 (1957).

119. N. F. Stanley, D. C. Dorman, J. Ponsford, and M. Larkin, *Australian J. Exptl. Biol. Med. Sci.*, **34**, 411 (1956).

120. G. J. Papaevangelou and J. S. Youngner, *Virology,* **15**, 509 (1961).

121. C. Wallis, J. L. Melnick, G. D. Ferry, and I. L. Wimberly, *J. Exptl. Med.*, **115**, 763 (1962).

122. W. A. Woods and F. C. Robbins, *Proc. Natl. Acad. Sci. U.S.*, **47**, 1501 (1961).

123. H. L. Hodes, H. D. Zepp, and E. Ainbender, *Virology*, **11**, 306 (1960).

123a. J. L. Delsal, P. Lepine, and V. Sautter, *Compt. Rend.*, **251**, 290 (1960).

124. A. C. Hollinshead, *Med. Exptl.*, **2**, 303 (1960).

125. A. Boeyé, *Virology*, **21**, 587 (1963).

126. J. Koza, *Virology*, **21**, 477 (1963).

127. W. D. McBride, *Virology*, **7**, 45 (1959).

5

MYXOVIRUSES

C. SCHOLTISSEK, R. DRZENIEK, AND R. ROTT

INSTITUT FÜR VIROLOGIE
JUSTUS LIEBIG-UNIVERSITÄT
GIESSEN, GERMANY

5-1. Introduction

The myxovirus group includes those RNA containing viruses which have a specific affinity for certain mucoproteins (1) and which contain as an integral component of the particle a virus-induced enzyme which splits sialic acid from sialic-acid-containing substrates. This enzyme has been named sialidase (2) or neuraminidase (3, 4). In accordance with this definition a number of morphologically related viruses, which are sometimes tentatively included in this group, will not be dealt with in this chapter, e.g., measles, Rinderpest, distemper, respiratory syncytial, or the group of the leucosis viruses of chicken and mice.

The myxoviruses have a more complex structure than viruses containing only nucleic acid and protein. Their central core consists of a long helical ribonucleoprotein component, the RNP-antigen. This RNP-antigen is contained in an envelope consisting of proteins, lipids, and carbohydrates of viral and of host-cell origin. The envelope carries the immunogenic, hemagglutinating, and sialidase activities of the virus particle (5).

The ability to agglutinate a wide spectrum of red cells, i.e., hemagglutination, a phenomenon first described simultaneously by Hirst (6) and by McClelland and Hare (7) using influenza virus, is commonly used as a rapid and simple procedure for the quantitation of myxoviruses.

The myxoviruses play an important role as causative agents of diseases in men and animals, e.g., influenza and other respiratory diseases, mumps, and fowl pest. The complex structure of the virion is of great interest to the student of biology. Biological investigations on the site of synthesis of viral components, the rate of synthesis, and the formation of the mature virus particle are aided by the many parameters available. In this respect, myxoviruses offer more possibilities than viruses of simpler structure.

5-2. Classification

On the basis of their morphological and biological properties, myxoviruses can be divided into two groups: influenza and parainfluenza viruses (8). The taxonomy of the main types of these viruses is shown in Table 5-1.

The grouping of myxoviruses into influenza and parainfluenza

TABLE 5-1
Classification of Myxoviruses

Influenza group, subgroups	Parainfluenza group, subgroups
Influenza A:	Newcastle disease
(Human influenza viruses A,	Mumps
A_1, A_2; swine-, horse-, duck-,	Parainfluenza 1 (Sendai, hemadsorption virus 2)
avian influenza virus; fowl	Parainfluenza 2 (Croup-associated virus, simian
plague virus)	myxovirus SV_5, parainfluenza
Influenza B	2 virus of chicken)
Influenza C	Parainfluenza 3 (Hemadsorption virus 1, bovine
	parainfluenza 3 virus)
	Parainfluenza 4

viruses is based on the differences summarized in Table 5-2. Marked differences in the size of the viria were revealed by electron microscopic and ultracentrifugational studies. Influenza viruses are smaller (80–120 mμ) and more uniform in size than parainfluenza viruses (150–250 mμ). Only parainfluenza viruses are known to have hemolytic activity. The diameter and structure of the RNP-antigens exhibit the most significant differences. The RNP-antigen of members of the influenza group is about 90 Å in width, and the loosely bound helix

TABLE 5-2
Differences between Influenza and Parainfluenza Viruses

	Influenza	Parainfluenza	References
Structure			
Particle size	800–1200 Å	1500–2500 Å	8
Filaments	Common	Unusual	8
RNP-antigen, diameter	90 Å	170 Å	19, 20, 195
RNP-antigen, length	?	1 μ	56, 57
RNP-antigen, structure	Double helix	Single helix	196
RNA, sedimentation coefficient	38S (?)	57S	59, 60, 62
Multiplication			
Sensitivity to actinomycin	Sensitive	Not sensitive	92, 108, 109
RNP-antigen, site of accumulation	Nucleus	Cytoplasm	97, 131
Eosinophilic cytoplasmic inclusions	−	+	8
Genetic			
Von Magnus phenomenon	+	−	156
Rate of recombination	High	Low	9, 186
Multiplicity reactivation	High	Low	190, 191–193
Stepwise inactivation	+	−	164, 165

seems to be double stranded. On the other hand, the RNP-antigen of the parainfluenza viruses is a single helix about 170–180 Å in width. The highest sedimentation coefficient of the influenza virus RNA found was 38S. But this determination is equivocal. The RNA of parainfluenza viruses has a sedimentation coefficient of 57S. It is noteworthy that none of these ribonucleic acids with high molecular weight show any infectivity.

Influenza viruses do not multiply in the presence of actinomycin, which is known to interfere with the DNA-dependent RNA synthesis. Parainfluenza viruses, however, and most of the other RNA containing viruses are not inhibited by this antibiotic. Another significant difference was revealed by studies with fluorescent antibodies: Only the RNP-antigen of influenza viruses accumulates in the nucleus of the host cell.

The two groups show different rates of genetic recombination and multiplicity reactivation. Influenza viruses exhibit the so called "von Magnus phenomenon," and they can be inactivated stepwise (Sec. 5-5, B). These differences in the genetic properties may be an expression of the different structure of the viral RNA (9).

Influenza- and parainfluenza viruses have to be treated as distinctly different species in spite of a superficial morphological similarity. The only thing common to both groups is the presence of sialidase. By definition these two groups are therefore myxoviruses.

The influenza viruses are subdivided into three subgroups (A, B, C). This subdivision is based on the serologic differences of the RNP-antigen. Members of a given subgroup differ only in their surface antigens, but have a common RNP-antigen (1).

In the parainfluenza virus group the antigenic relationship between the various subgroups is not as well defined as with influenza viruses. Nevertheless, the antigenic makeup of these viruses provides the basis for the creation of several subgroups, i.e., viruses belonging to one subgroup share a common serologic behavior (10), i.e., all show some cross reactions with at least one other member of the group (11).

5-3. Composition and Structure

A. PHYSICO-CHEMICAL PROPERTIES OF THE VIRION

Physico-chemical studies of myxoviruses were conducted mainly with human influenza virus, fowl plague virus, and NDV. The particle

TABLE 5-3
Chemical Composition of Myxoviria in Per cent

Virus	C	N	P	RNA	Protein	Carbo-hydrates	Lipids
Influenza A	53.2	10.0	0.97	0.7–1.0	60–70	6–13	18–37
	(197)	(197)	(197)	(198, 199)	(197)	(200)	(75)
Fowl plague				1.8	60–70		25
				(52a)	(52a)		(52a)
NDV				0.5	65	7	27
				(201)	(202)	(202)	(202)

weights as determined by sedimentation and diffusion measurements were reported to be 150–280 × 10⁶ for influenza A and 800 × 10⁶ for NDV (12).

A brief summary of the major chemical constituents of several representative examples of myxoviruses is presented in Table 5-3. Since no reliable criteria for the purity of myxoviruses exist and since the virus particles incorporate up to 40% host material (13, 14), these values are rather rough estimates.

B. MORPHOLOGY

Electron microscopic examinations revealed that myxoviruses are in general spherical particles with variations in size and shape (15). This might be one of the reasons that it is not possible to crystallize these viruses. Besides spherical particles, dumbell-shaped and filamentous forms occur in influenza virus preparations. This pleomorphism may depend on the strain and the cultural conditions (16). The appearance of NDV particles is affected by the ionic strength. At a low ionic strength the particles have a spherical shape; at a high ionic strength there are predominantly extended forms (17, 18).

At a higher resolution the myxoviruses appear as tiny thorn apples after staining with phosphotungstic acid (Fig. 5-1). The envelope, which is not rigid, consists of a distinct membrane (60–100 Å thick) from which cylindrical projections arise which are 80–100 Å long and 20 Å wide (19, 20). These spike-like structures seem to be spaced fairly regularly at intervals of about 70–100 Å over the virus surface (15, 21). Like viruses with cubic symmetry containing only nucleic acid and protein the envelope of influenza viruses is made up of subunits surrounded by either five or six neighboring subunits (Fig. 5-1).

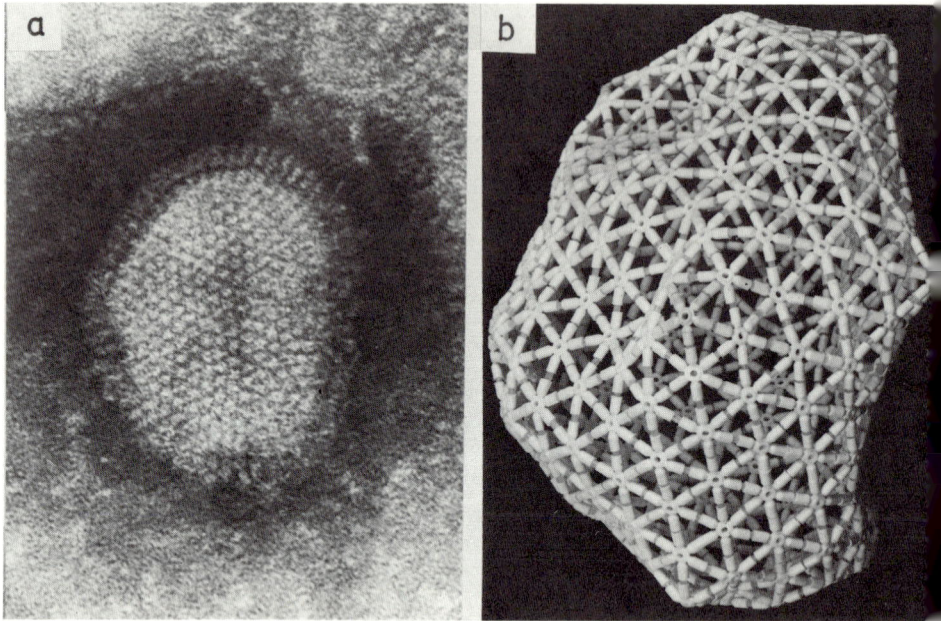

FIG. 5-1 Surface structure of influenza virus. (a) PTA-staining; (b) model. (Reprinted from Ref. *22*, p. 110, by courtesy of University Press, Cambridge, United Kingdom.)

While the distribution of "fives" and "sixes" follows definite rules in isometric viruses, e.g., the adenovirus, it is random for the influenza virus envelope (*22–24*).

It is not certain whether or not the observed "spikes" are in reality protrusions of the viral envelope. It is known that phosphotungstic acid binds tightly to proteins but not to carbohydrates or lipids. For this reason the surfaces of myxoviruses do not necessarily have to have spikes but may possess a lamellar-like structure consisting of phosphotungstic acid-binding protein which alternates with substances such as lipids and carbohydrates which do not bind the stain. This interpretation is supported by the finding that sialidase activity can be inhibited reversibly by minute amounts of phosphotungstic acid (*25*). After removal of sialidase, which as a particle should be large enough to be visible in the electron microscope, the surface structure of the virus particle is not visibly changed (*26*).

In electron micrographs of disrupted myxovirus particles, long, rod-like structures in a spiral arrangement may occasionally be seen which have the typical appearance of the RNP-antigen. From such pictures a model of the general structure of myxoviruses has been developed (*27*) (Fig. 5-2).

C. COMPONENTS

Myxoviruses can be disintegrated by ether or Tween-ether treatment into two distinct components: the hemagglutinin and the RNP-antigen. Such preparations are no longer infective (*28–31*).

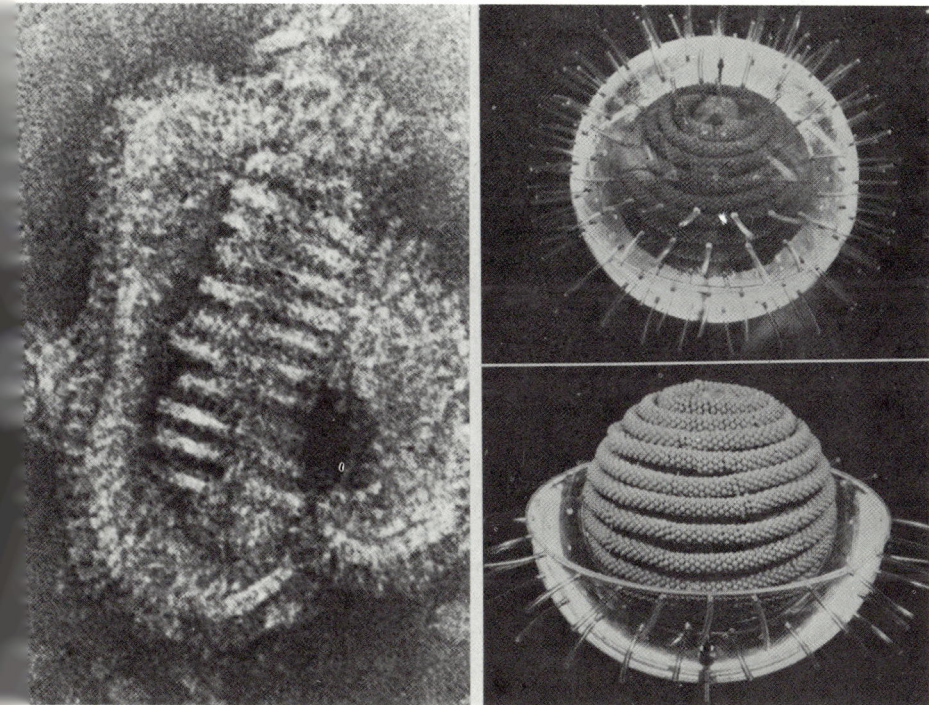

FIG. 5-2. Left: Electron micrograph of partially disrupted influenza virus. By courtesy of Dr. A. P. Waterson. Right: Schematic model constructed to show a possible arrangement of components forming myxovirus particles. (Reprinted from Ref. *27*, p. 355, by courtesy of Academic Press Inc., New York, London.)

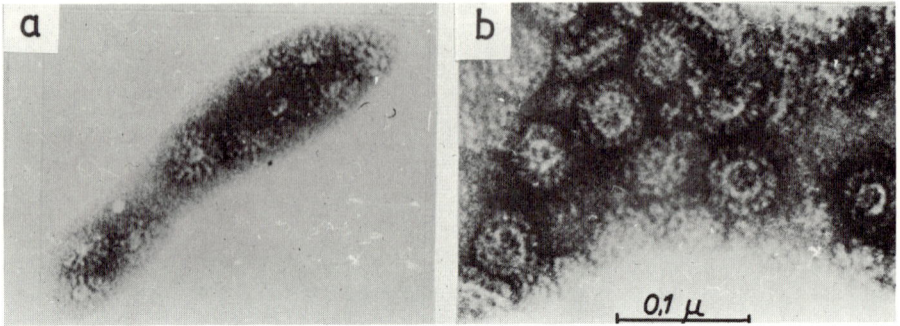

FIG. 5-3. Hemagglutinin of fowl plague (left) and Newcastle disease virus (right). PTA-staining. (Reprinted from Ref. *54*, p. 28, by courtesy of J. and A. Churchill Ltd., London.)

1. *Hemagglutinin*

Fragments of the virus surface which show the same spike-like protrusions as the envelope of the intact virus particle, carry the surface antigens, and carry out the hemagglutinating and enzymatic activities. These fragments have been called hemagglutinin. Depending upon the method of disintegration the hemagglutinin varies in size and shape (Fig. 5-3). It consists of proteins, carbohydrates, and lipids; it is free of RNA, but still contains the host-specific components (see below).

After removal of lipids by chloroform/methanol from purified influenza viruses a protein fraction can be solubilized by 8 m urea and dithiothreitol or by 67% acetic acid. The component(s) obtained have a sedimentation coefficient of 2S (*32, 33*). These components reaggregate after dialysis and form distinct aggregates with S_{20} values between 4S–20S. They possess the antigenic properties of the viral envelope. The largest aggregate(s) (20S) hemagglutinate guinea pig erythrocytes, but have no sialidase activity (*33*).

With detergents it was possible to dissociate influenza viruses into smaller subunits and then to separate the sialidase and hemagglutinating activities to a certain degree by electrophoresis on cellulose acetate strips (*34, 35*). The smallest component which still has hemagglutinating activity is reported to have a sedimentation coefficient of about 15S.

In some cases a further dissociation with sodium dodecylsulfate resulted in component(s) free of hemagglutinating activity, but which were readily adsorbed to erythrocytes. This component(s) was called *hemadsorbin* (*35*).

2. *Sialidase*

The sialidase can be separated from the envelope of myxoviruses by treatment with proteolytic enzymes, with butanol, or with detergents (*26*, *34*, *37–42*). The smallest, enzymatically active entity of influenza viruses which is free of other envelope antigens has a sedimentation coefficient of 9S. This corresponds to a molecular weight of about 200,000 (*43*).

Antibodies against the isolated sialidase do not inhibit hemagglutination. Therefore, hemagglutinin and sialidase seem to be two distinct antigens (*43*, *44*). Both antigens are type specific (*43*, *45*, *45a*). This holds for influenza as well as for parainfluenza viruses.

Generally, sialidase (neuraminidase; N-acetylneuraminate glyco-hydrolase EC 3.2.1.18) is an enzyme, which hydrolyzes the α-ketosidic linkage between sialic acids and other carbohydrate moieties. Besides myxoviruses this enzyme was found also in different bacteria, organs of birds and mammals (for review, see Ref. *4*, 1966; Ref. *49*, 1966). The substrate specificity of viral sialidase differs from that of *Vibrio cholerae sialidase*. *V. cholerae* enzyme hydrolyzes 2 → 3′ as well as 2 → 6′ linkages of sialic acid with carbohydrates. Viral neuraminidase only splits 2 → 3′ linkages easily (Fig. 5-4). Both types of substrates combine, however, with the viral enzyme (*36*, *42*, *50*).

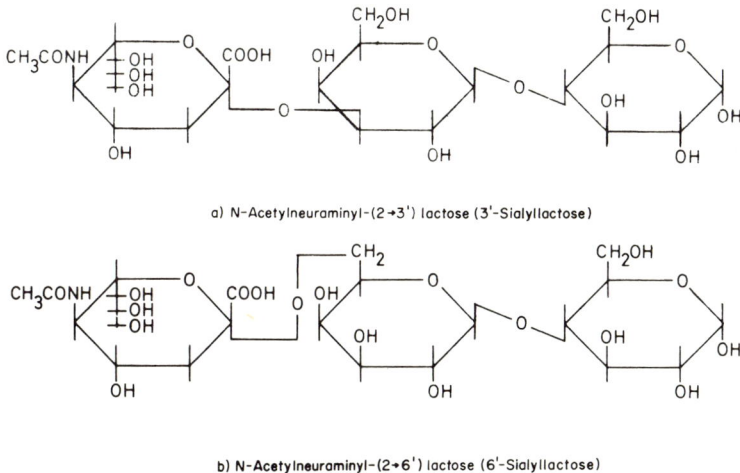

a) N-Acetylneuraminyl-(2→3') lactose (3'-Sialyllactose)

b) N-Acetylneuraminyl-(2→6') lactose (6'-Sialyllactose)

FIG. 5-4. Isomeric substrates of sialidase.

The pH optimum of myxovirus sialidase is generally between pH 4.5–7.0. It varies for different viruses and depends upon the substrate used (*43, 46–48*). The Michaelis constants range from 2×10^{-4} to 2×10^{-3} *m* (*43, 46*). Recently it has been shown that divalent cations (*46a*) or polycations (*42*) are necessary for optimal enzymatic activity.

Effective inhibitors against the viral sialidase were found recently. All of them have polyanionic character, e.g., phosphotungstic acid, RNA, DNA, dextran sulfate, heparin, porcine submaxillary mucine, congo-red, and trypan-red. The *V. cholerae* enzyme is only slightly affected by these substances. Enzymes of different myxoviruses are inhibited to varying degrees (*25, 51*).

3. *RNP-Antigen*

The other virus-specific component obtained after ether treatment, the RNP-antigen, is the carrier of the genetic material of myxoviruses. Several names have been used synonymously for the same component: internal soluble (s) and gebundenes (virus-bound) (g) antigen, or nucleocapsid. The "soluble" antigen found in infected cells is identical with the RNP-antigen of the virion (*52–54*).

The RNP-antigen can be detected as a virus-specific entity only by serological methods. Either hyperimmune sera against the purified RNP-antigen or reconvalescent sera obtained from an influenza strain of the same subgroup which have envelope antigens different from the virus under investigation are used for tests. It is evident that with the latter serum all antigens common to the strains under investigation will react. It is not yet known whether the RNP-antigens are the only common antigens. For the complement-fixation test (CF-test) mainly reconvalescent sera were used; for fluorescent antibody studies hyperimmune sera were used. (In the following it will be specified which test has been used for the quantitation of the RNP-antigen.)

The RNP-antigen can be detected serologically only after the virion has been disrupted. It contains 10–15% RNA in the case of influenza viruses and 4–5% in the case of parainfluenza viruses (*52a, 55, 56*). The length of the RNP-antigen of parainfluenza viruses has been determined to be 1 μ (*56, 57*). It has been shown with Sendai virus that besides viria containing a RNP-antigen of this length, larger virus particles contain up to 20 multiples of this unit (*57*).

The relative amount of RNA in the RNP-antigen of parainfluenza viruses is about the same as in tobacco mosaic virus (TMV). Since the RNP-antigen of these viruses is about three times as long as TMV and the molecular weight of their RNA is also three times larger (see below), it may be assumed that the arrangement of the RNA is similar in both cases (*56*).

The RNP-antigen of influenza viruses is more flexible and the helical structure is not as clearly resolved as in the parainfluenza viruses. After release from the viria only pieces with a maximal length of 150–200 mμ are found (*20, 29*) (Fig. 5-5).

4. *Ribonucleic Acid*

The genetic material of myxoviruses is a single-stranded ribonucleic acid (RNA): the nucleic acid is RNase sensitive and has a base composition (Table 5-4) typical of single-stranded RNA (*58–61, 95*).

Marked differences in the molecular weights of RNA from parainfluenza and influenza viruses were found. From NDV, a RNA with a molecular weight of 7.5×10^6 ($S_{20} = 57S$) was isolated using detergents and phenol. This RNA corresponds to a single molecule per virus particle; however, it is not infectious. Besides this RNA of high molecular weight, a RNA fraction with a lower sedimentation coefficient of about 4S has been found in extracts from purified virus. The base composition of this RNA, however, is indistinguishable from that of cellular RNA (*62, 63*).

The size of the influenza virus RNA has not been definitely determined. Isolated influenza RNA shows two peaks in the ultracentrifuge: one sedimenting with 7S, the other with about 16S–20S (*59–61, 64, 65*). On some occasions a RNA of higher molecular weight (38S) was found which might be an aggregate (*59*).

It has been mentioned that the RNP-antigen of influenza viruses can be isolated only in short pieces and never as continuous filament

TABLE 5-4
Base Composition of Myxovirus RNA

Virus	C	A	G	U	References
NDV	23.0	23.8	23.8	29.4	*62, 203*
	23.2	20.1	25.4	31.2	
Influenza/PR8 (19S-RNA)	23.5	23.1	20.5	33.0	*59, 61*
	24.1	22.9	19.0	34.1	
Fowl plague	25.2	23.1	19.8	32.0	*95*

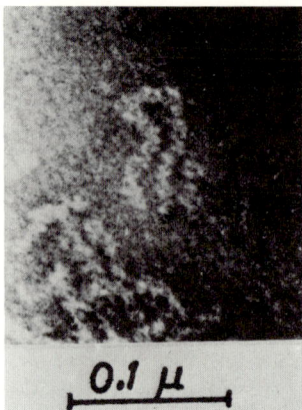

(a)

0.1 μ

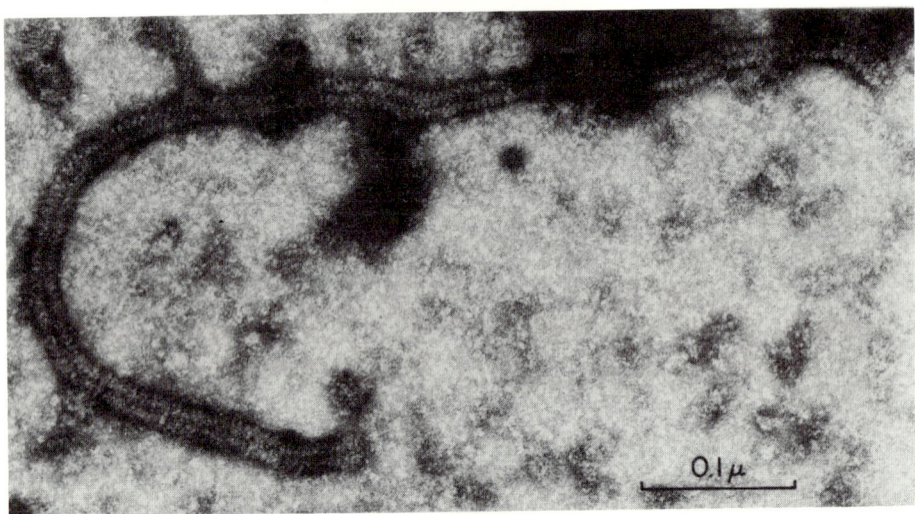

(b)

0.1 μ

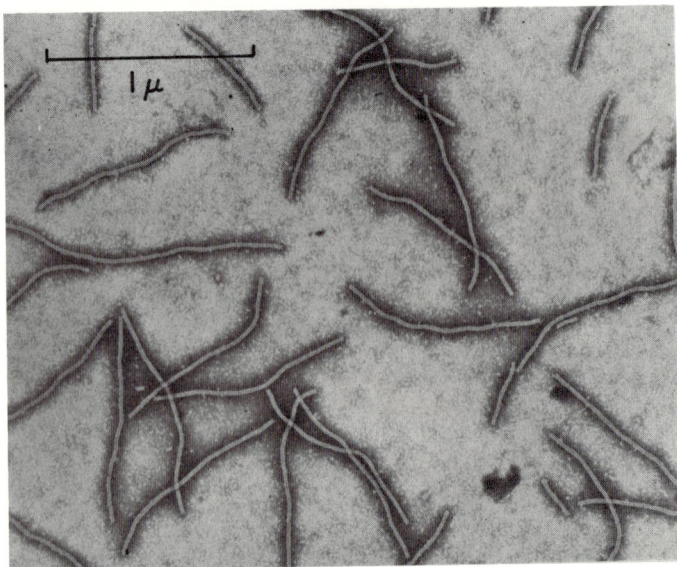

1 μ

(c)

like NDV-RNP antigen. This property of the RNP-antigen of influenza viruses might be the reason for the difficulties encountered in isolating a RNA of high molecular weight. It could be possible that the disruption of the viral envelope by the deproteinizing agents liberates the RNP-antigen so quickly that, because of the rapid unfolding of the RNP-antigen, only fragments can be obtained. If this would be the case, the addition of TMV-RNA as a marker of the RNA stability is not justified.

The observation that in all preparations at least the major part of influenza virus RNA has rather low sedimentation coefficients has led to the conclusion that this RNA exists in preformed pieces or has preferential breakage points. This assumption is supported by genetic experiments (9, 66) as will be discussed later. Many attempts have been made to isolate infectious RNA from influenza virus or from influenza-virus-infected cells. The results reported to date are confusing and contradictory. (For literature see Ref. 65.) There does not exist a method which yields reproducibly infectious RNA from influenza viruses.

5. Host-Specific Components

Serologic studies gave the first evidence that besides virus-specific material, host-specific components, so-called normal components, are integral constituents of myxoviruses (13, 14). These components cannot be removed from the virus particles by physico-chemical means without destroying their infectivity (14, 67, 68, 68a). Even after ether disruption the purified hemagglutinin still contains the host components (67, 69). Only extensive treatment of the virus with methanol separates some of the normal components from the virus-specific surface antigens (70).

Some of the host components could be characterized by their antigenic properties. Depending on the host in which the virus was propagated, various blood group substances, Forssman and mononucleosis antigens, were incorporated into the virion (67, 68, 71, 71a). By a single passage in a heterologous host the antigens derived from the original host were lost (72, 73).

Chemical studies revealed that the composition of the carbohy-

---

Fig. 5-5. RNP-antigen of influenza (a) and parainfluenza (b), (c) virus. [(a) Reprinted from Ref. 20 by courtesy of Verlag der Zeitschrift für Naturforschung, Tübingen. (b), (c) By courtesy of Dr. P. W. Choppin.]

drates and the lipids of the myxoviruses correspond to that of the host cell (*69, 73–77*).

Some of the viral properties are related to the presence of certain host components. Thus the buoyant densities of NDV depend on the hosts in which it was propagated (*78*). Although a pronounced hemolytic activity is exhibited only by the parainfluenza viruses such as NDV, Sendai, and mumps, it has been found that also this property is host dependent (*72*). The hemagglutinating activity, shown to be virus specific, can be inhibited by antibodies directed against host components (*13, 69*). This effect is most probably due to the close neighborhood of the two antigens on the virus surface.

5-4. Multiplication

A. Adsorption and Penetration

The ability of myxoviruses to adsorb to erythrocytes and to agglutinate them, i.e., the hemagglutination phenomenon, was widely used to investigate the mechanisms of adsorption of these viruses to cell surfaces. Results of experimental studies have revealed that electrostatic forces are responsible for the attachment of myxovirus particles to the cells. Myxoviruses do not adsorb in isotonic medium free of ions. By addition of electrolytes, however, adsorption can be obtained. Adsorbed viruses dissociate from the cells at higher salt concentrations without thereby splitting off sialic acid. If, on the other hand, sialic acid has been removed from the cell surface (*79*), myxoviruses do not adsorb under any conditions. Influenza viruses adsorbed to red cells will, after the sialic acid has been split off by the viral enzyme, elute from the cell surface. By the removal of sialic acid— a negatively charged molecule—the charge of the cell surface is drastically changed. In this sense sialic acid is essential for adsorption (for review see Refs. *80* and *83*).

The formation of the sialic acid–sialidase complex, which is necessary to hydrolyze the sialic acid of the cell surface, does not seem to be essential for the adsorption process. Sialidase activity can be inhibited by specific antibodies against sialidase without interfering with hemagglutination and infectivity (*81*). Heat-treated myxoviruses (indicator virus) can still adsorb to red cells but do not elute (*82*). Although these results favor the concept that sialidase is not involved in adsorption, they are equivocal, since sialidase is not completely destroyed under the conditions used.

After incubation of NDV, mumps, or Sendai viruses with erythrocytes above 20°C a portion of the adsorbed virus particles cannot be removed by viral or bacterial sialidase. The mode by which this virus portion is irreversibly bound is not known (*83*).

An active engulfment (pinocytosis) has been proposed as the mechanism responsible for the *entry* of myxoviruses into the host cell (*84*). The pinocytosis theory is supported by electron microscopic investigations which revealed intact NDV particles in the cytoplasm as early as 30 min after inoculation of chick embryo tissue culture cells (*85*). Disintegration of the virus particles must take place within the cytoplasm of the host cell. An interaction between the lipid structures of the viral envelope and lipid structures of the host cell has been claimed to be responsible for the disintegration (*86*). Attempts to demonstrate the liberation of the RNP-antigen after infection have been performed using ^{32}P-labeled influenza viruses (*87, 88*). In these experiments a ^{32}P-containing fraction which sediments more slowly than the intact viria was found 30–180 min after infection; and this fraction can be precipitated with RNP-antigen-specific antiserum. Besides this inner component, RNA was also demonstrated, probably released from the RNP-antigen.

B. ECLIPSE PHASE

Information about the events occurring between penetration of the virus and the appearance of new myxoviruses was obtained mainly from investigations of influenza A, NDV, and Sendai virus-infected cells. It was shown that the two groups of myxoviruses have completely different modes of reproduction.

1. *Influenza Viruses*

a. *Early Proteins*. As with other RNA-containing viruses the synthesis of a RNA complementary to the viral RNA is necessary for the replication of myxoviruses. Since *cellular* RNA is synthesized only on a DNA template, one must assume that for the replication of viral RNA no cellular enzymes are available. Thus protein synthesis should precede the synthesis of new viral RNA. This has indeed been shown with the aid of specific inhibitors of protein synthesis. The RNA synthesis of fowl plague virus can be inhibited by large amounts of *p*-fluorophenylalanine (FPA) when it is added up to 2 hr after infection but not when added later. This shows that the production of the major part of the protein(s) necessary for viral RNA synthesis ["early protein(s)"] has been completed 2 hr after infection and that it is stable throughout the remainder of the infectious cycle (*90, 91*). A virus-induced RNA-dependent RNA polymerase has been found in the microsomal fraction of cells infected with an influenza A virus (*89*). This enzyme functions only in the presence of all four nucleoside triphosphates and is not inhibited by actinomycin. It is not yet known whether the RNA-dependent RNA polymerase is the only "early protein" necessary for the synthesis of viral RNA.

b. *Viral RNA.* Influenza viruses do not multiply in the presence of actinomycin (*92*) which interferes specifically with DNA templates (*93*). For this reason this antibiotic cannot be used to unravel the viral RNA synthesis as is usually done with other RNA viruses. Therefore a chemical method was employed for the characterization of influenza RNA (*94*). This method takes advantage of (a) after digestion by RNase the oligonucleotide pattern of fowl plague virus RNA is completely different from that of cellular RNA and (b) in addition to cellular RNA no significant amounts of RNA other than that found in purified viria are synthesized in infected cells (*61, 94, 95*). In contrast to parainfluenza, only negligible amounts of RNA complementary to RNA of influenza particles is demonstrable (*60, 61*). It was found that the synthesis of viral RNA starts between 1–2 hr after infection. The maximum is reached 3 hr after infection (p.i.) (*94*). The highest sedimentation coefficient of viral RNA found in infected cells was 18S (*61*). Only minute amounts of double-stranded RNA were isolated from these cells (*60, 61*).

Autoradiographic investigations of fowl plague infected L-cells suggest that the viral RNA is synthesized in the nucleus (*96*).

c. *RNP-Antigen.* The protein moiety of the RNP-antigen of fowl plague virus is synthesized at the same time as the viral RNA. It has been demonstrated that as early as 2 hr after infection, when RNP-antigen can first be detected serologically, nearly all of the newly synthesized viral RNA can be precipitated by RNP antiserum (*94*). This implies that both components of the RNP-antigen are assembled immediately after synthesis.

When the RNP-antigen becomes detectable by the fluorescent antibody technique, it is located only in the nucleus. Later in the multiplication cycle it is also found in the cytoplasm (Fig. 5-6) (*97*).

Under certain conditions the RNP-antigen can be retained in the cell nucleus: (a) if large amounts of FPA are added 2 hr after in-

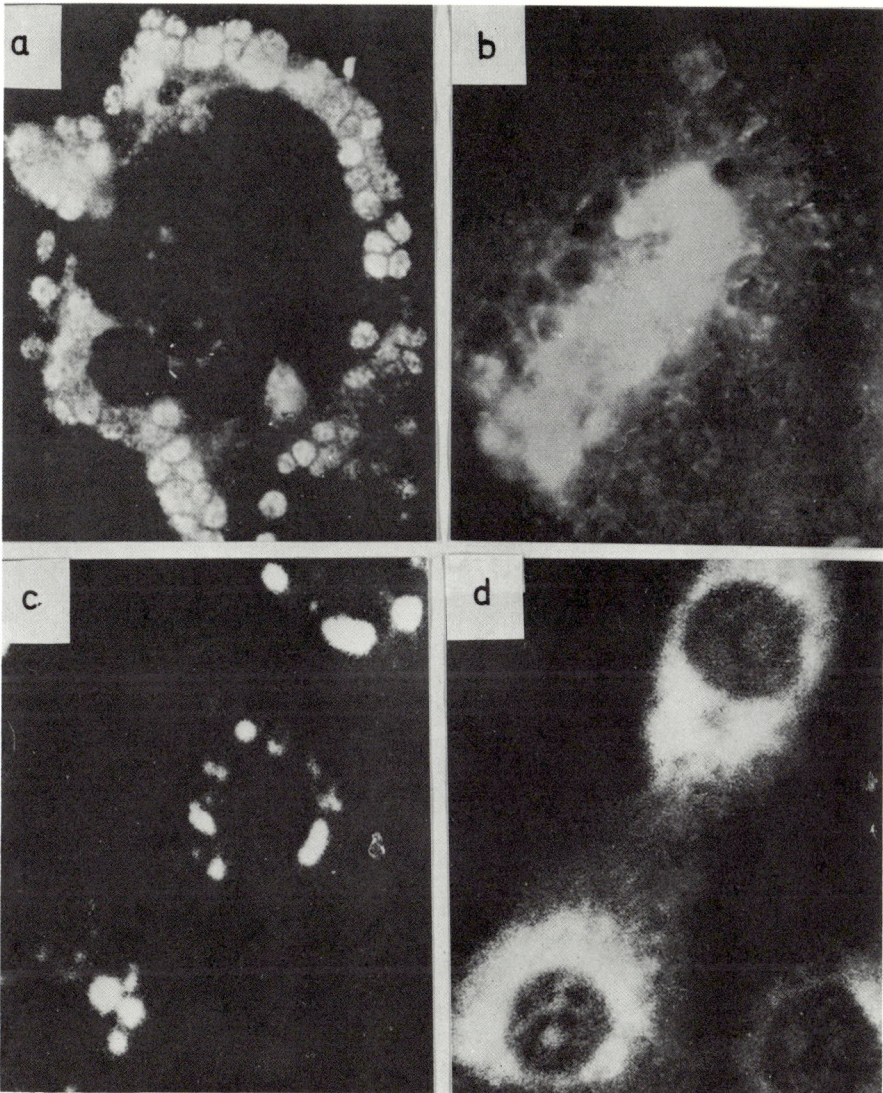

FIG. 5-6. Intracellular appearance of myxovirus components demonstrated by fluorescent antibody technique. (a) RNP-antigen and (b) hemagglutinin of fowl plague virus grown in multinucleated chicken giant cells; (c) RNP-antigen and (d) hemagglutinin of NDV grown in KB cells. (Reprinted from Ref. *125*, p. 31, by courtesy of Butterworths Scientific Publications, London.)

fection, when only the RNP-antigen is being synthesized and no other viral components (98); (b) under conditions of undiluted passages (von Magnus phenomenon, see below) (99); (c) in L-cells, when an abortive infectious cycle occurs (see below) (100).

The RNP-antigen which is synthesized in the presence of large amounts of FPA (addition 2 hr p.i.) is not incorporated into those virus particles formed after the removal of the block by phenylalanine (90, 91).

The site of synthesis of the RNP-antigen cannot be deduced from fluorescent antibody studies since this technique demonstrates only the site of accumulation. Recent studies using autoradiography have revealed a virus-induced protein with a high arginine/leucine or arginine/lysine ratio, which was tentatively identified as the RNP-antigen by using different doses of FPA and by determining the time course of its synthesis. After a pulse length between 10–30 min with ^{3}H-arginine this viral component was found in the cell nucleus. Using a shorter pulse, 5 min, the localization of the virus-specific protein was not possible. Therefore, it cannot be excluded that the RNP-antigen is not synthesized in the nucleus (101) (Fig. 5-7).

d. *Hemagglutinin.* A component which has the biological activities of the hemagglutinin is first demonstrable 1 hr after the RNP-antigen has appeared. This suggests that the production of the RNP-antigen and of the biologically active hemagglutinin are separate processes. This view is supported by experiments where the effect of FPA on influenza virus multiplication was investigated. When large amounts of FPA were added to the infected cultures 2 hr after infection, the RNP-antigen was produced in normal amounts, but no hemagglutinating activity, envelope antigen, nor infectious virus could be detected. When FPA was added between 3–6 hr p.i., hemagglutinin, envelope antigens, as well as infectious virus were produced. The amounts increased the later the compound was added (98, 102).

When a moderate dose of FPA is added immediately after infection, the synthesis of the "early protein" proceeds normally without a concomitant synthesis of hemagglutinin (103). Thus, the production of the viral envelope is most sensitive to the action of FPA.

e. *Sialidase.* The synthesis of influenza virus sialidase is directed by the viral genome. This is indicated by the fact that antibodies against the virion or the isolated enzyme will inhibit only the sialidase activity of the homologous strain (43–45a). Recombination experi-

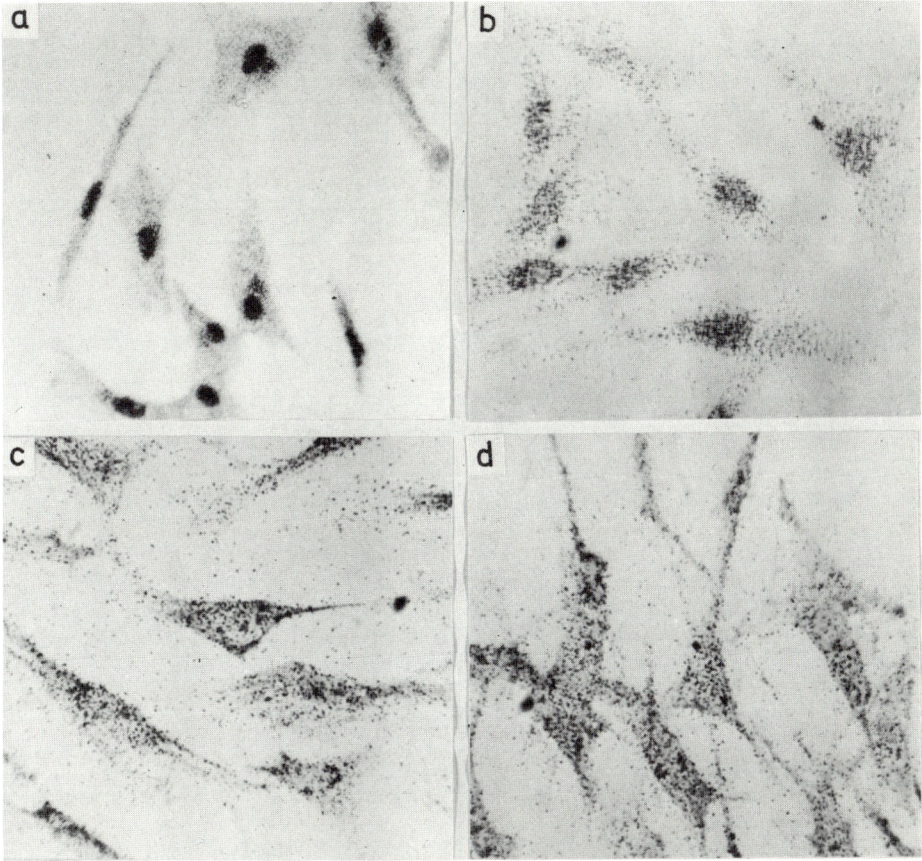

Fig. 5-7. Autoradiographs of fowl plague-(a), (b) and Newcastle disease virus (c)-infected secondary chick fibroblasts (3×10^5 cells/culture). (a) ³H-arginine, 7.5 μCi/culture, pulse from 2.5–3 hr p.i.; (b) ³H-leucine, 15 μCi/culture, pulse from 2.5–3 hr p.i.; (c) ³H-arginine, 5.0 μCi/culture, pulse from 6.0–6.5 hr p.i.; (d) noninfected control, pulse same as (a). Exposure time: 3 weeks. (By courtesy of Dr. H. Becht.)

ments with two different influenza strains produced a stable antigenic hybrid, which possessed the hemagglutinin and the RNP-antigen of one parent strain and the sialidase of the other (104).

The enzyme appears in measurable amounts 2–3 hr after infection with influenza A viruses. By extrapolation it has been estimated that its synthesis may start as early as 2 hr p.i. The major portion (60%)

TABLE 5-5

Effect of Actinomycin on the Labeling of the Acid Soluble Pool and
RNA of NDV-Infected and Noninfected Cells[a]

	μg/culture actinomycin	Counts per minute in	
		RNA	TCA-extract
Not infected	0	25,500	221,000
	0.3	1,480	66,000
	1.5	528	64,000
Infected	0	69,000	424,000
	0.3	7,600	294,000
	1.5	5,800	300,000

[a] Secondary chick fibroblast cultures (2×10^6 cells) on glass were either infected with NDV (multiplicity of $\sim$10) or were not infected. Four hours postinfection the actinomycin was added. The ³H-uridine pulse (5 μCi/culture) was given from 7 to $7\frac{3}{4}$ hr postinfection.

of the sialidase activity first appears, free of the hemagglutinating activity, in the "post microsomal fraction" (105).

2. Parainfluenza Viruses

a. *Early Proteins.* As with influenza viruses the existence of an "early protein(s)" has been demonstrated by using inhibitors of protein synthesis. Within the first 4 hr p.i., a stable "early protein(s)" is synthesized (106, 107).

b. *Viral RNA.* The course of virus-specific RNA synthesis can be determined in the case of parainfluenza viruses with the aid of actinomycin since these viruses multiply in the presence of this antibiotic (92, 108–112).

In NDV-infected cells newly synthesized actinomycin-resistent RNA can first be detected 3 hr after infection (106, 107). An analysis of this newly synthesized RNA revealed that it exists in various pieces of different molecular weights, the highest sedimentation coefficient being 57S. In contrast to influenza viruses most of this RNA is complementary (—strand) to the RNA found in NDV particles. Only small amounts of a double-stranded RNA with a thermal transition point of about 85°C is found in NDV-infected cells. Since most of the newly induced viral RNA is complementary to the RNA isolated from the virion, it is not certain how much of the double-stranded RNA is due to an artifact during the isolation (113–115).

The site of RNA synthesis was studied using autoradiography. Using short pulses of ³H-uridine, grains were found exclusively in the nucleus (116) in NDV-infected and actinomycin-treated cells. With pulses longer than 30 min, labeled viral RNA is demonstrable mainly, in the cytoplasm close to the nuclear membrane (116–118). These results were interpreted to show that the parainfluenza virus RNA is synthesized in the cell nucleus and then rapidly released into the cytoplasm where it accumulates in the immediate neighborhood of the nucleus (116).

This interpretation has to be taken with reservation, since it has been shown that actinomycin has a marked effect on the phosphorylation of labeled uridine in primary or secondary chick fibroblasts (119–121). As can be seen in Table 5-5 actinomycin reduces the phosphorylation of ³H-uridine in noninfected chick fibroblasts drastically, while in infected cells the radioactivity in the TCA-soluble pool is similar to noninfected cells without the antibiotic. Thus, the percentage of labeled cellular RNA synthesized in NDV-infected cells in the presence of actinomycin is much higher as might be expected from the data of the corresponding noninfected control. It is well known that actinomycin preferentially inhibits the synthesis of ribosomal *and* tRNA. Thus, the residual, labeled RNA in NDV-infected cells using a short pulse could be mainly cellular mRNA which moves relatively quickly to the cytoplasm.

c. *Proteins of the Virus Particle.* In contrast to influenza viruses, the virus-specific proteins were demonstrated in parainfluenza-virus-infected cells by fluorescent antibodies only in the cytoplasm (129–131) (Fig. 5-6). In most cases the RNP-antigen accumulates in distinct sites of the cytoplasm located in the immediate neighborhood of the nucleus. The hemagglutinin antigen, on the other hand, is spread over the whole cytoplasm. An arginine-rich virus-specific component, as found by autoradiography in fowl-plague-infected cells, was not demonstrable with NDV (101). All virus-specific proteins appear at about the same time in NDV-infected cells, yet 1 to 2 hr later than in influenza-virus-infected cells.

The formation of *NDV-sialidase* seems also to be directed by the viral genome since its synthesis is resistent to the action of actinomycin (109).

3. The Role of DNA in Myxovirus Multiplication

For the multiplication of parainfluenza viruses a *de novo* synthesis of DNA (*122*) or the function of cellular DNA is not necessary (*92, 95, 108–110*). This has been shown by investigations with substances which interfere with DNA synthesis or by the use of actinomycin D. Different opinions exist, however, whether cellular DNA plays a significant role in the multiplication of influenza virus. Inhibitors for DNA synthesis, like aminopterin (*90*), FUDR and BUDR (*111*), JUDR, and cytosine arabinoside (*123*) do not prevent influenza virus multiplication. On the other hand, actinomycin D inhibits their reproduction as already mentioned. Furthermore, if the cells are first treated with uv light (*95, 111, 112*) or mitomycin C (*95, 124*), which are known to destroy the cell genome, and then infected with influenza virus, no multiplication is found although NDV grows normally in those cells. From these results one may conclude that *de novo* synthesis of DNA is not necessary for influenza virus replication. It may be possible, however, that the already existing cellular DNA is involved in one of the multiplication steps (*92, 112*). An alternative explanation is offered by Rott and Scholtissek (*125*). From the following it is suggested that in the cell nucleus a RNA-degrading factor is set free by pretreatment with uv light, actinomycin, or mitomycin. (a) The effect of actinomycin is dependent on the multiplicity of infection (*103*). (b) The effect of low doses of actinomycin on cellular RNA synthesis can be reversed, but not the effect on the multiplication of influenza viruses (*111*). (c) The capacity to synthesize viral RNA decreases with time when the antibiotic is added 2 hr p.i. (*95*). This factor destroys the early viral RNA. The recent finding that the double-stranded RNA of influenza virus synthesized up to $1\frac{1}{2}$ hr after infection disappears again when actinomycin is added at this time, is in agreement with the finding mentioned above and lends further support to this explanation. When the antibiotic was given $3\frac{1}{2}$ hr p.i. the already synthesized "replicative form" remained intact (*126*).

The different effects of proflavine on NDV and fowl plague multiplication are noteworthy. Proflavine, in concentrations which inhibit DNA-dependent RNA synthesis, inhibits fowl plague multiplication but not the multiplication of NDV. Higher doses of proflavine, which interfere with protein synthesis, also prevent NDV multiplication. If these larger amounts of proflavine, however, are added after the "early protein" of NDV has already been synthesized, the NDV–

RNA synthesis proceeds normally, but infectious virus is not found (*127, 128*).

C. VIRUS MATURATION AND RELEASE

The newly synthesized viral components are transported to the marginal area of the cell where they seem to be finally assembled in protrusions of the cell wall into new infective virus particles. This is supported by various observations. Hoyle (*132, 133*), for example, observed filamentous microvilli bulging from infected cells. Investigations with fluorescent antibodies (*97*), ferritin conjugated antibodies (*134*), and direct electron microscopic observations (*135*) showed convincingly that these protrusions contain virus-specific material (Fig. 5-8). A feature of this altered cell surface is the specific adsorption of erythrocytes (hemadsorption) (*136, 137*). It is believed that parts of this modified membrane then form the viral envelope together with virus-specific material. Recent studies have suggested that the addition of vitamin A to the host cell can alter the phenotype of myxoviruses by producing changes in the packing of the molecular lipid leaflet (*138*). The maturation process is temperature dependent. Although at 25°C all known fowl plague virus-specific components are synthesized *slowly* yet in normal amounts, no infectious virus is formed (*139*).

Sialidase seems to play an important part in the release mechanism of myxoviruses. Usually, it takes about 1 hr for the particle to be released, but after a pretreatment of the infected cells with sialidase this time can be reduced to a few minutes (*140*). Furthermore, it has been shown that α-amino-p-methoxyphenylmethane sulfonic acid (AMPS) interferes with the release of newly formed influenza viruses. This effect can be abolished with sialidase obtained from *V. cholerae* (*141*). Finally, sialidase specific antiserum inhibits the release of fowl plague virus from infected cells (*81*).

5-5. Abnormal Cycles of Multiplication

A. CELL-TYPE-DEPENDENT ABNORMAL MULTIPLICATION

1. *Abortive Cycle*

After the eclipse of the virion in the host cell, the viral genetic material normally induces the formation of the viral components, and finally infectious virus is formed. In various different host cells

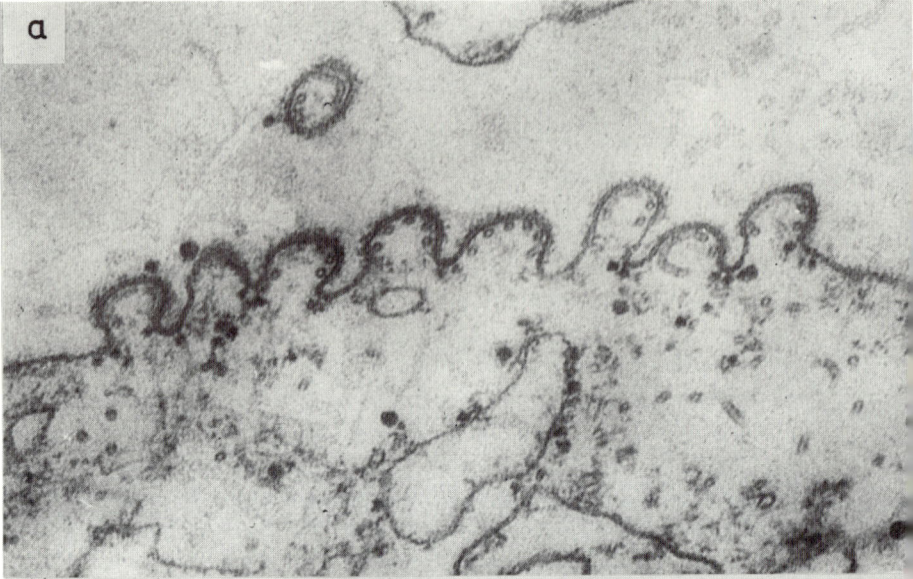

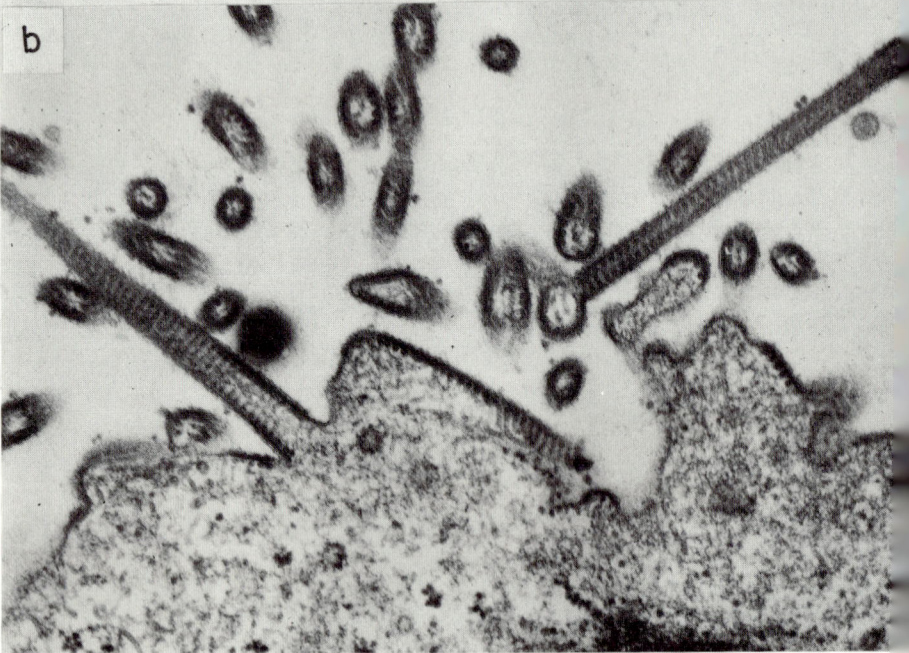

FIG. 5-8. Maturation of parainfluenza 2 virus at the cell periphery. (Reprinted from Ref. *135*, pp. 416, 417, by courtesy of Academic Press Inc., New York, London.)

the viral replication can proceed only to certain stages. This type of host-dependent interruption is called an abortive cycle.

The first example of such an abortive cycle was found in mice infected with a nonneurotropic strain of influenza virus (142). It was shown that infectious virus was not formed in these animals. Hemagglutinin, however, was produced. The higher the inoculated dose of the virus, the higher was the amount of newly synthesized hemagglutinin. After infection of HeLa-, L-, or BHK 21 cells with influenza A viruses, RNP-antigen and hemagglutinin are produced, but significant amounts of infectious virus are not synthesized (100, 143–146). Similar results have been obtained with NDV in different cell systems (131, 147, 148).

2. Persistent Infection

In most systems the infected host cells finally die, but there are cases where the cells survive although they are persistently infected. When the BHK 21/C13 line of hamster fibroblasts, for example, was infected with a parainfluenza virus type 2, all cells were found to contain virus-specific antigens, but they could nevertheless be subcultured for at least 40 passages. The virus yield was not higher than 10^{-4} plaque forming units per cell in 4-day-old subcultures (149). Many other cases of persistent virus infections of continuous cell lines are known. In many of these instances the balance between virus and host is controlled by the presence of extraneous factors such as added *antiviral antibody,* while in others the virus itself appears to limit the extent of the infection by inducing interferon in the infected cells. (For more detailed information, see Refs. 150–152.)

One interesting example of the latter type is the infection of L-cells with the egg-grown Victoria strain of NDV. Upon exposure of L-cells to high multiplicities of the virus all cells became abortively infected. They were found to produce only small amounts of viral antigen and little, if any, infectious virus. The cells yielded large amounts of interferon and the cultures survived. When the virus was adapted to L-cells and reisolated in eggs, it yielded in L-cells normal amounts of viral antigens and infectious virus, but no detectable interferon, and the cultures were destroyed (153).

Mumps virus infection of human conjunctiva cells in a minimal medium resulted in infectious progeny and cell destruction. If, however, the growth conditions for these cells were such that the cells were able to divide, a persistent infection was found. The cells could be cloned with the same efficiency as uninfected cells, and the virus yield was very low, about 1% of the cells showed hemadsorption, but 85–95% of the cells had in their cytoplasm detectable viral antigens. These cells could be easily superinfected with vesicular stomatitis virus, but a measurable resistance to Sendai virus or NDV was demonstrable. After the medium had become exhausted, or the cultures had been placed at 33°C instead of 37°C, 50–90% of the cells adsorbed erythrocytes and the level of infective virus in the medium rose about 10-fold. This suggests that under suboptimal conditions the equilibrium is shifted from mainly nonproducer cells to mainly virus producers (154).

B. Virus-Dependent Abnormal Multiplication

1. Von Magnus Phenomenon

If influenza virus is passaged with a multiplicity of infection higher than 1 (155), increasing amounts of incomplete viruses are formed and released from the host cell (v. Magnus phenomenon) (156, 157). While an abortive infection is found generally with all types of myxo-viruses, the von Magnus phenomenon is restricted to influenza viruses. The incomplete viruses have a surface structure similar to infectious virus, they are immunogenic and produce, to a certain degree, homologous interference (158–160). This phenomenon cannot be attributed to the action of interferon since with an increasing number of un-diluted passages less interferon is produced (161). The incomplete virus particles contain less RNA and RNP-antigen, show a lower infectivity/HA-ratio, and contain more lipid than complete viruses (156, 158, 162). The RNA incorporated into the incomplete virus particles is very similar to viral RNA. Both have almost the same oligonucleotide pattern and their specific radioactivities after labeling with ^{32}P are alike (162).

It has been suggested (162) that the von Magnus incomplete viruses are incomplete in a genetical sense, i.e., they have lost some of their RNA during these passages. Since no multiplicity reactivation has been observed, this loss of RNA should progress from one given end of the macromolecule. The concept of a sequential loss of RNA during passages at high multiplicity is supported by the finding that the first capacity which is lost is the one to produce infectious virus. Thereafter, the production of hemagglutinin and sialidase is lost and the last capacity to disappear is the ability to produce the RNP-anti-gen (CF-test) (*163*).

2. Incomplete Virus Formation by Partially Inactivated Influenza Virus

A further characteristic feature of influenza viruses is their ability to produce virus-specific components after partial destruction of their genetic material. This is in contrast to the parainfluenza virus NDV, where each hit of the viral genome inactivates the virus completely. Even viral subunits are no longer produced (*164, 165*). Cells infected with fowl plague virus which had been treated with an ethylene iminoquinone (Bayer A 139) or hydroxylamine, produce to a different

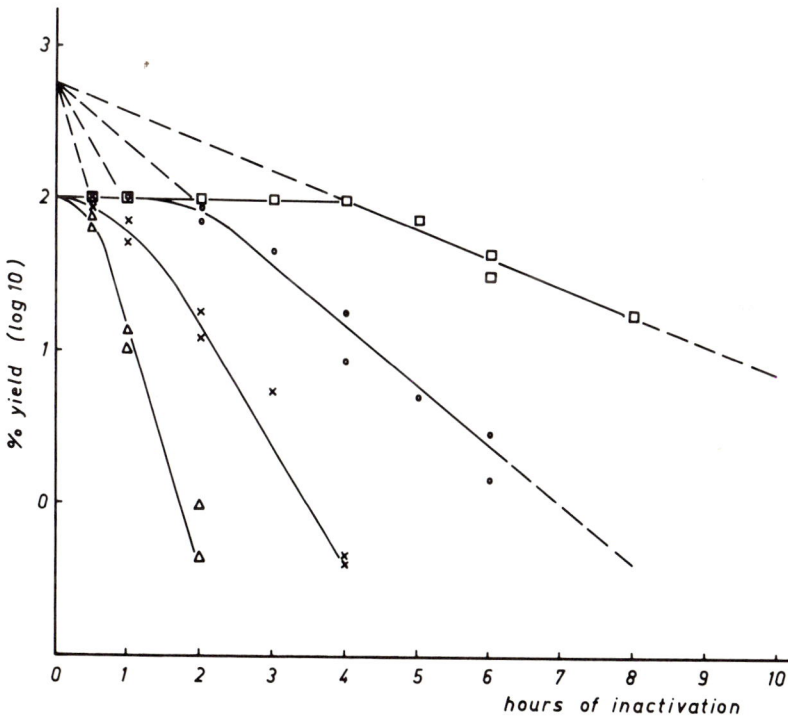

Fig. 5-9. Stepwise inactivation of fowl plague virus. △ = plaque forming units; ✕ = hemagglutination titer; ○ = sialidase activity; □ = RNP-antigen titer (CF-test). The curve for viral RNA synthesis follows that of the RNP-antigen (*164*). (Reprinted from Ref. *165*, p. **174**, by courtesy of Academic Press Inc., New York, London.)

Scheme 1

Scheme 2

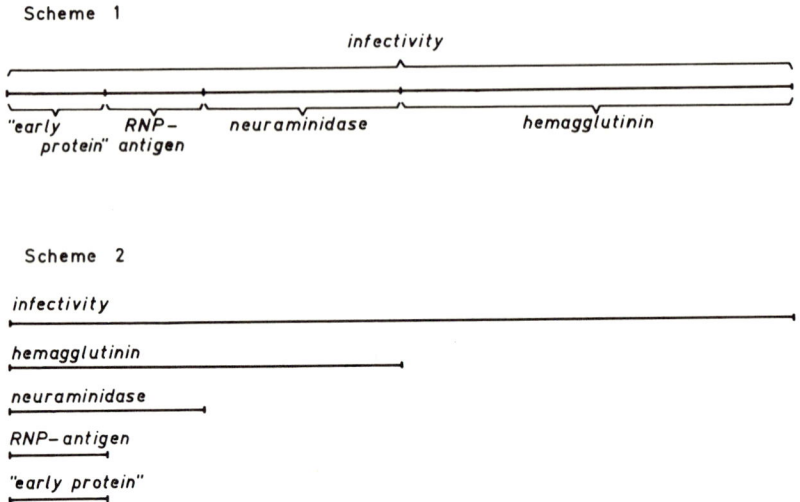

Fig. 5-10. Schematic arrangements of targets. For details see text.

extent infectious virus, hemagglutinin, viral sialidase, RNP-antigen (CF-test), and viral RNA, depending on the length of the inactivation of the virus, as shown in Fig. 5-9. The agents mentioned react exclusively with the genetic material and leave the viral proteins intact. The capacity to induce the synthesis of hemagglutinin is twice as resistant as the infectivity, while the capacity to induce sialidase is four times as resistant, and the capacity to induce RNP-antigen (CF-test) and the "early protein" is ten times as resistant. The inactivation of each function follows a single-hit kinetic.

From Fig. 5-9 the target size for the different activities relative to that of infectivity ($= 100\%$) can be determined. Thus the target size for the hemagglutinin is about 50%, the target size for sialidase is about 25% of that of infectivity, etc. Assuming that the total RNA is necessary for infectivity the various targets can be arranged along the viral genome in two different ways (Fig. 5-10). In *scheme 1* the various targets are arranged one after the other. The inactivation of each gene occurs independently with consequent loss of function and the different levels of resistance would be an expression of a varying target size. In *scheme 2* an overlapping arrangement of targets is shown which may represent the sequence necessary for the function of the different components. A hit into the gene coding for the "early protein(s)" will abolish also the ability to synthesize all the other virus-specific components, since viral RNA will not be multiplied and the other viral components cannot be synthesized in amounts detectable by the tests used. A hit into the gene coding for the hemagglutinin, however, will not abolish the ability to synthesize the "early protein(s)," RNP-antigen, or sialidase, but will abolish the ability to produce hemagglutinin and viria.

The RNP-antigen in these studies has been tested using reconvalescent sera (CF-test) as mentioned above. If the RNP-antigen was measured using fluorescent antibodies of a hyperimmune serum against RNP-antigen, and if the percentage of cells showing fluorescence was determined, the corresponding target size was the same as that for sialidase. With the CF-test the amount of the *not-enveloped* RNP-antigen of the total cell population is measured, while with the other test the *number* of cells synthesizing the RNP-antigen is counted. Thus, the total amount of *free* RNP-antigen in cells in which no envelope is synthesized might be higher than in those infected with active virus (*166*).

The target size of the antigenic component of the envelope, different from that of sialidase, has also been found to be the same as sialidase (*166*). Three different components have the same target size. This could mean that either the corresponding genes are located within the same target or that each of these genes belongs to a different target of equal size. Thus a compromise between the two extreme situations as proposed in scheme 1 and scheme 2 is probably correct.

5-6. Interaction between Virus Growth and Cell Metabolism

Although, as a rule, in picornavirus-infected cells cellular RNA and protein syntheses are inhibited by virus-specific protein (*167*), the myxoviruses leave the metabolisms of the infected cells unaltered —at least in the early stages of the multiplication cycle.

With fowl plague virus it has been demonstrated that the viral RNA synthesis is superimposed on the cellular RNA synthesis even at the maximum of viral RNA formation. The same has been found in NDV-infected cells. The protein and DNA-synthesis only decreases slowly during influenza A and NDV multiplication (*90, 107, 168, 169*). Most of these conclusions are derived from experiments where the incorporation of isotopes into the corresponding macromolecules was measured. Recently, it has been shown that after infection of chick fibroblasts with fowl plague virus the precursor pool is increased (*170*). *In vitro* studies revealed that after infection uridine, cytidine, and guanosine kinase activities are increased but not the one of adenosine kinase (*171*). The relatively late inhibition of the cellular metabolism may be due to cytotoxic activity of a virus component. Biochemical studies of the cytopathic effect (CPE) of an influenza virus have shown that such a component is indeed synthesized and has a target size similar to sialidase (*170*). The observation that many myxoviruses—e.g., parainfluenza 2 and 3 (*47, 172*) —multiply in various cells without destroying them, in spite of the formation of sialidase is evidence that sialidase itself is not responsible for the cytopathic effect. The intactness of cells was shown by superinfection with poliovirus, which give a normal yield of parainfluenza- as well as poliovirus (*172a*).

The CPE should not be confused with the cytotoxic effect which is induced by large amounts of virus particles per cell (*173*). In contrast to the CPE, which is a consequence of the synthesis of virus-specific material, the cytotoxic effect can also be caused by completely inactivated viruses (*174*).

5-7. Polykaryocytosis

A special kind of virus-host cell interrelationship is the induction of syncytia [i.e., multinuclear giant cells (polykaryocytes)]. Among the myxoviruses only the parainfluenza viruses induce polykaryocytosis (*175, 176*). This phenomenon has gained high significance for the formation of cell hybrids (*177*). In these experiments high multiplicities of uv-inactivated Sendai virus have been used. This phenomenon can be observed also at low multiplicities of infection late in the infectious cycle (*176*). Little is known about the mechanism of polycaryocytosis. Fusion is prevented by a pretreatment of the cells with *V. cholerae* sialidase or by periodate. The ability of NDV to cause polykaryocytosis is abolished only by agents which hydrolyze lipids in the virus membrane (*176*). These results indicate that an interaction of both the viral and the cell surface is involved in this phenomenon.

5-8. Genetic Interactions

The genetic interactions between viruses can be classified into those between two active viruses (recombination), between an active and inactivated virus (cross-reactivation), and between two inactivated viruses (multiplicity reactivation).

A. RECOMBINATION

Recombination of animal viruses was first reported for influenza A viruses (*178*). Much additional evidence has been accumulated since then and has been summarized exhaustively elsewhere (*9, 179–183*). The frequency of recombination of influenza viruses is higher as one would expect from the amount of RNA present in the virus particles. If closely related serotypes like the WS and WSN strain were used, the recombination frequences found ranged from 10–97%. The yield of recombinants parallels the input multiplicities employed (*182*). A most elegant recombination of two influenza A viruses was found where sialidase and hemagglutinin of two different strains were exchanged (*104*). Recombination was demonstrated among a wide variety of strains of influenza type A of human as well as of animal origin (*184, 185*), and among strains of influenza B, but not between A and B strains. The high rate of recombination found with certain markers might be due to an exchange of large pieces of RNA. However, the low rate of recombination of some other markers, which falls within the same range as found with other RNA-containing animal viruses, including parainfluenza viruses (*186*), may involve crossing over within these pieces (*9*).

B. CROSS REACTIVATION AND MULTIPLICITY REACTIVATION

1. *Cross Reactivation*

Cross reactivation is used synonymously with marker rescue and is a consequence of genetic recombination between a partially inactivated and an active parent virus. This phenomenon has been found with a variety of different serotypes of influenza viruses of human and animal origin. As a criterium of cross reactivation the capacity of plaque formation on a certain host cell was chosen, the plaque forming parent was carefully irradiated by uv (*185, 187*). This cross reactivation was found only within influenza A viruses, but not between influenza A and B viruses.

2. *Multiplicity Reactivation*

Multiplicity reactivation, a genetic interaction between two inactivated viruses, was found with influenza viruses by Henle and Liu

(*188*) and was confirmed by Kilbourne (*189*). In the meantime this phenomenon was reinvestigated using various virus host systems and different inactivation methods. Ultraviolet-inactivated influenza A virus, grown in pieces of allantois membranes at a multiplicity of above 1, yielded a higher progeny as one would have expected from the degree of inactivation (*190*). With fowl plague virus inactivated either by Bayer A 139 or hydroxylamine, a marked multiplicity reactivation was observed (*191, 192*).

A significant multiplicity reactivation of inactivated NDV could not be demonstrated (*191, 193*). The unusual properties of influenza viruses, e.g., the great difficulty in obtaining viral RNA as a single piece of high molecular weight, the von Magnus phenomenon, the ability to inactivate the virus stepwise, the high rates of genetic recombination, of marker rescue, and of multiplicity reactivation are consistent with the proposal (*9*) that the viral RNA either exists within the virion as distinct subunits or has preferential breaking points. Both possibilities would facilitate an exchange of genetic material.

C. PHENOTYPIC MIXING

Phenotypic mixing occurs in cells infected with at least two different virus strains or even virus types. In such a case the nucleic acid of one strain will be enclosed by the protein(s) of the other strain. The phenotype of such progeny particles is then not genetically fixed.

Myxoviruses as unrelated as NDV and influenza A virus may combine parts of their envelope (*186, 194*). In this manner a mosaic envelope is built containing both parental hemagglutinins (for a review see Ref. *181*). Since the structure of the myxovirus envelope is not as regular as the capsid of icosahedrical viruses and since their subunits are produced in a common pool in the doubly infected cells, it is not too surprising that they are incorporated into the virion at random.

ADDENDUM

After this chapter had been finished, several important contributions appeared; they are briefly considered here.

In a comparative study of the lipids of Newcastle disease (NDV) and Sendai virus grown in the same host, about 50% more unsaturated fatty acids were found in Sendai virus compared to NDV, while NDV contained more saturated fatty acids. This data may be significant for the site of maturation of these two viruses within the host cell (*204*).

Viral sialidase could be split off from influenza A2 virus by a protease from *Bacillus subtilis*. After this treatment, about 40% of total mass had been removed and the surface structure of the virus was markedly changed (*205, 206*). Sialidases isolated from a variety of influenza A viruses of human and avian origin could be classified by serological means into several groups. No cross-reactivity was found between these enzymes and influenza B virus sialidase (*207, 208*).

Highly purified sialidase isolated from influenza viruses showed ring-like structures under the electron microscope with a diameter between 80 and 90 Å. Purified hemagglutinating components, however, had a rod-like appearance with a diameter of about 60 Å and a length of 210 to 230 Å, which is similar to the spikes of the viria (*209*). There was a considerable difference in substrate specificities between myxo-virus sialidases. The NDV enzyme hydrolyzed $2 \rightarrow 8'$ linkages of sialic acids easily, whereas the fowl plague virus enzyme did not (*210*).

The RNP-antigen isolated from NDV has been reinvestigated by Kingsbury and Darlington (*211*). Its length appeared to be 1.3 to 1.4 μ rather than the values described recently (*56*). Its RNA had a S_{20} value of 50S. The RNA and the RNP-antigen of parainfluenza virus SV 5 had essentially the same parameters as those demonstrated for NDV (*212*).

Using polyacrylamid gel electrophoresis and sucrose density gradients, more evidence has been accumulated during the past year that the RNA of influenza virus exists in distinct pieces (*213–215*). An analysis of the replicative form of RNA isolated from influenza infected cells showed a corresponding size distribution of the pieces (*216*). The RNA of von Magnus-incomplete influenza virus was even more heterogenous with a higher proportion of smaller pieces (*214*). This is in support of a concept proposed by Rott and Scholtissek (*99*), where it was proposed that von Magnus—incomplete viruses have lost parts of RNA starting from one given end.

In equine influenza virus-infected cells about 90% of the newly synthesized viral RNA was plus-strand, while only about 10% was minus-strand. About 75% of this minus-stranded RNA did not exist in a double-stranded form inside the cell (*217*).

The RNA-dependent RNA polymerase induced after infection with fowl plague virus has been studied in more detail. This enzyme is first detectable 1.5 hr after infection. Its activity increases steadily during the total infectious cycle. It could be inhibited by low doses of polyanions like dextran sulfate or polyvinyl sulfate (218). The product of the reaction consists of 85 to 100% single-stranded RNA which has a base sequence complementary to the parental viral RNA (219). A corresponding enzyme has been found in NDV-infected cells. The NDV enzyme, however, is not sensitive to polyanions (220).

Influenza A2 virus multiplied normally in monkey kidney cells but showed an abortive infectious cycle in human diploid cells. In these cells all viral subunits were produced, but no infectious virus. The RNP-antigen left the nucleus normally (221).

The multiplication of NDV could be inhibited by superinfection with poliovirus in the presence of guanidine. Guanidine normally has no effect on cellular protein synthesis. However, it prevents multiplication of poliovirus, but the inhibition of cellular protein synthesis after infection with poliovirus is not affected (222).

Temperature-sensitive mutants of influenza A viruses have been isolated (223, 224). With the WSN-strain different mutants showed a frequency of recombination between 5 and 20% (223). Using antisera of low avidity in the overlay medium, antigenic mutants of influenza viruses have been selected which showed differences in the peptide maps of their hemagglutinating proteins (225).

ACKNOWLEDGMENTS

We thank Drs. M. A. Koch and J. E. Rodriguez for their criticism and help in preparing the manuscript. We are indebted to Dr. P. W. Choppin, Dr. W. Schäfer, Dr. H. Becht, and Dr. A. P. Waterson for their generous supply of photographs.

REFERENCES

1. C. H. Andrewes, F. B. Bang, and F. M. Burnet, Virology, 1, 176 (1955).
2. R. Heimer and K. Meyer, Proc. Natl. Acad. Sci. U.S., 42, 728 (1956).
3. A. Gottschalk, Biochim. Biophys. Acta, 23, 645 (1957).
4. A. Gottschalk, Glycoproteins. Their Composition, Structure and Function, Elsevier, Amsterdam–London–New York, 1966, p. 628.
5. J. G. Cruickshank, Ciba Foundation Symposium, Cellular Biology of Myxovirus Infections, Churchill, London, 1964, p. 5.
6. G. K. Hirst, Science, 94, 22 (1941).

7. L. McClelland and R. Hare, *Can. J. Public Health,* **32,** 530 (1941).

8. A. P. Waterson, *Nature,* **193,** 1163 (1962).

9. G. K. Hirst, *Cold Spring Harbor Symp. Quant. Biol.,* **27,** 303 (1962).

10. C. H. Andrewes, F. B. Bang, R. M. Chanock, and V. M. Zhdanov, *Virology,* **8,** 129 (1959).

11. M. K. Cook, B. E. Andrews, H. H. Fox, H. C. Turner, W. D. James, and R. M. Chanock, *Am. J. Hyg.,* **69,** 250 (1959).

12. G. Schramm, *Biochemie der Viren,* Springer, Berlin, 1954, p. 224.

13. C. A. Knight, *J. Exptl. Med.,* **80,** 83 (1944).

14. K. Munk and W. Schäfer, *Z. Naturforsch.,* **6b,** 372 (1951).

15. R. W. Horne and P. Wildy, *Advan. Virus Res.,* **10,** 101 (1962).

16. E. D. Kilbourne and J. S. Murphy, *J. Exptl. Med.,* **111,** 387 (1960).

17. F. B. Bang, *Proc. Soc. Exptl. Biol. Med.,* **63,** 5 (1946).

18. W. Schäfer, G. Schramm and E. Traub, *Z. Naturforsch.,* **4b,** 157 (1948).

19. L. Hoyle, R. W. Horne, and A. P. Waterson, *Virology,* **13,** 448 (1961).

20. A. P. Waterson, R. Rott, and W. Schäfer, *Z. Naturforsch.,* **16b,** 154 (1961).

21. L. Hoyle, R. W. Horne, and A. P. Waterson, *Virology,* **17,** 533 (1962).

22. J. D. Almeida and A. P. Waterson, *J. Gen. Microbiol.,* **46,** 107 (1967).

23. J. Archetti, A. Jemolo, D. Steve-Bocciarelli, G. Arangio-Ruiz, and ·F. Tangucci, *Arch. Ges. Virusforsch.,* **20,** 133 (1967).

24. T. H. Flewett and K. Apostolov, *J. Gen. Virology,* **1,** 297 (1967).

25. R. Drzeniek, *Nature,* **201,** 1205 (1966).

26. H. Noll, T. Aoyagi, and J. Orlando, *Virology,* **18,** 154 (1962).

27. R. W. Horne and P. Wildy, *Virology,* **15,** 348 (1961).

28. L. Hoyle, *J. Hyg.,* **50,** 229 (1952).

29. W. Schäfer and W. Zillig, *Z. Naturforsch.,* **9b,** 779 (1954).

30. Y. Hosaka, Y. Hosakawa, and K. Fukai, *Biken's J.,* **2,** 367 (1959).

31. R. Rott, *Zentr. Veterinaermed.,* **12b,** 74 (1965).

32. E. A. Eckert, *J. Bacteriol.,* **92,** 1430, 1907 (1966).

33. E. A. Eckert, *J. Virol.,* **1,** 920 (1967).

34. W. G. Laver, *Virology,* **20,** 251 (1963).

35. W. G. Laver, *J. Mol. Biol.,* **9,** 109 (1964).

36. M. L. Schneir and M. E. Rafelson, Jr., *Biochim. Biophys. Acta,* **130,** 1 (1966).

37. L. W. Mayron, B. Robert, R. J. Winzler, and M. E. Rafelson, Jr., *Arch. Biochem. Biophys.,* **92,** 475 (1961).

38. R. Drzeniek and R. Rott, *Z. Naturforsch.,* **18b,** 1127 (1963).

39. V. W. Wilson and M. E. Rafelson, Jr., *Biochem. Prep.,* **10,** 113 (1963).

40. M. Reginster, *J. Gen. Microbiol.,* **42,** 323 (1966).

41. J. T. Seto, R. Drzeniek, and R. Rott, *Biochim. Biophys. Acta,* **113,** 402 (1966).

42. R. Drzeniek, Habilitation thesis, Giessen, 1968.

43. R. Drzeniek, J. T. Seto, and R. Rott, *Biochim. Biophys. Acta,* **128,** 547 (1966).

44. J. T. Seto, K. Okuda, and Y. Hokama, *Nature,* **213,** 188 (1967).

45. G. L. Ada, P. E. Lind, and W. G. Laver, *J. Gen. Microbiol.,* **32,** 225 (1963).

45a. M. E. Rafelson, Jr., M. Schneir, and V. W. Wilson, Jr., *Arch. Biochem. Biophys.,* **103,** 424 (1963).

46. M. E. Rafelson, Jr., V. W. Wilson, Jr., and M. Schneir, *Presbyterian St. Luke's Hosp. Med. Bull.*, **1**, 34 (1962).

46a. V. W. Wilson, Jr. and M. E. Rafelson, Jr., *Biochim. Biophys. Acta*, **146**, 160 (1967).

47. R. Drzeniek, K. Bögel, and R. Rott, *Virology*, **31**, 725 (1967).

48. H. Tozawa, M. Homma, and N. Ishida, *Proc. Soc. Exptl. Biol. Med.*, **124**, 734 (1967).

49. M. E. Rafelson, Jr., M. Schneir, and V. W. Wilson, Jr., *The Amino Sugars*, Vol. II. B., Academic Press, New York, 1966, p. 171.

50. R. Drzeniek, *Biochim. Biophys. Res. Commun.*, **26**, 631 (1967).

51. H. Becht and R. Drzeniek, *J. Gen. Virol.*, **2**, 333 (1968).

52. W. Schäfer, *Ciba Foundation Symposium, The Nature of Viruses*, Churchill, London, 1957, p. 91.

52a. W. Zillig, W. Schäfer, and S. Ullmann, *Z. Naturforsch.*, **10b**, 199 (1955).

53. R. Rott, A. P. Waterson, and I. M. Reda, *Virology*, **21**, 663 (1963).

54. R. Rott and W. Schäfer, *Ciba Foundation Symposium, Cellular Biology of Myxovirus Infections,* Churchill, London, 1964, p. 27.

55. R. Rott, *Newcastle Disease Virus* (R. P. Hanson, ed.), University of Wisconsin Press, Madison, 1964, p. 133.

56. R. W. Compans and P. W. Choppin, *Proc. Natl. Acad. Sci. U.S.*, **57**, 949 (1967).

57. Y. Hosaka, H. Kitano, and S. Ikeguchi, *Virology*, **29**, 205 (1966).

58. W. Schäfer and E. Wecker, *Arch. Exptl. Veterinaermed.*, **12**, 418, (1958).

59. H. O. Agrawal and G. Bruening, *Proc. Natl. Acad. Sci. U.S.*, **55**, 818 (1966).

60. M. W. Pons, *Virology*, **31**, 523 (1967).

61. P. H. Duesberg and W. S. Robinson, *J. Mol. Biol.*, **25**, 383 (1967).

62. P. H. Duesberg and W. S. Robinson, *Proc. Natl. Acad. Sci. U.S.*, **54**, 794 (1965).

63. D. W. Kingsbury, *J. Mol. Biol.*, **18**, 195 (1966).

64. M. K. Cook, cited by G. K. Hirst, Ref. *9*.

65. P. Davies and R. D. Barry, *Nature*, **211**, 384 (1966).

66. R. W. Simpson, *Ciba Foundation Symposium, Cellular Biology of Myxovirus Infections,* Churchill, London, 1964, p. 187.

67. G. F. Springer and H. Tritel, *Science*, **138**, 687 (1962).

68. P. Isacson and A. E. Koch, *Virology*, **27**, 129 (1965).

68a. R. M. Franklin, *Progr. Med. Virol.*, **4**, 1 (1962).

69. A. Harboe, *Acta Path. Microbiol. Scandinav.*, **57**, 317 (1963).

70. R. Drzeniek, E. Reichert, and R. Rott, *Zentr. Veterinaermed.*, **13B**, 260 (1966).

71. R. Drzeniek, M. S. Saber, and R. Rott, *Z. Naturforsch.*, **21b**, 254 (1966).

71a. W. Smith, G. Belyavin, and F. W. Sheffield, *Proc. Roy. Soc., London*, Ser. B., **143**, 504 (1955).

72. T. Matsumoto and K. Maeno, *Virology*, **17**, 563 (1962).

73. R. Rott, R. Drzeniek, M. S. Saber, and E. Reichert, *Arch. Ges. Virusforsch.*, **19**, 273 (1966).

74. G. L. Ada and A. Gottschalk, *Biochem. J.*, **62**, 686 (1956).

75. M. Kates, A. C. Allison, D. A. J. Tyrell, and A. T. James, *Cold Spring Harbor Symp. Quant. Biol.*, **27**, 293 (1962).

76. E. Wecker, *Z. Naturforsch.*, **12b**, 208 (1957).
77. W. G. Laver and R. G. Webster, *Virology*, **30**, 104 (1966).
78. W. Stenbach and D. P. Durand, *Virology*, **20**, 545 (1963).
79. F. M. Burnet and J. D. Stone, *Australian J. Exptl. Biol. Med. Sci.*, **25**, 227 (1947).
80. A. Cohen, *Mechanism of Virus Infection* (W. Smith, ed.), Academic Press, New York, 1963, p. 153.
81. J. T. Seto and R. Rott, *Virology*, **30, 731** (1966).
82. T. Francis, Jr., *J. Exptl. Med.*, **85,** 1 (1947).
83. S. G. Anderson, *The Viruses*, Vol. III. (F. M. Burnet and W. M. Stanley, eds.), Academic Press, New York, 1959, p. 21.
84. S. Fazekas de St. Groth, *Nature*, **162,** 294 (1948).
85. M. Mussgay and J. Weibel, *Virology*, **16, 506** (1962).
86. L. Hoyle, *Cold Spring Harbor Symp. Quant. Biol.*, **27,** 113 (1962).
87. L. Hoyle and W. Frisch-Niggemeyer, *J. Hyg.*, **53,** 474 (1955).
88. E. Wecker and W. Schäfer, *Z. Naturforsch.*, **12b,** 483 (1957).
89. P. P. K. Ho and C. P. Walters, *Biochemistry*, **5,** 231 (1966).
90. C. Scholtissek and R. Rott, *Z. Naturforsch.*, **16b,** 663 (1961).
91. C. Scholtissek and R. Rott, *Nature*, **191,** 1023 (1961).
92. R. D. Barry, D. R. Ives, and J. G. Cruickshank, *Nature*, **194,** 1139 (1962).
93. E. Reich, R. M. Franklin, A. J. Shatkin, and E. L. Tatum, *Science*, **134,** 556 (1961).
94. C. Scholtissek and R. Rott, *Z. Naturforsch.*, **16b,** 109 (1961).
95. R. Rott, M. S. Saber, and C. Scholtissek, *Nature*, **205,** 1187 (1965).
96. C. Scholtissek, R. Rott, P. Hausen, H. Hausen, and W. Schäfer, *Cold Spring Harbor Symp. Quant. Biol.*, **27,** 245 (1962).
97. P. M. Breitenfeld and W. Schäfer, *Virology*, **4,** 328 (1957).
98. T. Zimmermann and W. Schäfer, *Virology*, **11,** 676 (1960).
99. R. Rott and C. Scholtissek, *J. Gen. Microbiol.*, **33,** 303 (1963).
100. R. M. Franklin and P. M. Breitenfeld, *Virology*, **8,** 293 (1959).
101. H. Becht, *J. Gen. Virol.*, **4,** 215 (1969).
102. H. Becht, personal communication.
103. D. O. White, H. M. Day, E. J. Batchelder, I. M. Cheyne, and A. J. Wansbrough, *Virology*, **25,** *289* (1965).
104. W. G. Laver and E. D. Kilbourne, *Virology*, **30,** 493 (1966).
105. A. Noll, *Cold Spring Harbor Symp. Quant. Biol.*, **27,** 256 (1962).
106. D. E. Wilson and P. LoGerfo, *J. Bacteriol.*, **88,** 1550 (1964).
107. C. Scholtissek and R. Rott, *Nature*, **206,** 729 (1965).
108. D. W. Kingsbury, *Biochem. Biophys. Res. Commun.*, **9,** 156 (1962).
109. R. Rott and C. Scholtissek, *Z. Naturforsch.*, **19b,** 316 (1964).
110. A. Granoff and D. W. Kingsbury, *Ciba Foundation Symposium, Cellular Biology of Myxovirus Infections*, Churchill, London, 1964, p. 96.
111. R. D. Barry, *Virology*, **24,** 563 (1964).
112. D. O. White and I. M. Cheyne, *Virology*, **29,** 49 (1966).
113. F. Sokol, *Second Meeting of the Federation of European Chemical Societies, Vienna, 1965*, Verlag der Wiener Medizinischen Akademie, Vienna, 1965.
114. D. W. Kingsbury, *J. Mol. Biol.*, **18,** 204 (1966).
115. M. A. Bratt and W. S. Robinson, *J. Mol. Biol.*, **23,** 1 (1967).

116. A. G. Bukrinskaya, O. Burducca, and G. K. Vorkunova, *Proc. Soc. Exptl. Biol. Med.*, **123**, 236 (1966).
117. E. F. Wheelock, *Proc. Soc. Exptl. Biol. Med.*, **114**, 56 (1963).
118. R. Schneider, Doctoral thesis, Tübingen (1967).
119. C. Scholtissek, *Biochim. Biophys. Acta*, **145**, 238 (1967).
120. J. J. Skehel, A. J. Hay, D. C. Burke, and L. N. Cartwright, *Biochim. Biophys. Acta*, **142**, 430 (1967).
121. H. Becht and C. Scholtissek, unpublished results.
122. E. H. Simon, *Virology*, **13**, 105 (1961).
123. R. Rott and C. Scholtissek, unpublished results.
124. D. P. Nayak and A. F. Rasmussen, Jr., *Virology*, **30**, 673 (1966).
125. R. Rott and C. Scholtissek, *Mod. Trends Med. Virol.*, **1**, 25 (1967).
126. M. W. Pons, *Virology*, **33**, 150 (1967).
127. C. Scholtissek and R. Rott, *Nature*, **204**, 39 (1964).
128. V. Stakhanova and C. Scholtissek, *J. Gen. Virol.*, **1**, 571 (1967).
129. M. I. Traver, R. L. Northrop, and D. L. Walker, *Proc. Soc. Exptl. Biol. Med.*, **104**, 268 (1960).
130. E. F. Wheelock and I. Tamm, *Virology*, **8**, 532 (1959).
131. I. M. Reda, R. Rott, and W. Schäfer, *Virology*, **22**, 422 (1964).
132. L. Hoyle, *J. Hyg.*, **48**, 277 (1950).
133. L. Hoyle, *J. Hyg.*, **52**, 180 (1954).
134. C. Morgan, R. A. Rifkind, and H. M. Rose, *Cold Spring Harbor Symp. Quant. Biol.*, **27**, 57 (1962).
135. R. W. Compans, V. V. Holmes, S. Dales, and P. W. Choppin, *Virology*, **30**, 411 (1966).
136. J. Vogel and A. Shelokov, *Science*, **126**, 358 (1957).
137. P. J. Marcus, *Cold Spring Harbor Symp. Quant. Biol.*, **27**, 351 (1962).
138. H. A. Blough, D. R. Weinstein, D. E. M. Lawson, and E. Kodicek, *Virology*, **33**, 459 (1967).
139. R. Rott and C. Scholtissek, *J. Gen. Virol.*, **3**, 239 (1968).
140. H. J. F. Cairns and P. J. Mason, *J. Immunol.*, **71**, 38 (1953).
141. W. W. Ackermann and H. F. Maassab, *J. Exptl. Med.*, **100**, 329 (1954).
142. R. W. Schlesinger, *Proc. Soc. Exptl. Biol. Med.*, **74**, 541 (1950).
143. G. Henle, A. Girardi, and W. Henle, *J. Exptl. Med.*, **101**, 25 (1955).
144. F. Deinhardt and G. Henle, *J. Immunol.*, **79**, 60 (1957).
145. V. Ter Meulen and R. Love, *J. Virol.*, **1**, 626 (1967).
146. K. B. Fraser, *J. Gen. Virol.*, **1**, 1 (1967).
147. A. M. Prince and H. S. Ginsberg, *J. Exptl. Med.*, **105**, 177 (1957).
148. W. C. Wilcox, *Virology*, **9**, 30, 45 (1959).
149. K. B. Fraser and J. Anderson, *J. Gen. Microbiol.*, **44**, 47 (1966).
150. H. S. Ginsberg, *Progr. Med. Virol.*, **1**, 36 (1958).
151. A. Isaacs, *Advan. Virus Res.*, **10**, 1 (1963).
152. D. L. Walker, *Progr. Med. Virol.*, **6**, 111 (1964).
153. J. E. Rodriguez, V. Ter Meulen, and W. Henle, *J. Virol.*, **1**, 1 (1967).
154. D. L. Walker and W. Hinze, *J. Exptl. Med.*, **116**, 739, 751 (1962).
155. R. D. Barry, *Virology*, **14**, 389 (1961).
156. P. v. Magnus, *Advan. Virus Res.*, **2**, 59 (1954).
157. R. Rott and W. Schäfer, *Z. Naturforsch.*, **15b**, 691 (1960).

158. A. Isaacs, *The Viruses,* Vol. III (F. M. Burnet and W. M. Stanley, eds.), Academic Press, New York, 1959, p. 111.

159. K. Paucker, A. Birch-Anderson, and P. v. Magnus, *Virology,* **8,** 1 (1959).

160. R. Rott and W. Schäfer, *Z. Naturforsch.,* **16b,** 310 (1961).

161. G. Schloer, to be published.

162. R. Rott and C. Scholtissek, *J. Gen. Microbiol.,* **33,** 303 (1963).

163. C. Scholtissek, R. Drzeniek, and R. Rott, *Virology,* **30,** 313 (1966).

164. C. Scholtissek and R. Rott, *Nature,* **199,** 200 (1963).

165. C. Scholtissek and R. Rott, *Virology,* **22,** 169 (1964).

166. H. Becht and C. Scholtissek, unpublished results.

167. R. Bablanian, H. J. Eggers, and I. Tamm, *Virology,* **26,** 100 (1965).

168. M. Rosenbergova, *Acta Virol.,* **9,** 55 (1965).

169. D. P. Bolognesi and D. E. Wilson, *J. Bacteriol.,* **91,** 1896 (1966).

170. C. Scholtissek, H. Becht, and R. Drzenick, *J. Gen. Virol.,* **1,** 219 (1967).

171. C. Scholtissek, unpublished results.

172. K. V. Holmes and P. W. Choppin, *J. Exptl. Med.,* **124,** 501 (1966).

172a. P. W. Choppin and K. V. Holmes, *Virology,* **33,** 442 (1967).

173. P. M. Cooke, *Am. J. Med. Sci.,* **241,** 383 (1961).

174. R. Rott and G. Müller, *Arch. Ges. Virusforsch.,* **17,** 139 (1965).

175. Y. Okada, *Biken's J.,* **1,** 103 (1958).

176. A. Kohn, *Virology,* **26,** 298 (1965).

177. H. Harris and J. F. Watkins, *Nature,* **205,** 640 (1965).

178. F. M. Burnet and P. E. Lind, *Australian J. Sci.,* **12,** 109 (1949).

179. F. M. Burnet, *The Principles of Animal Virology,* Academic Press, New York, 1960.

180. E. D. Kilbourne, *Progr. Med. Virol.,* **5,** 81 (1963).

181. F. Fenner and J. F. Sambrook, *Ann. Rev. Microbiol.,* **18,** 47 (1964).

182. R. W. Simpson, *Ciba Foundation Symposium, Cellular Biology of Myxovirus Infections,* Churchill, London, 1964, p. 187.

183. E. D. Kilbourne, F. S. Lief, J. L. Schulman, R. I. Jahiel, and W. G. Laver, *Perspectives Virol. Symp. N.Y.* **1967–V,** 87.

184. H. R. Staiger, *Virology,* **22,** 419 (1964).

185. B. Tumova and H. G. Pereira, *Virology,* **27,** 253 (1965).

186. A. Granoff, *Cold Spring Harbor Symp. Quant. Biol.,* **27,** 319 (1962).

187. R. W. Simpson and G. K. Hirst, *Virology,* **15,** 436 (1961).

188. W. Henle and O. C. Liu, *J. Exptl. Med.,* **94,** 305 (1951).

189. E. D. Kilbourne, *J. Exptl. Med.,* **101,** 437 (1955).

190. R. D. Barry, *Virology,* **14,** 298 (1961).

191. C. Scholtissek, R. Rott, and W. Schäfer, *Z. Naturforsch.,* **17b,** 222 (1962).

192. W. Schäfer and R. Rott, *Z. Hyg. Infektionskrankh.,* **148,** 256 (1962).

193. R. D. Barry, *Nature,* **103,** 96 (1962).

194. A. Granoff and G. K. Hirst, *Proc. Soc. Exptl. Biol. Med.,* **86,** 84 (1954).

195. R. Rott and W. Schäfer, *Virology,* **14,** 298 (1961).

196. J. D. Almeida, A. P. Waterson, and E. W. L. Fletcher, *Nature,* **206,** 1125 (1965).

197. A. R. Taylor, *J. Biol. Chem.,* **153,** 675 (1944).

198. G. L. Ada and B. T. Perry, *Australian J. Exptl. Biol. Med. Sci.,* **32,** 453 (1954).

199. L. H. Frommhagen and C. A. Knight, *Virology,* **2,** 430 (1956).

200. G L. Ada and A. Gottschalk, *Biochem. J.,* **62,** 686 (1956).

201. R. Rott, *Zentr. Veterinaermed.,* **12B,** 74 (1965).

202. R. Cunha, M. L. Weil, D. Beard, A. R. Taylor, D. G. Sharp, and J. W. Beard, *J. Immunol.,* **55,** 69 (1947).

203. D. W. Kingsbury, *J. Mol. Biol.,* **18,** 204 (1966).

204. H. A. Blough and D. E. M. Lawson, *Virology,* **36,** 286 (1968).

205. F. Biddle, *J. Gen. Virol.,* **2,** 19 (1968).

206. A. P. Kendal, F. Biddle, and G. Belyavin, *Biochim. Biophys. Acta,* **165,** 419 (1968).

207. C. K. J. Paniker, *J. Gen Virol.,* **2,** 385 (1968).

208. R. G. Webster and H. G. Pereira, *J. Gen Virol.,* **3,** 201 (1968).

209. R. Drzeniek, H. Frank, and R. Rott, *Virology,* **36,** 703 (1968).

210. R. Drzeniek, to be published.

211. D. W. Kingsbury and R. W. Darlington, *J. Virol.,* **2,** 248 (1968).

212. R. W. Compans and P. W. Choppin, *Virology,* **35,** 289 (1968).

213. M. W. Pons and G. K. Hirst, *Virology,* **34,** 385 (1968).

214. P. H. Duesberg, *Proc. Natl. Acad. Sci. U.S.,* **59,** 930 (1968).

215. R. D. Barry and P. Davies, *J. Gen. Virol.,* **2,** 59 (1968).

216. M. W. Pons and G. K. Hirst, *Virology,* **35,** 182 (1968).

217. D. P. Nayak and M. A. Baluda, *J. Virol.,* **2,** 99 (1968).

218. C. Scholtissek and R. Rott, *J. Gen. Virol.,* **4,** 125 (1969).

219. C. Scholtissek, *Biochim. Biophys. Acta,***179,** 389 (1969).

220. C. Scholtissek and R. Rott, *J. Gen. Virol.,* to be published.

221. H. Závadová, V. Vonka, L. Kutinová, and E. Tučková, *J. Gen. Virol.,* **2,** 341 (1968).

222. Y. Ito, H. Okazaki, and N. Ishida, *J. Virol.,* **2,** 645 (1968).

223. R. W. Simpson and G. K. Hirst, *Virology,* **35,** 41 (1968).

224. T. E. Medvedeva, G. I. Alexandrova, and A. A. Smorodintsev, *J. Virol.,* **2,** 456 (1968).

225. W. G. Laver and R. G. Webster, *Virology,* **34,** 193 (1968).

6

THE REOVIRUSES

C. J. GAUNTT AND A. F. GRAHAM

THE WISTAR INSTITUTE OF ANATOMY AND BIOLOGY
PHILADELPHIA, PENNSYLVANIA

6-1. Introduction

Reovirus (*respiratory enteric orphan* virus) is a term originally coined by Sabin (*1*) in 1959 to describe a group of viruses that had first been classified as ECHO type 10. Inclusion of the reoviruses in the ECHO group in 1954 was based on their isolation from human stools, growth in primary monkey kidney cell cultures, lack of association with any specific disease and lack of pathogenicity for newborn mice, guinea pigs, rabbits, and monkeys (*1, 2*). It soon became apparent, however, that the ECHO type 10 group differed from all other enteroviruses in cytopathogenic effect in cell culture (*3, 5*), in agglutination of human type O erythrocytes (*6*), and in having a greater size (*7*). More recently, the reoviruses have been shown to have a genome of double-stranded RNA and are, in this respect, in a unique position among mammalian viruses.

Three serotypes of reoviruses, designated types 1, 2, and 3, are recognized through differences in extent of neutralization by specific antisera (*2, 8*) and by hemagglutination inhibition (HI) tests (*1, 9*). Type 2 strains are further classified into four subtypes according to the HI assay (*10*). A common complement-fixing antigen is shared by all three serotypes (*10*). No significant morphological differences have been found among the three types of reovirus (*12–18*).

Reoviruses are almost ubiquitous in their distribution in natural hosts. Evidence for infection is derived by direct isolation of the virus or the detection of antibody by neutralization, HI inhibition, or complement fixation tests. Antibody to one or another of the three types of virus is common in bats, birds, camels, cats, cattle, chickens, dogs, guinea pigs, hares, horses, marsupials, mice, pigs, rabbits, rats, reptiles, and sheep (*19–21*). Infection of humans with reoviruses occurs on all continents and in most races (*19*). There is no obvious difference between the human and animal strains (*19, 21*). Reoviruses have not yet been associated with specific clinical illnesses in man, although it is likely that infection occurs early during childhood as evidenced by the presence of antibody and by isolation of virus from both healthy children and those with mild clinical illnesses (*1, 2, 19, 22, 23*).

6-2. Virus Assay

Reoviruses have a wide host range in cultures of mammalian cells, having been propagated in primary monkey kidney cells (*24, 25*), human fetal diploid kidney cells, baby hamster kidney cells, a grivet monkey cell line (BSC-1), HeLa cells (24) and L-cells (12) among others.

The plaque method, first introduced by Rhim and Melnick (25, 26), is now the generally used procedure for virus assay. A number of different cell types can be used and, in general, plaques can be enumerated 4–8 days after inoculation of monolayers (12, 27–29). A comprehensive study has been made of the relative plating efficiencies of nine different virus strains on several different types of cell monolayer (27). Titers varied widely with the particular strain of virus and type of cell. In some cases the ratio of infectious virus particles to total particles can be high as shown by Rhim, Smith, and Melnick (15) who plated a strain of type 1 virus on layers of monkey kidney cells and found that there were 11 particles per plaque-forming unit. Considering that this virus suspension contained an excess of "coreless" forms and had a tendency to form small aggregates the authors felt that the actual ratio of particles to plaque-forming units was probably close to unity.

Another assay method for reovirus depends on infection of cell layers on coverslips and, after 24 hr, staining the cells with fluorescein-labeled reovirus antibody (27, 30). Specifically stained cells are then enumerated under a fluorescence microscope and provide an estimate of the amount of virus in the original inoculum. The relationship between virus concentration and number of infectious units is linear (30), as it is with the plaque assay.

Some strains of reovirus are associated with what appears to be a protein inhibitor since the virus titer of a lysate can often be markedly increased by treatment with chymotrypsin before assay. This phenomenon has been observed with strains of all three types of reovirus (27). Treatment with pancreatin (28) or incubation at 55°C with 2 M magnesium chloride (31) also enhanced the infectivity of some strains of virus.

6-3. Properties and Structure of the Virion

A. PHYSICAL PROPERTIES

Complete virions have been variously reported to have a buoyant density of 1.32 (32), 1.36 (17), 1.38 (33–35, 37) and 1.41 (36) g/cm³ when centrifuged to equilibrium in density gradients of cesium chloride. Type 3 virions were found to have a sedimentation coefficient of 630S and a diffusion constant of 8.3×10^{-8} cm^2 sec and from these data the mass of the virion was calculated to be 70×10^6 daltons (*34*). This must be regarded as a minimum value for the molecular weight in view of uncertainties in measurement of the diffusion constant.

Resistance to inactivation by heat varies markedly among strains of reovirus. Thus for a type 3 strain, in a serum-containing medium at approximately pH 7, inactivation of infectivity was exponential at 37, 45, and 56°C, the half-lives being 157, 33, and 1.6 min, respectively (*12*). A strain of type 1 virus is probably considerably more resistant (*15*), being inactivated at 4, 24, and 37°C with half-lives of 3.7 days, 2.0 days, and 19 hr at pH 7.5. In fact, it is relatively easy to select a reovirus mutant more resistant than the wild type to inactivation at 50°C (*38*).

Reovirus is reported to be stable to changes in pH in the region pH 2.2 to 8.0 (*19, 39*), but the region of pH stability is also undoubtedly a function of the specific strain of virus (*38*).

Infectivity is not reduced after treatment for 1 hr at 20°C with 1% hydrogen peroxide, 1% phenol, 3% formaldehyde, 20% lysol (*19*), or to treatment with diethylether (*1, 2, 12, 19*) but is inactivated by 70% ethanol (*19*) and potassium periodate (*40*). Magnesium ion has a 2-fold effect on reovirus depending on the temperature. A 4- to 8-fold increase in infectivity was found when type 1 virus was heated to 50°C in 2 M MgCl$_2$ (*24, 31*), whereas freezing the virus to -30°C at this salt concentration led to extensive inactivation (*31*). Several other divalent cations did not enhance the infectivity of reovirus at 50°C but increased the extent of inactivation at -30°C over that observed in their absence (*31*). The action of various enzymes on the infectivity of reovirus will be considered later.

B. MORPHOLOGY

Early measurements of particle size by filtration through gradocol membranes suggested a diameter of 70 to 75 mμ for the reovirus virion (*1, 7, 8*). More recent estimates by electron microscopy of reovirus from ammonium acetate suspensions (*12*), of negatively stained virus (*13, 15, 16, 18, 32*), and of thin sections of infected cells (*11, 16, 41, 42*) are in agreement with the filtration results, although it has been claimed that one type 1 strain has a diameter of approximately 60 mμ (*14, 17*).

The viral particle is an icosahedron with 5:3:2 symmetry, the capsid probably being composed of 92 prismatic, elongated capsomeres (*12–14, 17, 18, 43*). These capsomeres are five or six sided and about 7.5 mμ in diameter with a hollow center 3 to 4 mμ across. An alternative suggestion has been made that the capsid may be comprised of 18 morphological units with 92 holes on the surface (*44*). The capsomeres are arranged on an inner shell which is approximately 50 mμ in diameter and contains the RNA genome of the virus (*17, 18, 45*). Some success has been achieved in stripping the capsomeres from the inner shell by treatment of purified virus with sodium pyrophosphate (*45, 46*), trypsin (*17, 45*), and by sonication (*45*). The inner shell thus exposed is composed of regularly arranged protein subunits (*17, 18, 45*) and may have a sievelike character with 92 holes corresponding to the hollow centers of the 92 capsomeres (*17*).

C. Composition

Reoviruses are comprised almost exclusively of protein and ribonucleic acid. The virus does not contain any essential lipid as the infectivity is not affected by extraction with diethyl ether (*1, 2, 12, 19*). DNA could not be detected in purified virus preparations by the diphenylamine test (*34, 37*) and, moreover, multiplication of the virus was not reduced by inhibitors of DNA synthesis such as fluorouracil- or bromouracildeoxyriboside (*12*) or 1-β-D-arabinofuranosylcytosine (*48*). The hemagglutinin of reovirus is thought to contain an oligosaccharide and the evidence for this supposition will be discussed in the next section.

Early cytochemical studies (*12, 32, 49*) which suggested that the viral nucleic acid was RNA were rapidly confirmed by direct chemical analysis of purified virus. Such analyses showed that 14.6% of the viral mass was RNA (*34*). Calculations from the buoyant density of the particle in cesium chloride gave a value of 14% RNA in the virus (*17, 37*). The major portion of viral RNA is judged to be double stranded by a number of criteria as described in a later section (6-4).

There is still considerable uncertainty about the molecular weight of the viral genome. Calculations based on the molecular weight of the virus particle and its content of RNA gave a value of 10×10^6 daltons (*34*). This is a minimum value only; the actual molecular weight of the genome determined by this method is uncertain because of inaccuracies in measurement of the diffusion constant of the virus.

Molecular weights have also been calculated from the lengths of RNA observed in electron micrographs of disrupted virions. However, as discussed in more detail below, reovirus RNA has an extraordinary tendency to fragment on its release from virions and it is difficult to be sure of the exact length of the genome. Among the short pieces, occasional longer threads of RNA are seen with lengths of 5–7.7 μ (50–53). The former length would be equivalent to a molecular weight of approximately 12×10^6; the latter to approximately 18×10^6 daltons, assuming 2.39×10^6 daltons/μ (50) and that the full length of RNA is double stranded.

In addition to the major double-stranded RNA (dsRNA) component of the virion, it has been found recently (54, 55, 126) that 15–25% of the total RNA is comprised of an adenine-rich (A-rich), single-stranded RNA. The molecular weight of this material is between 5000 (55) and 20,000 daltons (54). Its function is not yet known although it is virtually certain that the genetic information of the virus is encoded only in the double-stranded portion of viral RNA (36, 56–58).

D. Viral Hemagglutinin

Human type O erythrocytes are uniformly agglutinated by all types of reovirus in the temperature range of 4–37°C (19, 59, 60). Hemagglutinin has been associated with two types of particles isolated from infected cells and separated from each other by gradient sedimentation in cesium chloride (33, 37). The more dense particle was infectious; the lighter was not and probably represented coreless viral capsids. In addition to human type O erythrocytes, all three types of reovirus will agglutinate human type A, AB, and B erythrocytes, the order of reactivity being A > AB > O > B (61).

Bovine erythrocytes are useful in differentiating type 3 virus from types 1 and 2 (62) since these cells are agglutinated only by strains of type 3 virus and not by the other two types (19, 62, 63). Agglutination occurs only at 4°C and not at room temperature.

Studies on the hemagglutination reaction have given some insight into the nature of the cellular receptors for reovirus. Treatment of human type O erythrocytes with trypsin (40, 63, 64) or periodate (40) resulted in inactivation of the receptor sites for all three types of virus. Hemagglutination was unaffected by treatment of cells with *Vibrio cholerae* filtrate or purified neuraminidase (6, 22), with parachlormercuribenzoate (PCMB) or N-ethyl maleimide (*40*) or with ovomucin, a glycoprotein inhibitor of influenza virus hemagglutination (*63*). Thus, agglutination of human erythrocytes by the reoviruses does not seem to involve neuraminic acid although the cellular receptors may be glycoprotein in nature. On the other hand, agglutination of ox erythrocytes by type 3 virus does appear to involve neuraminic acid receptors on the cells since the interaction is prevented by treatment of the cells with ovomucin or with *V. cholera* filtrates (*63*). After attachment to ox cells, reovirus is not released again as are the myxoviruses from agglutinated chicken erythrocytes (*40*).

As far as the viral surface is concerned, sulfhydryl groups are apparently located on the capsids of all three types of virus and are essential for hemagglutination since the reaction was prevented by treatment of the virus with PCMB (*40, 63, 76*) and the inhibition was reversed by the action of reduced glutathione (*40*). Moreover, treatment of virus preparations with periodate (*70*) or β-glucosidase (*64*) destroyed the ability to hemagglutinate suggesting that the capsids contained oligosaccharides required for the virus–erythrocyte interaction. Gelb and Lerner (*103*) found that N-acetyl-D-glucosamine would bind to all three types of reovirus and inhibit their ability to agglutinate human group O erythrocytes. These latter results suggested that hemagglutination might occur by a nonenzymatic interaction of a glycoprotein moiety in the viral capsid and a component of the erythrocyte surface which contains N-acetyl-D-glucosamine.

Various inhibitors of reovirus hemagglutination are contained in human sera (*65*) animal sera (*63*), tissue culture fluids (*66*), and saliva (*67*). Fetal calf serum does not contain inhibitors of the hemagglutination reaction (*63*) and can therefore be used in the medium to support the growth of reovirus in cell cultures.

E. Chemical Analysis of Viral RNA

The early base analyses on reovirus RNA by Gomatos and Tamm (*34, 68, 69*) clearly showed a purine-to-pyrimidine ratio close to unity and provided evidence that the RNA was double stranded. Later analyses on purified, whole virus (*36, 55*) indicated that this was not the entire picture. As presented in Table 6-1, the purine-to-pyrimidine ratio of RNA in the whole virus is actually much greater than unity, with an adenine content of more than 40 mole%. The high adenine content is provided by a single-stranded RNA component of the virion.

TABLE 6-1

Base Composition of Reovirus RNA

Viral RNA	C	A	G	U	$\dfrac{A + G}{C + U}$	References
Whole virus						
Type 1	17.6	46.7	16.0	19.6	1.7	55
Type 2	19.1	41.2	18.3	21.6	1.5	55
Type 3	19.0	42.5	17.9	20.9	1.5	55
	16.9	37.7	17.5	27.9	1.2	36
Double-stranded component						
Type 1	24.9	26.8	22.8	25.5	1.0	55
Type 2	24.4	26.0	24.2	25.4	1.0	55
Type 3	24.8	26.3	23.3	25.5	1.0	55
	21.9	27.5	21.9	28.7	1.0	36
	20.5	29.7	19.3	30.5	1.0	34
	22.0	28.0	22.3	27.9	1.0	69
Single-stranded component						
Type 1	1.4	89.7	1.0	8.0	9.7	55
Type 2	1.0	88.7	0.9	9.3	8.7	55
Type 3	1.3	88.0	0.6	10.1	7.8	55
	1.4	87.8	0.3	10.5	7.4	54

This single-stranded component can be readily separated from the bulk of viral RNA by chromatography on a column of methylated bovine albumin-kieseleguhr (MAK) (55) or by electrophoresis through columns of polyacrylamide gels (54, 55). Its base composition is shown in Table 6-1. When the A-rich component is removed, the remainder of viral RNA has the base composition shown in Table 6-1 with a purine-to-pyrimidine ratio of unity, in agreement with the early analyses of Gomatos and Tamm (34).

Fifteen to twenty-five per cent of the total RNA of the reovirus virion is estimated to be comprised of the single-stranded component which contains nearly 90% of its weight as adenine (54, 55). Its properties will be described in a later section.

F. Composition of Viral Protein

Practically nothing has been published on the proteins of reovirus. Work in the authors' laboratory (35) on the degradation of purified virus with sodium dodecyl sulfate, mercaptoethanol and urea, and analysis of the resulting products by electrophoresis on polyacryl-amide gel, suggests that the virion contains at least five major poly-peptide components. There are probably several other minor components as well. More recent work using similar methods of analysis has indicated that virions of all three reovirus serotypes contain three major and four minor components (130).

6-4. Structure of Viral RNA

A. Double-Stranded Nature of the Genome

Most physical studies have been carried out on the bulk of viral RNA that remains after removal of the A-rich component since the extraction and purification steps ordinarily employed in obtaining the RNA would not have retained the A-rich material. The evidence that this major part of viral RNA is double stranded is several-fold.

1. *Base Composition*

As shown in Table 6-1 the major fraction of viral RNA contains equimolar amounts of A and U on the one hand and G and C on the other. The requirement for complementarity in base ratios is thus satisfied.

2. *Buoyant Density*

The buoyant density of reovirus RNA, when centrifuged to equi-librium in density gradients of Cs_2SO_4, is approximately 1.61 g/cm^{-3} (71) in agreement with the values reported for the double-stranded RNA replicating forms of poliovirus (72) and MS2 bacteriophage (73). Single-stranded RNA has a buoyant density of approximately 1.66 g/cm^{-3} (77) under similar conditions.

3. *Thermal Denaturation*

Reovirus RNA undergoes thermal transition in the temperature range of 70–105°C with a hyperchromicity of approximately 35% at 260 mμ (34, 36, 69, 71). The relationship between thermal transition and ionic concentration is shown in Fig. 6-1. Low concentrations of Mg^{2+} (2.5 $\times$ 10^{-3} M) have a marked effect in raising the thermal transi-tion point (71) but the dependence of Tm on Mg^{2+} concentration has not yet been worked out. While addition of formaldehyde to ribosomal

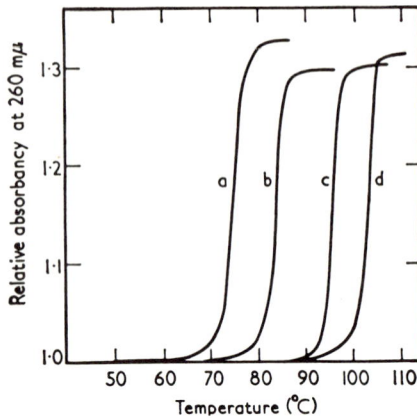

Fig. 6-1. Thermal transition of double-stranded RNA of reovirus at different ionic strengths. (a) $0.01 \times$ SSC, (b) $0.1 \times$ SSC, (c) $1.0 \times$ SSC, (d) $10.0 \times$ SSC. SSC is $0.15\,M$ NaCl and $0.15\,M$ sodium citrate. (Reprinted from Ref. *36* by courtesy of Academic Press Inc.)

RNA markedly changes the ultraviolet adsorption spectrum, it effects little change in the spectrum either of reovirus RNA or of DNA. The presence of formaldehyde does, however, considerably reduce the Tm of reovirus RNA (*31*).

As will be described in more detail later, the reovirus genome breaks into fragments on extraction and these fragments fall into a trimodal distribution of sizes. Each of the three size classes of RNA can be separated from the others by several methods (*36, 56*). When each class of RNA is denatured, either by heating or by treatment with dimethylsulfoxide, the rate at which they can be reannealed follows second-order kinetics (*56*). These observations provide additional evidence that reovirus RNA has a double-stranded structure of two complementary chains held together in much the same manner as the double helix of DNA.

4. *Effect of Ribonuclease*

The marked resistance of reovirus RNA to digestion by pancreatic ribonuclease was first described by Gomatos and Tamm (*34*), and the problem of its enzymatic degradation has been more thoroughly investigated by Shatkin (*71*) and Bellamy et al. (*36*). In summary, the extent to which reovirus RNA is digested by pancreatic ribonuclease depends on ionic concentration, species of ion, and on the concentration of ribonuclease. At a concentration of 1.0 μg/ml of ribonuclease, reovirus RNA is almost totally converted to acid-soluble products in 30 min at 37°C when the medium is 0.015 M in Na; but no more than 10% of the RNA is digested when the ion concentration is 0.15 M Na$^+$ or 0.01 M Mg^{2+} (36). Even at these higher ionic concentrations, 10 μg/ml of enzyme can cause complete hydrolysis of the viral RNA. Reovirus RNA is resistant to digestion by deoxyribonuclease, exonucleases I and III of $E.$ $coli,$ and spleen phosphodiesterase; it is degraded to acid-soluble fragments by venom phosphodiesterase, micrococcal nuclease (71), and by a double-stranded RNA nuclease isolated from $E.$ $coli$ (78). After heat denaturation of reovirus RNA, it is degraded by traces of ribonuclease ($34,$ 71) and by spleen phosphodiesterase but not by deoxyribonuclease (71).

The behavior of reovirus RNA with ribonuclease is consistent with a tightly hydrogen-bonded, double-stranded structure where many or all of the bonds are broken at low ionic strength or high temperature.

5. X-Ray Diffraction

X-ray diffraction data led Langridge and Gomatos (79) to postulate that reovirus RNA is a right-handed, double-stranded helix with a pitch of 30 Å and ten nucleotides per turn, with the base-pairs tilted at 75° to 80° to the helix axis. Thus, the translational value per nucleotide residue was 3 Å, which lead Gomatos and Stoeckenius (50) to calculate that a 1 μ length of reovirus RNA would have a molecular weight of 2.39×10^6 daltons. More recent consideration ($80,$ 123) has slightly altered the original interpretation of the X-ray diffraction data and it now appears likely that the double helix may have 11 rather than ten nucleotide pairs per turn. If the latter interpretation proves to be correct the translational distance between nucleotides will have to be reduced accordingly.

B. FRAGMENTATION OF THE VIRAL GENOME

1. Electron Microscopy of Viral RNA

The first electron micrographs of viral RNA were published by Gomatos and Stoeckenius (50) and Kleinschmidt et al. (52). They were made of RNA prepared from virus by phenol extraction and demonstrated a nonrandom distribution of lengths between 0.1 and 1.0 μ. If the viral genome had the molecular weight of 10^7 daltons as calculated from chemical analysis of purified virus (*34*), the RNA filaments seen by electron microscopy should have been uniformly close to 5 μ in length (*50, 52*). The unexpectedly low sedimentation values obtained for viral RNA (*50, 81, 83*) were consistent with the short lengths seen by electron microscopy. Clearly, the viral genome was composed of numerous short lengths of RNA or had been fragmented during its preparation.

Later work with improved methods of RNA extraction (*53*) showed a trimodal distribution of fragments with maxima at 0.35, 0.6, and 1.1 μ The assignment of molecular weights to these lengths will depend on the value finally accepted for the unit length of double-stranded RNA. Employing the weight of 2.39×10^6 daltons/μ, obtained from X-ray diffraction data by Gomatos and Stoeckenius (*50*), the above lengths are equivalent to weights of 8.3×10^5, 1.4×10^6, and 2.6×10^6 daltons, respectively. The relative numbers of filaments in each of the three classes which would comprise a viral genome were reported to be, respectively, 4–6, 3, and 2 (*53*).

Very gentle methods of releasing RNA from virions resulted in some longer filaments of RNA being seen in electron micrographs. Granboulan and Niveleau (*51*) have observed filaments of 5 μ in length, and lengths up to 7.7 μ have been seen by Dunnebacke and Kleinschmidt (*53*). Whatever the correct length may be, these observations strongly suggest that the viral genome exists as a continuous chain of RNA and that the shorter fragments result from breakage during extraction.

2. *Characterization of Viral Double-Stranded RNA Components*

Sedimentation analysis of double-stranded viral RNA, either in the analytical ultracentrifuge (*71*) or in sucrose gradients (*36, 56*) revealed three components with sedimentation coefficients of approximately 10.5S, 12S, and 14–15S. There is still some uncertainty in translating sedimentation rates of RNA into molecular weights, but two laboratories, using different methods, are in good agreement in assigning molecular weights of $0.8 \times 1 \times 10^6$, 1.4–1.5×10^6, and 2.3–2.5×10^6 daltons to the three classes of viral RNA (*54, 56*). These estimates are also in agreement with the molecular weights calculated from the contour lengths of viral RNA segments observed in electron micrographs (*50, 53*), as described in the previous section.

The three classes of viral RNA can be separated from each other by chromatography on MAK columns (*56*) or by electrophoresis on polyacrylamide gels (*36, 57*). The fraction of material which is eluted earliest from a MAK column (dsRNA-1, double-stranded RNA-1) represents the slowest moving component in sedimentation analysis, the second eluted fraction (dsRNA-2) represents the fraction that sediments at an intermediate rate, and the third column fraction (dsRNA-3) is the fastest sedimenting component. Thus the double-stranded RNA fragments are eluted from the column in order of increasing size as might be expected, or in the same order as they migrate on gel electrophoresis (*36, 57, 85*).

Separation of the double-stranded viral RNA fragments is most efficiently obtained by electrophoresis on polyacrylamide gels although extraction of the separated fragments from the gel is generally far from quantitative. Figure 6-2 illustrates the electrophoretic analysis of a ^{14}C-labeled preparation of viral RNA. The dsRNA-1 fraction has separated into three well-resolved components and the

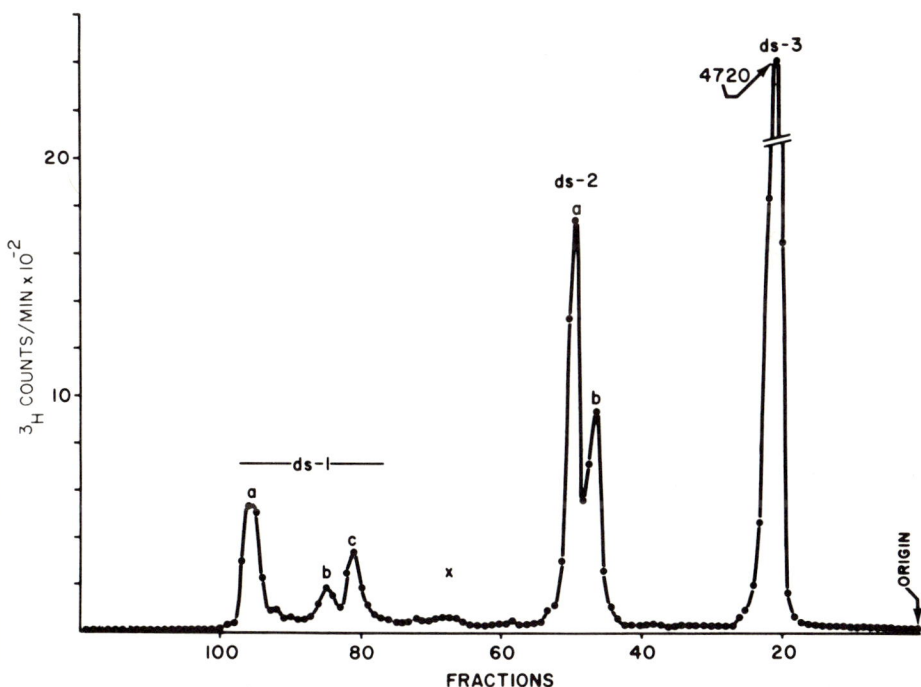

FIG. 6-2. Separation of the segments of viral RNA by electrophoresis on a column of polyacrylamide–agarose gel (*74*). ds-1 is dsRNA-1, ds-2 is dsRNA-2, ds-3 is dsRNA-3.

dsRNA-2 fraction into two. From earlier analyses the relative amounts of the three classes, dsRNA-1:dsRNA-2:dsRNA-3 were estimated to be approximately 1:1.3:2 giving an average molar ratio of 1.6:1.1:1. The relative frequencies of the segments dsRNA-1:dsRNA-2:dsRNA-3 were suggested tentatively to be 1.5:n:n (36, 57). If $n = 2$, the dsRNA-1 fraction should contain three pieces, dsRNA-2 should contain two, and dsRNA-3 should contain two pieces to give a sum of seven fragments with a total molecular weight of 10^7 daltons. From more recent analyses (82) such as that presented in Fig. 6-2 it is felt that a more likely distribution of the various segments is two dsRNA-1a, one dsRNA-1b, one dsRNA-1c, two dsRNA-2a, one dsRNA-2b, and three dsRNA-3. This interpretation follows from the observation that the amount of dsRNA-1a is about twice that of dsRNA-1b or 1c, and that dsRNA-2a is about twice the amount of dsRNA-2b. The resulting molecular weight of the double-stranded portion of the viral genome would be approximately 14.5×10^6 daltons. That this latter distribution is correct is indicated by the appearance of ten bands when type 3 virus RNA is analyzed by polyacrylamide gel electrophoresis and the resulting gels are stained with methylene blue (74, 82). Thus, dsRNA-1b, 1c, and dsRNA-2b (Fig. 6-2) are single segments of the viral genome.

As stated above, reovirus RNA is highly resistant to the action of low concentrations of RNase at salt concentrations of 0.1 M or greater. When heated to 100°C in 0.01 M NaCl for 3 min and quickly chilled, the RNA becomes almost completely sensitive to ribonuclease, suggesting that separation of the complementary strands has occurred (34, 56, 71, 75, 83, 84). Denaturation can also be readily accomplished at room temperature by dissolving the viral RNA in 90% dimethylsulfoxide (36, 56, 75) according to the method of Katz and Penman (88). When the RNA is denatured, three ribonuclease-sensitive RNA components can be demonstrated by sucrose-gradient sedimentation with sedimentation coefficients of approximately 14S, 18.5S, and 25S (36, 56, 75). These single-stranded RNA components are derived through the denaturation of dsRNA-1, dsRNA-2, and dsRNA-3, respectively (36, 56).

Denatured double-stranded RNA can be efficiently reannealed by heating it at 72.5°C in 0.3 M NaCl. Under these conditions the kinetics of reannealing have been found to be second order, indicating that the renaturation process is limited by frequency of collision of the complementary strands and, thus, that the strands were completely separated by denaturation (56).

3. *Origin of the Double-Stranded RNA Fragments*

Whether or not the fragments of double-stranded RNA arise as a result of random breaks in a continuous linear thread has posed a major problem. Data obtained by electron microscopy suggested that the breaks were not random; the lengths of the fragments and their frequency distribution were inconsistent with what would be expected from shear fragmentation as pointed out by Dunneback and Kleinschmidt (*53*) and, moreover, these two characteristics were unaltered by wide variations in technique of extracting viral RNA either from virions or from the infected cell (*36, 53, 56*). The question was answered conclusively by Watanabe and Graham (*56*) who separated the three classes of double-stranded RNA from each other, denatured them and hybridized them in all possible pairs. Lack of homology between any of the pairs indicated that the fragments arose by breakage of viral RNA at specific weak spots in the molecule. Evidence will be presented in a later section that the points of breakage function also as full stops during transcription of the viral genome into messenger RNA in infected cells.

C. THE ADENINE-RICH RNA COMPONENT OF THE VIRUS

While the original base analysis of reovirus RNA (*34, 69*) indicated complementary base ratios consistent with a double-stranded form, later analyses of whole virus showed that it had an unusually high adenine content (Table 6-1). It was then found that purified virions contained a single-stranded polynucleotide component, in addition to the double-stranded RNA, which had the base composition shown in Table 6-1. Upwards of 90% of the bases in this component are adenine; the traces of cytosine and guanine are considered to be impurities and probably not integral components of the polynucleotide (*54, 55*). Types 1, 2, and 3 reoviruses all contained the adenine-rich material when grown in L-cells (Table 6-1). Type 3 virus grown in HeLa cells (*55*), mouse, hamster, and human cells (*54*) also contained it. The amount of adenine-rich component is variously put at 15–20% (*54*) to 25% (*55*) of the double-stranded viral RNA. It is synthesized only in infected cells at about the same time as double-stranded viral RNA (*54*). The adenine-rich material could be separated from double-stranded RNA of the virus by electrophoresis on polyacrylamide gels, where it migrated at an even faster rate than 4S cellular RNA (*54, 55*) or by chromatography on MAK columns, where it was eluted by 0.2 to 0.6 M NaCl in comparison to double-stranded RNA which was

eluted by 0.8 M NaCl (55). In sucrose-gradient sedimentation analyses the adenine-rich material sedimented at a rate less than 4S (55,
126).

Adenine-rich RNA was degraded by snake venom phosphodiesterase
but not attacked by spleen phosphodiesterase, and only to a very
limited extent or not at all by pancreatic ribonuclease (54, 55). On
heating it underwent a gradual hyperchromic shift of no more than
14% and it would hybridize with polyuridylic acid. Bellamy and
Joklik (54) consider the chain length to be less than 60 nucleotides.
Shatkin and Sipe (55) have suggested a molecular weight of approximately 4000.

The function of this adenine-rich polynucleotide is not yet known.
It has been considered a candidate for a linker in holding together the
double-stranded RNA segments of the viral genome. However, since
each virion contains at least 100 molecules of the adenine-rich component there is a considerable excess over that needed to link together
the few double-stranded segments (54, 55). Moreover, highly sensitive
hybridization procedures indicated that no binding occurred between
double-stranded RNA of the virus and the adenine-rich component
(54). There is thus no evidence as yet to suggest that the material
serves in a linker capacity.

6-5. Replication of Reovirus

A. One-Step Growth Cycle

The length of the growth cycle depends on the strain of virus used
and the particular host cell, and a variety of such systems have been
employed (12, 15, 30, 84–86). One of the popular current combinations utilizes type 3 reovirus and suspension cultures of L-cells; a
growth curve for this system is shown in Fig. 6-3. In the experiment
illustrated in Fig. 6-3 the eclipse period was approximately 6 hr
although, in general, its duration depends on multiplicity of infection
as shown by Silverstein and Dales (87). The rise period was about
9 hr. Burst sizes of 2000 PFU/cell are readily obtained and, since the
number of noninfectious particles may exceed that of infectious particles by a factor of ten in viral lysates (15), the total number of virus
particles formed per cell can be of the order of 20,000. Most of the
newly formed virus remains cell associated and even after lysis of the
cells adheres tenaciously to the cellular debris.

B. ADSORPTION, PENETRATION, AND UNCOATING

Adsorption of reovirus to a variety of cell types in monolayer culture has been studied (*12, 15, 86, 89*). Generally there is insufficient information to calculate adsorption rate constants and the original papers must be referred to for the exact conditions. For suspension cultures

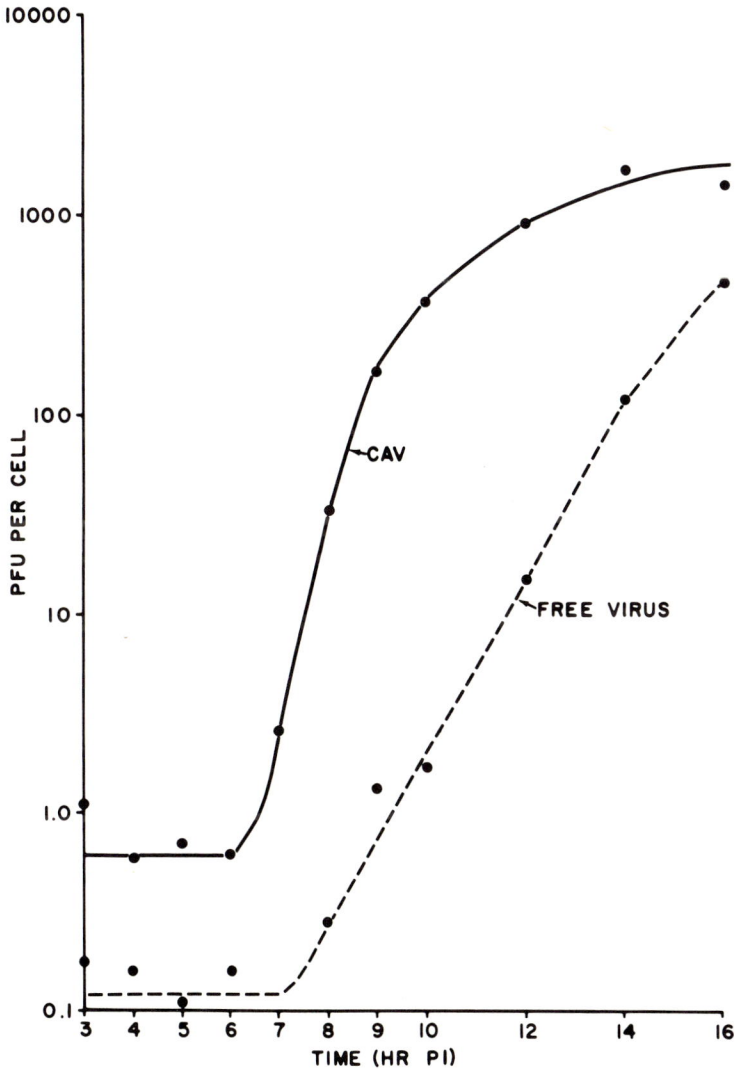

FIG. 6-3. One-cycle-growth curve of type 3 reovirus in a suspension culture of L-cells (*38*). CAV is cell-associated virus.

of L-cells it can be calculated from published data (*81*) that type 3 reovirus is adsorbed with a $k = 1.2 \times 10^{-8}$ min^{-1}. The rates of adsorption of virus to L-cells are similar at 4 and 37°C.

Penetration of reovirus into L-cells occurs, or is facilitated by phagocytosis (*16, 45, 87*). Within 15 to 20 min after adsorption and localization of the virus in phagocytic vacuoles, the vacuoles become concentrated in the centrosphere region of the cell where they fuse with lysosomes or with other vacuoles which emanate from the Golgi apparatus (*45, 87*). About 30 min after commencement of penetration, some 60% of the viral inoculum appears inside lysosomes and by 60 min most of the particles, which were taken singly into the cell, appear as groups of particles in the lysosomes (*87*).

Stripping of the coat from reovirus by enzymes in the lysosomes has been studied by biochemical and electron microscopic means (*87*). Degradation of the virus coat commences about 20–30 min after penetration, is 25% complete in 2 hr, and 50% in 10 hr, after which no further coat protein is broken down. Breakdown of viral coat protein was assessed by the release of acid-soluble fragments from virus which had been labeled with ^{14}C-leucine and purified prior to adsorption (*87*). Phagocytosis and stripping of the coat protein can occur in complete absence of new protein synthesis in the infected cell (*87, 90*) and are presumably carried out by enzymes and structures normally present in the uninfected cell.

In addition to measuring the uncoating process by digestion of the coat protein, Silverstein and Dales (*87*) studied the release of the viral RNA. When cells were infected with purified virus labeled in its RNA, no breakdown of the RNA occurred up to 10 hr after infection: None of the viral RNA appeared in an acid-soluble form, and none of it could be digested with pancreatic ribonuclease or "double-strandase" (*78*). Apparently any uncoated genomes were conserved, being resistant to action of the lysosomal nucleases and were tightly bound in the cellular particulate fraction (*45, 87*). Nevertheless, the uncoating process did release double-stranded RNA from the parental virus. The cytoplasmic particulate fractions obtained from cells after infection were broken up and the products chromatographed on columns of silicic acid coated with methylated albumin. Free parental RNA was found. In fact, the amount of demonstrable, free parental RNA at different times after infection coincided with the amount of coat protein that was degraded in the infected cell. Thus, 1–1.5 hr after infection 15–20% of the viral protein was hydrolyzed and 10–

15% of the viral RNA had been released, and this process of uncoating continued until at least 5 hr after infection. These processes are summarized in Fig. 6-4 (*87*).

C. Sites of Viral Synthesis

The immediate fate of uncoated parental viral RNA is not entirely clear although it is probably transported from lysosomes to other cytoplasmic compartments (*45*). In any event, no parental RNA is found free in the cytoplasm; it remains firmly bound to the cyto-

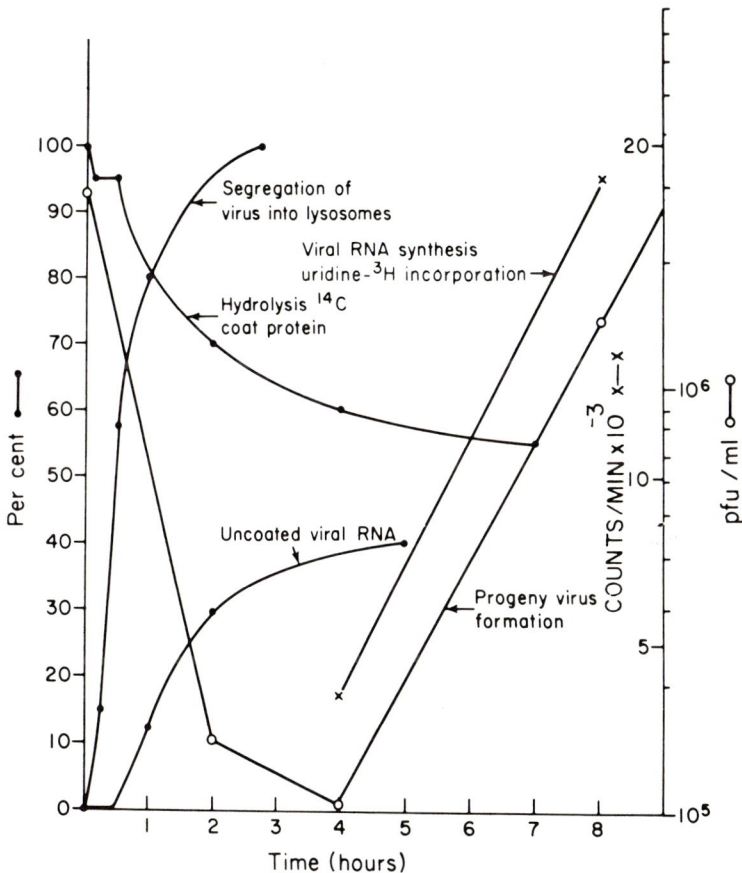

Fig. 6-4. Sequence of events in viral eclipse and development. (Reprinted from Ref. *87* by courtesy of The Rockefeller University Press.)

plasmic matrix. On extraction of the cytoplasmic components of the
infected cell, the parental RNA sediments with the large particle
fraction (87).

Reovirus components are synthesized in the cytoplasm in association with particulate elements which have been loosely termed "factories" (87). A series of cytological studies has been carried out on the
development of reovirus: by acridine orange staining (12, 32, 86),
fluorescent antibody analysis (12, 32, 49, 86, 91), ferritin-conjugated
antibody analysis (45), radioautographic analysis (45, 87), and electron microscopy (16, 45, 47, 92, 93). In summary, these studies showed
that virus can be assembled throughout the cytoplasm and develops
into large, sometimes crystalline aggregates embedded in a matrix of
fine filaments. However, the major concentration of viral components
occurred in a perinuclear position in association with the spindle
tubules (30, 87, 91, 92, 94). Experiments with antimitotic agents such
as colchicine, which disrupts the spindle tubules, changed the site of
virus synthesis but not the rate of virus development (92, 94) indicating that involvement of the cellular mitotic apparatus is not obligatory
for replication of reovirus.

D. Release of Virus

While formation of new virus is almost complete in 16–18 hr in a
one-cycle-growth experiment, release of virus into the medium is
generally relatively slow. Virus that is released is partly aggregated
(15, 17, 31) and the remaining cell-associated virus is very firmly held
by the cellular debris (47). This firm binding between virus and
cellular components is of some advantage in concentrating virus prior
to purification, but necessitates some rather vigorous treatment in
releasing the virus. Hence most virus purification procedures have
steps involving sonication and treatment with proteolytic enzymes,
detergents, and fluorocarbon (34, 36, 47, 81, 95).

It is presumably the tendency of virus to combine with cellular
protein that leads to enhancement of infectivity on treatment of cell
lysates with various proteolytic enzymes such as chymotrypsin (28)
or by heating to 55°C in 2 M $MgCl_2$ (31). The ability of proteolytic
enzymes to convert "potentially infectious virus" to infectious virus has
been thoroughly studied by Spendlove and his collaborators (24, 27,
96) and depends markedly on the type of virus, type of cell, method
of virus assay, and how long after infection the progeny virus is
harvested.

E. BIOCHEMISTRY OF VIRAL REPLICATION

1. Effect of Reovirus on Synthesis of Host Macromolecules

It was first shown by Gomatos and Tamm (97) that infection of L-cells with reovirus, type 3, had no early effect on the synthesis of cellular RNA, DNA, and protein. In fact, the first discernible effect was a suppression of cellular DNA synthesis commencing some 9–10 hr after infection. By 15 hr, when virus synthesis was almost complete, cellular DNA synthesis was inhibited by 80%. Essentially similar results were later obtained by Loh and Soergal (48, 98) with type 2 virus and by Kudo and Graham with type 3 virus (81). The latter workers found no inhibition of ribosomal RNA or nuclear RNA synthesis up to 12 hr after infection.

2. Effect of Metabolic Inhibitors on Reovirus Multiplication

Evidence that reovirus was an RNA-containing virus came initially from experiments with inhibitors of DNA synthesis (12). It was shown that fluoro- and bromodeoxyuridine (12) and cytosine arabinoside (48) markedly inhibited the synthesis of DNA in infected cells without affecting the yield of virus. It was clear from these experiments that the virus did not contain DNA and that synthesis of new cellular DNA was not required for viral replication. In fact, such results, combined with the observation that viral inclusions in infected cells stained orthochromatically pale green with acridine orange, led to the initial suggestion that reovirus contained a double-stranded RNA (12).

Continuing synthesis of cellular RNA in reovirus-infected cells makes it almost impossible to distinguish the formation of virus-specific RNA, and the early observation that virus synthesis was suppressed by actinomycin D (12) seemed to eliminate this very useful agent as a selective inhibitor of cellular RNA synthesis. Later it was found that certain concentrations of actinomycin D (0.5 μg/ml) effectively inhibited cellular RNA synthesis without affecting the synthesis of virus-specific RNA, viral antigen, or mature virus (81, 98–100). In the presence of actinomycin D, virus-specific RNA synthesis was shown to commence at approximately 5 hr after infection (81, 84, 98, 100, 102). The nature of this virus-specific RNA will be discussed in more detail in the next section.

Synthesis of virus-specific RNA requires the synthesis of early protein. This requirement was shown by adding protein inhibitors such as puromycin (*101*) or cycloheximide (*84, 102*) to cultures at various times after infection and determining whether virus-specific RNA was formed. If the inhibitor was added at any time before 5 hr postinfection there was virtually no synthesis of virus-induced RNA. However, when added at later times, synthesis of virus-specific RNA could take place, indicating that the new protein started to be synthesized at about 5 hr after infection.

3. *Nature of the Virus-Specific RNA Synthesized in Reovirus Infected Cultures*

Virus-specific RNA synthesized in infected cells sediments broadly in sucrose gradients, from approximately 6S–35S, and is partially resistant to the action of ribonuclease (*81, 84*). The ribonuclease-resistant portion sediments at an average rate of 15S. It sediments in an identical manner to viral RNA in sucrose gradients (*81*) and is eluted from columns of methylated bovine albumin-kieselguhr (*81, 84*) in the same position as viral RNA, at a concentration of 0.8 *M* NaCl; It migrates in exactly the same way as viral RNA upon electrophoresis through polyacrylamide gel (*74*) and has the same base composition as viral RNA (*84*). By all these criteria the double-stranded RNA formed in infected cells is viral progeny RNA.

In general, the amount of viral dsRNA synthesized in infected cells is about 20–40% of the total virus-specific RNA (*58, 81, 84, 102*). The remainder of the newly formed RNA is highly sensitive to the action of ribonuclease and is single-stranded RNA (ssRNA) (*81, 84*). This ssRNA is found associated with polyribosomes, hybridizes efficiently with denatured viral RNA, and does not self-hybridize (*56, 58, 83–85*). A possibility that some of this material might be a precursor of dsRNA has been largely eliminated (*102*) and it is now fairly certain that it is entirely messenger RNA (mRNA) copied off the dsRNA during infection as will be described in more detail later.

It was mentioned in the previous section that inhibitors of protein synthesis, such as puromycin and cycloheximide, if added early in the course of infection, prevent synthesis of both single- and double-stranded RNA. If the inhibitor is added later than 6 hr after infection, synthesis of single-stranded RNA goes on unabated, but double-stranded RNA formation is rapidly inhibited (*84, 101, 102, 124*). These results have suggested that two new proteins may be formed in infected cells, one associated with synthesis of dsRNA and the other with synthesis of ssRNA.

Interferon depresses the synthesis of both classes of reovirus-specific RNA and reduces the yield of virus if added postinfection (*38*). The response of this virus to the action of interferon is similar to that found for mengovirus (*38*). The uridine analog, 6-azauridine, has also been shown to inhibit the synthesis of all reovirus-specific RNA (*125*).

4. *Transcription of the Viral Genome.*

Single-stranded RNA formed in infected cells in presence of actinomycin D can be isolated either from polyribosomes or directly from infected cells by phenol extraction. When analyzed by sedimentation through sucrose gradients (*46, 56, 58*) (Fig. 6-5) or by acrylamide gel electrophoresis (*57, 58*) (Fig. 6-6), the RNA separates clearly into three size-classes having molecular weights of approximately 0.4×10^6, 0.7×10^6, and 1.2×10^6 daltons (*56*). These three classes of ssRNA have been termed ssRNA-1, ssRNA-2, and ssRNA-3, respectively (*56*). They have base compositions similar to each other as shown in Table 6-2 (*46*).

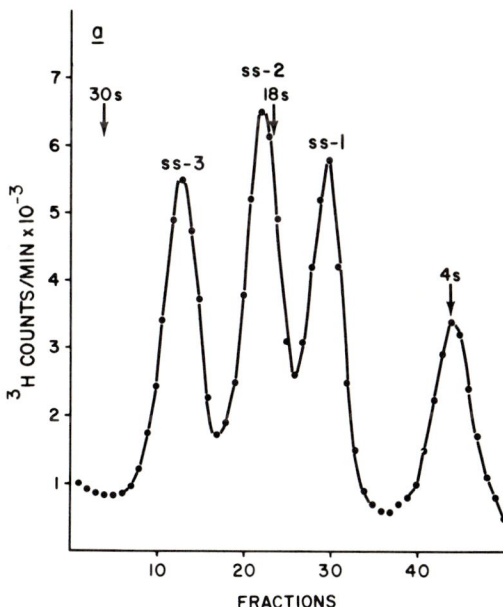

FIG. 6-5. Sucrose-gradient sedimentation analysis of virus-induced, single-stranded RNA from reovirus infected cells (*74*). ss-1 is ssRNA-1, ss-2 is ssRNA-2, ss-3 is ssRNA-3. Direction of sedimentation is from right to left.

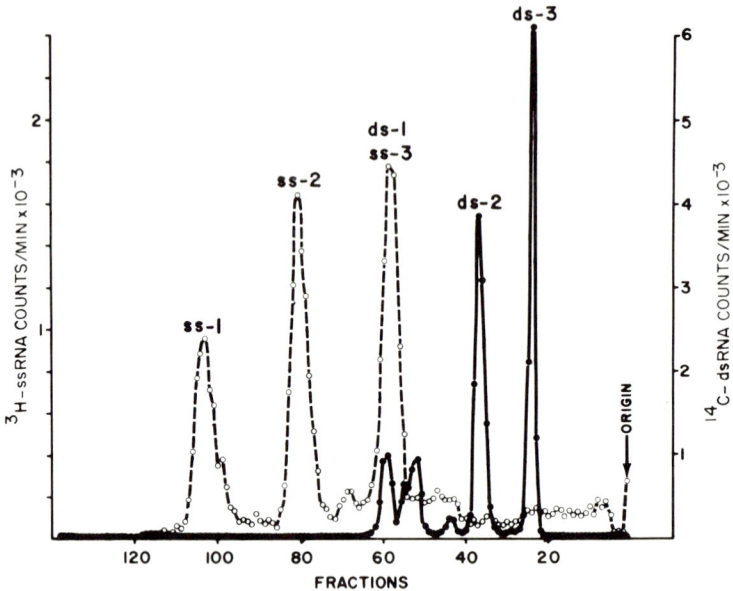

FIG. 6-6. Separation by electrophoresis on a polyacrylamide–agarose gel column of single- and double-stranded RNA isolated from reovirus-infected L-cells (74).

Each class of ssNRA corresponds almost exactly in sedimentation rate to one class of viral dsRNA after denaturation (56, 58), suggesting that each ssRNA is similar in size to one strand of the dsRNA fragment. Moreover, when the three classes of ssRNA were separated from each other and hybridized with the three classes of denatured dsRNA it was found that ssRNA-1 hybridized uniquely and efficiently with dsRNA-1, ssRNA-2 with dsRNA-2, and ssRNA-3 with dsRNA-3 (57, 58). An analysis illustrating this point is shown in Fig. 6-7: Each

TABLE 6-2

Base Composition of the Three Classes of Single-Stranded
RNA Isolated from Infected L-Cells (46)

	ssRNA-1	ssRNA-2	ssRNA-3
C	23.2	24.0	24.7
A	28.8	28.1	27.9
G	19.8	19.0	18.9
U	28.1	28.5	28.1

class of ssRNA was hybridized with dsRNA and the products were separated by electrophoresis on columns of polyacrylamide gels (57). Each ssRNA migrated exclusively with a corresponding length of dsRNA showing that the hybridization was unique and that the homology was complete throughout the length of the ssRNA. It was concluded that each segment of dsRNA acts as an individual template in the infected cell and is transcribed from end-to-end into a messenger

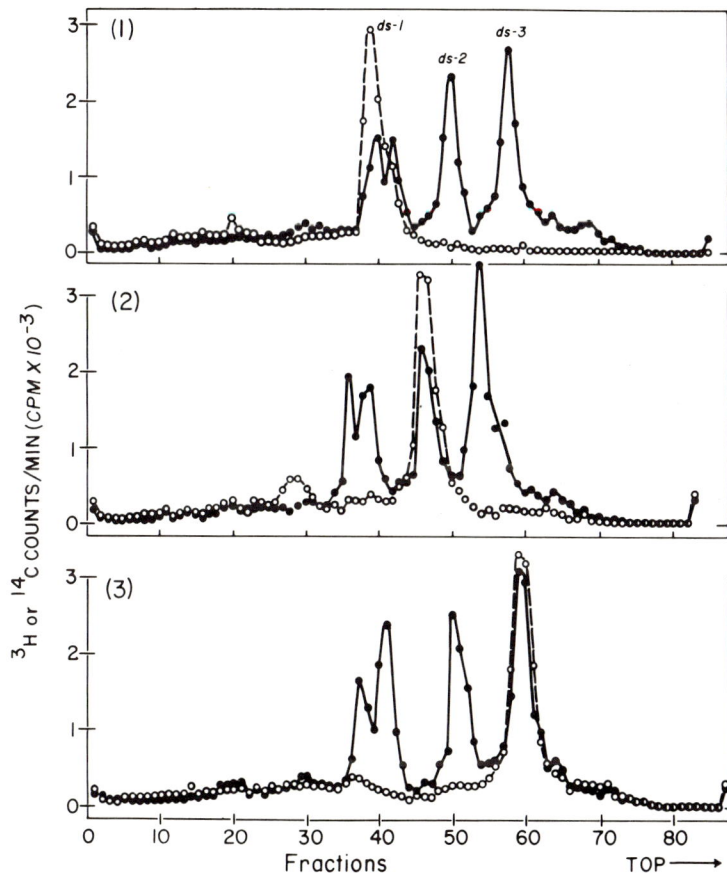

FIG. 6-7. Hybridization of the three species of ssRNA from infected cells with viral dsRNA and analysis of the products by acrylamide gel electrophoresis. (1) ssRNA-1-^{3}H with dsRNA-^{14}C; (2) ssRNA-1-^{3}H with dsRNA-^{14}C; (3) ssRNA-3-^{3}H with dsRNA-^{14}C. (O– – – –O) is ^{3}H; (●———●) is ^{14}C. (Reprinted from Ref. 57 by courtesy of Proceedings of The National Academy of Sciences.)

RNA molecule (*57, 58*). Regardless of how the segments of dsRNA
may be held together in the virion or infected cell, the "linkers" have
the effect of full stops in transcription of the genome.

Throughout the course of infection the three classes of ssRNA,
found in L-cells infected with type 3 virus, are formed in almost equal
amounts (*85*). This result was obtained from sedimentation analyses
of ssRNA obtained from cells that had been pulse labeled with ^{3}H-
uridine at different times after infection. Since the relative amounts of
dsRNA-1:dsRNA-2:dsRNA-3 are 1:1.3:2 (*57*) some segments of the
viral genome must be transcribed more frequently than others.
Furthermore, when L-cells are infected with type 1 reoviruses it was
found that ssRNA-3 was formed in much smaller quantity than
ssRNA-1 and ssRNA-2 (*85*). Both sets of results indicate that some
measure of regulation is exerted at the level of transcription of the
viral genome.

Sedimentation analysis of the three classes of ssRNA as described
in the previous paragraph was not sufficiently sensitive to determine
whether some of the messenger RNAs were being transcribed "early"
and some "late." Since each ssRNA class probably contained more
than one mRNA, changes in the relative rates of synthesis of
the mRNA's within a given class would go unnoticed. This problem was approached by Watanabe, Millward, and Graham (*82*)
who hybridized ssRNA, removed from infected cells at various times,
with denatured dsRNA and quantitatively determined the amount of
hybrid under each dsRNA peak obtained by polyacrylamide gel electrophoresis. Figure 6-8 shows such an analysis and it can be seen
that at early times in infection there is less ssRNA-1*b*, 1*c*, and ssRNA-
2*b* made than at later times, suggesting a temporal control over the
formation of these three messenger RNA's. Furthermore, when cycloheximide was added at the time of infection to block protein synthesis, ssRNA-1*b*, 1*c* and 2*b* were not formed at all (Fig. 6-9). The
conclusion is that there are messenger RNA's in the classes ssRNA-1*a*,
2*a*, and 3 which are transcribed from the parental viral genome by a
preexisting enzyme, while ssRNA-1*b*, 1*c*, and 2*b* are synthesized later
and are classified as "late" messengers.

Recently it has been shown (*127, 128*) that purified reovirions contain an RNA polymerase that transcribes ssRNA from the double-
stranded viral genome *in vitro*. The ssRNA thus transcribed does not
correspond in size to any of the ssRNA classes formed in the infected
cell but it does hybridize efficiently with viral dsRNA. Since this virion

polymerase is clearly a candidate to perform the transcription of early messenger RNA's in the infected cell, its further study is a matter of considerable importance.

F. Reovirus RNA as Template for Polymerases *In Vitro*

Several reports have suggested that reovirus RNA can serve as a template for both RNA polymerase (*104, 105*) and DNA polymerase (*106*) isolated from *E. coli*. However, later work suggested that the observed priming activity was not a property of the dsRNA *per se* but resulted from a small amount of cellular DNA carried through the

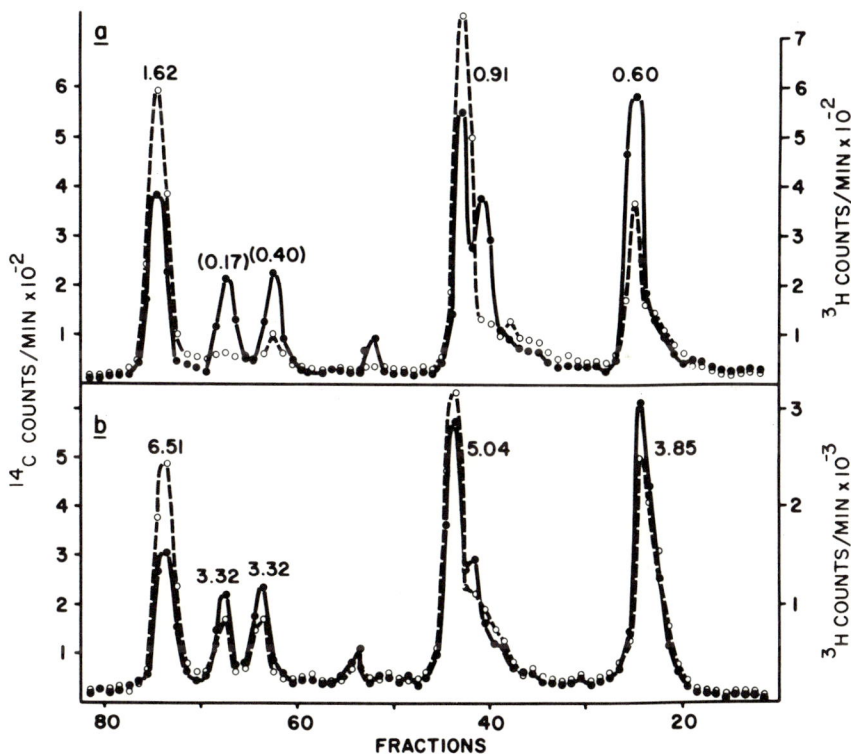

Fig. 6–8. Analysis of ssRNA extracted at different times from reovirus-infected L-cells. (a) ssRNA was labeled with ³H-uridine 4–6 hr after infection, hybridized with dsRNA-¹⁴C, and analyzed by acrylamide gel electrophoresis. (b) ssRNA was labeled with ³H-uridine 10–12 hr after infection and analyzed as in (a). (○----○) is ³H; (●———●) is ¹⁴C. The numbers over the peaks are ³H/¹⁴C (*82*).

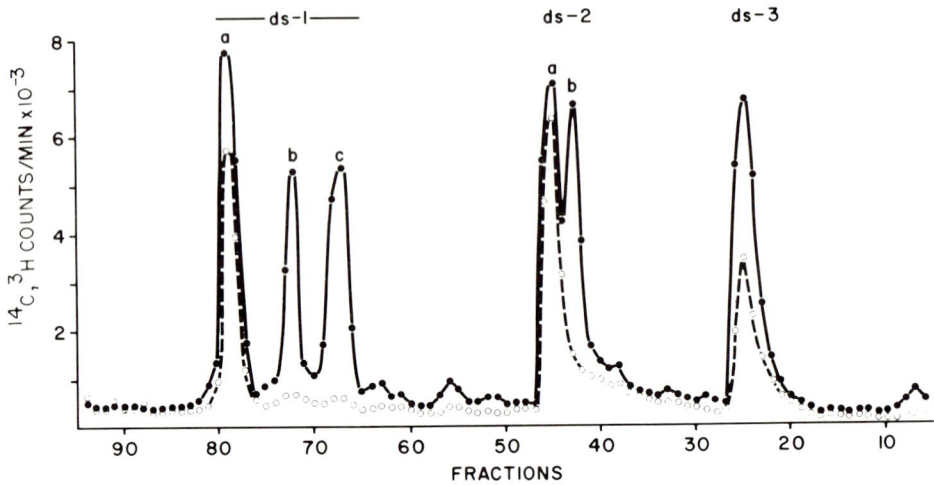

Fig. 6-9. Analysis of the ssRNA formed in infected cells in absence of early protein synthesis. Cells were infected at time zero and cycloheximide was added. ssRNA was labeled 4–8 hr after infection, extracted, hybridized with dsRNA–^{14}C, and analyzed by gel electrophoresis. (O- - - -O) is ^{3}H; (●———●) is ^{14}C (82).

purification of the virus and extraction of the RNA (71). Reovirus dsRNA that had been purified by sedimentation to equilibrium in a cesium chloride gradient was devoid of priming activity (71). This explanation satisfactorily accounts for most of the *in vitro* synthesis achieved with the *E. coli* polymerases and the observation that such synthesis was blocked by actinomycin D. It should be mentioned that the earlier authors (104) had, in fact, tested the effect of crystalline DNase on the template activity of reovirus RNA for RNA polymerase and found only a small reduction.

Recently, a new enzyme complex has been discovered in reovirus-infected L-cells that catalyses the *in vitro* synthesis of both single- and double-stranded RNA (107). The nature of the ssRNA product is unknown, it does not hybridize with viral RNA, but the double-stranded product is viral RNA. This enzyme commences to form at approximately 5 hr after infection and has not yet been obtained free from the dsRNA primer with which it is firmly associated. Probably, the dsRNA product of this enzyme reaction represents the *in vitro* completion of dsRNA strands attached to the complex whose synthesis had already been commenced in the infected cells.

Gomatos (129) has recently described the isolation of an RNA polymerase from infected cells that catalyses the synthesis of ssRNA. Three classes of ssRNA are formed *in vitro* and these correspond in sedimentation rate to the three classes ssRNA-1, ssRNA-2, and ssRNA-3 formed in infected cells. Moreover, in hybridization tests of the *in vitro* product with viral RNA, each of the three classes of product was competed for only by the corresponding class of ssRNA formed *in vivo*. Thus the enzyme seems to be faithful in transcribing the viral genome *in vitro*. Whether this enzyme described by Gomatos is identical or not to the RNA polymerase contained in purified virions (*127, 128*) is not yet known.

6-6. Other Double-Stranded RNA Viruses

Three other viruses in addition to reoviruses have so far been shown to possess double-stranded RNA as their genetic material, wound tumor virus (*69, 108, 109*), rice dwarf virus (*110*), and the cytoplasmic polyhedrosis virus of silkworm (*111*). Very little is yet known about the latter virus. Rice dwarf virus is approximately 70 mμ in diameter, while wound tumor virus is somewhat smaller at 60 mμ diameter (*112, 114*). There are 92 capsomeres in the capsids of both wound tumor virus (*112, 113*) and rice dwarf virus (*114*) as in reovirus. Rice dwarf virus contains 11% RNA (*115*) and has a G + C content of 44% (*110*) compared to a G + C content of 37.7% for wound tumor virus (*69*). Both viruses multiply in the cytoplasm of infected cells (*116–119*) and can multiply in their insect vectors, the leaf hopper (*119, 120*).

The close similarities in size, structure, and type of RNA in wound tumor virus and reovirus have prompted several serological studies to determine whether there was any cross reaction. An early report (*121*) claimed that, in fact, the two viruses had a common complement-fixing antigen, although more recent work has been unable to substantiate the claim (*69, 122*).

ACKNOWLEDGMENT

We are indebted to our colleagues Dr. Y. Watanabe, Dr. J. Borsa, Dr. S. Millward and Mr. S. Newlin for their numerous critical comments during the writing of the manuscript. Work in the authors' laboratory was supported in part by Public Health Service research grant AI 0245 4 from the National Institute of Allergy and Infectious Diseases, by grants E148 and 149 from The American Cancer Society, and by grant GB364 from the National Science Foundation.

REFERENCES

1. A. B. Sabin, *Science*, **130**, 1387 (1959).
2. A. D. Macrae, *Ann. N.Y. Acad. Sci.*, **101**, 455 (1962).
3. G. D. Hsuing, *Proc. Soc. Exptl. Biol. Med.*, **99**, 387 (1958).
4. H. Malherbe and R. Harwin, *Brit. J. Exptl. Pathol.*, **38**, 539 (1957).
5. D. N. Shaver, A. L. Barron, and D. T. Karzon, *Am. J. Pathol.*, **34**, 943 (1958).
6. L. Dardanoni and P. Zaffiro, *Boll. Inst. Sieroterap. Milan*, **37**, 346 (1958).
7. A. B. Sabin, *Ann. N.Y. Acad. Sci.*, **67**, 250 (1957).
8. N. F. Stanley, D. C. Dorman, and J. Ponsford, *Australian J. Exptl. Biol. Med. Sci.*, **32**, 543 (1954).
9. L. Rosen, *Am. J. Hyg.*, **71**, 242 (1960).
10. J. W. Hartley, W. P. Rowe, and J. B Austin, *Virology*, **16**, 94 (1962).
11. J. M. Papadimitriou, *Am. J. Pathol.*, **50**, 59 (1967).
12. P. J. Gomatos, I. Tamm, S. Dales, and R. M. Franklin, *Virology*, **17**, 441 (1962).
13. C. P. Loh, H. R. Hohl, and M. Soergel, *J. Bacteriol.*, **89**, 1140 (1965).
14. L. E. Jordan and H. D. Mayor, *Virology*, **17**, 597 (1962).
15. J. S. Rhim, K. O. Smith, and J. L. Melnick, *Virology*, **15**, 428 (1961).
16. N. Anderson and F. W. Doane, *J. Pathol. Bacteriol.*, **92**, 433 (1966).
17. H. D. Mayor, R. M. Jamison, L. E. Jordan, and M. van Mitchell, *J. Bacteriol.*, **89**, 1548 (1965).
18. C. Vasquez and P. Tournier, *Virology*, **17**, 503 (1962).
19. N. F. Stanley, *Brit. Med. Bull.*, **23**, 150 (1967).
20. L. Rosen, *Ann. N.Y. Acad. Sci.*, **101**, 461 (1962).
21. N. F. Stanley, P. J. Leak, G. M. Grieve, and D. Perret, *Australian J. Exptl. Biol. Med. Sci.*, **42**, 372 (1964).
22. N. F. Stanley, *Nature*, **189**, 687 (1961).
23. N. F. Stanley, *Presbyterian St. Luke's Hosp. Med. Bull., Chicago*, **3**, 146 (1964).
24. R. S. Spendlove and F. L. Schaffer, *J. Bacteriol.*, **89**, 597 (1965).
25. J. S. Rhim and J. L. Melnick, *Texas Rept. Biol. Med.*, **19**, 851 (1961).
26. J. S. Rhim and J. L. Melnick, *Virology*, **15**, 80 (1961).
27. M. E. McClain, R. S. Spendlove, and E. H. Lennette, *J. Immunol.*, **98**, 1301 (1967).
28. C. Wallis, J. L. Melnick, and F. Rapp, *J. Bacteriol.*, **92**, 155 (1966).
29. C. Wallis, J. L. Melnick, and M. Bianchi, *Texas Rept. Biol. Med.*, **20**, 693 (1963).
30. R. S. Spendlove, E. H. Lennette, C. O. Knight, and J. N. Chin, *J. Immunol.*, **90**, 548 (1963).
31. C. Wallis, K. O. Smith, and J. L. Melnick, *Virology*, **22**, 608 (1964).
32. J. S. Rhim, L. E. Jordan, and H. D. Mayor, *Virology*, **17**, 342 (1962).
33. M. T. A. Fouad and R. Engler, *Z. Virusforsch.*, **21**, 706 (1966).

34. P. J. Gomatos and I. Tamm, *Proc. Natl. Acad. Sci. U.S.*, **49,** 707 (1963).

35. S. Newlin, unpublished.

36. A. R. Bellamy, L. Shapiro, J. T. August, and W. K. Joklik, *J. Mol. Biol.*, **29,** 1 (1967).

37. R. Engler and M. T. A. Fouad, *Arch. Ges Virusforsch.*, **20,** 29 (1967).

38. C. J. Gauntt, unpublished.

39. V. V. Hamparian, M. R. Hilleman, and A. Ketler, *Proc. Soc. Exptl. Biol. Med.*, **112,** 1040 (1963).

40. A. M. Lerner, J. D. Cherry, and M. Finland, *Virology*, **19,** 58, (1963).

41. P. Tournier and M. Plissier, *Presse Med.*, **68,** 683 (1960).

42. J. M. Papadimitriou, *Am. J. Pathol.*, **47,** 565 (1965).

43. H. D. Mayor and J. L. Melnick, *Science,* **137,** 612 (1962).

44. C. Vasquez and P. Tournier, *Virology*, **24,** 128 (1964).

45. S. Dales, P. J. Gomatos, and K. C. Hsu, *Virology*, **25,** 193 (1965).

46. P. J. Gomatos, *Proc. Natl. Acad. Sci. U.S.*, **58,** 1798 (1967).

47. P. J. Gomatos, *Procedures in Nucleic Acid Res.*, **4,** 493 (1965).

48. P. C. Loh and M. Soergel, *Nature,* **214,** 622 (1967).

49. V. Drouhet, *Ann. Inst. Pasteur*, **98,** 618 (1960).

50. P. J. Gomatos and W. Stoeckenius, *Proc. Natl. Acad. Sci. U.S.*, **52,** 1449 (1964).

51. N. Granboulan and A. Niveleau, *J. Miscroscopie*, **6,** 23, 1967,

52. A. K. Kleinschmidt, T. H. Dunnebacke, R. S. Spendlove, F. L. Schaffer, and R. F. Whitcomb, *J. Mol. Biol.*, **10,** 282 (1964).

53. T. H. Dunnebacke and A. K. Kleinschmidt, *Z. Naturforsch.*, **22b,** 159 (1967).

54. A. R. Bellamy and W. K. Joklik, *Proc. Natl. Acad. Sci. U.S.*, **58,** 1389 (1967).

55. A. J. Shatkin and J. D. Sipe, *Proc. Natl. Acad. Sci. U.S.*, **59,** 246 (1968).

56. Y. Watanabe and A. F. Graham, *J. Virology*, **1,** 665 (1967).

57. Y. Watanabe, L. Prevec, and A. F. Graham, *Proc. Natl. Acad. Sci. U.S.*, **58,** 1040 (1967).

58. A. R. Bellamy and W. K. Joklik, *J. Mol. Biol.*, **29,** 19 (1967).

59. J. S. Rhim, J. I. Kato, and W. Pelon, *Proc. Soc. Exptl. Biol. Med.*, **118,** 453 (1965).

60. M. Goldfield, S. Srihongse, and J. P. Fox, *Proc. Soc. Exptl. Biol. Med.*, **96,** 788 (1957).

61. M. Brubaker, B. West, and R. J. Ellis, *Proc. Soc. Exptl. Biol. Med.*, **115,** 1118 (1964).

62. H. J. Eggers, P. J. Gomatos, and I. Tamm, *Proc. Soc. Exptl. Biol. Med.*, **110,** 879 (1962).

63. P. J. Gomatos and I. Tamm, *Virology*, **17,** 455 (1962).

64. A. M. Lerner, E. J. Bailey, and J. R. Tillotson, *J. Immunol.*, **95,** 1111 (1965).

65. N. J. Schmidt, J. Dennis, M. N. Hoffman, and E. H. Lennette, *J. Immunol.*, **93,** 377 (1964).

66. N. J. Schmidt, J. Dennis, M. N. Hoffman, and E. H. Lennette, *J. Immunol.*, **93,** 367 (1964).

67. A. M. Lerner, E. J. Bailey, and M. Kofender, *J. Immunol.*, **96,** 59 (1966).

68. H. Kudo, unpublished results.

69. P. J. Gomatos and I. Tamm, *Proc. Natl. Acad. Sci. U.S.*, **50,** 878 (1963).

70. J. R. Tillotson and A. M. Lerner, *Proc Natl. Acad. Sci. U.S.*, **56,** 1143 (1966).

71. A. J. Shatkin, *Proc. Natl. Acad. Sci U.S.*, **54**, 1721 (1965).
72. J. M. Bishop, D. F. Summers, and L. Levintow, *Proc. Natl. Acad. Sci. U.S.*, **54**, 1273 (1965).
73. C. Weismann, P. Borst, R. H. Burdon, M. A. Billeter, and S. Ochoa, *Proc. Natl. Acad. Sci. U.S.*, **51**, 682 (1964).
74. A. J. Shatkin, J. D. Sipe, and P. Loh, *J. Virol.*, **2**, 986 (1968).
75. W. J. Iglewski and R. M. Franklin, *J. Virol.*, **1**, 302 (1967).
76. F. E. Buckland, *Nature,* **188**, 768 (1960).
77. R. L. Erikson, *J. Mol. Biol.*, **18**, 372 (1966).
78. H. D. Robertson, R. E. Webster, and N. D. Zinder, *Virology*, **32**, 718 (1967).
79. R. Langridge and P. J. Gomatos, *Science,* **141**, 694 (1963).
80. S. Arnott, M. H. F. Wilkins, W. Fuller, and R. Langridge, *J. Mol. Biol.*, **27**, 525 (1967).
81. H. Kudo and A. F. Graham, *J. Bacteriol.*, **90**, 936 (1965).
82. Y. Watanabe, S. Millward, and A. F. Graham, *J. Mol. Biol.*, **36**, 107 (1968).
83. L. Prevec and A. F. Graham, *Science,* **154**, 522 (1966).
84. A. J. Shatkin and B. Rada, *J. Virol.*, **1**, 24 (1967).
85. L. Prevec, Y. Watanabe, C. J. Gauntt, and A. F. Graham, *J. Virol.*, **2**, 289 (1968).
86. H. Oie, P. C. Loh, and M. Soergal, *Arch. Ges. Virusforsch.*, **18**, 16 (1966).
87. S. C. Silverstein and S. Dales, *J. Cell. Biol.*, **36**, 197 (1968).
88. L. Katz and S. Penman, *Biochem. Biophys. Res. Commun.*, **23**, 557 (1966).
89. J. S. Rhim and J. L. Melnick, *Texas Rept. Biol. Med.*, **19**, 851 (1961).
90. S. Dales, *Proc. Natl. Acad. Sci. U.S.*, **54**, 462 (1965).
91. R. S. Spendlove, E. H. Lennette, and A. C. John, *J. Immunol.*, **90**, 554 (1963).
92. S. Dales, *Proc. Natl. Acad. Sci. U.S.*, **50**, 268 (1963).
93. A. B. Jenson, E. R. Rabin, C. A. Phillips, and J. L. Melnick, *Am. J. Pathol.*, **47**, 223 (1965).
94. R. S. Spendlove, E. H. Lennette, J. N. Chin, and C. O. Knight, *Cancer Res.*, **24**, 1826 (1964).
95. A. M. Lerner, *Bacteriol. Rev.*, **28**, 391 (1964).
96. R. S. Spendlove, E. H. Lennette, C. O. Knight, and J. N. Chin, *J. Bacteriol.*, **92**, 1036 (1966).
97. P. J. Gomatos and I. Tamm, *Biochim. Biophys. Acta,* **72**, 651 (1963).
98. P. Loh and M. Soergal, *Proc. Natl. Acad. Sci. U.S.*, **54**, 857 (1965).
99. A. J. Shatkin, *Biochem. Biophys. Res. Commun.*, **19**, 506 (1965).
100. P. C. Loh and M. Soergel, *Proc. Soc. Exptl. Biol. Med.*, **122**, 1248 (1966).
101. H. Kudo and A. F. Graham, *Biochem. Biophys. Res. Commun.*, **24**, 150 (1966).
102. Y. Watanabe, H. Kudo, and A. F. Graham, *J. Virol.*, **1**, 36 (1967).
103. L. D. Gelb and A. M. Lerner, *Science,* **147**, 404 (1965).
104. P. J. Gomatos, R. M. Krug, and I. Tamm, *J. Mol. Biol.*, **9**, 193 (1964).
105. R. M. Krug, P. J. Gomatos, and I. Tamm, *J. Mol. Biol.*, **12**, 872 (1965).
106. P. J. Gomatos, R. M. Krug, and I. Tamm, *J. Mol. Biol.*, **13**, 802 (1965).
107. Y. Watanabe, C. J. Gauntt, and A. F. Graham, *J. Virology,* **2**, 869 (1968).
108. K.-I. Tomita and A. Rich, *Nature,* **201**, 1160 (1964).
109. L. M. Black and R. Markham, *Neth. J. Plant Pathol.,* **69**, 215 (1963).
110. K.-I. Miura, I. Kimura, and N. Suzuki, *Virology,* **28**, 571 (1966).

111. M. Fuke, *Protein, Nucleic Acid, Enzyme* (*Tokyo*), **12,** 521 (1967).

112. R. F. Bils and C. E. Hall, *Virology*, **17,** 123 (1962).

113. M. K. Brakke, A. E. Vatter, and L. M. Black, *Brookhaven Symp. Biol.*, BNL No. 6, 137 (1954).

114. K. Maramorosch and D. D. Jensen, *Ann. Rev. Microbiol.*, **17,** 495 (1963).

115. S. Toyoda, I. Kimura, and N. Suzuni, *Protein, Nucleic Acid, Enzyme* (*Tokyo*), **9,** 861 (1964).

116. V. C. Littau and L. M. Black, *Am. J. Botany*, **39,** 87 (1952).

117. A. N. Nagaraj and L. M. Black, *Virology*, **15,** 289 (1961).

118. E. Shikata, S. W. Orenski, H. Hurumi, J. Mitsuhashi, and K. Maramorosch, *Virology*, **23,** 441 (1964).

119. T. Fukushi, E. Shikata, and I. Kumura, *Virology*, **18,** 192 (1962).

120. L. M. Black, *Proc. Natl. Acad. Sci. U.S.*, **44,** 364 (1958).

121. C. Streissle and K. Maramorsch, *Science*, **140,** 996 (1963).

122. R. Gamez, L. M. Black, and R. MacLeod, *Virology*, **32,** 163 (1967).

123. S. Arnott, M. H. F. Wilkins, W. Fuller, and R. Langridge, *J. Mol. Biol.*, **27,** 535 (1967).

124. P. C. Loh and J. R. Crowley, *Proc. Soc. Exptl. Biol. Med.*, **125,** 1287 (1967).

125. B. Rada and A. J. Shatkin, *Acta Virol.*, **11,** 551 (1967).

126. T. Koide, I. Suzuka, and K. Sekiguchi, *Biochem. Biophys. Res. Commun.*, **30,** 95 (1968).

127. J. Borsa and A. F. Graham, *Biochem. Biophys. Res. Commun.*, **33,** 895 (1968).

128. A. J. Shatkin and J. D. Sipe, *Proc. Natl. Acad. Sci. U.S.*, **61,** 1462 (1968).

129. P. J. Gomatos, *J. Mol. Biol.*, **37,** 423 (1968).

130. P. C. Loh and A. J. Shatkin, *J. Virol.*, **2,** 1353 (1968).

7

RNA TUMOR VIRUSES

JOHN P. BADER

CHEMISTRY BRANCH
NATIONAL CANCER INSTITUTE
BETHESDA, MARYLAND

7-1. Introduction

The discovery of viruses around the turn of the century generated widespread curiosity about the possible viral etiology of human disease, and investigators were soon seeking a role for viruses in tumorigenesis. Such studies led to the discovery that lymphomatosis of fowl could be transmitted by a filterable agent (1), and a short time later Rous (2) found that solid tumors could be generated in fowl using extracts of certain spontaneous chicken tumors. While many investigations were made into the general biology of the malignancies caused by these viruses, the implications in human diseases were obscure, and results of studies in avian systems were considered of academic interest.

Eventually certain mammary tumors of mice were shown to be transmittable by virus (3), indicating that viruses could be important in producing mammalian malignancies. The complexities of the host–virus system discouraged extensive experimentation, however, and there still was no direct parallel with human cancer to stimulate research in this area. The demonstration that a lymphatic leukemia of mice was caused by virus (4) finally aroused the interests of clinicians and researchers, for the observation of a virus-induced mammalian cancer which was clinically similar to that seen in man, made the prospect of virus-caused human malignancies more tenable. All three types of tumor viruses discussed thus far were shown on subsequent biochemical investigation to contain RNA, exclusive of DNA, while oncogenic viruses containing only DNA were found later.

The RNA-containing tumor viruses can be generally classified on the basis of the types of neoplasms they produce, i.e., solid tumors or reticulo-endothelial malignancies (Table 7-1). Those causing

TABLE 7-1
Known or Suspected RNA-Containing Tumor Viruses

Avian sarcoma viruses
Avian leukosis viruses
Avian reticuloendotheliosis virus
Murine sarcoma viruses
Murine leukemia viruses
Mouse mammary tumor virus
Feline leukemia virus
Canine leukemia virus

solid tumors are: (a) avian sarcoma viruses, which include the various strains of Rous sarcoma virus (*2, 5*) and the Fujinami sarcoma virus (*6*); (b) murine sarcoma virus (*7, 8*); and (c) mouse mammary tumor virus (*9*). The first two of these viruses are characterized by their ability to produce tumors rapidly *in vivo*, and by the ability to convert normal cells in culture to malignant forms, the malignant conversion accompanied by easily recognizable morphological changes. Other types of solid neoplasms are attributed to RNA viruses, e.g., nephroblastoma and osteopetrosis of fowl, but the responsible agents have not yet been isolated.

The viruses causing reticulo-endothelial cancers are: (a) avian leukosis viruses (*10*), including viruses causing visceral lymphomatosis, myeloblastosis, and erythroblastosis; and possibly avian reticuloendotheliosis virus (*11*); (*b*) murine leukemia viruses (*12*), causing lymphatic and myeloid leukemias. Viruses associated with canine (*13*) and feline leukemias (*14*) also have been described, and probably can be included in the realm of the RNA tumor viruses. The leukemia viruses often propagate in animals or cell cultures with no easily recognizable effect on the infected cells, and they characteristically produce disease in animals only after a prolonged latent period.

The nucleic acid character of only a few of these viruses has been determined experimentally, but extrapolation to other viruses in these groups can be made on the basis of similarities in basic virus structure, antigenicity, pathogenesis, etc. The list of viruses presented in Table 7-1 is undoubtedly incomplete, and we can expect that other RNA-containing oncogenic viruses will be identified.

7-2. Structure

A. Morphology

The known RNA-containing oncogenic viruses are generally similar in physical, chemical, and morphological structure, although specific differences among groups have been established. These virus particles are usually found to be smaller than myxoviruses and structurally are quite distinct, although they are often classed as myxoviruses, or myxovirus-like. The virions are basically spherical, but aberrant forms are found in suboptimal environments. Estimates of viral diameters have ranged from 65–150 mμ, based on a variety of experimental methods, including filtration, sedimentation, and electron

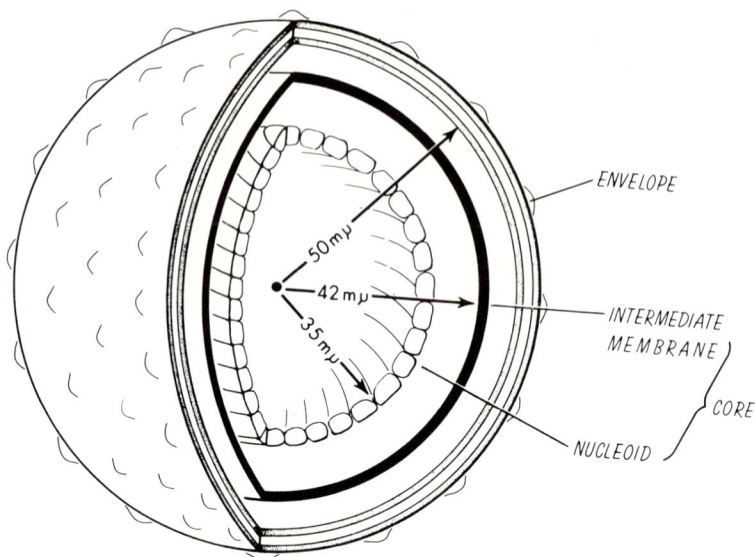

FIG. 7-1. Diagramatic representation of RNA tumor virus particle.

microscopy, but the most reproducible and accurate data from the electron microscope indicate diameters ranging 90–110 mμ. Estimates for the avian viruses (other than reticuloendotheliosis virus) generally tend to be smaller than those for murine viruses (*15–19*). A diagramatic representation of an RNA-containing tumor virus particle is given in Fig. 7-1.

1. *The Envelope*

Like many viruses, infectious particles are completed by budding at at the cell membrane (Fig. 7-2—No. 1, 4, 8), and the virus particle contains a protein–lipid membrane as its envelope. This membrane undoubtedly contains components of the plasma membrane, but antigenic elements specific for the virus are also present. The envelope is responsible for attachment of virus and its penetration of cells, and genetic differences in virus strains are reflected in the envelope character which is a determinant of infectivity.

Morphologically the envelope has the structure of the so-called "unit membrane," i.e., two peripheral layers of protein with a central layer of lipid. Examination of the viral surface by negative staining shows tiny projections, 5–10 mμ long, emanating from the membrane. These projections are characteristic of budding viruses and are not usually found on normal plasma membranes. Little is known about the role of these projections in infectivity or about their chemical nature.

2. The Intermediate Membrane (Core)

The electron microscope reveals a membrane intermediate between the envelope and a central electron dense nucleoid. This membrane has not been characterized chemically but is probably protein. The diameter of the structure enclosed by the intermediate membrane is approximately 85 mμ, indicating a volume approximately one-half that of the intact virion.

"Cores" of murine leukemia virus (20) or Rous sarcoma virus (21) have been isolated by stripping off the envelope with detergent and ether. The cores have structures compatible with that of the intermediate membrane and nucleoid. Resistance of the cores to such treatment indicates that lipid is not an important constituent of the intermediate membrane. Intracerebral injection of mice with cores derived from murine leukemia virus induced leukemia, demonstrating that a degree of infectivity is retained after ether treatment (22).

The cores have a buoyant density (1.26 g/cc) distinct from that of whole virus (1.16 g/cc) in K tartrate, K citrate, or Cs_2SO_4 gradients and are easily separable from virus. Ribonuclease releases ^{3}H-uridine from labeled viral cores, but not from whole virus, indicating that the cores are relatively porous (21).

Particles resembling leukemia virus cores are regularly found intracellularly in tissues containing mouse mammary tumor virus, and such particles may be the cores of the mammary tumor virus (Fig. 7-2–No. 7). Whether these constitute precursors to the final budding particles or represent residual degradation products of ingested virus is uncertain. Experiments involving particle diffusion rates indicate that these incomplete particles may be infectious as well as the completely enveloped particles (23).

The viral envelope and intermediate membranes are semipermeable giving virus particles properties of osmometers. When particles of murine leukemia virus or Rous sarcoma virus are suspended in hypotonic or isotonic medium during fixation with OsO_4, spherical particles are noted with the electron microscope (24, 25). If fixation is performed in hypertonic medium, the membranes collapse around the nucleoid and tailed particles are prevalent (8). This permeability is also reflected in the buoyant densities of the virus particles. In density

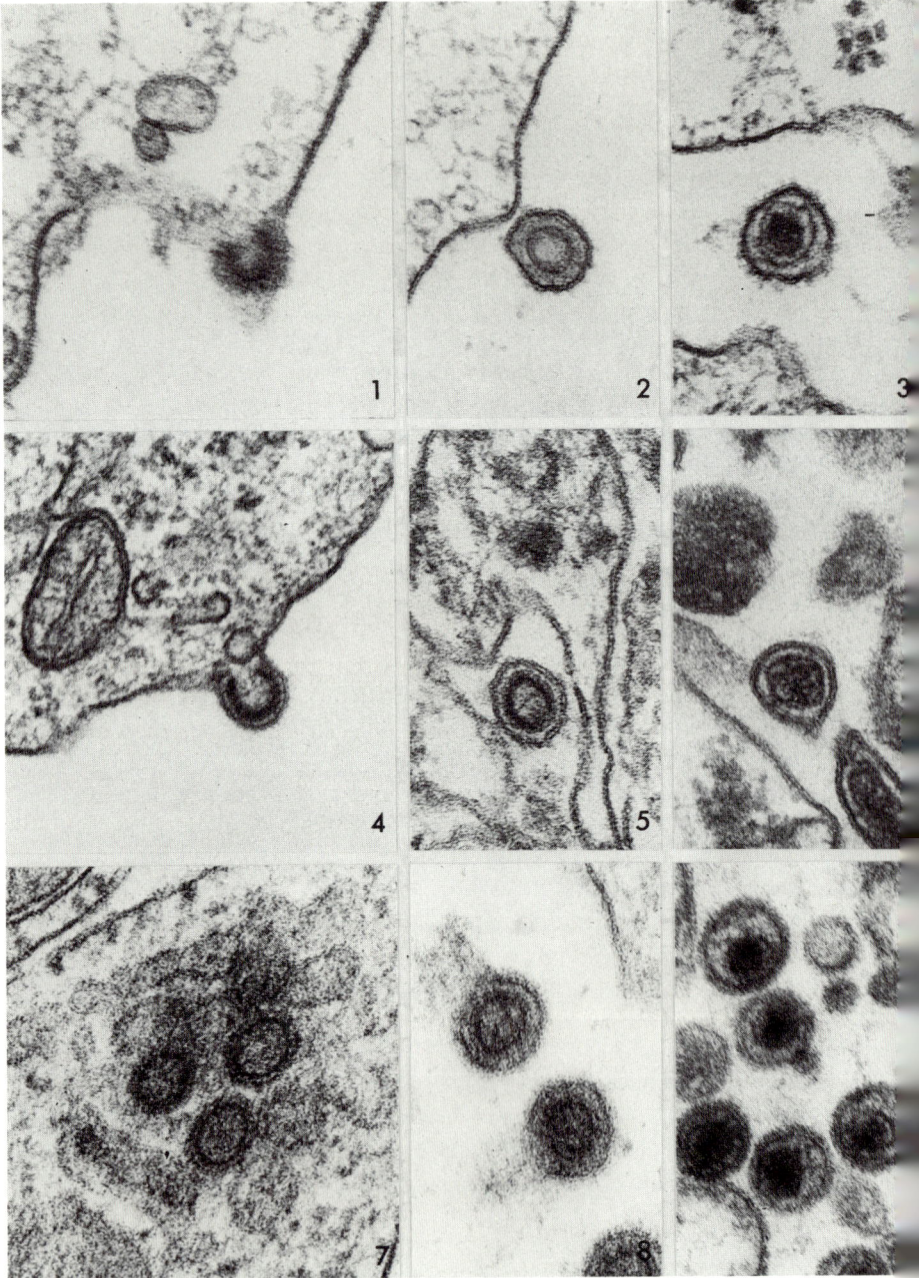

FIG. 7-2. RNA-containing tumor viruses ($\times$ 116,200). (1) Particle budding

gradients of high osmotic pressure (sucrose, K citrate, Cs_2SO_4, etc.) the particles reach equilibrium at about 1.16 g/cc (*26–28*). In Ficoll (polysucrose), which has a minimal osmotic effect, the viruses band at 1.08 g/cc (*28*). The lighter density in Ficoll indicates a greater hydration of the particle and is probably the density of the virus under physiological conditions.

Selective permeability to solutes is seen also with acridine dyes. These compounds react with nucleic acids and render them sensitive to visible light. Poliovirus remains resistant to visible light after addition of neutral red to the virus suspension (*29*), indicating that the poliovirus capsid prevents contact of the dye with the internal nucleic acid. However, suspension of Rous sarcoma virus in neutral red renders the virus sensitive to visible light (*30*), indicating the dye has passed the membranes and attached to the internal nucleic acid.

3. *The Nucleoid*

The nucleoid of the RNA-tumorigenic viruses contains the viral RNA, possibly complexed with protein (*31, 32*). The nucleoid of a virus particle recently released from a cell presents an electron-lucent center, with an electron-dense area, representing the RNA, at the periphery. These particles, often called "immature," are fully infectious. Particles of all of the groups under discussion look alike at this stage, and cannot be distinguished morphologically with confidence (Fig. 7-2–No. 2, 5, 8). This staining characteristic indicates that the viral nucleic acid is coiled into a hollow ball within the viral core, rather than bunched into a compact mass. Changes in general morphology eventually occur in extracellular virus and this change can be attributed to condensation of the nucleoid. After this change has occurred, viruses of the avian tumor, murine tumor, and mouse mammary tumor groups can be distinguished morphologically (Fig. 7-2–No. 3, 6, 9). It is probable that such "mature" particles are in fact

---

from chick embryo cell infected with mixture of Rous sarcoma virus (Bryan) + Rous-associated virus. (RSV-RAV₁). (2) "Immature" avian leukosis virus particle (RSV-RAV₁). (3) "Mature" avian leukosis virus particle (RSV-RAV₁). (4) Particle budding from rat cell infected with murine leukemia virus (Moloney str.). (5) "Immature" murine leukemia virus particle (Moloney str.). (6) "Mature" murine leukemia virus particle (Moloney str.). (7) Intracytoplasmic "A" particles in mammary tissue infected with mouse mammary tumor virus. (8) "Immature" mammary tumor virus particles in process of completion. (9) "Mature" mammary tumor virus particles. (Through the courtesy of Dr. A. F. Valentine and Dr. G. H. Smith, National Cancer Institute, Bethesda, Maryland.)

inactivated virus. The high sensitivity of these viruses to moderate heat suggests that the ribonucleoprotein breaks down and the nucleoid collapses, giving this "mature" form.

B. CHEMICAL CHARACTERIZATION

1. *General*

No single RNA-tumor virus has been chemically analyzed extensively and a general idea is constructed only by piecing together data from several laboratories on different viruses. As indicated previously, lipid, protein, and RNA are important constituents of these viruses. Lipid is found mainly in the envelope and constitutes about 35% of the dry weight of avian myeloblastosis virus (*33*), and 27% of mouse mammary tumor virus (*34*). Disintegration of viruses following treatment by phospholipases indicates the presence of phospholipid, a regular component of cellular membranes (*35, 36*). Lipids of murine leukemia viruses have been analyzed chemically (*37*), and 75–80% of the total lipid is phospholipid. Although these percentages are higher than those found for cellular lipids (about 60% phospholipid), the specific lipid components of virus are very similar to those of cells.

Proteins make up about 50% of the avian myeloblastosis particle (*33*), and all of the usual amino acids are represented (*38*). Analysis of a murine leukemia virus showed that at least two electrophoretically separable proteins are present (*39*), not a surprising finding, since at least two distinct antigens are found in murine (*40*) or avian leukosis (*41*) viruses.

About 2% of the total particle is RNA (*33*), and no significant amount of DNA has been reported for any of these viruses.

In addition to protein, lipid, and RNA, some carbohydrate moieties may be integral parts of viral structure. Glucosamine has been found in avian myeloblastosis virus (*33*), and the presence of this amino sugar plus mannose, galactose, and fucose in influenza virus suggests that other carbohydrates may be found. It is likely that at least some of these components are derived from host-cell membrane. Also, it is possible that such substances contribute to the antigenic character of the virus.

2. *The RNA*

These viruses contain the largest natural molecules of RNA found in nature. Molecules of molecular weight 10^7 have been extracted from avian sarcoma and leukosis, murine leukemia, and mouse mammary tumor viruses (*42–45*), and possibly murine sarcoma virus (*46*). While this large RNA constitutes most of the RNA in the virus particle (*47*), the evidence that this molecule is the entire viral genome is inconclusive. Attempts to attain full infectivity with RNA from various avian tumor viruses have yielded negative results (*30, 48*). An infectious RNA derived from tissues infected by a murine leukemia virus has been reported (*49*) and awaits confirmation. An RNA extracted from Rous sarcomas also has been reported to induce virus-specific antigen in chick embryo cells (*50*) but no virus production resulted and cells were not converted to malignant forms. The difficulty in finding infectious RNA consistently may be due to any of several factors including: (a) a possibility of two or more different molecules of RNA required for infectivity; (b) the difficulty of introducing such a large piece of RNA into the cell; (c) possible action of intracellular nucleases before the RNA can reach a proper site for replication.

Several lines of evidence indicate that the large RNA molecules found in these viruses are single stranded: (a) nucleotide base ratios (Table 7-2) of several viral RNA's are inconsistent with a double-stranded base-paired molecule (*42*), (b) the RNA is digested to small pieces by pancreatic ribonuclease (*26, 39*); (c) heating of extracted RNA changes its light ($\lambda = 260$) absorbency in a manner similar to other single-stranded RNA's (*51*); (d) sedimentation properties in solution are characteristically affected by changes in salt concentration (*42*).

Base ratio analyses of extracted viral RNA's have yielded widely varying results from different laboratories, even among identical virus strains. Analyses from a single laboratory, employing essentially identical techniques, have been made on preparations of three avian viruses [avian myeloblastosis virus, a mixture of Rous sarcoma and Rous-associated viruses, and "defective" Rous sarcoma virus (*28*)] and a murine leukemia virus (*42*). Comparison of the bases (Table 7-2) shows that there are slight differences among the various viruses, but the differences are probably not large enough to be of practical use in identification of the viral RNA type.

It can be calculated that the large molecule of viral RNA has sufficient nucleotides to code for over 20 proteins, assuming non-redundancy of genes. Thus far, however, only two antigens have been identified in virus or virus-infected cells, and of these, only one,

TABLE 7-2
Base Composition of Viral RNA's[a]

	Murine leukemia virus	Avian myeloblastosis virus	Rous sarcoma virus + Rous-associated virus (mixture)	Rous sarcoma virus
C	26.7	23.0	24.2	24.2
A	25.5	25.3	25.1	24.8
G	25.1	28.7	28.3	29.2
U	22.7	23.0	22.4	21.7

[a] References 37, 40.

the type-specific antigen of the viral envelope, can be directly related to the translation of viral genes.

3. Host-Cell Components of Virions

Molecular components of normal cells can certainly be found in virus particles, but at least some of these molecules cannot be considered structurally important or necessary for infectivity. Adenosine triphosphatase has been identified on the surface of particles of avian myeloblastosis virus (52) and murine leukemia virus (53). However, if these viruses are propagated in cells which produce minimal amounts of triphosphatase normally then little or no enzyme is seen associated with virus particles. Apparently the viruses acquire cellular enzyme in the process of budding from the cell membrane and it is likely that other enzymes of the infected cell surface are associated with the viral envelope.

An alkaline phosphatase (53) and a polynucleotide phosphorylase (54) have also been found in preparations of avian myeloblastosis. These enzymes are not usually associated with the cellular surface, and their presence in virus may represent adsorption to the viral envelope after completion of the virus.

Lipids of the viral envelope probably are host derived if we can extrapolate from other types of budding viruses (55, 56). The indicated chemical analyses probably reflect the cellular origin of viral lipids.

Intuition suggests that other structural components of the cell membrane may be found in viral envelopes, but experimental verification is wanting. Several reports have indicated that antibody elicited against normal tissues were capable of neutralizing viral activity (*57, 58*), but in related systems other workers observed contradictory results, or found such neutralization to be an accident of the system (*59–61*).

Some cellular RNA possibly is incorporated into virions. Sedimentation of extracted RNA from avian myeloblastosis virus revealed absorbency peaks in regions of 26S, 18S, and 5S material in addition to viral 70S RNA (*62*). The 5S material showed acceptor activity for amino acids and was concluded to be cellular transfer RNA. While such an inclusion of cellular RNA into virion is unprecedented, the reported association of cellular RNA with plasma membrane (*63*), and the presence of other host membrane materials in virions, make this an interesting possibility.

C. Physical Properties

A number of biophysical analyses have been done on RNA tumor viruses which give insight into the nature of the virus particles and lend support to biochemical investigations.

1. *Density*

As indicated previously, the buoyant densities of avian tumor and murine leukemia viruses are about 1.16 g/cc in density gradients of low molecular weight solutes (*11, 13, 24*). The density of the mammary tumor virus seems to be slightly higher (1.18–1.22 g/cc) (*64*). Variation in density within a single preparation of virus has been noted (*65*) and is a phenotypic property of the virions. The relatively low densities compared to other viruses are due in part to the large amount of lipid in the viral envelope but also are indicative of extensive hydration of the particles.

2. *Heat Sensitivity*

These enveloped viruses are more sensitive to heat than viruses with an external capsid, and special care must be given to retain infectivity. The half-life of viral infectivity at 37°C is about 3 hr for most of these viruses, although determinations ranging from less than 2 to more than 7 hr have been made. Analyses of inactivation show single-hit kinetics at various temperatures between 37 and 60°C (*66, 67*). An Arrhenius' plot of inactivation rate constants reveals a change in the activation energy at about 43°C. The activation energy below 43°C (19.9 kcal/mole) coincides with that required for degradation of RNA molecules in solution (68), suggesting that at the lower temperatures it is RNA which is disintegrating. Such an interpretation is consistent with the observed condensation of the nucleoid in "mature" virions discussed earlier. The relatively free permeability of water and certain solutes into the particles may be partially responsible for this inactivation. Conversely, the relative stability of viruses with an external capsid may be related to their impermeability (69).

3. Radiation Sensitivity

A notable feature of the avian tumor viruses is their relative resistance to ultraviolet light (70–72). About ten times as much irradiation is required for inactivation when compared to other budding RNA viruses. Structural differences possibly may account for this resistance, but an interesting hypothesis based upon "integration" of the viral genome into host metabolism has also been presented. Irradiation of virus particles with X rays results in similar inactivation rates for Rous sarcoma virus and Newcastle disease virus, indicating similar target sizes for the two viruses.

D. ANTIGENIC COMPOSITION

1. Type-Specific Antigens

Two general types of antigens are found in oncogenic RNA viruses type specific and group specific (gs). The type-specific antigens are located on the viral surface and constitute part of the envelope. The antigens appear to be identical with viral receptors which are important in initial stages of infectivity. The avian tumor group can be separated into three subgroups on the basis of (a) neutralization with type-specific antibody (73), (b) relative infectivity for genetically susceptible or resistant avian cells (74), (c) sensitivity to interference by prior infection with specific avian leukosis strains (75). The identical surface component may be the basis for grouping by these three methods.

Immunological typing of murine leukemia viruses has been accomplished by phenotypic mixing of leukemia virus with murine sarcoma virus, accompanied by the production of sarcoma virus "pseudotypes" (40). Antibody against the homologous leukemia virus prevents transformation of cells by the pseudotype, and on this basis strain differences have been identified.

A comparison of mammary tumor viruses isolated from three strains of mice revealed no significant immunological differences (76), but it is possible that more extensive studies will show differences. A single study showing alterations in reactivity of mammary tumor virus to antiserum after changing the animal host used for propagation of the virus (77) indicates that antigenic strain differences may also exist among mammary tumor viruses.

2. Group-Specific Antigens

The group-specific (gs) antigen of the avian tumor virus groups is similar for all of the RNA-containing avian tumor viruses thus far examined, and differs from that of the murine leukemia group. The antigen can be identified in cells infected by avian tumor viruses using serum from an avian sarcoma-bearing hamster (41). Such antiserum fails to neutralize infectious Rous sarcoma virus. Using complement-fixation tests, the identification of gs antigen in avian myeloblastosis virus required disintegration of virus with strong detergent or a weak detergent plus ether (78, 79). Also, experiments with fluorescent antibody directed against the avian gs antigen failed to demonstrate antigen on the surface of the Rous sarcoma virion (80). Staining occurred only if preceded by fixation. Thus, the gs antigen is probably an internal structural component of the virion.

The chemical nature of this internal antigen is unknown and data pertaining to the antigen is puzzling. Antibody to such antigen is found in hamsters (41) and other mammals bearing Rous sarcomas, but no gs antibody is formed in infected chickens, the natural host of these avian viruses. Also, an attempt to establish a relationship between the presence of gs antigen in chick embryos and infection of chick embryos with avian leukosis viruses was unsuccessful (81). Virus particles were found where no gs antigen was apparent, or gs antigen was plentiful in the absence of infectious virus or morphologically identifiable leukosis particles. These observations raise the interesting possibility that the gs antigen is similar or identical to a normal embryonic chicken component. Chickens perhaps are immunologically tolerant and fail to produce antibodies against this antigen, while other species respond to it.

A group-specific antigen has also been observed in the murine leukemia viruses by using immunoprecipitation (82) or complement-fixation techniques (40). This gs antigen is not found in mammary tumor or avian tumor viruses and presents a basis for distinguishing members of the tumor-virus groups.

Murine sarcoma virions can acquire murine leukemia type-specific antigens in a mixed infection of sarcoma and leukemia viruses, and the infectivity of the sarcoma virus may be augmented by this process. In the absence of leukemia virus, murine sarcoma virus and cells infected with the virus do not react with antibody against the murine leukemia group-specific antigen (83). Thus murine sarcoma viruses may be antigenically unrelated to the murine leukemias.

E. "DEFECTIVENESS" OF SARCOMA VIRUSES

Certain developments in investigations into Rous sarcoma virus suggested that this virus may be "defective." An examination of stock preparations of Rous sarcoma virus (Bryan "high titer" strain) revealed the presence of another avian leukosis virus (84), termed Rous-associated virus (RAV). Attempts to obtain Rous sarcoma virus devoid of RAV resulted in the isolation of transformed cells producing no apparent infectious virus, as determined by the standard infectivity techniques in chick embryo cells (85). Deliberate addition of RAV to such "nonproducing" cells resulted in production of infectious Rous sarcoma virus containing the type-specific antigen of RAV. Other leukosis viruses likewise were effective as "helper" viruses in inducing infectious Rous sarcoma virus, and the resulting transforming particles, in each case, could be neutralized by antisera against the specific helper virus employed, but not by antiserum against a heterologous leukosis virus (86). Properties of the helper virus envelope were also noted in the host range (87, 88) and sensitivity to interference (89, 90) of the derived Rous sarcoma virus. On the basis of these observations the Bryan "high titer" strain of Rous sarcoma virus was declared "defective," in that it was unable to provide for the synthesis of its own envelope and required a helper virus for completion of infectivity. Similar experiments with nonproducing transformed cells had been done with the Bryan "standard" strain of Rous sarcoma virus (91, 92). In this case, a hypothesis based on interaction between cellular and viral genomes was offered.

Subsequent experiments showed that interpretations of defectiveness or genomic interaction were unnecessary to account for nonproductive transformed cells. Examination of nonproducer Rous sarcoma cells by electron microscopy revealed that the cells produced particles morphologically identical to avian leukosis particles (93–95). The particles could be labeled with ³H-uridine, and the RNA extracted from the particles had characteristics of viral RNA (96). Addition of "nonproducer" culture fluids to Japanese quail cells resulted in transformation typical of Rous sarcoma virus, and certain selected chick embryo cells were also found to be susceptible to infection (97). Also, normally resistant chick embryo cells could be transformed by particles from "nonproducer" cells after introduction of irradiated Newcastle disease virus (42). The latter effect was interpreted as a trapping of the Rous particles by cells in the process of fusion caused by the second virus.

These results show clearly that the Bryan "high titer" strain of virus is not "defective" in the usual sense, since completely infectious progeny are produced by infected cells. Also, quantitative experiments involving radioactive labeling showed that about the same number of Rous sarcoma virus particles are produced in the presence or absence of "helper" virus (42). It can be concluded, therefore, that the deficiency of the Rous sarcoma virus, portrayed by a relatively inefficient infectivity for certain cells, is related to the composition of the viral envelope, and does not constitute a failure of viral genome to provide an envelope.

A similar deficiency has been observed for the murine sarcoma virus. Production of virus by infected cells has been found thus far only with concomitant infection by a murine leukemia virus (40, 98, 99). However, certain transformed rat cells containing the viral genome of murine sarcoma virus, but no leukemia virus, produced typical virus particles as determined by electron microscopy, buoyant density measurements, and the presence of high molecular weight RNA (46). Infectivity of these particles has not yet been demonstrable. However, deliberate infection of these cells by a murine leukemia virus results in infectious murine sarcoma virus (46, 99). Thus, the murine sarcoma virus appears analogous to "defective" strains of Rous sarcoma virus in that complete progeny are produced by infected cells, but these progeny are inefficient in reinfection.

7-3. Susceptibility

Successful infection of a cell by RNA-containing tumor viruses requires compatibility in two general areas. These are (a) the surface

phenomena determining initial acceptability of the virus by the cell
and (b) intracellular factors associated with the cells' metabolic activity. The mechanism by which the RNA tumor viruses enter the
cell is not fully known. One hypothesis suggests that the particle is
ingested intact (viropexis) (*100*), the envelope being stripped in the
cytoplasm, with release of the nucleoid. A more appealing mechanism involves coalescing of the viral envelope with the plasma membrane at the cell surface, with release of the viral core or nucleoid
into the cytoplasm (*101*). Whatever the mechanism, a specific interaction between viral and cellular receptor sites must first take place
at the cell surface; the envelope of the virus must be compatible with
the membrane of the cell for successful infection.

A. Susceptible and Resistant Cells

Several types of avian cells varying in susceptibility or resistance
to avian tumor viruses have been delineated (*55*). The response to
exposure to virus is noted in living animals, as well as cells derived
from them, and the susceptibility has a genetic basis (*102–104*). Indeed, genetic experiments show that susceptibility to subgroup B
avian tumor viruses is controlled by a single autosomal gene, with
susceptibility dominant over resistance. Cells derived from embryos
of susceptible or resistant chickens have been classified in a manner
similar to that for bacterial resistance to bacteriophage. Conversely,
the avian tumor viruses can be classified on the basis of their ability
to infect the various cell types, and this classification can be correlated with the antigenic subgroup of the virus discussed previously
(*73–75*).

The low infectivity of virus for resistant cells is not due to a failure
of virus to adsorb to these cells. Rous sarcoma virus is removed from
suspending medium as rapidly by resistant as by susceptible cells
(*103, 104*). However, the disappearance of infectivity of adsorbed
virus is delayed in resistant cells, indicating a difficulty of the virus in
attaining the eclipse phase.

Although certain avian sarcoma viruses can produce tumors in
mammals, these animals are relatively resistant to infection by the
avian viruses. A major factor in this resistance is an inconsonance of
the viral envelope and the cell membrane, a conclusion derived from
the following observations (*105*). A strain of Rous sarcoma virus
(Schmidt-Ruppin 50) effectively produces tumors in Syrian hamsters,

while another strain (RSV-RAV₁) is ineffective. A nontransforming avian leukosis virus (RAV50) with envelope properties similar to Schmidt-Ruppin 50 was grown as a mixed infection with RSV-RAV₁ and the resulting progeny injected into hamsters. Tumors developed in these animals, but not in those in which mixed virus was exposed to anti-RAV50 serum before injection. These results showed that phenotypic mixing of viruses had occurred with the RSV-RAV₁ acquiring the envelope, or important elements of the envelope, of RAV50. The acquisition of this envelope enabled the RSV to successfully combine with and to transform the hamster cells.

B. INTERFERENCE

It appears that the various viral subgroups bind to different surface sites on a given cell type, while virus strains within a given subgroup share common binding sites. Prior infection of cells with an avian leukosis virus prevents infection by a second virus of the same subgroup but not by a virus of another subgroup (75, 87, 88, 106). This type of homologous resistance can occur within minutes after exposure of cells to large amounts of the initial virus or may require several days if a small amount of virus is used (103, 104). It seems that the cellular receptors for the first virus are either covered by the material of the viral envelope or are somehow changed as to be incompatible with infection by a homologous subgroup member, but receptors are still available for a heterologous virus. Similar interference reactions have been observed among the murine viruses (107). Such interference among viruses has been used as a means of classifying virus subgroups described earlier. It perhaps is not surprising that members of the interference subgroups are also classed together when immunological or cellular resistance methods are employed, indicating that the same parameter is being measured by three different means. The primary events of infection with the RNA-containing oncogenic viruses are probably, in a general way, similar to other viruses which have a membrane-derived envelope at their outer surface. Unfortunately little is known yet about the biochemistry of these initial interactions.

C. THE ROLE OF HOST-CELL METABOLISM

The virus genome, having successfully gained access to the cellular cytoplasm, encounters a host metabolism which determines the fates of cell and virus. The metabolic events affecting the virus are obscure at this stage. A determining factor, however, is the stage of the cell cycle during which infection takes place. In experiments utilizing cells synchronized with respect to the mitotic cycle (108), the efficiency of infection of cells with Rous sarcoma virus was highest when virus was added prior to or during the S phase, the stage of synthesis of nuclear DNA. Under similar circumstances, infection with Newcastle disease virus occurred with equal efficiency at all stages of the cell cycle. It is possible that the Rous sarcoma virus utilizes factors involved in cellular DNA synthesis, which would be most active during the S phase, for synthesis of an essential virus constituent or precursor. Experiments to be discussed in a later section show a requirement for DNA synthesis for successful infection by Rous sarcoma virus. Successful infection may also require a single cellular division (109), perhaps because of the proximity of the S phase to mitosis and the general incapacity of cells to produce virus during mitosis (110). The absence of these specific metabolic events may result in an abortive infection. Once uncoated the exposed viral RNA may be vulnerable to inactivation by ribonuclease or other RNA-degrading substances, and the longer the viral genome must wait for the proper metabolic events to occur, the greater the chance of its destruction.

The early metabolic requirements for successful infection are reflected in various observations concerning growth of virus or transformation. Cell cultures which have grown to confluency (and where cell divisions are then infrequent) vary in their susceptibility to infection by murine sarcoma virus, depending upon treatment subsequent to exposure to the virus. If confluent cells are dispersed with trypsin within a few hours after exposure to virus and transferred into sparse populations capable of growth and division, the efficiency of infection is high compared with cultures initially containing sparse growing cells (111). However, if the confluent cultures are left for 24 hr before dispersal and transfer, successful infections are reduced 1000-fold. Under similar conditions of confluency, many other types of RNA viruses (e.g., Newcastle disease virus, influenza virus) infect cells without any requirement for cellular dispersal.

Similarly, cells which have been suspended in liquid medium for several days and have stopped dividing are less susceptible to infection with RSV than are suspended dividing cells (112). Such observations suggest that in the otherwise susceptible animal the tissues most likely to be infected by leukemia viruses, or infected and later transformed by sarcoma viruses, are those most active in cellular divisions. The neoplasms observed as a consequence of infection with RNA tumor viruses are consistent with this notion.

7-4. Metabolic Requirements for Infection

A. THE REQUIREMENT FOR DNA SYNTHESIS

Knowledge of intracellular events leading to viral replication mainly relies upon evidence obtained through manipulation of the metabolic pathways of the cell and the effect of such manipulation on virus growth. The evidence, therefore, can be deemed indirect. Nonetheless there are important distinctions to be made between the RNA-tumor viruses and other RNA-containing viruses. The most striking difference is a requirement for DNA synthesis. The growth of Rous sarcoma viruses, Rous-associated virus, and murine sarcoma virus is prevented by such antimetabolites as 5-iododeoxuridine or cytosine arabinoside, or by excess thymidine, if any of these substances is added to cells soon after infection (111, 113–116). These antimetabolites prevent DNA synthesis, or alter newly synthesized DNA, but have little immediate effect upon the synthesis of RNA or protein. The different mechanisms by which these and other virus-inhibitory DNA antagonists (117–119) affect DNA argue against the possibility that their effect on virus growth is secondary to inhibition of some other type of metabolic event. Also, under essentially identical conditions the growth of other RNA viruses (e.g., influenza or vesicular stomatitis viruses) is unaffected by the pressure of such DNA antagonists (108). The virus-inhibitory action of 5-iododeoxuridine is prevented by thymidine, and inhibition by cytosine arabinoside or excess thymidine is prevented by deoxycytidine. Thymidine and deoxycytidine are specific precursors of DNA. The bulk of these observations cannot easily be reconciled with a mechanism of virus growth independent of DNA.

The requirement for DNA synthesis is a transient one, as demonstrated by studies on the time course of virus growth in cell culture. Inhibition of DNA synthesis later than 8–12 hr after initial exposure to virus has little effect on the subsequent production of Rous sarcoma virus, although inhibition only during the first 8 hr after viral exposure protects the cells from infection (120) and abolishes the subsequent production of virus (116).

The requirement for DNA synthesis suggested that certain features of DNA-containing tumor viruses might obtain with the RNA-tumor viruses. The DNA-containing papovaviruses can induce DNA synthesis in nongrowing cells under specified conditions (*121, 122*). Under similar conditions, exposure of cells to Rous sarcoma virus or murine sarcoma virus has no stimulating effect on DNA synthesis (*111, 123*), and the infection is aborted for lack of this metabolism.

The results of several other reported studies can be explained by the transient requirement for DNA synthesis: (a) As was briefly mentioned before, cells which were transferred from a growing monolayer into suspension in liquid stopped dividing within a few days (*112*). Although viability was retained, DNA synthesis presumably was minimal. These cells produced Rous sarcoma virus only if infected during the initial suspension period, but if virus was added after cellular divisions diminished infection was unsuccessful. However, once infection was established, suspended cells continued to produce virus over several weeks.

(b) Likewise, chick embryo cells maintained in a serum-deficient medium stopped growing, although capabilities for growth and division remained (*116*). Synthesis of DNA decreased to a low level in such cells, while RNA and protein synthesis remained active. These cells could not be successfully infected by Rous sarcoma virus unless serum was added soon after exposure to virus and cultures resumed DNA synthesis. However, if infection had been established in growing cultures when cells were synthesizing DNA, virus production continued indefinitely after deletion of serum from the cell-culture medium. In both cases, (a) and (b), if the requirement for DNA synthesis was fulfilled, infection was successful, and once infection was established DNA synthesis was no longer necessary for virus production.

(c) An interpretation of the results of X-irradiation of infected and noninfected cells (*124*) also can be made on the basis of a transient requirement for DNA synthesis. Cells which were X-irradiated prior to infection failed to produce Rous sarcoma virus, but if infection was established before X-irradiation, cells continued to produce high levels of virus after treatment. X-irradiated cells stop dividing but retain a degree of viability, as evidenced by continuation of certain metabolic activities. Under these limiting conditions of growth which fail to provide Rous sarcoma virus with the essential DNA synthesis, other types of RNA viruses reproduce.

B. DNA AS POSSIBLE REPLICATIVE FORM

The nature of the DNA synthesized has not been determined. Perhaps it is cellular DNA which must be replicated so that some sort of integration of viral RNA can take place. Another possibility is that the single-stranded viral RNA serves as primer for a complementary strand of DNA and the replicative form of the virus contains at least one strand of DNA (*125*). This hypothesis is consistent with the reported observations and is supported by experiments with 5-bromodeoxyuridine (BUdR), which is incorporated into DNA as a substitute for thymidine (*117*). Cultivation of cells in the presence of BUdR for as long as 4 days, had little effect upon the subsequent yield of infectious Rous sarcoma virus, if the BUdR was removed before infection. However, if BUdR was retained in the medium, or if nontreated cells were exposed to BUdR immediately after exposure of the cells to Rous sarcoma virus, then virus reproduction was inhibited. These results indicate that it is the specific DNA synthesized during viral replication which is important.

Other experiments using partially synchronized cells (*109*) showed that the DNA required by Rous sarcoma virus can be made during the G_2 period of the mitotic cycle (the phase between nuclear DNA synthesis and mitosis). Inhibition of DNA synthesis during this time prevented subsequent virus growth but had little effect on mitosis, indicating that the viral DNA requirement is something other than nuclear DNA.

A new intracellular DNA rendered single stranded would be expected to attach to its complementary viral RNA to form a DNA–RNA hybrid under appropriate experimental conditions, and attempts to detect a new DNA by hybridization methods have been made. A single report (*126*) tendered data showing an increased binding of viral RNA to DNA from RSV-infected chick embryo cells compared to DNA from uninfected cells. These data are tenuous, however, and other laboratories have been unable to confirm the results (*123, 127*). One difficulty lies in the high degree of binding of viral RNA to noninfected chicken DNA which may obscure differences one might expect to see between infected and noninfected cells. This "normal" homology may play a role in the replication of virus, but its significance has yet to be determined.

The level of binding of viral RNA was found to be much less to hamster DNA than to chicken DNA (123), reducing the problem of nonspecific homology in hybridization experiments. Hamster tumors induced by Rous sarcoma virus contained DNA having a greater capacity than nontumor hamster DNA for binding viral RNA. The increased binding was specific for rapidly sedimenting viral RNA, and differences in binding disappeared when slowly sedimenting RNA was used. This latter low-molecular weight RNA, although extracted from density-gradient purified virus, may represent nonviral RNA.

Further support for the hypothesis that DNA constitutes part of the replicative form of the virus is provided by experiments involving sedimentation of presumed RNA extracted from RSV-infected cells (128). A considerable fraction of this RNA sedimenting in the viral RNA region was found resistant to ribonuclease, in contrast to RNA extracted from virus particles or from noninfected cells. The addition of deoxyribonuclease decreased this resistance to ribonuclease, and it was concluded that a DNA–RNA hybrid molecule was present.

C. DNA Synthesis and Transformation

The requirement for DNA synthesis in virus growth is paralleled by a similar requirement in the transformation of cells by virus. Exposure of cells to cytosine arabinoside or 10 mM thymidine within 12 hr after addition of Rous sarcoma (120) or murine sarcoma virus (111) prevented subsequent transformation of cells after the inhibitors were removed. The addition of inhibitor for a similar interval before or after the critical 12-hr period had little effect on transformation.

Morpholigical transformation of cells after infection requires several days to develop under the usual conditions of cell culture. If cells are exposed to virus, incubated for 12 hr, and then DNA synthesis is continually suppressed by thymidine over the next five days, cells still become transformed (174). These cells normally have a generation time of greater than 14 hr, and the fixation of transformation occurred within this time. It was concluded that development of transformation could occur in the absence of DNA synthesis, although the initiation of transformation required it.

D. A Role for Cellular Enzymes

The synthesis of the DNA required early in infection is performed by cellular enzymes. Inhibition of protein synthesis by puromycin immediately after exposure of cells to RSV had little effect on infection if the puromycin was removed within 8 hr (*116*). Addition of cytosine arabinoside (a DNA inhibitor) after the puromycin likewise had little effect although earlier addition of cytosine arabinoside resulted in an abortive infection. The DNA, therefore, is made in the presence of puromycin. Existing enzymes are sufficient for infection, and the conclusion is drawn that cellular DNA-synthesizing enzymes are utilized. This result is consistent with the observation that cells are most efficiently infected during or just prior to the S (DNA synthesis) phase of the mitotic cycle (*108*) since it is at this stage that cellular DNA-synthesizing enzymes would be most active.

E. THE REQUIREMENT FOR FUNCTIONING DNA

The addition of actinomycin D, an inhibitor of DNA activity, to infected cells prevents subsequent production of Rous sarcoma virus (*114, 117, 119, 129*). In contrast to cytosine arabinoside which affects only the early stages of virus replication, actinomycin D is inhibitory at both early and late stages. This effect is due to the failure of actinomycin-treated cells to synthesize viral RNA (*130*). A similar inhibition of production of virus and synthesis of viral RNA has been noted with murine leukemia virus (*131, 132*). These observations indicate that the viral RNA, as well as whole virus depends upon the transcription of DNA for synthesis.

In this case, actinomycin could be acting either upon cellular DNA or upon a hypothetical viral DNA. It seems unlikely, however, that the actinomycin acts directly upon a double-stranded viral RNA if this be the replicative form of the virus. No other double-stranded RNA has been observed to bind or be otherwise affected by actinomycin.

The effects of ultraviolet irradiation (uv) on the growth of RSV contrast with the effects of X-irradiation (*70, 71, 133*), probably for the reason that the metabolic consequences of X-irradiation differ from those of uv. In addition to inhibition of DNA synthesis, uv has a striking immediate effect on cellular RNA synthesis, whereas RNA synthesis continues after X-irradiation of cells. Exposure of cells to a critical dose of uv resulted in inhibition of virus production whether irradiation occurred before or at any time after infection. Cells similarly exposed to uv before infection were fully capable of supporting the growth of vesicular stomatitis virus or Newcastle disease virus.

These results complement those of experiments using actinomycin D to show that, in addition to the transient requirement for DNA synthesis, functioning cellular DNA is necessary.

Further support for a role for functioning cellular DNA in virus production was provided by experiments utilizing 5-bromodeoxyuridine (BUdR) (133). The DNA synthesized in the presence of BUdR has some of its thymidine replaced by BUdR, and this substitution renders the DNA excessively sensitive to the effects of uv. Cells pretreated with BUdR and subsequently infected with Rous sarcoma virus and incubated in the absence of BUdR were hypersusceptible to the virus-inhibiting action of uv. These experiments show that, whatever the nature or function of the DNA synthesized early in the course of infection, a DNA-based host function is required for the continued virus production of infected cells.

7-5. Completion of Infectious Virus

A. EXTRUSION

Infectious particles of the RNA tumor viruses are probably completed and released soon after synthesis of viral components. Inhibition of synthesis of nucleic acids by actinomycin or uv quickly decreases the rate of appearance of infectious Rous sarcoma virus (117, 133). Likewise, inhibition of protein synthesis by puromycin prevents further production of virus (116). It appears that after synthesis, viral components are not maintained in a pool for later incorporation into virus particles. A firm conclusion about this must await the biochemical identification of these components intracellularly.

Virus particles are completed by extrusion from the plasma membrane (134–137). Other than the budding particles, no structures which can be identified as virus specific, and which can be shown to be part of the replicative process, are seen in the cell by electron microscopy. Both the intermediate membrane and nucleoid of the particle first appear in the early stage of budding, and the blebbing viral envelope is continuous with plasma membrane.

Sections of mouse mammary tumors regularly contain intracytoplasmic structures similar to viral cores (nucleoid and intermediate membranes, also called "A" particles) when viewed in the electron microscope (*137*). Some consider these to be precursors of the completed mammary tumor virus, but incomplete intermediate membrane and nucleoid are often seen in apposition to the blebbing cell membrane, suggesting that the mechanism of viral synthesis is similar to the other RNA viruses. A more likely explanation for "A" particles is that the structures are virus particles which have been divested of the viral envelope upon reentry into the cell. However, they may even be unrelated to mammary tumor virus *per se,* or may be products of aberrant synthesis of mammary tumor virus.

B. PLASMA MEMBRANE AND VIRAL ENVELOPE

The identification of specific viral antigens on the viral envelope demonstrates unequivocally that the viral envelope is not identical with the host's cellular membrane from which it buds. What is not clear is the extent to which the virus imparts its antigenic character on the cell membrane. Viral structural antigens may be confined to specific areas of the membrane anticipating viral emergence or they may be more widespread.

If a cell is coincidentally infected with two strains of avian tumor virus [e.g., RSV (Bry) + RAV$_1$, or RSV (Bry) + RAV$_2$], progeny virus may be phenotypically mixed (*90, 138*). That is, within a single particle, the genome of one strain may be enclosed in an envelope containing elements of the other strain. The demonstration of this type of interaction indicates that virus-specific substances occur at cellular sites other than that required for completion of a homologous virus particle. This also suggests that part of the substance of the viral envelope is synthesized at a site removed from that of the specific RNA which is to be enveloped. Evidence for a generalized distribution of viral antigen on the surface of cell membrane is available for infections with myxoviruses (*139, 140*), and this may be a general attribute of cells infected with budding-type viruses.

An earlier section was addressed to the presence of host materials as integral parts of virus structure. Conversely, viral antigens apparently may become integral parts of the cellular membrane. Rous sarcomas in chickens regularly regress when strongly antigenic Rous-associated virus is produced along with Rous sarcoma virus (*86*). However, if only the weakly antigenic Rous sarcoma particles are produced, regressions are less likely.

C. Production of Antigens

Type-specific and group-specific antigens can be detected in the cytoplasm of infected cells using fluorescent antibody (141–146). In a few instances fluorescent staining was noted also in the nucleus, but the fluorescence was meager and irregular. Antigen synthesized in the nucleus may rapidly move into the cytoplasm, as was suggested by observations on infected cells incubated at subnormal temperature. Nonetheless, an antigen analogous to the intranuclear "T" antigen of papovaviruses has not been identified in cells infected with RNA tumor viruses. Whether any virus-specific materials are synthesized in the nucleus or even whether the viral genome has an intranuclear phase, are important considerations in understanding virus growth.

D. Non-Virus-Producing Mammalian Tumor Cells

Mammalian cells infected with avian tumor viruses usually do not produce infectious virus (147, 148). Certain hamster tumors induced by Rous sarcoma virus produce no virus particles as determined by infectivity measurements, by electron microscopy, and by radioactive labeling techniques, although the presence of the viral genome can be demonstrated (46, 149). Similarly, cells from a hamster tumor containing the genome of murine sarcoma virus produce no virus particles (46, 150). Budding extrusions are absent in such cells, and there is no microscopic evidence of virus structures. The viral genomes are present in these tumor cells since mixed cultivation with susceptible avian cells (or murine cells in the case of murine sarcoma virus) results in the transfer of viral genome. Addition of the appropriate leukosis virus results in production of the sarcoma virus. Also, the group-specific antigen of the avian viruses is made in non-virus-producing mammalian Rous tumors, indicating that some of the viral genome is active in producing structural components. This host control over the completion of virus presents an intriguing problem, and the mechanism of production of the RNA-containing tumor viruses may be elucidated by investigations in this area.

7-6. Cytopathology

Viruses are often identified by their effects on cells, and standard techniques for the study of many viruses (e.g., poliovirus, vaccinia virus) depend on their ability to kill cells. The RNA-containing tumorigenic viruses generally do not kill the cells they infect. Often no visible microscopic change occurs in infected cells, even though large numbers of virus particles are in production. Cells may divide for unlimited generations after infection, the virus having no readily recognizable effect on cellular metabolism. Thus, mice may produce large amounts of leukemia virus for many months before developing leukemia (12), or virus-bearing chickens similarly may not develop lymphomatosis until adults, although productive infection was manifest in embryonic stages of development (151, 152).

A. THE ROLE OF VIRUS IN PRODUCING LEUKEMIA

The role of the leukemia virus, whether murine or avian, in inducing leukemia is not clear. It is obvious that all cells infected with leukosis virus do not become leukemic cells. Cultures of susceptible cells may be grossly infected with leukemia viruses, producing large quantities of infectious virus, yet the cells resemble nonleukemic cells morphologically (153–156). Indeed, it is not yet certain that the leukemia virus must infect the cell which eventually becomes the malignant cell. Cells in virus-induced leukemias resemble leukemic cells of other etiology, and the possibility exists that the leukosis virus in vivo produces its effect in other ways, perhaps on other cells, allowing uninfected cells to become malignant.

Certain types of chicken tissue cultures contain transformed cells after exposure to avian myeloblastosis virus (157). The number of resulting myeloblastoid cells is related to the dose of infecting virus, which suggests that transformation is directed by the virus. However, persistence of the viral genome in these and other virus-induced leukemia cells has not yet been established as a requirement for malignancy. An understanding of viral leukemogenesis depends upon the resolution of this problem.

Likewise, the specific role of the mammary tumor virus in producing tumors is obscure. That the virus is necessary for tumor production is not questioned, but the necessity for persistence of viral genome in the malignant cell requires experimental verification.

B. THE ROLE OF SARCOMA VIRUSES IN TRANSFORMATION

In contrast to the leukosis viruses, the avian and murine sarcoma-inducing viruses have obvious effects on the cells they infect. Cells in culture regularly become morphologically transformed after exposure to these viruses (*158, 159*), and a single particle of Rous sarcoma virus is sufficient to produce such transformation in a cell (*160*). All cells transformed by Rous sarcoma virus carry the viral genome. The transformed cells either can be shown to produce infectious virus or if no infectious virus is produced, viral genome can be transferred to another cell type which will produce virus (*161*). This latter method usually involves the mixed cultivation of transformed cells and chicken cells, and may rely upon a degree of cytoplasmic exchange between the heterologous cells. Similarly, hamster cells transformed by murine sarcoma virus produce no virus, but the viral genome can be demonstrated by transfer experiments with mouse cells (*150*).

Conversely, any cell producing Rous sarcoma virus is committed to malignancy and if not already morphologically transformed, will become so. Hamster cells transformed by Rous sarcoma virus, but producing no virus, may, upon division, segregate nontransformed cells (*162*). This suggests that the intracellular viral genome is not an inseparable part of the cellular genome. Also, a newly infected cell may segregate a noninfected daughter cell if division occurs before infection is established (*163, 164*). However, once the infection of Rous sarcoma virus in chick embryo cells is fixed, all cellular progeny are transformed. It is clear that the sarcoma viruses play a direct role in cellular transformation, and that the functioning viral genome is responsible for the intracellular events leading to malignancy.

It is possible that the entire viral genome is not required for transformation. X-irradiation of Rous sarcoma virus resulted in inactivation of transforming activity at a slightly decreased rate compared to inactivation of ability to reproduce infectious progeny (*165*). The results indicated that only a small portion of the viral genome could bear tampering and retain transforming activity. Possibly only late functions of the viral genome dealing with particle completion are expendable.

The kind of morphological transformation elicited by Rous sarcoma virus is determined by the viral genome and varies with different strains of the virus. Clear distinctions in morphological type are seen in cell cultures infected with the Bryan "high titer," the Bryan "standard," and the Schmidt-Ruppin strains of Rous sarcoma virus. These viruses all produce "rounded" cell types distinguishable from each other by the size of the transformed cells and the morphology of isolated colonies of cells. These are also distinct from cells transformed by the Harris strain, or a derived "morph^f" mutant of the Bryan "standard" strain, both of which induce a fusiform morphology in the cells they infect (*166*). The variety of cell types found after transformation by Rous sarcoma viruses makes one question if there is a common biochemical feature of Rous sarcoma cells which determines their malignancy. On the other hand, the contrast in morphology may be a reflection of quantitative differences in a specific substance rather than qualitative biochemical differences among transformed cells.

C. BIOCHEMISTRY OF TRANSFORMED CELLS

1. *Increased Hyaluronic Acid*

From a biochemical standpoint, Rous sarcoma virus-transformed cells exhibit a few notably different features from their noninfected counterparts. One is the increased production of acid mucopolysaccharide by these cells. Long noted as a characteristic of Rous sarcomas *in vivo* (*167*), this production of acid mucopolysaccharide also was observed *in vitro* (*168*), and an increase in the enzyme, hyaluronic acid synthetase was held responsible (*169*). This enzyme has all the characteristics of the usual hyaluronic acid synthetase of chick embryo cells and increased hyaluronic acid probably is the result of stimulation of the normal enzyme in virus-infected cells. Nonetheless, the increased production of hyaluronic acid in these cells may be directly responsible for the conversion to malignancy, and such a possibility demands attention.

2. *Size*

Cells transformed by Rous sarcoma virus are visibly larger than normal chick embryo cells in culture, and measurements of cell volume, and RNA and protein content, confirm this increase in size (*170*). The relation of size, in this case, to malignancy is not known. The chromosomes, qualitatively and quantitatively, are not notably different in transformed cells (*160*), so increased size cannot be attributed merely to the increased content of cellular genetic material.

3. *Glycolysis*

Glycolysis (production of lactic acid) is a common feature of tumor cells and a theory of cancer etiology is based upon the premise that tumor cells have a characteristic change in glucose metabolism which is responsible for malignancy (*171*). Indeed, an increase in glycolysis has been reported for Rous sarcoma cells in culture compared to non-infected cells (*172*). Rous cells also showed increased levels of the enzyme hexokinase compared to uninfected controls (*126*). However, carefully controlled experiments revealed that such increases are concomitant with higher growth rates of Rous cells (*173*). The standard conditions of cell culture used in most experiments of this type selectively allowed the transformed cells to grow while growth of non-infected cells was limited. Under conditions of equal optimal growth, glycolysis of normal and transformed cells was equivalent. Conditions which restrained growth of both cell types inhibited glycolysis in both transformed and control cultures. It appears that the general demands of cellular growth for energy are reflected in observed rates of glycolysis, whether the cells be transformed or not.

7-7. Conclusions

The RNA-containing tumor viruses are a category of microorganisms, unique with respect to structure, to the cytopathology of infection, and to metabolic requirements for infection. The genome of these viruses is completely RNA, with at least one very large molecule. One would be tempted to apply the basic findings on the replication of other RNA viruses to the RNA-tumor viruses were it not for the obvious differences in cytopathology, and the interpolation of synthesis of DNA in the replicative cycle. Other viral RNA's can replicate new RNA directly in the absence of any DNA synthesis, and the consequence of virus production is always cell death. Cell death is not the usual result of infection with RNA-tumor viruses, but other biological changes often occur.

The alteration of cellular morphology and biological responsiveness observed in viral-induced malignancies can *a priori* be attributed to a modification of cellular genetic activities. The prospect of physical interaction of viral genome, or the replicative form of the viral nucleic acid, with cellular genome, seems a possible attendant to a disturbance of cellular genes, although one can easily suppose other mechanisms. The introduction of a DNA intermediate in viral reproduction is an appealing proposition since it provides a reasonable mechanism for genomic interactions, precedented by observations on bacteriophage lysogeny.

An ultimate understanding of viral-induced tumors depends on the resolution of the specific metabolic and structural changes which render the cell malignant. Whether the prime product of gene activity responsible for malignancy derives from the viral genome directly, or is a product of cellular genes activated by virus-specific substances will be a matter of major interest in investigations into the carcinogenesis of RNA-containing viruses.

REFERENCES

1. C. Ellermann and O. Bang, *Zentr. Bakteriol. Parasitenk.*, **4**, 595 (1908).

2. P. Rous, *J. Exptl. Med.*, **13**, 397 (1911).

3. J. J. Bittner, *U.S. Public Health Rept.*, **54**, 1590 (1939).

4. L. Gross, *Proc. Soc. Exptl. Biol. Med.*, **76**, 27 (1951).

5. H. R. Morgan, *Natl. Cancer Inst. Monograph*, **17**, 392 (1964).

6. A. Fujinami and S. Hatano, *Gann*, **23**, 67 (1929).

7. J. J. Harvey, *Nature*, **204**, 1104 (1964).

8. J. B. Moloney, in *Some Recent Developments in Comparative Medicine*, Academic Press, London, 1966, p. 251.

9. J. J. Bittner, *U.S. Public Health Rept.*, **54**, 590 (1939).

10. J. W. Beard, *Advan. Cancer Res.*, **7**, 1 (1953).

11. R. F. Zeigel, G. H. Thielen, and M. J. Tweihaus, *J. Natl. Cancer Inst.*, **37**, 709 (1966).

12. J. B. Moloney, *Federation Proc.*, **21**, 19 (1962).

13. S. Yazgan and A. L. Chapman, *Virology*, **32**, 537 (1967).

14. W. F. Jarrett, E. M. Crawford, W. B. Morton, and F. Davis, *Nature*, **202**, 568 (1964).

15. W. Bernhard, C. Oberling, and P. Vigier, *Bull. Cancer*, **43**, 407 (1956).

16. F. Hagenau, A. J. Dalton, and J. B. Moloney, *J. Natl. Cancer Inst.*, **20**, 633 (1958).

17. A. J. Dalton, *Natl. Cancer Inst. Monograph*, **22**, 143 (1966).

18. R. F. Zeigel, R. L. Tyndall, T. E. O'Connor, E. Terter, and B. V. Allen, *Natl. Cancer Inst. Monograph*, **22**, 237 (1966).

19. M. J. Lyons and D. H. Moore, *J. Natl. Cancer Inst.*, **35**, 549 (1965).

20. G. de Thé and T. E. O'Connor, *Virology*, **28**, 713 (1966).

21. J. P. Bader and N. R. Brown, in preparation.

22. G. P. Shibley, F. E. Durr, G. Schlidlovsky, B. S. Wright, and R. Schmitter, *Science*, **156**, 1610 (1967).

23. D. H. Moore, E. Y. Lasfargues, M. R. Murray, C. D. Haagensen, and E. C. Pollard, *J. Biophys. Biochem. Cytol.*, **5**, 85 (1959.

24. E. de Harven and C. Friend, *Virology*, **23**, 119 (1964).

25. J. P. Levy, M. Boiron, D. Silvestre, and J. Bernard, *Virology*, **26**, 146 (1965).

26. W. S. Robinson, A. Pitkanen, and H. Ribin, *Proc. Natl. Acad. Sci. U.S.*, **54**, 219 (1965).

27. T. E. O'Connor, F. J. Rauscher, and R. F. Zeigel, *Science*, **144**, 1144 (1964).

28. S. Oroszlan, L. W. Johns, Jr., and M. A. Rich, *Virology*, **26**, 638 (1965).

29. J. N. Wilson and P. D. Cooper, *Virology*, **21**, 135 (1963).

30. J. P. Bader, unpublished observations.

31. M. A. Epstein, *Nature*, **181**, 1808 (1958).

32. F. Padgett, V. Kearns-Preston, H. Voelz, and A. S. Levine, *J. Natl. Cancer Inst.*, **36**, 465 (1966).

33. R. A. Bonar and J. W. Beard, *J. Natl. Cancer Inst.*, **23**, 183 (1959).

34. M. J. Lyons and D. H. Moore, *J. Natl. Cancer Inst.*, **35**, 549 (1965).

35. R. A. Bonar, V. Heine, D. Beard, and J. W. Beard, *J. Natl. Cancer Inst.*, **30**, 949 (1963).

36. F. Padgett and A. S. Levine, *Virology*, **27**, 633 (1965).

37. M. Johnson and P. T. Mora, *Virology*, **31**, 230 (1967).

38. R. H. Purcell, R. A. Bonar, D. Beard, and J. W. Beard, *J. Natl. Cancer Inst.*, **28**, 1003 (1962).

39. P. H. Duesberg and W. S. Robinson, *Proc. Natl. Acad. Sci. U.S.*, **55**, 219 (1966).

40. R. J. Huebner, *Proc. Natl. Acad. Sci. U.S.*, **58**, 835 (1967).

41. R. J. Huebner, D. Armstrong, M. Okuyan, P. S. Sarma, and H. C. Turner, *Proc. Natl. Acad. Sci. U.S.*, **51**, 742 (1964).

42. W. S. Robinson, H. L. Robinson, and P. H. Duesberg, *Proc. Natl. Acad. Sci. U.S.*, **58**, 825 (1967).

43. J. Harel, J. Huppert, F. Lacour, and L. Harel, *Compte Rend.*, **261**, 2266 (1965).

44. F. Galibert, C. Bernard, P. Chenaille, and W. Boiron, *Compte Rend.*, **261**, 1771 (1965).

45. P. T. Mora, V. W. McFarland, and S. W. Luborsky, *Proc. Natl. Acad. Sci. U.S.*, **55**, 438 (1966).

46. A. F. Valentine and J. P. Bader, *J. Virol.*, **2**, 224 (1968).

47. L. V. Crawford and E. M. Crawford, *Virology*, **13**, 227 (1961).

48. R. Weil, A. Bendich, and C. M. Southam, *Proc. Soc. Exptl. Biol. Med.*, **104**, 670 (1960).

49. J. B. Moloney, *Uni. Intern. Contra. Cancrum.*, **19**, 250 (1963).

50. F. G. Rabotti and M. K. Cook, *Natl. Cancer Inst. Monograph*, **17**, 619 (1964).

51. P. T. Mora, V. W. McFarland, and S. W. Luborsky, *Natl. Cancer Inst. Monograph*, **22**, (1966).

52. G. de Thé, *Natl. Cancer Inst. Monograph*, **17**, 651 (1964).

53. G. de Thé, *Natl. Cancer Inst. Monograph*, **22**, 169 (1966).

54. J. Riman and B. Thorell, *Biochem. Biophys. Acta*, **40**, 565 (1960).

55. M. Kates, A. C. Allison, D. A. Tyrell, and A. T. James, *Cold Spring Harbor Symp. Quant. Biol.*, **27**, 433 (1962).

56. E. R. Pfefferkorn and R. L. Clifford, *Virology*, **23**, 217 (1964).

57. W. E. Gye and W. J. Purdy, *Brit. J. Exptl. Pathol.*, **14**, 250 (1933).

58. C. R. Amies and J. G. Carr, *J. Pathol. Bacteriol*, **49**, 497 (1938).

59. H. Rubin, *Virology*, **2**, 545 (1956).

60. T. Borsos, *J. Natl. Cancer Inst.*, **20**, 1215 (1958).

61. M. Rychlikova and J. Svoboda, *Virology*, **10**, 545 (1959).

62. R. A. Bonar, L. Sverak, D. P. Bolognesi, A. J. Langlois, D. Beard, and J. W. Beard, *Cancer Res.*, **27**, 1138 (1967).

63. E. R. Burka, W. Schreml, and C. J. Kick, *Biochemistry,* **6,** 2840 (1967).

64. P. H. Duesberg and P. B. Blair, *Proc. Natl. Acad. Sci. U.S.,* **55,** 1490 (1966).

65. L. V. Crawford, *Virology,* **12,** 143 (1960).

66. R. A. Bonar, D. Beard, G. S. Beaudreau, D. G. Sharp, and J. W. Beard, *J. Natl. Cancer Inst.,* **18,** 831 (1957).

67. R. M. Dougherty, *Virology,* **14,** 371 (1961).

68. J. Eigner, H. Boedtker, and G. Michaels, *Biochim. Biophys. Acta,* **51,** 165 (1961).

69. J. Youngner, *J. Immunol.,* **78,** 282 (1957).

70. H. Rubin and H. M. Temin, *Virology,* **7,** 75 (1959).

71. H. Rubin, *Virology,* **11,** 28 (1960).

72. W. Levinson and H. Rubin, *Virology,* **28,** 533 (1966).

73. R. Ishizaki and P. K. Vogt, *Virology,* **30,** 375 (1966).

74. P. K. Vogt and R. Ishizaki, *Virology,* **26,** 664 (1965).

75. P. K. Vogt and R. Ishizaki, *Virology,* **30,** 368 (1966).

76. P. B. Blair, *Proc. Soc. Exptl. Biol. Med.,* **103,** 188 (1960).

77. J. J. Bittner, H. M. Hirsch, J. D. Ross, and R. Gabrielson, *Cancer Res.,* **19,** 918 (1959).

78. E. A. Eckert, R. Rott, and W. Schäfer, *Virology,* **24,** 426 (1964).

79. E. A. Eckert, R. Rott, and W. Schäfer, *Virology,* **24,** 434 (1964).

80. G. Kelloff and P. K. Vogt, *Virology,* **29,** 377 (1966).

81. R. M. Dougherty and H. S. DiStefano, *Virology,* **29,** 586 (1966).

82. G. Geering, L. J. Old, and E. Boyse, *J. Exptl. Med.,* **124,** 753 (1966).

83. J. P. Bader and P. Sarma, unpublished observations.

84. H. Rubin and P. K. Vogt, *Virology,* **17,** 184 (1962).

85. H. Hanafusa, T. Hanafusa, and H. Rubin, *Proc. Natl. Acad. Sci. U.S.,* **49,** 572 (1963).

86. H. Hanafusa, T. Hanafusa, and H. Rubin, *Proc. Natl. Acad. Sci. U.S.,* **51,** 41 (1964).

87. H. Hanafusa, *Virology,* **25,** 248 (1965).

88. P. K. Vogt, *Virology,* **25,** 237 (1965).

89. T. Hanafusa, H. Hanafusa, and H. Rubin, *Virology,* **22,** 643 (1964).

90. H. Hanafusa, and T. Hanafusa, *Virology,* **28,** 369 (1966).

91. H. M. Temin, *Cold Spring Harbor Symp. Quant. Biol.,* **27,** 407 (1962).

92. H. M. Temin, *Virology,* **20,** 235 (1963).

93. H. M. Dougherty and H. S. DiStefano, *Virology,* **27,** 351 (1965).

94. H. S. DiStefano and R. M. Dougherty, *Virology,* **27,** 360 (1965).

95. D. Courington and P. K. Vogt, *J. Virol.,* **1,** 400 (1967).

96. H. L. Robinson, *Proc. Natl. Acad. Sci. U.S.,* **57,** 1655 (1967).

97. P. K. Vogt, *Proc. Natl. Acad. Sci. U.S.,* **58,** 801 (1967).

98. J. W. Hartley and W. P. Rowe, *Proc. Natl. Acad. Sci.,* **55,** 780 (1966).

99. R. C. Y. Ting, *Proc. Soc. Exptl. Biol. Med.,* **126,** 778 (1968).

100. S. Fazekas de St. Groth, *Nature,* **162,** 294 (1948).

101. L. Hoyle, *Cold Spring Harbor Symp. Quant. Biol.,* **27,** 113 (1962).

102. L. B. Crittenden, W. Okazzaki, and R. Reamer, *Virology,* **20,** 541 (1963).

103. F. T. Steck and H. Rubin, *Virology,* **29,** 628 (1966).

104. F. T. Steck and H. Rubin, *Virology,* **29,** 642 (1966).

105. H. Hanafusa and T. Hanafusa, *Proc. Natl. Acad. Sci. U.S.,* **55,** 532 (1966).

106. H. Rubin, *Proc. Natl. Acad. Sci. U.S.*, **46**, 1105 (1960).

107. P. Sarma, M. P. Cheong, J. W. Hartley, and R. J. Huebner, *Virology*, **33**, 180 (1967).

108. J. P. Bader, in *The Molecular Biology of Viruses* (J. S. Colter and W. Parenchych, eds.), Academic Press, New York, 1967, p. 697.

109. H. M. Temin, *J. Cell. Physiol.*, **69**, 1 (1967).

110. P. I. Marcus and E. Robbins, *Proc. Natl. Acad. Sci. U.S.*, **50**, 1156 (1963).

111. Y. Nakata and J. P. Bader, **2**, 1255 (1968).

112. J. P. Bader, *J. Cell. Physiol.*, **70**, 301 (1967).

113. J. P. Bader, *Natl. Cancer Inst. Monograph*, **17**, 781 (1964).

114. J. P. Bader, *Virology*, **26**, 253 (1965).

115. E. E. Force and R. C. Stewart, *Proc. Soc. Exptl. Biol. Med.*, **116**, 803 (1964).

116. J. P. Bader, *Virology*, **29**, 444 (1966).

117. J. P. Bader, *Virology*, **22**, 462 (1964).

118. H. M. Temin, *Virology*, **23**, 486 (1964).

119. P. Vigier and A. Goldé, *Virology*, **23**, 511 (1964).

120. J. P. Bader, *Science*, **149**, 757.

121. R. Dulbecco, L. H. Hartwell, and M. Vogt, *Proc. Natl. Acad. Sci. U.S.*, **53**, 403 (1965).

122. M. Hatanaka and R. Dulbecco, *Proc. Natl. Acad. Sci. U.S.*, **56**, 736 (1966).

123. J. P. Bader, in *Subviral Carcinogenesis* (Y. Ito, ed.), "Nissha" Printing, Kyoto, Japan, 1966, p. 144.

124. H. Rubin and H. M. Temin, *Virology*, **7**, 75 (1959).

125. H. M. Temin, *Natl. Cancer Inst. Monograph*, **17**, 557 (1964).

126. H. M. Temin, *Proc. Natl. Acad. Sci. U.S.*, **52**, 323 (1964).

127. L. Harel, J. Harel, F. Lacour, and J. Huppert, *Compt. Rend.*, **263**, 616 (1966).

128. A. M. Prince and W. A. Adams, *Virology*, **30**, 151 (1966).

129. H. M. Temin, *Virology*, **20**, 577 (1963).

130. W. S. Robinson, in *The Molecular Biology of Viruses* (J. S. Colter and W. Parenchych, eds.), Academic Press, New York, 1967, p. 681.

131. P. H. Duesberg and W. S. Robinson, *Virology*, **4**, 742 (1967).

132. R. E. Bases and A. S. King, *Virology*, **32**, 175 (1967).

133. J. P. Bader, *Virology*, **29**, 452 (1966).

134. F. Hagenau and J. W. Beard, in *Tumors Induced by Viruses: Ultrastructural Studies* (A. J. Dalton and F. Haganau, eds.), Academic Press, New York, 1962, p. 1.

135. E. de Harven, in *Tumors Induced by Viruses: Ultrastructural Studies* (A. J. Dalton and F. Haganau, eds.), Academic Press, New York, 1962, p. 183.

136. A. J. Dalton, in *Tumors Induced by Viruses: Ultrastructural Studies* (A. J. Dalton and F. Haganau, eds.), Academic Press, 1962, p. 207).

137. D. H. Moore, in *Tumors Induced by Viruses: Ultrastructural Studies* (A. J. Dalton and F. Haganau, eds.), Academic Press, 1962, p. 113.

138. P. K. Vogt, *Virology*, **32**, 708 (1967).

139. C. Morgan, R. A. Rifkind, and H. M. Rose, *Cold Spring Harbor Symp. Quant. Biol.*, **27**, 57 (1962).

140. H. Duc-Nguyen, H. M. Rose, and C. Morgan, *Virology*, **28**, 404 (1966).

141. R. A. Malmgren, M. A. Fink, and W. Mills, *J. Natl. Cancer Inst.*, **24**, 995 (1960).

142. R. C. Mellors and J. S. Munroe, *J. Exptl. Med.*, **112**, 963 (1960).

143. W. F. Noyes, *Virology*, **12**, 488 (1960).

144. P. K. Vogt and H. Rubin, *Virology*, **13**, 528 (1961).

145. M. A. Fink and R. A. Malmgren, *J. Natl. Cancer Inst.*, **31**, 1111 (1963).

146. F. E. Payne, J. J. Solomon, and H. G. Purchase, *Proc. Natl. Acad. Sci. U.S.*, **55**, 341 (1966).

147. J. Svoboda, *Natl. Cancer Inst. Monograph*, **17**, 277 (1964).

148. P. Vigier and L. Montagnier, in *Subviral Carcinogenesis* (Y. Ito, ed.), "Nissha" Printing, Kyoto, 1966, p. 156.

149. P. S. Sarcoma, W. Vass, and R. J. Huebner, *Proc. Natl. Acad. Sci. U.S.*, **55**, 1435 (1966).

150. R. J. Huebner, J. W. Hartley, W. P. Rowe, W. T. Lane, and W. I. Capps, *Proc. Natl. Acad. Sci. U.S.*, **56**, 1164 (1966).

151. H. Rubin, *Proc. Natl. Acad. Sci. U.S.*, **46**, 1105 (1960).

152. H. Rubin, *Cold Spring Harbor Symp. Quant. Biol.*, **27**, 441 (1962).

153. H. Rubin, *Virology*, **13**, 200 (1966).

154. B. S. Wright and J. C. Lasfargues, *Natl. Cancer Inst. Monograph,* **22**, 685 (1966).

155. R. A. Manaker, P. C. Strother, A. A. Miller, and C. V. Piczak, *J. Natl. Cancer Inst.*, **25**, 1411 (1960).

156. H. Ginsburg and L. Sachs, *Virology*, **13**, 380 (1961).

157. M. A. Baluda and I. E. Goetz, *Virology*, **15**, 185 (1961).

158. R. A. Manaker and V. Groupé, *Virology*, **2**, 838 (1966).

159. J. W. Hartley and W. P. Rowe, *Proc. Natl. Acad. Sci. U.S.*, **55**, 780 (1966).

160. H. M. Temin and H. Rubin, *Virology*, **6**, 669 (1958).

161. D. Simkovic, N. Valentova, and V. Thurzo, *Neoplasma*, **9**, 104 (1962).

162. I. Macpherson, *Science*, **148**, 1731 (1965).

163. G. W. Trager and H. Rubin, *Virology*, **30**, 266 (1966).

164. G. W. Trager and H. Rubin, *Virology*, **30**, 275 (1966).

165. A. Goldé and R. Laterjet, *Compt. Rend.*, **262**, 420 (1966).

166. H. M. Temin, *Virology*, **10**, 182 (1960).

167. L. N. Loomis and A. W. Pratt, *J. Natl. Cancer Inst.*, **17**, 101 (1956).

168. S. Ericksen, J. Eng, and H. R. Morgan, *J. Exptl. Med.*, **114**, 435 (1961).

169. N. Ishimoto, H. M. Temin, and J. L. Strominger, *J Biol. Chem.*, **241**, 2052 (1966).

170. A. Goldé, *Virology*, **16**, 9 (1962).

171. O. Warburg, *Science*, **123**, 309 (1956).

172. H. R. Morgan and S. Ganapathy, *Proc. Soc. Exptl. Biol. Med.*, **113**, 312 (1963).

173. T. L. Steck, S. Kaufman, and J. P. Bader, *Cancer Res.*, **28**, 1611 (1968).

174. Y. Nakata and J. P. Bader, *Virology*, **36**, 401 (1968).

8

BIOCHEMISTRY OF ADENOVIRUS INFECTION

HAROLD S. GINSBERG

DEPARTMENT OF MICROBIOLOGY
SCHOOL OF MEDICINE
UNIVERSITY OF PENNSYLVANIA
PHILADELPHIA, PENNSYLVANIA

8-1. Introduction

With the initial isolations of adenoviruses (*1, 2*) it was clear that important etiologic agents of acute respiratory diseases in man had been uncovered. Their inapparent association with tonsils and adenoids also suggested that they may serve as useful viruses to study latent infections (*1*). It was then unforeseen, however, that these DNA-containing viruses would prove to be valuable tools to investigate the directed biosynthesis of viral macromolecules and to probe the imposition of viral controls on the synthesis of host macromolecules. The implications of these studies are broad, for adenoviruses not only produce lytic and abortive infections but also induce tumors in animals (*3–5*) and effect malignant transformation of cells *in vitro* (*6, 6a*). Hence, ultimately the data obtained may expose the differential mechanisms by which a virus kills cells or conversely induces their uncontrolled proliferation.

Representatives of four families of DNA-containing viruses which infect animals (i.e., poxviruses, herpesvirus, adenoviruses, and papovaviruses) have been studied in detail [most recently reviewed generally by Green (*7*)]. Although all of these viruses follow the same general pattern of biochemical events in their replication cycle, the marked structural differences of the virions have set certain biosynthetic demands which are reflected in specific biochemical differences noted. This chapter confines itself to a critical discussion of the biochemical events which in a productive infection culminate in the appearance of new infectious adenovirus particles while concomitantly blocking the production of host DNA, RNA, and protein, and to a brief consideration of those biosynthetic steps which occur normally and those which are restricted in abortive adenovirus infections.

8-2. Background Matters

A. MORPHOLOGICAL AND CHEMICAL STRUCTURE

The numerous immunological types of adenoviruses of man and other animals have similar morphological and chemical features which are detailed elsewhere in this volume. These structural features will be briefly summarized here, however, to facilitate the discussion which follows. The virions are naked icosahedrons, 650–800 Å in diameter, composed of a protein capsid consisting of 252 distinct capsomers (*8*) and a core containing the DNA viral genome (*9*). Contrary to the original concepts that the capsid of an isometric virion is assembled from numerous structural subunits of a single type of protein (*10*), the capsids of adenoviruses consist of several different macromolecular protein structures (*11*). (1) The hexons are hollow polygonal structures, 70–85 Å in diameter; they are distributed on the edges and triangular faces of the icosahedron, and each has six neighbors (*11*, *12*). (2) The pentons consist of a base about 70–80 Å in diameter and a fiber of variable length; they are the 12 vertex units of the icosahedron, and each is the center of a 5-fold axis of symmetry (*13*). (3) The penton fiber (*11*, *13*), which varies in length depending upon the adenovirus type (*11*, *13–17*), appears on morphological (*11*, *13*, *14*, *17*) and immunological grounds (*14*, *18*, *19*) to consist of at least two distinct proteins, the string-like structure and a terminal knob. In addition to the proteins of the capsid, there appears to be an internal protein (*20–23*) associated with the viral DNA.

The DNA's extracted from purified virions of several selected serological types of human adenoviruses are characteristic linear, double-stranded molecules of $20–25 \times 10^6$ daltons (*24–26*). The DNA contents of the 31 human adenoviruses studied range from 11.6–13.5% of the weight of the virion, and the guanine + cytosine content varies from 48–60% of the total base composition (*24*). Green and his colleagues have made an intriguing correlation between DNA structure (i.e., its size and base composition) and the oncogenic potential of the viruses in newborn hamsters (*24*, *27*).

B. Cell Cultures for Studies

The data to be reviewed in subsequent sections were derived from experiments made with monolayer and suspension cultures of cell lines, most often HeLa or KB cells. The choice of conditions for cell growth were based mainly on deductive reasoning rather than experimental evidence demonstrating the value of one over the other. The major arguments bearing on the selection follow.

1. Suspension Cultures of Exponentially Dividing Cells

Cells growing logarithmically in complete medium are synthesizing nucleic acids and protein at an optimum rate; the enzymes for synthesizing polymers and generating energy are maximally operational; and the physiological state of the cells is well defined. These optimum conditions for cell division should be the appropriate conditions to obtain maximum production of viral components and mature virions as well as to measure the greatest biochemical changes in virus- infected cells.

2. *Monolayer Cultures under Maintenance Conditions*

Since adenovirus infection stops cells from dividing (*28*), it is more appropriate to compare the biochemical events in infected cells with those in uninfected cells which also are not actively multiplying. These conditions are valid because the physiological condition of the cell can be controlled with chemically defined media and the findings are highly reproducible.

The data derived from studies done in cell cultures of the two types indicate that both arguments are valid and that each type of cell culture has distinct advantages for investigation of different aspects of infection. (1) Cells in spinner cultures can be handled more conveniently, and repeated sampling from a single culture vessel gives more reproducible results than using individual monolayer cultures unless at least three to six cultures are used for each experimental point. (2) Optimum synthesis of host as well as viral macromolecules and maximum yield of virus occur in suspended cell cultures. (3) The profound influence of viral infection on the synthesis of host macromolecules can be investigated more effectively in exponentially dividing cells. (4) Synthesis of viral macromolecules can be measured in either proliferating or nondividing cells, but the changes, observed when compared to uninfected cells, are more dramatic in the latter type cell cultures. (5) An increase in synthesizing enzymes during the viral eclipse period [e.g., aspartic transcarbamylase (*29*) and thymidine kinase (*30, 31*)] is detected only in nondividing cells which have low base-line activities of enzyme levels.

8-3. The Process of Infection

Because the majority of biochemical studies have been carried out with types 2 and 5 adenoviruses, the characteristics of the productive infection will be summarized with these agents which have similar biological and chemical characteristics.

A. A One-Step Growth Curve

The initial cycle of multiplication of type 5 virus in a spinner culture of KB cells is shown in Fig. 8-1. This one-step growth curve, however, is similar in its general features for all adenoviruses studied. (1) The eclipse period is relatively long, ranging from 12–18 hr, depending upon the virus examined, the host cell employed, and the method of cell propagation (there is not any evidence that fundamental differences exist in the mechanisms of multiplication of various adenoviruses). (2) The incremental period is comparatively short, reaching maximum viral yield in 24–30 hr after infection. (3) More than 90% of the virus remains intracellular, even when cytopathic changes are marked (*32*). When infection is initiated in monolayer cell cultures the eclipse period of the multiplication cycle is usually 2–3 hr longer and the viral yield per cell is approximately 1–10% less than that observed in spinner cultures (*28, 32*).

B. ATTACHMENT, PENETRATION, AND INTRACELLULAR UNCOATING OF VIRAL GENOME

Virions attach to susceptible cells over a period of 2–6 hr depending upon the viral and cell concentration (*21, 22, 28, 32*). The vertex subunits, the pentons each with its fiber end-organ, probably accomplish the virus–cell association since the fiber, but not the hexon, can combine with susceptible cells (*33*). Characterization of the cell's attachment receptors, however, has not yet been made. Virions are rapidly engulfed after this attachment (*34*), and eclipse of infectivity associ-

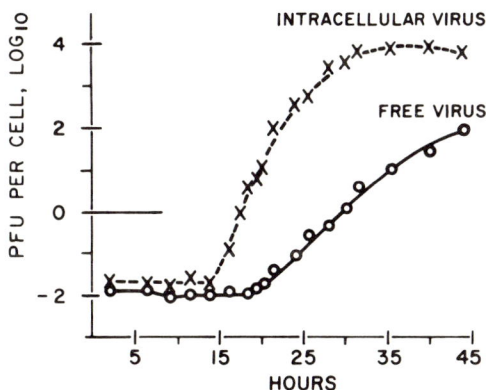

FIG. 8-1. Multiplication of type 5 adenovirus in KB cells. Cells growing exponentially were infected with virus at a multiplicity of 100 PFU/cell; 2 hr after virus was added, cells were washed two times with cold medium, resuspended in Eagle's MEM supplemented with 5% calf serum, and incubated at 36°C. At the indicated times samples were removed, the cells and medium were separated by centrifugation, and each fraction was assayed.

ated with shedding of the viral capsid follows almost immediately
(*21*).

The term uncoating has been used to denote the process of exposing the viral genome so that it can be transcribed and replicated. In
fact, the process actually measured only describes the conversion of
viral DNA from resistance to deoxyribonuclease (intact virus) to
sensitivity to the enzyme (*21, 22*), rather than complete freedom of
the DNA from viral protein or the capacity of the viral genome to be
transcribed or replicated. Perhaps virion disruption or disassembly
would be the more appropriate term to describe the phenomenon actually assayed, but to avoid confusion "uncoating" will be used in this
discussion. "Uncoating" of the adenovirus genome begins almost immediately after penetration of the virion into the cell's cytoplasm
(*21, 22, 37*), and unlike the "uncoating" of vaccinia DNA (*35, 36*),
protein synthesis is unnecessary to accomplish this step (*21, 22, 37*).
Displacement of the pentons with a loss of 5–6% of the viral protein
appears to be the initial event of "uncoating" (*37*). After maximum
"uncoating," i.e., conversion of 85% of the viral DNA to a state in
which it is DNase-sensitive, the viral DNA is still associated with
25–50% of viral protein (*21, 22*). During these events neither the uncovered viral DNA nor the released viral proteins become acid-soluble
(*21, 37*).

At the initiation of the "uncoating" studies it was anticipated that
they would reveal at least a partial explanation for the long interval
between infection of the cell and initiation of the biosynthetic steps
which produce new viral particles. Since "uncoating" as described
requires no more than 60 min, this anticipation has certainly not yet
been fulfilled. However, the data still leave unanswered critical questions, which prevent an understanding of when the viral genome can
be transcribed and replicated. (1) Can the genome which is available
to DNase be transcribed? There is evidence suggesting that the vaccinia virus core contains DNA which cannot react with nuclease but
can still be transcribed to a limited extent (*38, 39*). (2) What is the
nature of the protein still associated with viral DNA? It seems possible that the structure remaining is the nucleoprotein core, and that the
major part of the residual protein is an internal protein (*23*). (3)
Does the DNA-protein complex that remains after capsid proteins
are removed undergo additional uncoating, perhaps in the nucleus
where the viral DNA is replicated (*40*)? If this step is essential, perhaps like vaccinia virus, there is limited transcription of the viral

genome to make an "uncoating enzyme" or the viral core induces protein synthesis to accomplish the task.

There remains one major ambiguity which is inherent to all studies of this type, regardless of the experimental techniques employed. A high multiplicity of infection is required for these experiments and a population of viral particles, which may appear to react homogeneously, is measured. It is, however, impossible to ascertain whether a small per cent of the virions is unique. Therefore, although the major proportion of viral DNA after "uncoating" is still in the form of a nucleoprotein structure with a sedimentation coefficient about one-half that of intact virus, in all experiments a small amount of free viral DNA is detected (21, 22). It is of course possible that this minority population of DNA molecules could be responsible for infection. Present methods do not permit our clarifying this annoying uncertainty.

8-4. Biosynthesis of Viral Macromolecules

When cytochemical and biochemical studies on replication of adenoviruses began (41–43) it was not self-evident that the pieces of a virus composed of protein and DNA would be synthesized according to a single general pattern. However, the following sequence of biosynthetic events has now been demonstrated with all DNA-containing viruses which have been adequately investigated, although variations in details are naturally imposed by the varying complexities of different virions and the site of their multiplication within the cell.

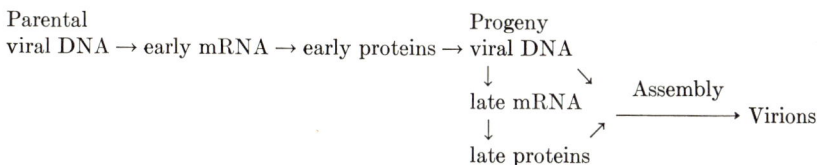

Parental   Progeny
viral DNA → early mRNA → early proteins → viral DNA
                                              ↓   ↘   Assembly
                                         late mRNA ─────────────→ Virions
                                              ↓   ↗
                                         late proteins

This section will summarize the experimental data which describe each step in the production of adenovirus particles. The time at which a particular macromolecule is synthesized may appear to vary in different studies, but these differences can usually be attributed to the cell system or multiplicity of infection employed; therefore, they are considered trivial, and they will be ignored in this discussion.

A. Virus-Specific RNA

Specific inhibitors of RNA synthesis, 6-azauridine and 5-fluorouridine, were used to demonstrate that RNA, as predicted, is obligatory for the propagation of a DNA-containing virus (44). These pyrimidine analogs, when added at varying intervals after infection, also furnished a sensitive tool to measure the biosynthesis of the *last* species of RNA demanded for viral propagation. The analog studies, as well as pulse-labeling RNA with ^{32}P, indicated that essential RNA was made during the last half of the eclipse period, and that final synthesis of the required RNA was completed shortly after the initial mature viral particles were assembled (44).

The pulse-labeled RNA has the following characteristics: (1) its base composition during the latter portion of the eclipse period approaches that of viral DNA, implying that it is virus-specific RNA (mRNA) (45); (2) it hybridizes with purified denatured viral DNA (46, 47); (3) it is synthesized, as detected by the hybridization techniques employed (48), about 6 to 7 hr before the assembly of infectious virions; and (4) it has a rapid rate of synthesis which continues for about 10 hr before its production becomes markedly reduced (47).

It should be noted that the mRNA, initially detected by hybridization (46, 47), and the RNA required for viral replication, identified with pyrimidine analog (44), are synthesized simultaneously and they must certainly be the same species of RNA, the mRNA related to synthesis of viral capsid proteins, i.e., "late" proteins (described in Sec. 8-4, B). But protein synthesis is also mandatory before viral DNA can be replicated (49, 50), which implies that "early" mRNA must likewise be made. Either the host or viral genome could serve for the transcription of these early informational macromolecules. A choice of these alternatives has come from recent studies by Green and colleagues (51) who, using the more sensitive hybridization technique of Gillespie and Spiegelman (52) in competition experiments, identified an "early" virus-specific mRNA and showed that "early" mRNA is made throughout the viral growth cycle, whereas "late" mRNA is synthesized only during the period in which viral capsid proteins are produced. Of course, these studies do not identify different early mRNA's, and therefore do not demonstrate that all early mRNA's are transcribed during the entire period of viral propagation.

Synthesis of a low molecular weight, 5S RNA, is also induced in cells infected with type 2 adenovirus (*53*), but its role in infection is obscure. Indeed, although DNA synthesis is required for the appearance of 5S RNA there is no evidence available that it is transcribed from either the host or viral genome, for it does not hybridize with either DNA species (*46*).

B. "EARLY" PROTEINS

Following synthesis of the "early" messenger RNA's and prior to replication of the viral genome, several proteins are made, at least some of which are required for synthesis of the viral DNA. The necessity for and the synthesis of these so-called "early" proteins were demonstrated using: (1) an amino acid analog, *p*-fluorophenylalanine (*49*); (2) enzyme assays (*29, 30*); and (3) immunological techniques (*23, 54–57*).

On the basis of extensive evidence indicating that a number of virus-specific enzymes appear during the early portion of the eclipse period of T-even bacteriophage multiplication it was anticipated that DNA-containing animal viruses would also induce the synthesis of similar enzymes. This forecast was substantiated: (a) Addition of *p*-fluorophenylalanine before synthesis of viral DNA inhibits replication of the viral genome and prevents formation of infectious virus (*49, 50*), suggesting that production of functional proteins, presumably enzymes, is a prerequisite for biosynthesis of viral DNA. (b) Infection of monolayers of HeLa or monkey kidney cells results in increased activities of aspartic transcarbamylase (*29, 30*), thymidine kinase (*30, 31*), deoxycytidylate deaminase (*30*), and DNA polymerase (*30, 31*) during the expected interval of the viral eclipse period; all require protein synthesis for the increase in activity to be manifest. Enzyme activities are not increased, however, in spinner cultures (*58, 59*), possibly because the enzymes are being made at close to their maximal rate in cells dividing exponentially prior to infection. It is generally thought that the enzymes which increase are different from host enzymes and that they are viral gene products. But with aspartate transcarbamylase (ATCase), although the enzyme purified from infected cells has a pH optimum, maximal velocity (V_{max}), and Km for aspartate distinctly different from the ATCase from uninfected cells, it is not a virus-coded enzyme but rather a host enzyme with altered characteristics resulting from a release from feedback inhibition: (1) Addition of CTP to ATCase purified from infected cells inhibits enzyme activity *in vitro* and changes the enzyme's properties so that they are the same as the enzyme from control cells; and (2) heating ATCase from uninfected cells, to remove its feedback inhibitor-sensitive structure, converts the enzyme so that it assumes the characteristics of ATCase from infected cells (*29, 60*).

The thymidine kinase, deoxycytidylate deaminase, and DNA polymerase from adenovirus-infected cells have not been characterized either enzymatically or immunologically. Therefore, one does not have any direct evidence on which to surmise whether the information to synthesize these enzymes in infected cells is encoded in the viral or the host genome. However, enzymes related to DNA synthesis also increase in cells infected with herpesviruses (*61–64*) and poxviruses (*58, 65–69*), and the characteristics of a few of them have been shown to be unique. On the basis of a small amount of genetic, biochemical, and immunological evidence (*63, 64, 68, 70, 71*) and a large amount of "hunch," it is generally believed that they are virus-specific and not virus-induced host enzymes. Since even fewer data are available, it cannot be concluded whether or not the enzymes which are increased in adenovirus-infected cells (except for ATCase which is clearly a modified host protein) are encoded in the viral DNA.

At least two candidates for "early" proteins, the T antigen (*54–56, 72*), and the P antigen (*23*), have been identified immunologically in lytic infections. Before discussing the similarities and differences in the characteristics of these antigens and some meager data concerning their possible functional roles it is necessary to clarify their identities and names.

In the cells of tumors induced by types 12, 18, and 31 adenoviruses, an antigen can be detected by complement fixation or immunofluorescence using sera from tumor-bearing hamsters (*74*). The antigens are immunologically similar or perhaps identical to those found in tumors produced by the three closely related adenoviruses [(*75*), see (*76*) for review]. Other more weakly oncogenic adenoviruses also induce the synthesis of a similar antigen which is immunologically unrelated to the antigens induced by types 12, 18, and 31 adenoviruses. During the early stages of productive infections with the viruses, a corresponding antigen is also made (*54–57, 72, 73*); hence, the term T antigen to imply its relationship to the antigen of tumor cells. A complexity of this apparently simple picture emerged with the finding that immunofluorescent techniques uncover an intranuclear antigen in cells infected with any type of human adenovirus, oncogenic or non-oncogenic, using sera from hamsters bearing tumors initiated by type 12 or 18 adenovirus (54). This cross-reacting antigen, which cannot be detected by the complement-fixation assay (54), is probably distinct from the T antigen identified in tumor cells, but the appropriate studies have not been done to establish this difference. Temptation beckons to give the cross-reacting antigen a new symbol but this will be suppressed. However, for convenience and clarity in the following discussion, and *not for permanent usage,* the cross-reacting antigen will be called Tx to distinguish it from the tumor T antigen and to suggest its unknown nature.

Russell and Knight discovered the P antigen with an antiserum from rabbits immunized with an extract of cells obtained 90 hr after infection with type 5 adenovirus in the presence of arabinosylcytosine to block synthesis of DNA and hence production of capsid proteins (23). But the anti-P serum detects not only the P antigen, a protein which becomes enclosed within the capsid, but also the Tx antigen, and possibly one or more other antigens (23, 77). The major antigen, the P antigen, measured by the antiserum employed is distinct from the T antigen, a nonvirion antigen (78, 79), and it appears to be different from the Tx antigen (77). Unfortunately, purified P or Tx antigens have not been prepared so that firmer data are not yet available.

These antigens have properties which make them acceptable as "early" proteins. (1) They are synthesized early in the viral eclipse period, 8–12 hr before the assembly of progeny virions (56, 57, 77). (2) They are made 3–6 hr before replication of viral DNA (56, 57). (3) Their synthesis is independent of and can proceed in the absence of viral DNA biosynthesis (23, 56, 72, 73). (4) They are manufactured 4–9 hr before the capsid proteins (hexons and pentons) (56, 57, 77), "late" proteins. (5) They are immunologically distinct from the capsid proteins (23, 56, 78).

Although little can be said about the nature or the function of the broad cross-reacting Tx antigen, the major T antigen has been purified and characterized. It is a heat labile, small, acidic protein with a sedimentation coefficient of 2.58 $S_{w,20}$ and an isoelectric point of approximately 5.0 (78, 79). The function of the T antigen either in tumorigenesis or in productive infections is unknown, and only one unequivocal fact can be stated concerning a role it does not fulfill: It is not the viral internal protein, nor is it a constituent of the virion (77).

In contrast, the P antigen can be identified as an internal structural component of the virion: Antiserum will not react with intact viral particles, but as the virions disintegrate on standing at 4°C the P antigen becomes accessible to specific antibodies (*23*). Recent evidence indicates that the P antiserum reacts with an internal protein which appears to be an arginine-rich macromolecule having the characteristics of a histone (*80*).

The high arginine content of the P antigen internal protein suggests a relationship between it and the "maturation protein" described by Rouse and Schlesinger (*81, 81a*). Their experiments, inspired by earlier studies on the arginine requirement for viral propagation (*82–84*), demonstrated that multiplication of type 2 adenovirus has an absolute requirement for arginine in the culture medium; that in the absence of arginine infectious virus does not appear, although viral DNA and late proteins are synthesized; and that addition of arginine late in infection induces the formation of mature virions without delay (*81*). Additional confidence in the notion that the major P antigen and the maturation protein are one and the same is garnered from the finding that the P antigen, like the "maturation protein," apparently is not made in arginine-depleted cells (*85*). It may be deduced from the above considerations that the viral DNA and capsid proteins are made sequentially but require the internal protein before their assembly into virions can be accomplished, perhaps to permit proper folding of the viral DNA within the capsid. Another rate limiting step in the maturation, i.e., assembly of the virion, however, may be the synthesis of the penton base which occurs after the hexon and fiber proteins are made (*57, 77*). By the time this volume is published the data to confirm or refute these notions will undoubtedly be available.

C. VIRAL DNA

With the appearance of the early proteins, preparations are completed for replication of the viral genome. Biosynthesis of viral DNA begins in the nuclei 4–7 hr before final assembly of the virions (*86–89*). Viral DNA is made *de novo* from the nucleotide pools of the cells and the medium, and although infection interrupts synthesis of host-cell DNA (to be discussed below), the host DNA is not broken down (*59, 86*).

The time course of production of viral DNA has been measured with the specific inhibitor of DNA synthesis, 5-fluoro-2-deoxyuridine

(*88, 89*) as well as by direct chemical measurements (*86, 87*). Once synthesis of viral DNA is initiated it proceeds at a relatively rapid rate for 8–12 hr and culminates in an accumulation of viral DNA which approximately doubles the total content of DNA in the infected cell (*42, 86, 87*). Hence, the number of viral DNA equivalents synthesized is large, but only about 10% of that made is packaged into viral particles (*42*). The large excess of unused viral DNA becomes a constituent of the characteristic inclusion bodies of adenovirus-infected cells (*41*).

D. CAPSID PROTEINS

Following the replication of the viral genome, the well-ordered series of biosynthetic steps continues toward the production of the viral capsid proteins, hexons, penton fibers, and penton bases (*49, 57, 77*). This regular program of events is dependent upon each of the preceding steps, for without synthesis of viral DNA and its transcription into late messenger RNA's capsid proteins are not made (*44, 88*). It is striking that the infecting parental genome cannot be transcribed to make late proteins—progeny DNA is essential for this task.

Production of viral capsid proteins is initiated approximately 2 hr after synthesis of viral DNA begins and 2 hr before the assembly of the first detectable new virions (*49, 50, 88*). But the three capsid proteins apparently are not simultaneously synthesized: Hayashi and Russell, utilizing complement-fixation and immunofluorescent techniques with antisera specific for the purified capsid proteins, reported that the hexon and fiber antigens appear first and that the penton base is detected about 2 hr later (*77*) (it is possible, however, that the differences in times of synthesis were more apparent than real and were only due to variations in the qualities of the antisera employed).

With immunocytologic techniques the capsid proteins can only be detected within nuclei where they accumulate in great amounts throughout their period of synthesis (*90–92*). These findings suggested a unique opportunity to study an intranuclear mechanism for synthesis of an easily identifiable protein since there is still considerable uncertainty whether any host-cell proteins can be made in the nucleus. Contrary to expectation, however, adenovirus capsid proteins, were shown to be synthesized on characteristic cytoplasmic polyribosomes; after synthesis they are transported rapidly into the nucleus. (1) Thomas and Green (*93*) detected virus-specific RNA on cytoplasmic polyribosomes of infected cells and showed that amino

acids are incorporated into acid-precipitable material on these structures (the nature of the peptides made was not studied, however). (2)
Velicer and Ginsberg (94) to overcome the severe handicap that nuclei
from adenovirus-infected cells are leaky, used autoradiographic techniques to show unequivocally that the viral proteins are made in
the cytoplasm of infected cells, and that they are rapidly transported
into the nuclei. (3) The latter investigators also demonstrated that
virus-specific, immunologically reactive polypeptide chains with a sedimentation coefficient of approximately 3S are synthesized rapidly on
200S polyribosomes (94).

A lag of approximately 2 hr intervenes between the initial synthesis of capsid proteins and final assembly of infectious virions (49).
The delay in utilization of the completed proteins may result from a
late synthesis of either the penton bases (77) or the arginine-rich internal protein, or "maturation protein," which is essential for the assembly of viral particles (23, 81, 85). [The appropriate experiments
to ascertain precisely when the "maturation" protein is made have
not yet been reported (see discussion above).] It is striking that utilization of the capsid proteins produced, like the viral DNA, is wholly
inefficient since only 10–15% is assembled into virions, and the remainder accumulates as wasteful excess to form intranuclear inclusion bodies (11, 41, 95, 96). The ineffectual utilization of structurally
normal viral proteins and DNA may occur because (1) the "maturation protein" or the penton base is made in limiting amounts; or (2)
the hexon and fiber proteins associate with viral DNA so that only
a relatively small proportion of these reactants are available for assembly as the penton base or maturation protein is synthesized. There
is evidence to support the latter possibility: (a) The hexon and fiber
proteins can combine with viral DNA (59); (b) both proteins and
particularly the fiber, are avidly complexed with DNA when extracted
from cells (97); and, (c) the unassembled viral DNA can be selectively extracted in 0.15 M NaCl (42) because the fiber protein confers
this unusual solubility to the deoxyribonucleoprotein (59).

8-5. Effects of Adenovirus Infection on Host Cells

Depending upon the host cells employed, adenoviruses may initiate
productive or abortive infections. The biosynthetic events of productive infections are described in the preceding sections; their effects
on the host-cell's machinery are now detailed. Although a full dis-

cussion of abortive infections and their consequences on the host cells is not within the scope of this essay, a brief consideration of their characteristics appears warranted.

A. PRODUCTIVE INFECTION

Cells cease to divide within 10–12 hr after infection (87), and gross and microscopic cytopathic changes soon appear (41, 90, 98). The characteristic rounding and clumping of infected cells, often in grape-like clusters, may be induced by the penton antigen, since this protein can produce similar cytopathic alterations when added to cell cultures alone (96, 99, 100). The most characteristic pathologic hallmark of adenovirus infections, however, is the development of large intra-nuclear basophilic inclusion bodies and crystalline-like structures (41, 90, 101). Specific viral subunits are the ingredients of these bodies: (1) Feulgen stain demonstrates the presence of DNA (41); (2) immunofluorescence proves the presence of specific viral antigens (90); and (3) electron microscopic examination reveals the basophilic crystals to be composed of viral particles arranged in crystalline lattices (102–104). Thus, the large basophilic inclusions are the pathological evidence for the accumulations of viral subunits, DNA and proteins, which are made in extravagant excess of those assembled into viral particles (see above).

Despite the production and accumulation of viral DNA and protein beginning 6–10 hr after infection, the overall rates of syntheses of these macromolecules remain unchanged for about 40 hr as compared to uninfected cells in spinner cultures (Figs. 8-2a and 8-2b) (47, 87). In contrast, total RNA synthesis is linear and unaltered in infected cells for only 20–24 hr, after which its production rapidly decreases (Fig. 8-2c) (47). Since division of infected cells ceases 10–12 hr after infection, and viral subunits continue to accumulate, the cell mass increases, and the DNA and protein per cell approximately doubles in 48 hr (86, 87). Consideration of these data immediately leads to the conclusion that biosyntheses of host macromolecules must be profoundly affected as a consequence of infection.

1. Host DNA Synthesis

Host and viral DNA can be readily seprarated by chromatography on methylated albumin–kieselguhr (MAK) columns (47) and by equilibrium centrifugation in cesium chloride (47, 89) on the basis of

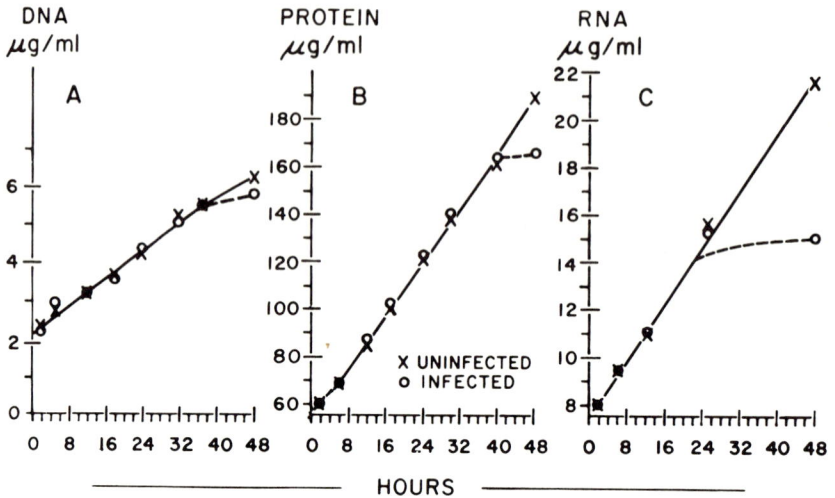

Fig. 8-2. Biosynthesis of DNA, protein, and RNA in uninfected and infected KB cells in spinner cultures. Cells in one culture were infected at a multiplicity of 100 PFU/cell; infected and uninfected cells were thereafter prepared as described in Fig. 8-1. [Modified from H. S. Ginsberg, L. J. Bello, and A. J. Levine (47)].

their differences in content of guanine + cytosine (GC). In cells infected with type 5 adenovirus, replication of host DNA begins to decline 6–10 hr after infection, depending upon the multiplicity of infection, and its production is totally blocked 2–4 hr later (Figs. 8-3 and 8-4) (47). Thus, the replication of host DNA is inhibited just before or at the time that biosynthesis of viral DNA commences (86–88). But the synthesis of neither host nor viral DNA is necessary for the switch-off of host synthesis to occur: When FUdR is added to uninfected and infected cells to stop all DNA synthesis, and 10 hr later the block is reversed with excess thymidine, only viral DNA is made in the infected cells, whereas host DNA synthesis resumes immediately in the uninfected controls (40).

Whether protein synthesis is essential to switch-off replication of host DNA is a pressing problem which cannot be answered satisfactorily because the compounds used to stop protein synthesis experimentally (i.e., cyclohexamide and puromycin) also halt production of DNA to a similar degree (47). Additional attention must be paid to this critical question in order to arrive at the mechanism by which adenovirus infection selectively blocks synthesis of host but not of viral DNA.

2. Host Protein Synthesis

Since the linear synthesis of total protein remains unaltered in virus-infected cells during the period in which viral proteins are being made at a maximal rate (Fig. 8-2b), it follows that production of host proteins must be reduced when the viral proteins are synthesized. This deduction was substantiated experimentally using two different procedures (with identical results) to measure host protein synthesis: (1) differential immunological precipitation of ^{14}C-labeled proteins; (2) assay of six host enzymes [fumerase, phosphoglucose isomerase, lactic dehydrogenase, acid and alkaline phosphatases, and deoxyribonuclease (47, 105)]. By 15–20 hr after infection, synthesis of host proteins is reduced whereas viral proteins are made at a maxi-

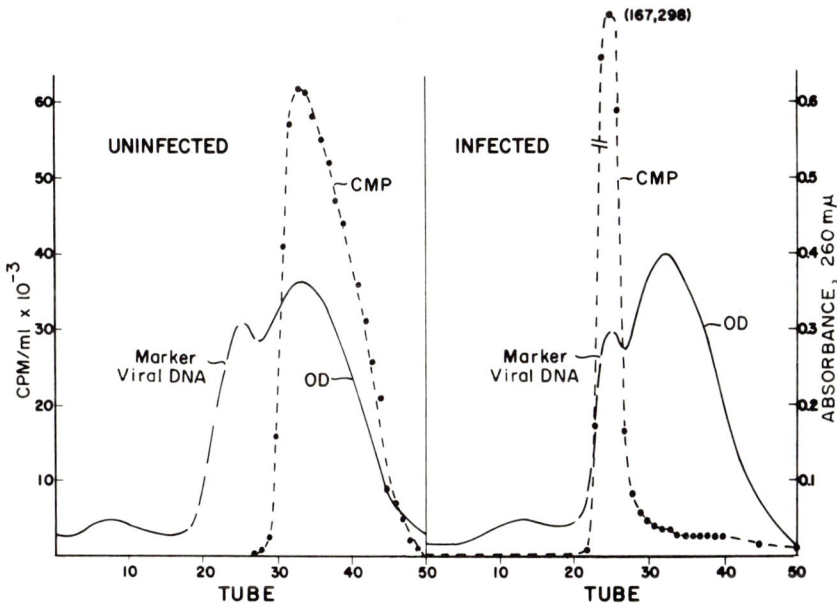

FIG. 8-3. Synthesis of host-cell DNA in uninfected KB cells and in cells 10–12 hr after infection with type 5 adenovirus. One-half of a culture of exponentially growing cells was infected with 100 PFU/cell. Ten hours after infection, ^{3}H-thymidine was added to the uninfected and infected spinner cultures, and 2 hr later DNA was extracted from cells of both cultures. To each sample 50 μg of unlabeled viral DNA was added as an OD marker, and each mixture was chromatographed on MAK columns using a linear gradient from 0.50 to 0.85 M NaCl at pH 6.8. The absorbance (260 mμ) was determined with a recording spectrophotometer, and the ^{3}H content was assayed in an automatic scintillation spectrometer. [From H. S. Ginsberg, L. J. Bello, and A. J. Levine (47).]

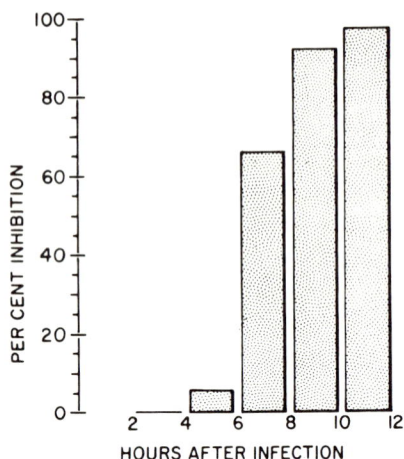

FIG. 8-4. Time at which host-cell DNA was inhibited in KB cells infected with type 5 adenovirus. Exponentially growing cells were infected with an input viral multiplicity of 200 PFU/cell. At the times indicated ^{3}H-thymidine was added to an aliquot of the culture for a 2-hr interval, after which DNA was extracted and chromotographed on an MAK column as described in Fig. 8-3.

mal rate until 20–25 hr postinfection (Fig. 8-5) (*105*). Thus, the blockade of host protein synthesis is not initiated until after the production of viral capsid proteins begins, and the major proportion of viral proteins is made during a period in which the biosynthesis of host protein is retarded by 75–85%. It is striking that when FUdR totally blocks replication of viral DNA, viral mRNA and viral capsid proteins are not made and synthesis of host proteins is not halted (*47*, *105*). These results introduce obvious questions: (1) Does viral infection interrupt transcription or translation of the host genome? and (2) Is viral mRNA or capsid proteins responsible?

3. *Host RNA Synthesis*

That over-all synthesis of host RNA is rapidly diminished 20–24 hr after infection is clear from the data summarized in Fig. 8-2. But if the production of a minor species such as host mRNA were blocked much earlier it might not be reflected in the determination of total RNA synthesis since virus-specific RNA's are made during this period. Direct measurements of the species of RNA complementary to host DNA, which presumably contains mRNA, by hybridization techniques demonstrated that transcription of the host genome does not begin to decline until about 15 hr after infection with type 5 adenovirus (Fig. 8-6) (*47, 106*). Hence, production of host protein is blocked at approximately the same time as the synthesis of the RNA from which it is translated (Fig. 8-7). This result is incompatible with the notion that host protein synthesis is affected during transcription of the host genome; translation of the host mRNA must be the step at which protein synthesis is interrupted.

4. *Possible Mechanisms for Inhibition of Host Macromolecules*

Central to a consideration of the means by which syntheses of host macromolecules are interrupted is the degree of specificity which permits production of viral subunits while the biosynthesis of similar

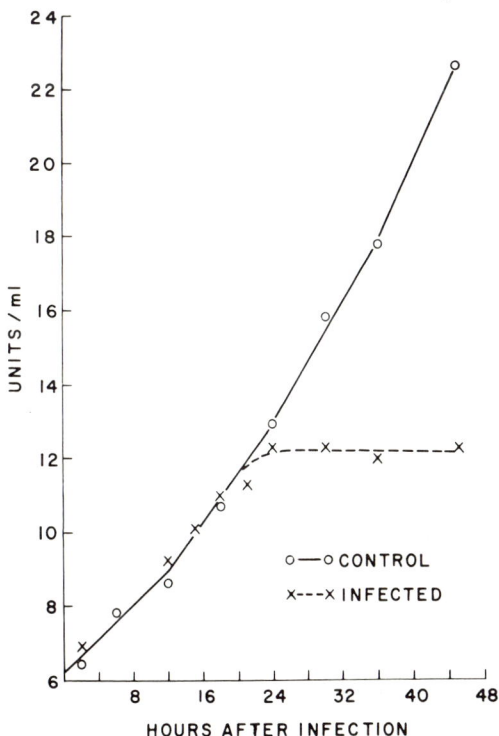

FIG. 8-5. Effect of type 5 adenovirus infection on synthesis of fumerase in spinner cultures of KB cells. Exponentially growing cells were infected with 200 PFU/cell of type 5 adenovirus and incubated at 36°C. At the times indicated, samples of uninfected and infected cells were centrifuged, washed, suspended in 0.01 *M* glycyl-glycine buffer (pH 7.4), and disrupted by sonication. [From L. J. Bello and H. S. Ginsberg (*105*).]

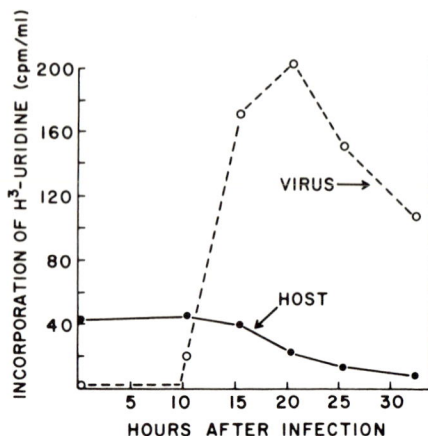

FIG. 8-6. The rate of synthesis of RNA complementary to host DNA in spinner culture of KB cells infected with type 5 adenovirus. For comparison, and to indicate the specificity of the hybridization technique, data are summarized to show the synthesis of virus-specific mRNA in the same cells. Cells infected at an input multiplicity of 200 PFU/cell were labeled with ^{3}H-uridine for 1-hr periods at the indicated times after infection. RNA was extracted with hot phenol and mixed with purified, denatured host or viral DNA for hybridization. [From L. J. Bello and H. S. Ginsberg (*106*).]

host materials are restricted. The mechanisms are not obvious for any of the controls observed, but several possibilities can be suggested which may lead to eventual solutions. But prior to suggesting models for the control of synthesis of host macromolecules in infected cells, it is desirable to make a brief summation of the intracellular and *in vitro* effects of the viral capsid proteins on these biosynthetic reactions: (1) Fiber, but not hexon, protein when added to cell cultures inhibits multiplication of adenoviruses, polioviruses, and vaccinia virus (*33, 107*); (2) highly purified fiber protein has a profound inhibitory effect on the biosyntheses of DNA, RNA, and protein in KB cells, although its adsorption to and apparent entry into cells is slow and inefficient (*33*); (3) hexon protein cannot attach to KB cells (*33*); and (4) the hexon as well as the fiber proteins complex with DNA *in vitro* and inhibit the activities of both DNA-dependent RNA polymerase and DNA polymerase (*59*).

Replication of host DNA stops relatively early in the infectious process when production of viral DNA begins; the time host DNA ceases to be made is dependent upon the multiplicity of infection (*47*).

However, for the switch-off to occur, neither the synthesis of viral DNA nor the continued replication of host DNA is mandatory. It is tempting to postulate that the viral capsid proteins, when shed from the DNA-protein core, complex with host DNA, inhibit polymerase activity, and therefore block further host DNA synthesis (as shown experimentally) (47, 59). Viral DNA replication would consequently be free to proceed unhindered because the limited number of capsid protein molecules dissociated from the parental virions would complex with host DNA before the hypothetical second stage of uncoating the viral DNA were completed. However, it is precisely the limited number of molecules of capsid proteins initially available from the infect-

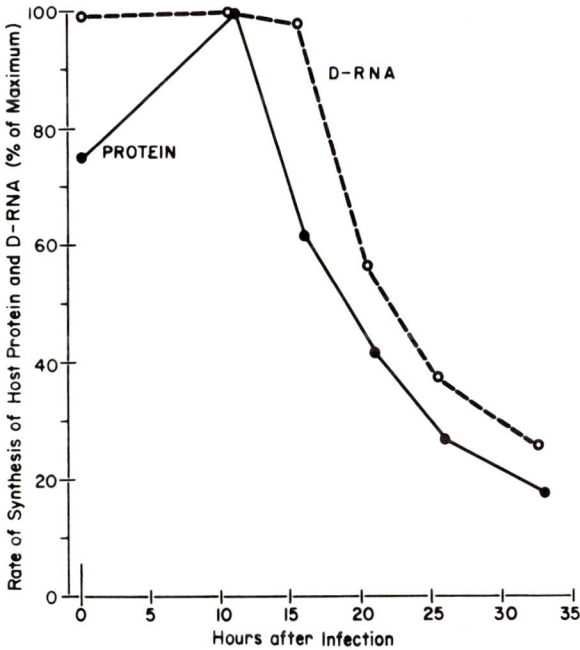

FIG. 8-7. Comparison of the rates of synthesis of host protein and RNA complementary to host DNA in spinner cultures of KB cells infected with type 5 adenovirus. Cells infected at a viral multiplicity of 200 PFU/cell were labeled with ^{14}C-valine for 2-hr periods at the times indicated. Viral proteins were precipitated with specific antiserum, and the number of acid-precipitable counts remaining in the supernatant fluid was considered to be host protein. The data for RNA complementary to host DNA (presumably containing host mRNA) were taken from experiments presented in Fig. 8-6. [From L. J. Bello and H. S. Ginsberg (106).]

ing virions which may restrict the potential role of these proteins in halting host DNA synthesis since the size of the host DNA molecule is great and the number of initiation points for replication are apparently many. Other possible mechanisms to explain the blockade of host DNA replication are: (1) The viral capsid proteins or the viral DNA-protein core reacts with the nuclear membrane to induce a process similar to contact inhibition which stops DNA synthesis; or (2) a basic protein which has great specificity for host DNA is made as an "early" protein either as a host response to the virus or from information encoded in the viral genome.

To halt translation of host proteins without affecting the synthesis of viral proteins demands a highly specific reaction. This fine selectivity can be most easily explained as a simple competition phenomenon based upon the findings that: (1) The viral mRNA is made at a rate 5 to 10 times that of host DNA-like RNA when synthesis of host proteins is reduced (Fig. 8-5) (*47, 106*), and (2) interruption of host protein synthesis is avoided if viral mRNA is not synthesized. From these data it is postulated that viral mRNA has a quantitative selective advantage over the host mRNA and therefore competes successfully for the limited number of intranuclear ribosomal subunits available for transport into the cytoplasm; consequently, only the viral mRNA is translated.

Cessation of the biosynthesis of host DNA-like RNA occurs when transcription of virus-specific RNA and replication of viral DNA also begin to decline (*47, 88, 106*). The concomitant reduction in syntheses of the three different macromolecules may then result from nondiscriminatory reactions, and all could be most readily explained as a consequence of over-production and excessive accumulation of viral capsid proteins. A large number of viral protein molecules are available late in infection and it seems inevitable that they would complex with the highly charged nucleic acids and prevent the syntheses of both RNA and DNA. It is perhaps ironic that just as the host cell receives its final oppressive blow, i.e., a harsh block of the remaining syntheses of vital macromolecules, propagation of its parasite, the virus, approaches its end because viral DNA also cannot replicate.

B. ABORTIVE INFECTIONS

An abortive infection is defined, for the purpose of this discussion, as a virus-cell interaction in which the virion enters the cell and is uncoated, but the process of viral multiplication is restricted so that the expected final product, i.e., infectious viral particles, cannot be fabricated. Thus, operationally, adenoviruses may be considered host-dependent conditionally lethal mutants: In a *permissive* cell the virus induces a "productive" infection; in a *restrictive* cell, an "abortive" infection ensues. Several different abortive adenovirus infections have been investigated, but biochemically most information has been gained in the study of restricted infections with strains of human adenoviruses in African green monkey kidney cells (AGMK). The abortive infection of AGMK cells by adenoviruses has been reviewed extensively elsewhere (*81a, 108, 109*), and it will be discussed here only as it pertains to the subject of this essay.

Infection of AGMK cells with simian virus 40 (SV_{40}) (*110*) or with SV_{15} simian adenovirus (*111*) prior to or concomitant with adenovirus infection converts the abortive adenovirus response into a productive infection. Thus, it is probably not an accident that abortive infections of AGMK cells with human adenoviruses are being extensively studied, for these intriguing findings add another parameter that facilitates investigation of the biochemical defect which will not permit the adenovirus genome to fulfill its natural productive function. The recognition that SV_{40} and SV_{15} viruses can enhance the multiplication of human adenoviruses leads to the predictions that (1) either the viral genome requires for its full expression some cell function which is supplied in permissive but not in nonpermissive cells, or the nonpermissive cell represses a cistron of the viral genome thus imposing a restriction upon it; and (2) the helper virus supplies the missing function as a viral gene product or by derepressing a cistron of the host or viral genome. Enhancement of adenovirus multiplication by SV_{40} virus has been most thoroughly investigated and the discussion to follow will be concerned mainly with the data which have emerged from these studies.

Operationally SV_{40} virus may be considered to complement the restricted adenovirus to permit complete viral propagation. Formally, however, the SV_{40} enhancement of adenovirus replication is not analogous to complementation since the cooperating viruses are not genetically related, and the enhancing virus is itself not defective. The enhancement phenomenon has also been termed "nonreciprocal complementation" (*109*), but this name also may be misleading since it carries with it genetic implications. Perhaps the title "helper function" may be less restrictive. The biochemical evidence available does not permit us to identify precisely the defect which imposes the limitation upon adenovirus propagation in AGMK cells. However, the known data are beginning to focus on the adenovirus deficiency and these will be considered. A few pertinent biological facts of the restricted and enhanced adenovirus infection will be summarized before the biochemical data are discussed.

As with most host-dependent conditionally lethal mutants, the restriction of adenovirus infection in AGMK cells is not absolute; i.e., some productive viral multiplication occurs and the quantity of virus made is dependent upon both the viral type and the multiplicity of infection (109, 112–114). But the viral yield is always markedly depressed as compared to infection of permissive cells (e.g., KB cells) and SV_{40} virus enhances adenovirus multiplication 100- to 1000-fold (112–115). In contrast, propagation of the "good Samaritan," SV_{40} virus, is depressed while serving its helper function (112). Inhibition of the helper's multiplication is not necessarily a general phenomenon, however, for the simian adenovirus SV_{15}, which also enhances replication of human types of adenoviruses in AGMK cells, multiplies without constrictions along with type 2 adenovirus (111).

Investigation of each event in the program of adenovirus multiplication (see Sec. 8-4) has revealed that several biosynthetic reactions proceed normally in the restricted infection. Early proteins such as T antigen (115) and thymidine kinase (116) are made and viral DNA is synthesized (117, 118). Recent quantitative studies in our laboratory confirmed the earlier results and showed that the quantity of viral DNA replicated is approximately the same in restricted and SV_{40}-enhanced infections (112). Moreover, Baum et al. have found that biosynthesis of viral mRNA was only enhanced about 2-fold in an SV_{40}-adenovirus double infection in which production of infectious adenovirus was increased 1000-fold (119).

Restricted replication of adenovirus in AGMK cells, must, therefore, result from defective translation of one or more viral proteins. That synthesis of capsid proteins is limited in the abortive infection is apparent but it is unknown whether the failure to make these late viral proteins is due to a primary or secondary defect. From the data available it is reasoned that the translation defect may result from: (1) a virus-induced or constitutive cell-specific differential inhibitor of translation—perhaps a translation inhibitor protein similar to that apparently induced by interferon (120); or (2) the absence of one or more essential tRNA's which are not coded for by adenoviruses or whose synthesis cannot be induced by these agents in AGMK cells. The helper function performed by SV_{40} virus would then be to repress the inhibitor or to induce the appearance of the essential tRNA.

It is striking that the enhancement of adenoviruses is a true interplay between the involved viruses and not a simple interaction which is mutually beneficial to both participants. The final result of the double infection finds the multiplication of the adenovirus enhanced, but synthesis of SV_{40} DNA is markedly depressed, replication of the host DNA is not stimulated, and propagation of SV_{40} virions is suppressed (112). Since replication of the SV_{40} genome is restricted in the doubly infected cells, it follows that the parental SV_{40} genome must supply an early function which transforms an abortive to a productive adenovirus infection. In fact, the entire SV_{40} DNA molecule is not mandatory to supply the missing function since the defective SV_{40} genome in the adenovirus-SV_{40} "hybrid particle" can also serve as the helper (121).

It is of special interest to note that the deleterious effect of the "helped" on the "helper" is not a unique phenomenon confined to adenovirus-SV_{40} virus double infections. Co-infection with an adenovirus is demanded for multiplication of the defective adeno-associated viruses (AAV) (122, 123), but propagation of the adenovirus is inhibited while the AAV is rescued (123–127). The ungrateful manner in which the benefactor is treated in these examples may be viewed as presumptive evidence that adenoviruses are at least sociologically even though not genetically related to AAV and SV_{40} virus.

8-6. Prospects

The fine structure of adenoviruses is now well described and in broad outline the biosynthesis of its component parts is understood in molecular terms. In concluding this essay it seems appropriate to indicate some areas where data are either inconclusive or absent and to mark those exciting, partially explored fields of regulation which future investigators will probably cultivate.

(1) Although the virion's fibers appear to serve as the organs of attachment to host cells, the chemical nature of the cell receptors is unknown.

(2) Intracellularly the capsid proteins are shed rapidly to make the viral DNA susceptible to DNase, but the viral genome remains asso-

ciated with viral protein, probably an internal protein. Whether this partially uncoated DNA can replicate, and if not, how and where the final stages of uncoating occur deserves investigation.

(3) The T and P antigens appear and the activities of several enzymes increase prior to replication of viral DNA. But the functions of these proteins and which of the enzymes are products of the viral genome is ambiguous. In particular, the role of the ubiquitous thymidine kinase is unclear, since it is not required for normal thymidine synthesis.

(4) Only messages for "early" proteins are made from the infecting parental viral genome, and the mRNA for the capsid proteins ("late" proteins) can only be transcribed from progeny DNA molecules. Investigation of the mechanisms by which sequential transcription of the parental and progeny genomes is regulated excites the imagination of many molecular biologists and the answers when obtained should be pertinent not only to virology but also to the problems of differentiation as well.

(5) Formation of the virion appears to be accomplished by self-assembly of the subunits when the bits and pieces reach a critical concentration. However, the discovery of a protein which appears to be essential for the assembly of previously made viral DNA and proteins suggest that the so-called "maturation protein" either actively promotes assembly or is a component essential for the pieces to fit perfectly into the icosahedral virion.

(6) Biosynthesis of host macromolecules is sequentially halted in productive adenovirus infections. Only hypotheses and theoretical models have been presented thus far to explain the highly selective controls involved. Of central interest are data to tell us whether the infecting viral genome must impose itself on and overcome the host cell's controls for the elaboration of infectious virus to ensue.

(7) Preliminary data suggest the possible type of biochemical deficiency which imposes a restriction on multiplication of adenoviruses in African green monkey kidney cells and permits coinfection with SV_{40} virus to abrogate the restriction of adenovirus replication. The deficiency which causes an abortive adenovirus infection and the mechanism of the inhibitory effect of adenovirus on its helper SV_{40} virus must be clearly explained in biochemical terms.

(8) Many but not all adenoviruses can transform cells and induce tumors in newborn hamsters, rats, and mice. An intriguing correlation has been made between the oncogenic potentiality of adenoviruses and

the proportion of guanine and cytosine (G + C) in their DNA; viruses with the lowest proportion of G + C have the greatest oncogenicity. It is considered possible that because a portion of the viral DNA has an adenine–thymidine-rich region similar to that of the host's DNA, a piece of the viral genome can be readily inserted and integrated into the host genome so that both can replicate in perpetuity together. This attractive postulate requires proof and its mechanism must be elucidated.

It now seems possible to predict with confidence that the biosynthetic steps in adenovirus multiplication will be known in detail in the near future. It may also be predicted that great attention will be directed to the intriguing investigations of mechanisms for the selective controls of the viral biosynthetic pathways and the manner by which the virus inserts its controls upon the cell. The data derived from these studies not only will complete the picture of adenovirus replication but also will offer molecular mechanisms for cell killing and malignant cell transformation. It seems unlikely that these goals can be attained with biochemical tools alone—it is probable that solutions to these problems will require excursions into the unexplored genetics of adenoviruses.

ACKNOWLEDGMENTS

The personal investigations summarized in this chapter were partially supported by United States Public Health Service grants Al-03620 and Al-05731, and conducted under the sponsorship of the Commission on Acute Respiratory Diseases, Armed Forces Epidemiological Board (supported by the Office of the Surgeon General, Department of the Army).

I am grateful to R. Walter Schlesinger and W. C. Russell for furnishing me preprints of their unpublished papers.

REFERENCES

1. W. P. Rowe, R. J. Huebner, L. K. Gilmore, R. H. Parrott, and T. G. Ward, *Proc. Soc. Exptl. Biol. Med.*. **84,** 570 (1953).
2. M. R. Hilleman and J. R. Werner, *Proc. Soc. Exptl. Biol. Med.*, **85,** 183 (1954).
3. J. J. Trentin, Y. Yabe, and G. Taylor, *Science*, **137,** 835 (1962).
4. R. J. Huebner, W. P. Rowe, and W. T. Love, *Proc. Natl. Acad. Sci.*, **48,** 2051 (1962).

5. A. J. Girardi, M. R. Hilleman, and R. E. Zewickey, *Proc. Soc. Exptl. Biol. Med.*, **118**, 15 (1965).

6. W. D. McBride and A. Wiener, *Proc. Soc. Exptl. Biol. Med.*, **115**, 870 (1964).

6a. J. H. Pope and W. P. Rowe, *J. Exptl. Med.*, **120**, 577 (1964).

7. M. Green, *Ann. Rev. Microbiol.*, **20**, 189 (1966).

8. R. W. Horne, S. Brenner, A. P. Waterson, and P. Wildy, *J. Mol. Biol.*, **1**, 84 (1959).

9. M. A. Epstein, S. J. Holt, and A. K. Powell, *Brit. J. Exptl. Pathol.*, **41**, 567 (1960).

10. F. H. C. Crick and J. D. Watson, *Nature*, **177**, 473 (1956).

11. W. C. Wilcox, H. S. Ginsberg, and T. B. Anderson, *J. Exptl. Med.*, **118**, 307 (1963).

12. U. Pettersson, L. Phillipson, and S. Höglund, *Virology*, **33**, 575 (1967).

13. R. C. Valentine and H. G. Pereira, *J. Mol. Biol.*, **13**, 13 (1965).

14. E. Norrby, *Virology*, **28**, 236 (1966).

15. E. Norrby and G. Wadell, *Virology*, **31**, 592 (1967).

16. H. Gelderblom, H. Bauer, H. Frank, and R. Wigand, *J. Gen. Virol.*, **1**, 553 (1967).

17. U. Pettersson, L. Philipson, and S. Hoglund, *Virology*, **35**, 204 (1968).

18. H. G. Pereira and M. V. T. de Figueiredo, *Virology*, **18**, 1 (1962).

19. E. Norrby and P. Skaaret, *Virology*, **32**, 489 (1967).

20. W. G. Laver, J. R. Suriano, and M. Green, *J. Virol.*, **1**, 723 (1967).

21. W. C. Lawrence and H. S. Ginsberg, *J. Virol.*, **1**, 851 (1967).

22. L. Philipson, *J. Virol.*, **1**, 868 (1967).

23. W. C. Russell and B. E. Knight, *J. Gen. Virol.*, **1**, 523 (1967).

24. M. Pina and M. Green, *Proc. Natl. Acad. Sci. U.S.*, **54**, 547 (1965).

25. A. J. van der Eb and L. W. van Kesteren, *Biochem. Biophys. Acta*, **129**, 441 (1966).

26. M. Green, M. Pina, R. Kimes, P. C. Wensink, L. A. MacHattie, and C. A. Thomas, Jr., *Proc. Natl. Acad. Sci U.S.*, **57**, 1302 (1967).

27. M. Green and M. Pina, *Proc. Natl. Acad. Sci. U.S.*, **51**, 1251 (1964).

28. M. Green and G. E. Daesch, *Virology*, **13**, 169 (1961).

29. R. A. Consigli and H. S. Ginsberg, *J. Bacteriol.*, **87**, 1034 (1964).

30. N. Ledinko, *Virology*, **28**, 679 (1966).

31. S. Kit, L. J. Piekarski, D. R. Dubbs, R. A. de Torres, and M. Anken, *J. Virol.*, **1**, 10 (1967).

32. H. S. Ginsberg, *J. Exptl. Med.*, **107**, 133 (1958).

33. A. J. Levine and H. S. Ginsberg, *J. Virol.*, **1**, 747 (1967).

34. S. Dales, *J. Cell Biol.*, **13**, 303 (1962).

35. W. K. Joklik, *J. Mol. Biol.*, **8**, 263 (1964).

36. W. K. Joklik, *J. Mol. Biol.*, **8**, 277 (1964).

37. J. S. Sussenbach, *Virology*, **33**, 567 (1967).

38. J. R. Kates and B. R. McAuslan, *Proc. Natl. Acad. Sci. U.S.*, **57**, 314 (1966).

39. J. R. Kates and B. R. McAuslan, *Proc. Natl. Acad. Sci. U.S.*, **58**, 134 (1967).

40. H. S. Ginsberg, unpublished work.

41. G. S. Boyer, C. Leuchtenberger, and H. S. Ginsberg, *J. Exptl. Med.*, **105**, 195 (1957).

42. H. S. Ginsberg and M. K. Dixon, *J. Exptl. Med.*, **109**, 407 (1959).

43. M. Green, *Virology,* **9,** 343 (1959).

44. J. F. Flanagan and H. S. Ginsberg, *J. Bacteriol.,* **87,** 977 (1964).

45. K. Köhler and T. Odaka, *Z. Naturforsch.,* **19b,** 331 (1964).

46. J. A. Rose, P. R. Reich, and S. M. Weissman, *Virology,* **27,** 571 (1965).

47. H. S. Ginsberg, L. J. Bello, and A. J. Levine, in *The Molecular Biology of Viruses* (J. S. Colter and W. Paranchych, eds.), Academic Press, New York, 1967, p. 547.

48. A. P. Nygaard and B. D. Hall, *Biochem. Biophys. Res. Commun.,* **12,** 98 (1963).

49. W. C. Wilcox and H. S. Ginsberg, *Virology,* **20,** 269 (1963).

50. H. Polasa and M. Green, *Virology,* **25,** 68 (1965).

51. M. Green, M. Pina, K. Fujinaga, S. Mak, and D. C. Thomas in *Perspectives in Virology,* Vol. VI (M. Pollard, ed.), Academic Press, New York, in press.

52. D. Gillespie and S. Spiegelman, *J. Mol. Biol.,* **12,** 829 (1965).

53. P. R. Reich, B. G. Forget, S. M. Weissman, and J. A. Rose, *J. Mol. Biol.,* **17,** 428 (1966).

54. J. H. Pope and W. P. Rowe, *J. Exptl. Med.,* **120,** 577 (1964).

55. M. D. Hoggan, W. P. Rowe, P. H. Black, and R. J. Huebner, *Proc. Natl. Acad. Sci. U.S.,* **53,** 12 (1965).

56. Z. Gilead and H. S. Ginsberg, *J. Bacteriol.,* **90,** 120 (1965).

57. W. C. Russell, K. Hayashi, P. J. Sanderson, and H. G. Pereira, *J. Gen. Virol.,* **1,** 495 (1967).

58. M. Green, M. Pina, and V. Chagaya, *J. Biol. Chem.,* **239,** 1188 (1964).

59. A. J. Levine and H. S. Ginsberg, *J. Virol.,* **2,** 430 (1968).

60. R. A. Consigli and H. S. Ginsberg, *J. Bacteriol.,* **87,** 1027 (1964).

61. S. Kit and D. R. Dubbs, *Biochem. Biophys. Res. Commun.,* **11,** 55 (1963).

62. H. M. Keir and E. Gold, *Biochem. Biophys. Acta.,* **72,** 263 (1963).

63. C. Hamala, T. Kamiya, and A. S. Kaplan, *Virology,* **28,** 271 (1966).

64. D. R. Dubbs and S. Kit, *Virology,* **22,** 493 (1964).

65. C. Jungwirth and W. K. Joklik, *Virology,* **27,** 80 (1965).

66. W. E. Magee, *Virology,* **17,** 604 (1962).

67. S. Kit, D. R. Dubbs, and L. J. Peirkarski, *Biochem. Biophys. Res. Commun.,* **8,** 72 (1962).

68. B. R. McAuslan and W. K. Joklik, *Biochem. Biophys. Res. Commun.,* **8,** 486 (1962).

69. T. Hanafusa, *Biken J.,* **4,** 97 (1961).

70. S. Kit and D. R. Dubbs, *Biochem. Biophys. Res. Commun.,* **13,** 500 (1963).

71. W. E. Magee and O. V. Miller, *Virology,* **31,** 64 (1967).

72. L. A. Feldman and F. Rapp, *Proc. Soc. Exptl. Biol. Med.,* **122,** 243 (1966).

73. H. Shimajo, H. Yamamoto, and C. Abe, *Virology,* **31,** 748 (1967).

74. R. J. Huebner, W. P. Rowe, H. C. Turner, and W. T. Lane, *Proc. Natl. Acad. Soc. U.S.,* **50,** 379 (1963).

75. R. J. Huebner, M. J. Casey, R. M. Chanock, and K. Schell, *Proc. Natl. Acad. Sci. U.S.,* **54,** 381 (1965).

76. R. J. Huebner, in *Perspectives in Virology,* Vol. V (M. Pollard, ed.), Academic Press, New York, 1967, p. 147.

77. K. Hayashi and W. C. Russell, *Virology,* **34,** 470 (1968).

78. Z. Gilead and H. S. Ginsberg, *J. Virol.*, **2**, 7 (1968).

79. Z. Gilead and H. S. Ginsberg, *J. Virol.*, **2**, 15 (1968).

80. W. C. Russell, W. G. Laver, and P. J. Sanderson, in press.

81. H. Rouse and R. W. Schlesinger, *Virology*, **33**, 513 (1967).

81a. R. W. Schlesinger, *Advan. Virus Res.*, **14**, 1 (1969).

82. V. H. Bonifas and R. W. Schlesinger, *Federation Proc.*, **18**, 560 (1959).

83. H. C. Rouse, V. H. Bonifas, and R. W. Schlesinger, *Virology*, **20**, 357 (1963).

84. V. H. Bonifas, *Arch. Res. Virusforsch.*, **20**, 20 (1967).

85. W. C. Russell and Y. Becker, *Virology*, **35**, 18 (1968).

86. H. S. Ginsberg and M. K. Dixon, *J. Exptl. Med.*, **113**, 283 (1961).

87. M. Green and G. E. Daesch, *Virology*, **13**, 169 (1961).

88. J. F. Flanagan and H. S. Ginsberg, *J. Exptl. Med.*, **116**, 141 (1962).

89. M. Green, *Cold Spring Harbor Symp. Quant. Biol.*, **27**, 219 (1962).

90. G. S. Boyer, F. W. Denny, and H. S. Ginsberg, *J. Exptl. Med.*, **110**, 827 (1959).

91. H. G. Periera, A. C. Allison, and B. Balfour, *Virology*, **7**, 300 (1959).

92. H. P. Stich, V. I. Kalnins, E. Mackinnon, and D. S. Yohn, *J. Ultra Structure Res.*, **19**, 556 (1967).

93. D. C. Thomas and M. Green, *Proc. Natl. Acad. Sci. U.S.*, **56**, 243 (1966).

94. L. F. Velicer and H. S. Ginsberg, *Proc. Natl. Acad. Sci. U.S.*, **61**, 1264 (1968).

95. H. G. Klemperer and H. G. Pereira, *Virology*, **9**, 536 (1959).

96. W. C. Wilcox and H. G. Ginsberg, *Proc. Natl. Acad. Sci. U.S.*, **47**, 512 (1961).

97. A. C. Allison, H. G. Pereira, and C. P. Farthing, *Virology*, **10**, 316 (1960).

98. H. S. Ginsberg, *Ann. N.Y. Acad. Sci.*, **67**, 383 (1957).

99. H. G. Pereira, *Virology*, **6**, 601 (1958).

100. S. F. Everett and H. S. Ginsberg, *Virology*, **6**, 770 (1958).

101. D. P. Block, C. Morgan, G. C. Godman, C. Howe, and H. M. Rose, *J. Biophys. Biochem. Cytol.*, **3**, 1 (1957).

102. L. Kjellen, G. Lagermalm, A. M. Svedmyr, and K. G. Thorsson, *Nature*, **175**, 505 (1955).

103. C. Morgan, C. Howe, H. M. Rose, and D. H. Moore, *J. Biophys. Biochem. Cytol.*, **2**, 351 (1956).

104. C. Harford, A. Hamlin, E. Parker, and T. van Ravensway, *J. Exptl. Med.*, **104**, 443 (1956).

105. L. J. Bello and H. S. Ginsberg, *J. Virol.*, **1**, 843 (1967).

106. L. J. Bello and H. S. Ginsberg, *J. Virol.*, **3**, 106 (1969).

107. H. G. Pereira, *Virology*, **11**, 590 (1960).

108. F. Rapp and J. L. Melnick, *Progr. Med. Virol.*, **8**, 349 (1966).

109. F. Rapp, in *The Molecular Biology of Viruses* (L. V. Crawford and M. G. P. Stoker, eds.), Cambridge, New York, 1968, p. 273.

110. G. T. O'Conor, A. S. Rabson, I. K. Berezesky, and F. J. Paul, *J. Natl. Cancer Inst.*, **31**, 903 (1963).

111. R. F. Naegele and F. Rapp, *J. Virol.*, **1**, 838 (1967).

112. M. Friedman, M. Lyons, and H. S. Ginsberg, in preparation.

113. A. S. Rabson, G. T. O'Conor, I. K. Berezesky, and F. J. Paul, *Proc. Soc. Exptl. Biol. Med.*, **116**, 187 (1964).

114. W. B. Beardmore, M. J. Havlick, A. Serafine, and I. W. McLean, Jr., *J. Immunol.*, **95**, 422 (1965).

115. L. A. Feldman, J. S. Butel, and F. Rapp, *J. Bacteriol.*, **91**, 813 (1966).

116. E. Bresnick and F. Rapp, *Virology*, **34**, 799 (1968).

117. P. R. Reich, S. G. Baum, J. A. Rose, W. P. Rowe, and S. M. Weissman, *Proc. Natl. Acad. Sci. U.S.*, **55**, 336 (1966).

118. F. Rapp, L. A. Feldman, and M. Mandel, *J. Bacteriol.*, **92**, 931 (1966).

119. S. G. Baum, W. H. Wiese, and P. R. Reich, *Virology*, **34**, 373 (1968).

120. P. I. Marcus and J. M. Salb, *Virology*, **30**, 502 (1966).

121. F. Rapp and M. Jerkofsky, *J. Gen. Virol.*, **1**, 311 (1967).

122. R. W. Atchison, B. C. Casto, and W. McD. Hammon, *Science*, **149**, 754 (1965).

123. M. D. Hoggan, N. R. Blacklow, and W. P. Rowe, *Proc. Natl. Acad. Sci. U.S.*, **55**, 1467 (1966).

124. K. Smith, W. P. Geble, and J. F. Thiel, *J. Immunol.*, **97**, 754 (1966).

125. B. C. Casto, R. W. Atchison, and W. McD. Hammon, *Virology*, **32**, 52 (1967).

126. W. P. Parks, J. L. Melnick, R. Rongey, and H. D. Mayor, *J. Virol.*, **1**, 171 (1967).

127. W. P. Parks, A. M. Casazza, J. Alcott, and J. V. Melnick, *J. Exptl. Med.*, **127**, 91 (1968).

THE BIOCHEMISTRY
OF POXVIRUS REPLICATION

B. R. McAUSLAN

ROCHE INSTITUTE FOR MOLECULAR BIOLOGY
NUTLEY, NEW JERSEY

9-1. Introduction

The scientific literature on poxvirus dates from about 1798 with Jenner's (*1*) classic work on smallpox vaccination. Interest shifted from clinical to epidemiological aspects of this virus with the dramatic application of myxomatosis to biological pest control in Australia in 1950. Interest has shifted again in the last few years to the molecular aspects of poxvirus replication. A number of intriguing questions peculiar to poxvirus replication remain unanswered but the most interesting problem in poxvirus studies is one central in biology today—an understanding of the mechanisms involved in the sequential expression of genes. As will be evident from consideration of studies on poxvirus RNA synthesis, several attributes of poxvirus make it a valuable system to unravel this important question.

This virus can be grown in large quantities, it is very stable and easy to purify. These properties made poxvirus one of the first animal viruses amenable to biochemical study. The entire replication cycle of the virus is carried out in the cell cytoplasm and as it replicates it gradually inhibits replication and expression of the host genome. With this system one can study the replication and expression of a DNA genome without many of the obvious problems in studying viruses that replicate in the nucleus.

Until recently, poxvirus was often referred to as the only DNA virus that replicated in the cytoplasm. We now have examples of other such viruses and these include African swine fever (2), frog virus (3), and the insect virus, Sericesthis iridescent virus (4). No doubt these viruses will exhibit differences in the details of their replication but one can expect that the experimental approaches devised for poxvirus will be directly applicable to a comparative study of other cytoplasmic DNA viruses.

An extensive review on the poxviruses has recently appeared (5). That review summarizes most of the information on the biology and biochemistry of poxviruses up to the beginning of 1967 and is useful background reading for this chapter. In this chapter I do not aim to present such an exhaustive treatise. Following a brief introduction to the poxviruses, I emphasize some current biochemical aspects and outline some experiments not described in previous reviews.

9-2. An Outline of the Poxvirus Group

Poxvirus is one of the major groups among the viruses of vertebrates. Of primary importance in their classification is the nucleic acid (double-stranded DNA of molecular weight about 160×10^6), morphology ("brickshaped" with dimensions about 300×200 mμ, composed of well-defined protein layers and a central nucleoid), and the occurrence of an internal antigen, the NP antigen, common to all poxviruses so far examined (6).

Several subgroups can be distinguished within the poxvirus group. Although first based primarily on host infection, the allocation to a subgroup is now based on the detailed external morphology of the virion as well as cross-protection and neutralization tests. Fenner (7) proposes that recombination may be applicable to subgrouping. Members of the vaccinia subgroup recombine to produce viable hybrids, but recombination is not demonstrable between them and members of other subgroups.

A partial list of the subgroups of poxviruses is given in Table 9-1.

TABLE 9-1

The Poxviruses

Subgroup	Members
Vaccinia	Vaccinia, cowpox, rabbitpox, ectromelia (mousepox), monkeypox, variola
Myxoma	Myxoma, rabbit fibroma, squirrel fibroma, hare fibroma
Orf	Orf, bovine papular stomatitis
Sheeppox	Sheeppox, goatpox
Avianpox	Fowlpox, canary pox, pigeon pox
Others not allocated	Yaba monkey virus, swinepox, molluscum contagiosum

An important property of the poxviruses as a group is nongenetic reactivation of heat-denatured poxvirus. This occurs with all members of the poxvirus group but with no other virus. This property is considered again elsewhere in this chapter. Most biochemical studies to date have been conducted with vaccinia strain WR, cowpox, and rabbitpox, and this chapter is concerned primarily with work conducted using these strains.

Members of the vaccinia subgroup will replicate in a large variety of cultured cell types. Cultured human cells of most types, L-cells, swine cells, monkey cells, baby hamster kidney cells, chick embryo fibroblasts, and the chorioallantoic membrane of hen eggs support replication of members of the vaccinia subgroup. However, not all members of the vaccinia subgroup will replicate in each of these cells. For example, L-cells do not support replication of myxoma or fowlpox; FL cells support myxoma and vaccinia growth but not fowlpox virus;

chicken embryo fibroblasts support replication of myxoma, vaccinia, and fowlpox (7a).

There are also differences in the extent of replication of different virus strains in various systems. Thus, vaccinia WR and rabbitpox multiply well in HeLa cells (about 200 pock-forming units per cell), whereas, in our experience, cowpox replicates poorly in this system (usually about 30 pock-forming units per cell). Rabbitpox and cowpox can be grown to very high titer on the chorioallantoic membrane of hen eggs; vaccinia WR and some rabbitpox mutants do not give good yields in this system and the pocks are barely discernible. Therefore, the selection of a virus-cell system depends on the particular experimental objects, the facilities available, and the magnitude of biochemical effects found in different systems. For example, cowpox has been the most useful in demonstrating induced enzymes because the magnitude of the induction in HeLa cells is usually higher than with other poxvirus strains studied. A comparison of several strains of the vaccinia subgroup made by Fenner and Woodroofe (8) shows the great differences between strains in their capacity to produce hemagglutinin. Several research groups in different areas of the United States (personal communications) have found unexpected difficulty in growing poxvirus in chick embryo fibroblasts. Usually members of the vaccinia subgroup grow to high titer in these cells, but several workers have also found that extremely low yields of rabbitpox virus are sometimes obtained from chick fibroblasts despite the fact that the virus still forms plaques in such fibroblasts.

Composition and Structure

A comprehensive study of the chemical composition of vaccinia virus has been made by Zwartouw (9), and other pertinent references are given in a recent review (5). The weight of a vaccinia virus particle is approximately 5.5×10^{-5} g. It consists of about 5% DNA, 2% lipid, 2% phospholipid, and the rest protein. Trace amounts of copper and riboflavin have been reported. These compounds are usually dismissed as being contaminants, but in view of the complexity of poxvirus one should at least be alert to the possibility that they do have a function in poxvirus replication. Even after purification of the virus by banding in density gradients (10–12) traces of DNAase, alkaline phosphatase, and ATPase can be detected. An ATPase activity has been localized within the virus particle (12a). A DNA-dependent RNA polymerase within the virion is discussed later. The amount of RNA in poxviruses is at most 0.2% (9, 11) and although it too is probably a contaminant, the presence of a low molecular weight RNA molecule cannot be ruled out.

Numerous electron microscopic studies on the structure of poxviruses have been conducted. These include thin sectioning with or without enzymatic digestion (13), negative staining with phosphotungstate (14), or controlled degradation with detergent coupled with electron microscopy (15).

An outline of the typical structure of the virion of the vaccinia, myxoma, or avian subgroups follows. Other subgroups exhibit different surface structure, but the internal structures seem similar for all groups studied.

The surface of most virus particles of a population is densely beaded. Such virions are referred to as M, for mulberry-like particles; the beads are formed by loops of threadlike double helices (14). In this form the virion is not permeable to phosphotungstic acid. But it is permeable to phosphotungstic acid in the so-called C (capsule) form which occurs naturally and to which the M form can be converted experimentally by treatment with enzymes, alkali, or fat solvents.

Within the outer layer or coat is a nucleoid filled with fibers composed of DNA and protein. The nucleoid wall is composed of an outer layer, which in cross section consists of cylindrical pegs about 50×100 Å, and an inner layer 45–50 Å thick which may be smooth or palisaded. The quality of published electron micrographs is not good enough to be sure of fine details. The nucleoid or nucleocapsid is sometimes called the "core" and sandwiched between the nucleoid and the outer coat are lateral bodies. The lateral bodies appear to be composed of a long filament(s) folded and oriented in the direction of the long axis of the particle. Cores with lateral bodies attached can be produced by controlled degradation of the virus with detergent and mercaptoethanol. The lateral bodies are readily digested with trypsin (15). Westwood et al. (14) demonstrate the envelope surrounding the core of the poxvirion. This is not to be confused with a loose, lipoprotein "envelope" that can be seen around some preparations of vaccinia virus (16) and around Orf virus (17). It is not clear what proportion of the virus population possesses this envelope or if it is present if the virions are purified from the cell-associated population. Orf virus also exhibits a distinct criss-cross pattern of threads or tubules on the surface of the virion. These can be clearly seen in the electron micrographs presented by Nagington and Horne (17).

Electron micrographs of a cross section of rabbit skin infected with

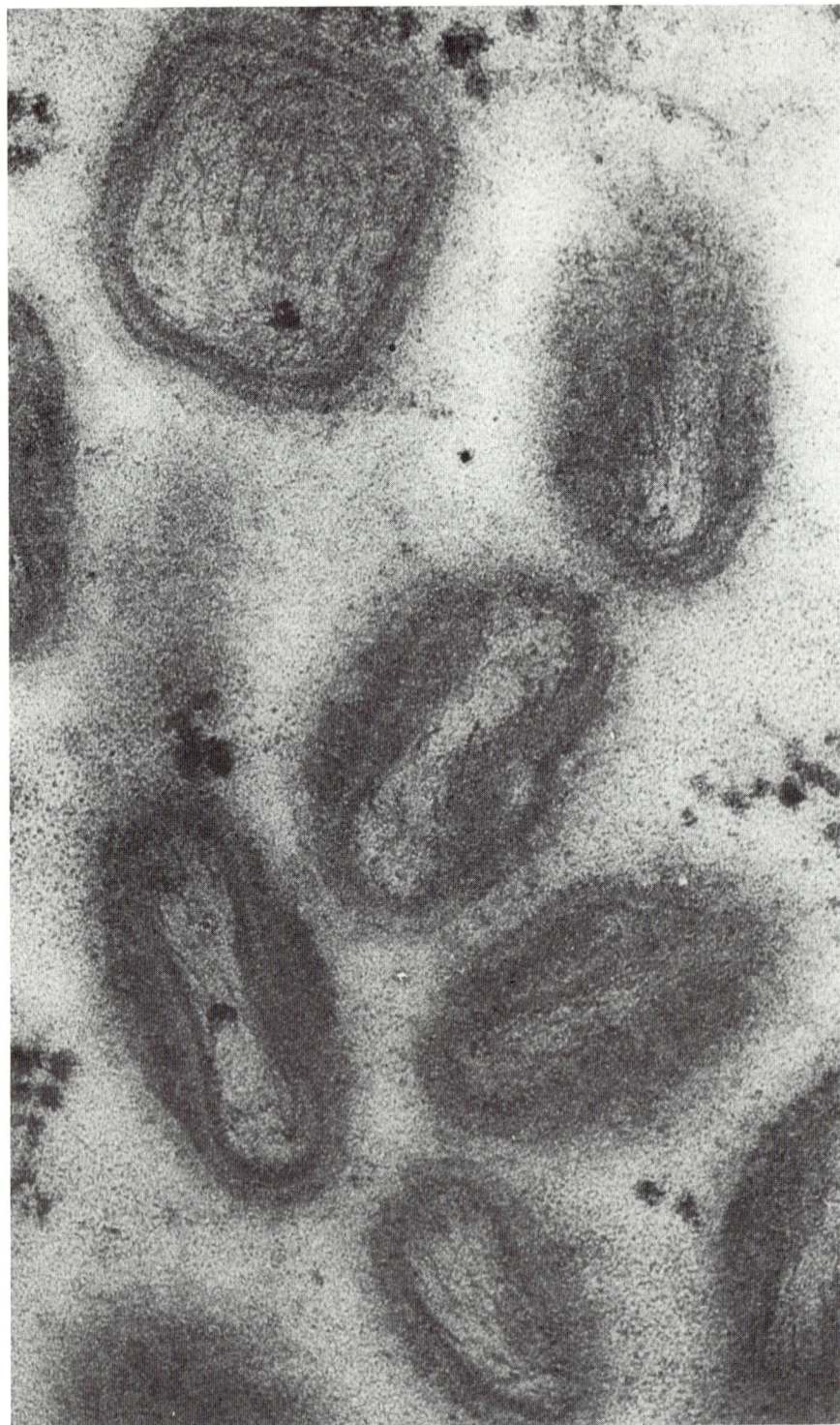

Fig. 9-1. Myxoma virus in rabbit skin. (Photograph supplied by Dr. F. Fenner, Australian National University.)

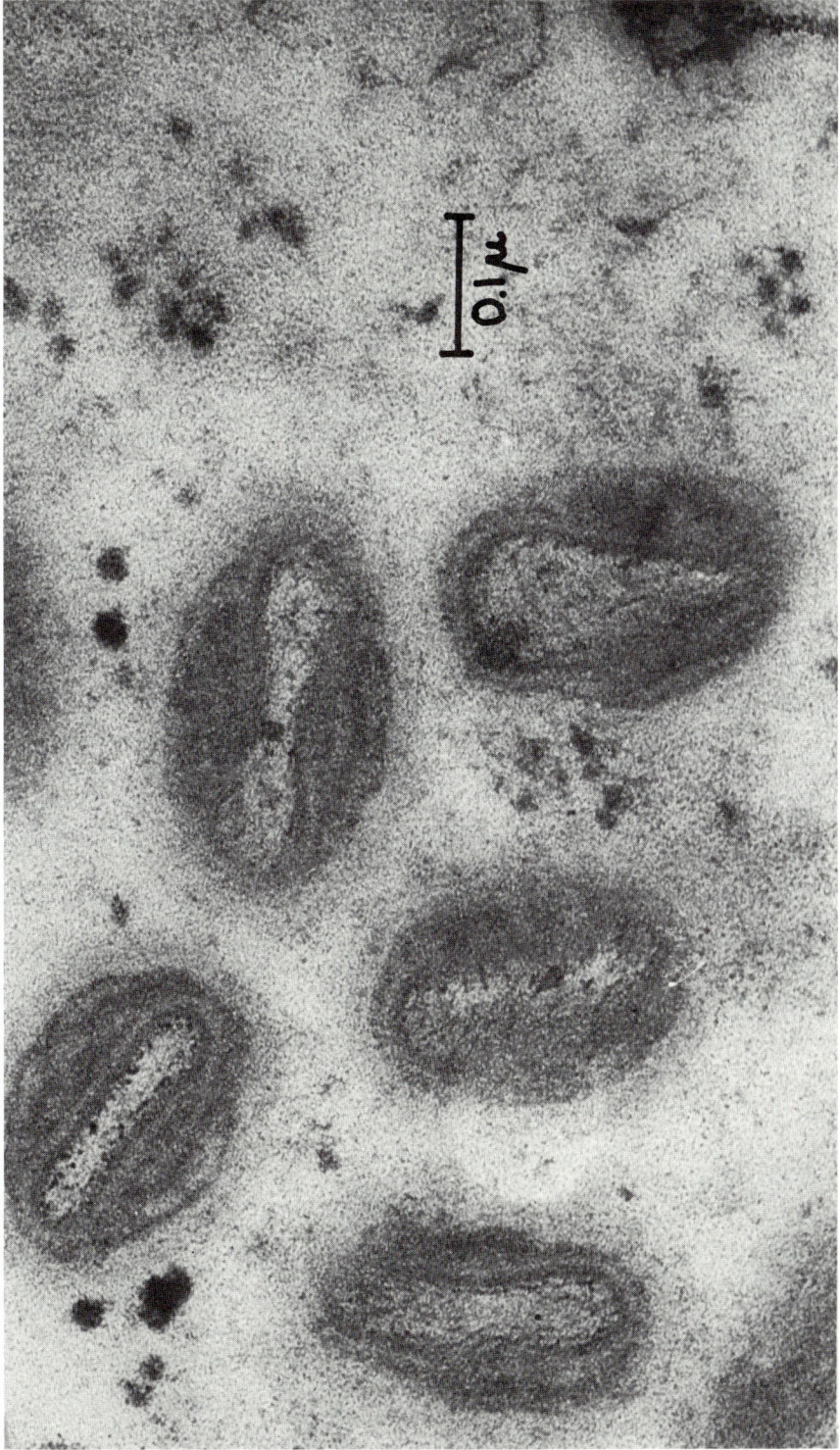

0.1μ

Fig. 9-1. (continued)

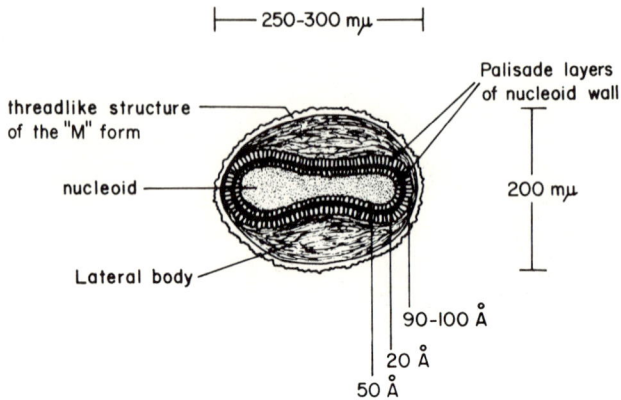

FIG. 9-2. Diagram of the "M" form of a poxvirion in cross section.

myxoma are shown in Fig. 9-1 and demonstrate the presence of the nucleoid.

A highly simplified diagram of the "M" form of a poxvirion is presented in Fig. 9-2.

For electron micrographs of the fine structure of the virion or immature forms of the developing virion, the papers of Westwood Easterbrook et al., or Dales (14, 15, 18) are highly recommended.

9-3. The Initiation of Infection and Uncoating

A convenient procedure for cell infection in biochemical studies is to suspend cells at a concentration of about 5×10^6 cells/ml of complete medium. The virus inoculum is added at a sufficiently high input (5–10 pock-forming units/cell) to infect all cells within a 15–30-min adsorption period. Following adsorption, excess virus is removed by centrifugation and the infected cells are rediluted to a concentration of about 5×10^5 cells/ml of growth medium. Within a few minutes after infection, virions become adsorbed to the outer membranes of the cells. The entire virion is engulfed by invagination of the membrane and transferred into phagocytic vacuoles. Our knowledge of these, and succeeding steps, comes mostly from the electron-microscope studies of Dales and co-workers (18–22), and has been supplemented by radiochemical studies on the fate of isotopically labeled virus (23–25).

Soon after the virus particles appear within phagocytic vacuoles, the outer membrane(s) is degraded and the virion is rapidly reduced to the core state. This coincides with a breakdown of the membrane of the vacuole and the cores appear free in the cytoplasm. The whole process takes about 20 min in L-cells. Compatible with this observation is the finding that a ^{32}P-labeled phospholipid-containing fraction of poxvirus is degraded soon after virus adosorption (23). In fact, the phospholipid of all particles becomes converted to acid soluble form (mainly glycerophosphate).

Using ^{14}C-leucine-labeled virus, Joklik (23) was able to see to what extent virus protein was degraded. Very little of the virus protein was completely degraded but up to 50% of the total protein was cleaved from the virion and appeared in a form still precipitable by trichloracetic acid. The loss of protein commences with imperceptible lag and is 50% completed by 2 hr. This must reflect breakdown of the outer virion coat within the phagosome.

The release of ^{32}P-labeled phospholipid and the degradation of virions to cores appearing in the cytoplasm is a constitutive process since it is unaffected by pretreating cells with inhibitors of protein synthesis (23). Heat-treated virus, for the most part, does not get out of the phagocytic vacuole where it is extensively degraded, although a very small proportion either does get out or survives in the vacuole in view of the phenomenon of the reactivability of heated virus.

Once in the cytoplasm the DNA of the cores gradually becomes susceptible to degradation by exogenous deoxyribonuclease. It is this process that has been termed "uncoating" (24, 25). In a series of elegant experiments that provided a convenient approach to the study of the early events of virus replication, Joklik (23) followed the kinetics of uncoating by determining the extent to which input viral DNA, labeled with ^{32}P, was susceptible to DNAase. Only 50–60% of the input virions ever become uncoated despite the fact that 100% of the input virus undergoes degradation of the phospholipid- containing layer. In cells inhibited with flurophenylalanine, which blocks protein synthesis, or with high concentrations of actinomycin D, which blocks RNA synthesis, uncoating does not take place. Cells arrested in metaphase or cells irradiated with ultraviolet light prior to infection do not exhibit appreciable capacity to uncoat virus. Such results led Joklik (25) to propose a model to explain uncoating. Figure 9-3 is a diagram of the Joklik uncoating hypothesis which is as follows:

After the intact virus enters the cells, a constitutive process breaks down the outer coat of the virus and releases an "inducer" protein (X).

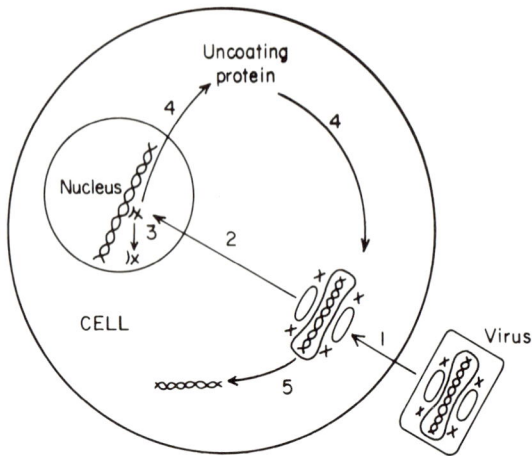

Fig. 9-3. An illustration of the Joklik uncoating hypothesis. For a description, see text. [Figure redrawn from W. K. Joklik, *J. Mol. Biol.*, **8**, 277 (1964) with permission of Academic Press Inc. (London).]

The virus "inducer" protein then derepresses the host gene(s) coding for uncoating protein (Steps 2 and 3). The uncoating protein is synthesized in the cytoplasm via messenger RNA (Step 4). Once formed the uncoating protein degrades the viral cores and releases naked viral DNA (Step 5). The main points of this hypothesis, which accounted for the facts known at that time, were that the host genome codes for an uncoating protein and that the viral genome in the core stage is unable to code for any functions.

This model could be applied to explain the phenomenon of the reactivability of poxvirus (*26*). Berry and Dedrick (*27*) noted that rabbits inoculated with a mixture of heat-inactivated virulent myxoma virus and infectious benign fibroma virus were killed and the active myxoma virus could be isolated from the animals. They considered this to be due to "transformation" of the fibroma virus into a myxoma virus by a component of the heated myxoma virus preparation. This phenomenon has since been demonstrated for other members of the poxvirus group. When a cell is mixedly infected with a heated (55°C for 120 min or 60°C for 12 min) poxvirus and another live poxvirus, the cell produces virus of both types. Reactivation was shown to be nongenetic since virus whose nucleic acid is selectively inactivated also has the capacity to reactivate heated virus. According to Joklik (*25*) heating denatures the inducer for uncoating protein and this inducer is provided by nonheated virus. Prior to this hypothesis, Abel (28) reported that she could treat heated (noninfectious but reactivable) vaccinia virus, *in vitro*, with extracts from cells infected with live virus and so produce an infectious "subviral entity." This subviral entity was susceptible to DNAase and the implication was that the extract contained a protein (uncoating enzyme) that removed the denatured protein from heated virus to yield infectious poxvirus DNA. A dramatic experiment following this was to infect *Bacillus subtilis* with infectious poxvirus DNA, produced *in vitro* by uncoating enzyme, and to report the replication of poxvirus in the bacterium (29).

The kinetics of the production of "uncoating enzyme" (28) show that it roughly follows the uncoating of virus particles as described by Joklik (24). About 4 hr postinfection, increase in uncoating enzyme was terminated and between 6 and 8 hr, when progeny virus appears, the uncoating enzyme was inactivated. These results pointed to an elaborate control of uncoating enzyme which conincides perfectly with the course of virus replication. Extracts of uninfected cells or cells infected with heated virus showed no reactivating capacity.

In collaboration with Abel, we attempted to confirm the *in vitro* uncoating experiments. No significant uncoating activity could be shown in extracts prepared exactly according to the Abel procedures or by numerous modifications of such procedures. In later work with highly radioactive virus, Degnen and McAuslan (29a) were unable to show any significant breadown of the protein coat or exposure of DNA by extracts of uninfected or infected cells. Furthermore, radioactive virus cores were remarkably stable in such "uncoating" extracts. A major problem encountered in uncoating experiments is the assay for "subviral entities"; the presence of some infectious particles in the heat-treated virus and in the uncoating extract gives a large background. Since the reported efficiency of reactivation is between 10^{-3} and 10^{-4} it was impossible to see significant increase above the background values. In the original report decoating activity was only described (28) as per cent of maximal activity. From such data one cannot calculate the actual number of plaque-forming units produced above background. Attempts to demonstrate the postulated uncoating enzyme should be repeated. It is difficult to control all parameters when assaying for an enzyme elicited in cultured animal cells and it is conceivable that we have not controlled conditions well enough to assay for uncoating enzyme.

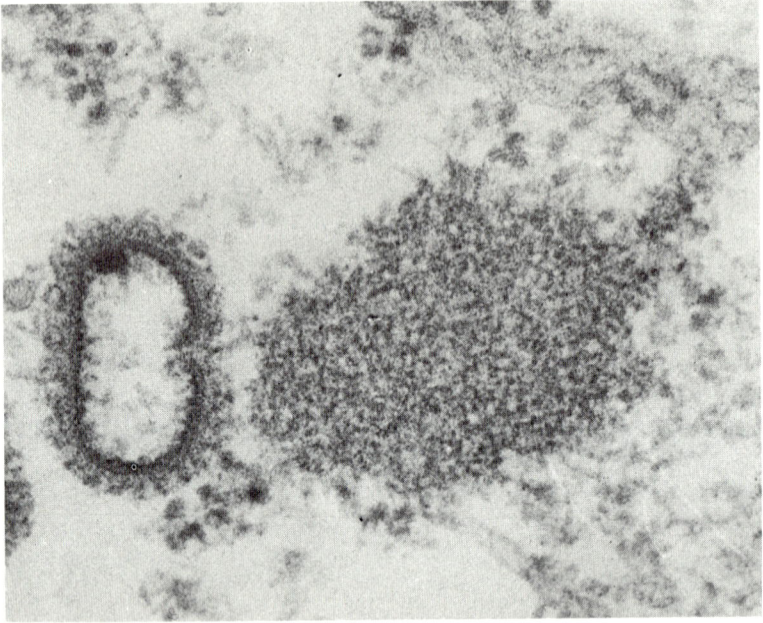

FIG. 9-4. Electron micrograph of a poxvirus core showing rupture of the wall and discharge of the nucleoid contents. Magnification ×130,000. [Photograph by courtesy of Dr. S. Dales, Rockefeller University. Reprinted from *Proc. Natl. Acad. Sci. U.S.*, **54**, 462 (1965), with permission of the publishers.]

Dales (*22*) has since prepared electron micrographs of infected cells showing that *in vivo*, release of viral DNA from the core does not involve complete digestion of the nucleoid wall. Cores appear to rupture and discharge their contents into the cytoplasmic milieu while the wall of the core remains undegraded throughout the rest of the infection cycle. An electron micrograph showing discharge of core contents is shown in Fig. 9-4.

If there is a specific uncoating protein it does not need to digest the outer wall of the core by say, extensive proteolysis, but needs merely to weaken the wall in one section to facilitate rupture. The nature of the protein requirement for uncoating remains obscure, and one cannot say if a specific protein is needed or if it is just the act of protein synthesis that provided favorable conditions for rupture of the core and release of its contents.

Further clues to the uncoating process have come from studies of the effect of various inhibitors on the uncoating of the genome and its expression. Dales (*18*) pointed out that at low levels of actinomycin D (1 μg/ml) the synthesis of RNA in infected L-cells is reduced by over 90%, but penetration and release of material from the cores progresses normally. Further, at this concentratiton of antibiotic the induction in L-cells of a viral enzyme, thymidine kinase, was reduced only to one-half the normal level (*30*). That is to say, even though host function was almost abolished, the virus is uncoated and a viral function was expressed. This observation was largely ignored because one could not determine, from electron microscopy, what proportion of the virions was uncoated under such conditions and because it could be argued that the remaining 10% of host RNA synthesis could be responsible for formation of uncoating protein.

A more quantitative analysis of the effect of inhibitors on the formation of RNA in infected (vaccinia strain IHD) L-cells was undertaken by Munyon and Kit (*31*). They found that even though host RNA synthesis was inhibited 99% by actinomycin D, poxvirus infection stimulated incorporation of uridine into cytoplasmic RNA. More importantly they found that pretreatment of cells with cycloheximide, which completely inhibits protein synthesis, did not prevent the stimulation of uridine incorporation by vaccinia infection. Unfortunately they did not investigate the nature of the mRNA formed in the inhibited system but suggested that it might be poxvirus messenger RNA.

The Joklik uncoating hypothesis was based on the assumption that viral genes are not expressed while the virus is at the core stage. The work with the inhibited L-cell system, described above, raised some doubt about this. In addition, arguments that a host gene codes for a decoating protein were questionable (*32*) and so a reinvestigation of the entire uncoating phenomenon was undertaken (*32*). In agreement with the work of Joklik (*23*), we found that inhibitors of protein synthesis block uncoating of the virus and from the studies of Dales (*22*) concluded that virus was arrested at the core stage. On the other hand, in contrast to Joklik's work, it was found that ultraviolet irradiation of the virus only partially inhibited uncoating. This is in accord with the results of Dales and co-workers (*20, 21*). Cells were infected in the presence of actidione to completely inhibit protein synthesis and to allow the accumulation of cores. Under this condition, synthesis of RNA in the cell cytoplasm took place and the RNA was shown, by specific hybridization with poxvirus DNA, to be viral messenger RNA. Therefore, the viral genome can express itself while in the core stage, and therefore it is unnecessary to postulate derepression of a host function in order to make an uncoating protein. Similar conclusions were reached by Woodson (*33*). If there is a specific uncoating protein, it could be coded by the viral genome. One could argue that even though the viral genome is inaccessible to DNAase (i.e., coated) there could be a small region of the genome, resistant to pancreatic DNAase by virtue of say, a high purine content, that is exposed and uncoating protein is synthesized from this. At this stage of the investigation such elaborate arguments do not seem worth entertaining but certainly can't be dismissed. From consideration of the effects of interferon on mRNA synthesis in poxvirus infected cells we suggested that interferon blocks the uncoating process (*32*). This was subsequently proved by experiment (*94*).

Given that poxvirus-specific RNA is synthesized by cores under conditions where protein synthesis is arrested prior to infection, there remained the problem of the DNA-dependent RNA polymerase. The obvious alternatives were that the virus uses a pre-existing host enzyme or that the RNA polymerase is an integral part of the virion. A DNA-dependent RNA polymerase within purified poxvirus virions or in purified poxvirus cores has now been demonstrated (*34*).

In view of all biochemical and electron-microscope work, a tentative proposal for the initiation of poxvirus infection is a follows: The poxvirus particle possesses a DNA-dependent RNA polymerase closely associated with the viral DNA. Once the virion is taken into a phagosome, the outer coat is degraded by constitutive host enzyme(s) to produce the core. With the entry of the core into the cytoplasm, the polymerase of the core utilizes pre-existing nucleoside triphosphates to synthesize a class of early poxvirus messenger RNA. If there is a specific uncoating protein this is synthesized under the direction of poxvirus messenger RNA. After uncoating an additional class of viral genes (including, for example, that for DNA polymerase) becomes available for transcription. The failure of heated virus to replicate can now be explained as being due in part to denaturation of the RNA-polymerase.

9-4. Poxvirus DNA and Its Replication

A. COMPOSITION

There is little information on the properties of the DNA of any animal virus compared to what is known about bacteriophage DNA

(35). This situation is in part due to the role of bacterial viruses in the study of molecular genetics and from the ease of extracting bacteriophage DNA from its protein coat. Extraction and examination of the nucleic acid from highly purified preparations of virus has been undertaken by Joklik (36) and Randall (37). These carefully conducted studies establish the guanosine plus cytosine content of cowpox DNA to be 36% and that of fowlpox to be 35%. No unusual purine or pyrimidines in poxvirus DNA have been reported but in all probability there has been no diligent search for very small amounts of such bases.

By the criteria of hyperchromic shift upon heating DNA solutions, base composition and resistance to degradation by nuclease specific for single-stranded DNA, the bulk of the DNA extracted from poxvirus is double stranded. Although Pfau and McCrea (38) have reported the isolation of single-stranded DNA from poxvirus, this DNA represents only about 5% of the total DNA and furthermore this single-stranded DNA was obtained only from a fraction of particles having a lower density and specific infectivity than the rest of the particles.

It is highly unlikely that any single-stranded DNA is enclosed within poxvirus particles, and in the absence of any information on the composition and degree of homology with double-stranded poxvirus DNA speculation on this single-stranded DNA does not seem warranted.

The problems of gently releasing the viral DNA from its intimate association with protein has made it difficult to establish the form and molecular weight of the nucleic acid. Sodium dodecyl sulfate of phenol extraction alone yields little free nucleic acid. To prepare DNA in any quantity the currently favored method is to first extract the virus with alcohol–ether to remove lipids and then to treat it with 3% sodium dodecyl sulfate in the presence of high concentrations of mercaptoethanol. The latter cleaves dithiol bonds and leads to considerable disorganization of the protein structure. This allows more effective digestion by proteolytic enzymes such as pronase and following enzymatic digestion, the preparation is deproteinized by chloroform–butanol and phenol (36, 39).

Recently Easterbrook (15) has introduced the use of a nonionic detergent (Shell Nonidet P_{40}) together with mercaptoethanol for the controlled degradation of the virion. Together with electron microscopic studies this has provided valuable information on the substructure of the virion. Extended Nonidet treatment does lead to extensive liberation of DNA strands and should be pursued further as a tool for isolation of both DNA and viral proteins in quantity.

Despite precautions to minimize shear in the extraction of DNA, the molecular weight of poxvirus nucleic acid was determined by velocity sedimentation to be 80×10^6. On the other hand, the molecular weight calculated from the particle weight and DNA content is about 160×10^6 (36). This raises the question of whether the poxvirion contains one or two molecules. Of a large number of bacterial viruses studied, only a single DNA molecule per virion has been found and this has led to the expectation that virions of all species will contain a single molecule. If one assumes that the DNA of poxvirus is in B form with a weight-to-length ratio of 2×10^6 daltons/μ (40) the estimates of 160×10^6 would correspond to an over-all length of about 80 μ per virion. By gentle extraction of the Mill Hill strain of vaccinia with Nonidet P_{40} and mercaptoethanol, Easterbrook (41) has been able to demonstrate the presence of two-ended molecules up to 83 μ long. It seems likely that poxvirions do contain a single DNA molecule and that in the normal course of extraction these molecules are usually broken when released from associated proteins. Fowlpox is unusual in that its DNA is readily liberated from lipoprotein coats with sodium lauryl sulfate (42). Measurement of the contour length of DNA molecules liberated in this way, together with sedimentation measurements, shows the molecular weight of fowlpox DNA to be in the range of 200–240×10^6 daltons (43). Fowlpox DNA is, therefore, the largest viral DNA molecule isolated.

We know virtually nothing about the physical state of DNA during infection and it might be informative to conduct studies comparable to those of Frankel (44) on the physical state of replicating DNA of T-phages. In these studies, Frankel showed that the replicating pool of phage DNA exists in a form with a sedimentation coefficient many times greater than that of DNA in the completed virion.

B. DNA REPLICATION

Poxvirus DNA replicates extensively in the cell cytoplasm and one of the early effects of infection is the inhibition of host DNA synthesis. This has been clearly shown by autoradiography of infected monolayer cells (45–47). This technique also showed that viral DNA synthesis can be initiated in most cells of a population regardless of whether or not the cell is making DNA at the time of infection (48). The amount of viral DNA made is at most 50% of that of un-infected cells; the GC content (36%) is not greatly different from that of mamalian cells (about 42%) and no unusual bases have been shown. Until recently this made it difficult to follow DNA replication by chemical or physical methods. A significant advance in the method-ology for the study of DNA virus replication was Salzman's (49) imaginative use of the thymidine analogue, 5'-fluorodeoxyuridine (FUdR). Although familiar to the poxvirus cognoscenti, it seems worthwhile to briefly outline the principle of Salzman's procedure be-cause of its use in the study of other DNA viruses.

The analog FUdR is a potent and selective inhibitor of thymidylate synthetase and as a consequence of this action, blocks DNA synthesis. In Hela cells it causes a complete suppression of DNA synthesis and an immediate block in the incorporation of uridine-2-^{14}C into thy-midine. Since addition of fluorodeoxyuridine to HeLa cells at the time of infection completely suppresses viral replication, it follows that the pool of free thymidine or thymidylate available for DNA synthesis must be negligible.

Salzman added inhibitor at various times after infection and compared this time with the yield of virus obtained. Since any virus formed must have contained DNA synthesized prior to the time of addition of inhibitor, the yield of virus after 24 hr in the inhibited culture was a measure of the quantity of viral DNA synthesized prior to the time of fluorodeoxyuridine addition.

Although this technique gives the approximate time of initiation of DNA synthesis it does not necessarily give the time that synthesis is completed; it gives only the time at which an amount of DNA is formed sufficient to provide a full yield of virus. A much more quanti-tative and informative technique, now possible with the introduction of convenient techniques to fractionate cells into nucleus and cyto-plasm (50), is to follow the incorporation of radioactive thymidine into cytoplasmic (viral) DNA. The time course of incorporation of label into viral DNA is compared with the appearance of viable rab-bitpox in Fig. 9-5. In general, studies using this technique confirm Salzman's conclusions concerning the time of viral DNA synthesis.

Despite the convenience with which one can follow thymidine in-corporation into the cytoplasmic fraction of cells, it has been difficult to clearly show the kinetics of inhibition of host nucleic acid syn-thesis. This is because a significant fraction of viral DNA remains

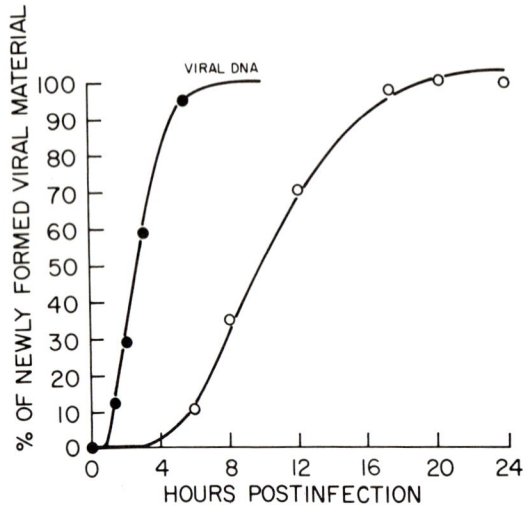

FIG. 9-5. Typical time course of viral DNA synthesis and appearance of viable rabbitpox in infected HeLa cell spinner cultures. Closed circles give the course of DNA synthesis and open circles the appearance of viable virus.

associated with nuclei after disruption and fractionation of cells. This problem has been recently overcome by the introduction of a technique, based on the selective renaturation of viral DNA, to increase the density difference between viral DNA and host DNA *(50a)*.

An important feature of the poxvirus system is that viral DNA synthesis precedes the appearance of progeny virus by about 3 hr and in fact is virtually completed by the time any significant yield of virus is formed. This contrasts with the sequence of events in the replication of FV 3 (another cytoplasmic replicating DNA virus) in which DNA synthesis commences shortly before the appearance of viral progeny and continues to be synthesized through the course of infection *(58)*.

Other compounds influencing DNA replication are bromouracil deoxyribose and mitomycin C.

1. Bromouracil Deoxyribose (BUdR)

In contrast to the action of fluorodeoxyuridine, BUdR permits DNA replication and is incorporated into poxvirus DNA. Easterbrook and Davern *(51)* showed that BUdR addition to infected cells could virtually abolish the infectivity of vaccinia virus progeny. Under the conditions used, there was 13% replacement of thymine by 5-bromouracil as determined from density determinations. Iododeoxyuridine can also be incorporated into vaccinia DNA (52) to the extent of 18% replacement of thymine by 5-iodouracil. The iodouracil-containing DNA either readily fragments during extraction from poxvirus or is packaged as fragments. BUdR-containing virus has an atypical morphology; such a virus cannot induce thymidine kinase, it does not reactivate nor is it reactivatable. On being taken up by the cell, BUdR-containing virions are prematurely uncoated, and their DNA is rapidly broken down as soon as it is uncoated (53).

2. Mitomycin C

This antibiotic in its reduced state can crosslink complementary DNA strands (54). DNA synthesis can be preferentially inhibited by mitomycin in both bacterial and animal systems. However, inhibition is not absolutely specific. High concentrations of mitomycin C will block RNA synthesis. If there is any degradation of DNA in mitomycin-treated animal cells, it takes place slowly and occurs late in the sequence of events. In general, viral DNA synthesis is remarkable in that mitomycin C does not inhibit it. This has been shown for T-even and T-odd bacteriophages (55), for pseudorabies virus (57), for FV 3 virus (58), and for lambda phage (56, 59). On the other hand, Reich and Franklin (60) noted that mitomycin C prevents infectious vaccinia virus formation. They reported that vaccinia DNA was not produced in the inhibited system and suggested that the block in DNA synthesis was due to mitomycin inhibition of uncoating of the input templates. Kajioka et al. (20) found that if mitomycin C was added to HeLa cells at the time of infection, viral DNA synthesis was "markedly depressed" but not abolished. They attributed the low rate of DNA synthesis to a delay in the action of mitomycin C.

Kit et al. (61) demonstrated that at a level of 15 μg mitomycin C/ml, the replication of total DNA in vaccinia infected cells was reduced to 5% of that in uninfected cells and the authors assume that viral DNA synthesis is effectively blocked. There are objections to all of these experiments. Reich and Franklin measured DNA synthesis at a time (6–8 hr postinfection) when one would expect viral DNA replication to be virtually completed so one is not sure if uncoating or DNA replication per se is blocked. The results of Kajioka et al., could be interpreted as showing either that uncoating or DNA synthesis is blocked. In Kit's experiments, DNA synthesis was measured very early (1–2 hr postinfection) and in the absence of appropriate experiments one is not sure if viral DNA synthesis in controls was proceeding maximally at this time. These experiments warranted reinvestigation because they implied that poxvirus DNA replication is really distinct from that of other DNA viruses in being sensitive to mitomycin C. This point has recently been investigated. In our experience, after pretreatment of cells with mitomycin C at levels that markedly depress host DNA synthesis, rabbitpox virus DNA synthesis takes place in the continuing presence of mitomycin C and is depressed only by 30% at most. At higher levels of mitomycin C (over 25 μg/ml) the action of the antibiotic is no longer selective and any inhibition of DNA synthesis could be a secondary effect of blocking RNA and protein synthesis. It may be that different poxviruses have different sensitivities to mitomycin but at present one should be cautious of accepting that poxvirus DNA synthesis is blocked by mitomycin C.

C. CONTROL OF DNA SYNTHESIS

The cessation of poxvirus DNA synthesis occurs at a time when the DNA that has just been replicated is still susceptible to DNAase (62). Therefore, synthesis is not arrested because DNA becomes encased within a protein coat nor is it arrested as a result of precursor depletion since host DNA synthesis, although markedly depressed, is not completely inhibited. It seems then that the process of viral DNA synthesis is controlled.

Synthesis of protein, other than that of any of the known enzymes of DNA synthesis, is required for the initiation of new rounds of bacterial and mamalian cell DNA replication *in vivo* (64).

For poxvirus DNA replication, protein synthesis is required contemporaneously with viral DNA synthesis since addition of cycloheximide or puromycin, both potent inhibitors of protein synthesis, at any time during DNA replication abruptly inhibits its synthesis. However, if viral DNA synthesis is blocked with fluorodeoxyuridine at the start of infection, the protein(s) essential for its synthesis accumulates. Thus, after removing FUdR, DNA synthesis will occur even if puromycin is then added (65).

The protein requirement must be an early viral function. It is a relatively stable protein since if allowed to accumulate for a limited time in the presence of fluorodeoxyuridine, it will support DNA synthesis where the thymidine analog is removed 4 hr later, and further protein synthesis inhibited.

It has also been possible to show that actinomycin D inhibits DNA synthesis indirectly by blocking the synthesis of a short-lived (45 min) messenger RNA for the protein in question. In contrast, the messenger RNA for two virus-elicited enzymes (thymidine kinase and polymerase) are stable (66, 67). The protein requirement for viral DNA replication thus differs from the known poxvirus-induced enzymes that might participate in DNA replication in the stability of its messenger and its apparently noncatalytic relation to DNA synthesis.

Ben-Porat and Kaplan (68) have shown that the enzyme system catalyzing the incorporation of thymidine into DNA in cellular extracts of pseudorabies-infected cells was stable after addition of puromycin; DNA synthesis *in vivo* was rapidly inhibited by the drug. These results show that for a DNA virus that replicates in the cytoplasm and for a DNA virus replicating in the nucleus, there is in each case a requirement for a protein distinct from known early enzymes; the requirement appears to function in a stoichiometric manner in the synthesis of viral nucleic acid. For convenience the protein requirement is referred to as replicator protein. In contrast to the poxvirus system, the replicator protein does not accumulate to any significant extent when DNA synthesis is prevented by FUdR in the pseudorabies system (68a) or in the FV 3 system (69). The poxvirus-infected cells thus assumes importance as a convenient system for studying this replicator protein.

The stoichiometric relationship between the protein requirement and DNA replication suggests that the protein(s) involved is a nonenzymatic structural protein of the viral chromosome or that it is somehow associated with some process of DNA replication other than the chemical aspects as we currently understand them. Of course an enzymatic function cannot be ruled out since the stoichiometric relation could, for example, result from the protein becoming attached to the chromosome, thus being unavailable for transfer for a second round of replication. By this mechanism, new enzyme would be required for initiation of each round of chromosome replication.

D. SITE OF DNA REPLICATION

Examination of infected cells by autoradiography shows that tritiated thymidine is incorporated into foci of DNA synthesis rather than scattered over the cell cytoplasm (70). These foci are sites where virus structural proteins can be demonstrated by fluorescent antibody staining (46), and this observation led to the idea that "factories" are set up in which the viral DNA replicates and the virus matures. Such factories have been most strikingly demonstrated by Randall and co-workers (71) using fowlpox-infected tissues. Fowlpox inclusions are clearly bounded by a limiting membrane and within these inclusions virus in all stages of maturity can be seen. The inclusions are stable structures and may be isolated during the infection cycle (72). There are, in fact, two types of inclusion bodies in cells infected with a number of poxvirus strains. These are referred to as A type or Marchal bodies and B type which is the site of DNA replication. The A type does not appear to be directly related to virus replication. They appear late in the course of virus replication and some strains of vaccinia virus do not produce A-type inclusions at all (70). The A-type inclusions have been isolated from ectromelia-infected cells (73) and shown to be predominantly protein with some lipid. The protein is unrelated to the NP antigen, soluble antigens, or hemagglutinin.

In a series of experiments designed to follow the fate of replicating viral DNA, Joklik and Becker (62) demonstrated that newly replicated viral DNA exists in the form of large aggregates, the size of which increases considerably between 2 and 4 after infection.

The aggregates do not appear to be bound by a membrane because of their stability to sodium deoxycholate but they are reversibly dissociated upon removal or addition of magnesium ions. Joklik and Becker (62) conclude that the aggregates are probably the "factories" observed by autoradiographic and electron microscopic studies.

In vitro DNA synthesis by these aggregates has been claimed (67) but attempts to confirm this in this author's laboratory suggest that the degree of incorporation of deoxynucleoside triphosphate cannot be considered significant. Now that it is possible to prepare extracts containing excess of the replicator protein(s) (65) it seems worthwhile re-attempting the demonstration of *in vitro* DNA synthesis by incorporating such protein(s) into the reaction system.

9-5. Expression of the Virus Genome

A. ENZYMES ELICITED BY POXVIRUS INFECTION

A search for poxvirus-elicited enzymes began at a time when little information was available on the intracellular distribution of thymidine-phosphorylating enzymes or DNA polymerase. These host enzymes were expected to be predominantly, if not exclusively, in the nucleus. The T-even bacteriophages were known to induce deoxynucleotide kinases and an obvious question was whether poxvirus induced enzymes to initiate viral DNA synthesis in the cytoplasm.

The enzyme activities augmented by T-even bacteriophages fall into two classes: (1) those not found in uninfected host cells but necessary for some phage function such as synthesis of a novel viral component, e.g., deoxycytidylate hydroxymethylase and (2) those duplicating activities of the type found in uninfected cells but possessing different physico-chemical properties to the pre-infection enzymes, e.g., thymidylate synthetase. [For a list of phage-induced enzymes, see Luria and Darnell (74).]

Enzymes of class 1 would not be expected for poxvirus although a novel poxvirus-induced nuclease recently found might fit this classification. So far, only seven poxvirus-induced enzymes have been reported and postinfection increases in these of usually 5–20-fold are routinely observed. The enzymes include thymidine kinase (75), DNA polymerase (67), and three deoxyribonucleases (67, 76, 77). Despite intensive searches no significant postinfection increases in the deoxynucleotide kinases, thymidylate synthetase, or other deoxynucleoside kinases have been found.

All the induced activities exhibit properties differing from those of the corresponding pre-infection activities of the host cell and are probably coded by the viral genome. The arguments for coding of the induced enzymes have been presented several times (5, 66). They are based on comparison of properties of pre- and postinfection activities, the detection of virus mutants unable to induce enzyme, and the fact that different DNA viruses induce the one enzyme activity having different properties for each virus (78). Such evidence has accumulated only for thymidine kinase. It can, of course, be argued that any virus-induced enzyme represents a derepressed or activated host-coded isozyme. A strong counter-argument is that Herpes virus induces a thymidine kinase that differs from both host cell and poxvirus-induced enzymes (66, 78). It is unlikely that there are three genes in the host genomes, one of which is derepressed by herpes and one by pox. One would still like to have evidence to show that mutations in the viral genome can be correlated with alterations in the properties of the induced enzymes as has been shown for phage enzymes. Given that bacteriophage can code for enzymes and considering the circumstantial evidence for the poxvirus system, it is more profitable in terms of planning other experiments to accept tentatively that poxvirus codes for thymidine kinase.

The only evidence cited against thymidine kinase being virus coded is Joklik's (5) argument that since uv-irradiated virus can elicity thymidine kinase (75) the enzyme cannot be coded by the viral genome; this was based on his finding that uv-irradiated poxvirus was poorly uncoated. It has since been shown (32) that the messenger for induced kinase synthesis is, in fact, made prior to the liberation of viral DNA from its protective protein coat (i.e., before the DNA is susceptible to DNAase).

An experiment demonstrating this was as follows. Cells were infected with virus in the presence of cycloheximide to prevent uncoating. After allowing time for stable messengers to accumulate, actinomycin D was added to block further messenger synthesis. Upon subsequent removal of cycloheximide, thymidine kinase synthesis commenced despite the presence of actinomycin D. Therefore, the kinase was made from messenger accumulated prior to uncoating and presumably from messenger specified by viral genes.

B. Properties and Function of the Induced Enzymes

1. Thymidine Kinase

Ten- to twenty-fold increases in this enzyme have been demonstrated with rabbitpox- or cowpox-infected HeLa cells. Using crude extracts, differences in Michaelis constants and temperature stability as compared to those of the pre-infection enzyme were demonstrated (66). These experiments were repeated with slightly less crude enzyme (5-fold purified) from vaccinia-infected cells with essentially the same results (78). More recently the enzyme has been purified over 300-fold from infected HeLa cells (78a) and shown to be reversibly inactivated by removal of some component present in uninfected HeLa cells. The inactivation or reactivation is accompanied by changes in sedimentation coefficient of the enzyme (*78b*). The enzyme is feedback inhibited by its distal product, deoxythymidine triphosphate; inhibition is reversed *in vitro* at high ATP concentrations. These properties of thymidine kinase indicate that it is a complex enzyme of the allosteric type.

Poxvirus mutants incapable of inducing thymidine kinase (TK⁻ mutants) can be isolated (*79*). When cells are mixedly infected with TK⁺ and TK⁻ mutants the amount of enzyme synthesized is dependent on the ratio of the two input multiplicities, not on the absolute multiplicity of either (*80*). The changes in sedimentation coefficient described above suggest that kinase is composed of subunits; competition for enzyme subunits could explain the observed results with mixed TK⁺ and TK⁻ infection. The possibility should not be overlooked that a subunit bearing the active site is coded by the host genome and another subunit activating the former is coded by the viral genome. TK⁻ mutants replicate well in cultured cells and TK⁻ cultured cells (*79*). Since kinase induction is gratuitous in these artificial systems, one can only suggest that if the viral genome does code for thymidine kinase there must be some survival advantage associated with possession of this gene. Certainly the induced kinase is a highly effective scavenger for small amounts of thymidine as shown in studies (*81*) of the reversal of FUdR inhibition with suboptimal amounts of exogenous thymidine; the rate of synthesis of viral DNA was maximal at thymidine levels that effectively controlled the rate of cellular DNA synthesis.

2. *Deoxyribonucleases*

The three poxvirus-induced nucleases are classified according to their substrate preference and pH optimum as "alkaline" (native DNA), "acid" (denatured DNA), and "neutral" (denatured DNA) nucleases. All viruses of the vaccinia group induce these three nucleases. Previous reports of failure to detect some of these nucleases in different systems (*77*) can be ascribed to the lability of the "alkaline" DNAase and the binding of the "acid" DNAase to a particulate fraction in the cell accompanied by a reduction in its measurable activity.

The "alkaline" DNAase acts exonucleolytically on native DNA at about pH 9 (*82a*). The corresponding pre-infection enzyme shows a preference for single-stranded DNA. "Neutral" DNAase acts optimally with heat-denatured DNA at pH 7.5; no studies on its mode

TABLE 9-2
Some Properties of "Acid" DNAase from
Normal and Infected HeLa Cells

Property	Virus-induced enzyme	HeLa enzyme
pH optimum	4–5	4–5
Substrate	Melted DNA	Native or melted DNA
Km (melted DNA)	7.7×10^{-6} g	5×10^{-6} g
Inhibition by sRNA	Completely inhibited	Partially inhibited depending on the physical state of the enzyme
Action	Exonucleolytic	Endonucleolytic
Response to SO_4^{2-} ions	Unaffected	Completely inhibited
Mg^{2+} requirement	None	None
Sedimentation coefficient, $SW_{20°}$	5.8	2.7

of action have been reported. The "acid" DNAase represents an activity that is not detectable at all in uninfected cells. It acts preferentially on heat-denatured DNA at about pH 4.5–5 and shows negligible activity towards native DNA. The enzyme has been purified over 3000-fold and its properties contrasted (Table 9-2) with those of an "acid" deoxyribonuclease purified from uninfected HeLa cells (82, 83).

In studying DNAases, reducing agents such as mercaptoethanol, used routinely in many enzyme assays, have been intentionally omitted. This is to avoid the possibility of artificially producing different molecular species of the one enzyme having different kinetic properties. This phenomenon occurs with the host "acid" endonuclease (82) and provides a good example of just how careful one must be before demonstration of a new viral induced enzyme can be claimed.

The wide variety of DNA viruses that induce deoxyribonucleases suggests a functional role for this activity. For example, the bacteriophages T2 and T4 induce oligonucleotide diesterase (84), an exonuclease closely associated with a polymerase activity (85), and an endonuclease acting on denatured DNA (86). Bacteriophage T5 induces a nuclease acting both endo- and exonucleolytically on both native and denatured DNA (87). Fifty-fold increases in a DNAase that attacks denatured DNA are found after infection of B. subtilis with phage SP3 (88). In addition to the nucleases described for poxvirus, there is evidence for the induction of DNAase by polyhedral virus infection of the pupae and larvae of Bombyx mori (89, 90); two "alkaline" DNAase activities are induced in herpes virus-infected

HeLa cells (77). Unsuccessful preliminary surveys have been made for adenovirus-induced nucleases. It may be that unusual assay conditions are required to demonstrate any such nucleases.

The plethora of both host- and virus-induced nucleases makes it difficult to suggest just how these nucleases can be used by the viruses. A possible role for nucleases in phage infection could be to provide precursors for viral DNA synthesis. Host DNA is degraded after T2, T4, or T5 infection, but in the case of the T-even phages, host DNA does not constitute a quantitatively important source of precursors. The nuclease responsible for the extensive and rapid degradation of host DNA after T5 infection has yet to be described. A T5 nuclease that can be demonstrated does not appear until over 90% of the host DNA is degraded. In contrast, poxvirus infection causes no significant degradation of host nucleic acid; induction of thymidine kinase as a result of a DNAase producing deoxynucleotides can be ruled out.

What evidence there is for a direct role of DNAase in viral DNA replication is largely circumstantial. For example, the phage T5-induced nuclease precedes and then closely parallels phage DNA synthesis; the synthesis of exonuclease can be closely correlated with lambda-phage replication. A recent review by Lehman (91) covers these phenomena in some detail and points out that if DNAases are to participate in chromosomal replication and recombination, they must possess an order of specificity yet to be demonstrated. At this time it is unlikely that the complete spectrum of DNAases induced by poxvirus or any other DNA virus is known and much more information on the details of the reactions, other than simply describing substrate preference, will be needed before any meaningful role for nucleases can be considered. A start has been made in the poxvirus system with "acid" DNAase and now that the enzyme can be readily purified it might be informative, using both cell fractionation and immunofluorescence techniques, to see if the enzyme is bound to, say, the viral "factories." A natural substrate for the enzymes, the single-stranded DNA described by Pfau and McCrea (38), might be present in infected cells. Perhaps the function of the enzyme is to monitor the system and ensure this material is largely removed before assembly of the virion takes place.

3. DNA-Dependent RNA Polymerase

Cytoplasmic extracts from actidione-treated uninfected cells possess negligible capacity to synthesize RNA. However, similar extracts from rabbitpox-infected HeLa cells show a marked capacity to incorporate precursors into RNA. Viral cores purified from such extracts also synthesize RNA *in vitro* (*34*). The synthesis of RNA is dependent on the presence of all four nucleoside triphosphates but is not stimulated by addition of native or denatured DNA primers. Actinomycin D, but not pancreatic DNAase, can apparently permeate the core and block RNA synthesis. The pH optimum of the reaction is about 9.0 to 0.05 M tris HCl. The reaction is strictly dependent on a divalent cation and $5 \times 10^{-3}\,M$ $MgCl_2$ is optimal. Phosphate ions in excess of $10^{-2}\,M$ completely inhibit the reaction. The requirement for an ATP generating system in the reaction stems, at least in part, from the presence of an ATPase associated with the core. High concentrations of ATPase are present in the crude cytoplasmic fraction of uninfected cells and the ATPase of the core is partially derived from this. The ATPase level of the core preparation is reduced as the cores are purified by passage through sucrose. Similarly the ATPase of the whole virion is reduced by repeated passage of the virus through 36% sucrose.

A low molecular weight heat-stable factor can be prepared from the cytoplasm of uninfected HeLa cells. This factor enhances the rate and extent of the RNA polymerase reaction. It does not function by inhibiting the ATPase activity of the core preparation (*91a*). In the presence of an ATP generating system and factor, the synthesis of virus-specific RNA *in vitro* by cores proceeds linearly for at least 2 hr. The product is of large molecular weight (8S–10S) which resembles the isolated RNA synthesized by cores *in vivo*. The newly synthesized RNA seems to be rapidly released from the core during the reaction.

Proof that the RNA polymerase is an integral part of the virion was obtained by partially degrading purified virus. The purified virions exhibit no capacity for *in vitro* RNA synthesis. However, if the virions are degraded by very mild trypsin digestion or if the surface of the virion is disrupted by treatment with high concentrations of mercaptoethanol, the virus becomes capable of RNA synthesis *in vitro*.

One might expect that if the virus codes for its own polymerase one could show synthesis of RNA polymerase during the course of infection. Recently we have detected large increases in RNA polymerase activity in cytoplasmic extracts of poxvirus infected cells (*34a*). In order to unmask this activity it is necessary to treat the extracts with mercaptoethanol and a nonionic detergent such as Shell Nonidet P_{40}. The increase in this activity parallels the increase in progeny virus and we are possibly measuring the RNA polymerase of newly matured virions. The use of detergent to demonstrate RNA polymerase in infected cells was also used by Kates et al. (34b). Their results differ in that polymerase activity could be demonstrated in particles considerably smaller than virions. It is conceivable that the assembly of subvirions to virions might be less efficient in the cell line they used and thus accounts for the broad distribution of polymerase activity in sucrose gradients.

In addition to the particulate RNA polymerase, poxvirus elicits an activity that synthesises poly AU when primed by the polymer deoxy AT (34a). Activity of this type is present in uninfected cells and increases as high as 30-fold by 4 hr postinfection. In contrast to the RNA polymerase just described, the dAT primed enzyme is clearly demonstrable only if the onset of viral DNA synthesis is inhibited. The induced dAT primed activity might represent an expansion of the preexisting activity. On the other hand, it more likely represents the soluble form of the RNA polymerase of virions. There is evidence (78a) that the dAT primed activity is synthesized under the direction of a stable messenger, but, unlike that of thymidine kinase, this messenger is not synthesized before viral DNA is uncoated.

4. *DNA Nucleotidyltransferase* (*DNA polymerase*)

Magee reported DNA polymerase activity to be increased after infection with vaccinia virus (92), but only very small increases were found. By first fractionating cells and assaying only the cytoplasmic extracts, Jungwirth and Joklik (67) demonstrated 2–4-fold increases in polymerase activity elicited by vaccinia WR infection of HeLa cells.

DNA polymerase activity is also increased in rabbit kidney cells infected by the oncogenic Shope fibroma virus, a member of the myxoma subgroup of poxvirus (93). In fact, two species of polymerase activities have been reported: one acting preferentially on native and the other on denatured DNA. High activities are obtained using native DNA; fibroma DNA is superior to calf thymus DNA as a primer. Despite claims of the discoverers there seem to be no significant differences in the kinetics of appearance of these two activities and it is quite likely that the activities are due to different physical states of the one enzyme.

In view of the preference for native fibroma DNA, this enzyme deserves consideration as a candidate for the role of poxvirus DNA replicase. Different assay conditions have been used by different investigators. Conditions necessary to show fibroma-induced polymerase activity for native DNA (high concentration of magnesium ions and high DNA concentrations) were not used in assaying for vaccinia polymerase. Magee (*92*) did show that if calf thymus DNA was replaced by salmon sperm DNA, native rather than denatured substrate was a better precursor for the induced activity. It might also be interesting to re-examine the properties of the vaccinia-induced polymerase to see if its preference for a particular primer state varies with the conditions of assay.

At the time of writing, a DNA replicase exhibiting preference or specificity for poxvirus DNA remains to be shown. Utilizing the "replicator" protein which itself might be the replicase described earlier, a search for a poxvirus replicase associated with viral "factories" appears warranted.

C. OTHER VIRAL PROTEINS

The molecular weight of DNA of the poxvirion is probably as high as 160×10^6 daltons. Potentially this contains sufficient genetic information to code for about 400 proteins of average molecular weight. If infected cells are disrupted and analyzed by the Ouchterlony gel diffusion technique, some 15–20 distinct antigens can be shown (*95, 96*). The precipitin bands formed in the gel diffusion test against hyperimmune or convalescent sera reflect the presence of excess viral structural proteins together with other proteins coded for by the virus. Until recently only four antigens had been examined in any detail. These were the hemagglutinin, NP antigen, LS antigen, and Appleyard's "protective antigen."

1. *Hemagglutinin* (*HA*)

Late after infection when about 20–40% of the mature virions have accumulated, hemagglutinin becomes detectable (*97*). This antigen appears only in cells infected with virus of the vaccinia subgroup but not with other poxviruses. Although some viruses of the vaccinia subgroup cause hemagglutinin, not all do (*98*). Vaccinia virus strain Lederle 7N (hemagglutinin positive) and rabbitpox RPu+ (hemagglutinin negative) differ in many characteristics and when crossed, produce progeny of all classes of HA+ and HA- recombinants (*99*). This suggests that the viral genome contains information only to cause synthesis of a hemagglutinin encoded by the host-cell genome. The hemagglutinin can be isolated as pleomorphic lipoprotein particles 50–65 mμ in diameter. About 74% of the total lipid is neutral lipid and the rest phospholipid (*100*). Its biological activity is destroyed by lecithinase (*101*). HA is not part of the virion from which it is separable by density-gradient centrifugation.

2. *NP Antigen*

Extraction of virions with dilute alkali (*102*) yields a nucleoprotein containing 6% DNA and about 50% of the mass of the virus. It is probably a mixture of internal and surface protein. All poxviruses yield this NP antigen which reacts with antisera to all poxvirus strains tested (*6*).

3. *LS Antigen*

This antigen has been purified (*102*) by isoelectric precipitation and shown to be a single-protein molecule of molecular weight 240,000. The molecule bears a heat-labile (L) and heat-stable (S) specificity which can be degraded independently. It probably constitutes the major viral-coded protein in supernatant fluids of vaccinia-infected rabbit skin. In cowpox-infected tissues it appears to occur in a non-diffusible state which can be rendered diffusible by trypsin treatment (*103*). While the location of this antigen is unknown, it is unlikely to be on the surface of the virion since inoculation of rabbits with LS antigen fails to stimulate production of neutralizing antibodies to produce active immunity (*104*).

4. *Appleyard's "Protective Antigen"*

Amongst the soluble antigens of rabbitpox-infected tissues there is one recognizable by its "serum-blocking" activity (*105, 106*). The antigen has a molecular weight of between 100,000–200,000 and reacts with rabbitpox-virus neutralizing antibody. Several of the other soluble antigens recognizable by gel-diffusion test have no serum-blocking activity.

Recently more attention has been directed towards the extraction and resolution of the protein of the virion by various techniques. Mechanical disintegration or extraction of highly purified preparations of vaccinia virus with alkaline buffers has yielded eight components reacting with hyperimmune antisera (*107, 108*).

Using antiserum against detergent-degraded virus, seven vaccinia-

induced soluble antigens were found and these could be divided into
two classes on the basis of their molecular weight (*109*). High molecular weight (HMW) species, of which there are three, elicit production
of neutralizing antibody; low molecular weight (LMW) antigens, of
which there are four, show negligible capacity of this sort. Antiserum
against purified virus degraded by detergent was able to combine with
antigen species of both molecular weight classes to fix complement,
so both HMW and LMW antigens are probably viral structural proteins produced in excess. Since the LMW antigens do not elicit formation of neutralizing antibody, they are thought to be internal
proteins associated with viral DNA (*110*).

It is obvious that as better ways of solubilizing the virion proteins
are found it will be possible to fractionate the virus structural proteins
and resolve them by polyacrylamide electrophoresis. A simple stepwise approach might be to start with purified viral "cores" and extract
for just one class of proteins such as histones so that one can study one
or two defined proteins and their course of synthesis within the cell.

9-6. Regulation of Gene Expression

A. Early and Late Events

Macromolecules synthesized in bacteria infected with the DNA
bacteriophage T4 appear in a definite chronological order. Soon after
infection, but prior to the synthesis of phage DNA, a number of
enzyme activities are elicited. Following the onset of viral DNA
synthesis, "early" enzyme synthesis is arrested (*110a, 110b*) and synthesis of "late" proteins such as lysozyme and phage coat proteins are
initiated. This highly organized scheme of development is further
shown by genetic mapping in which clustering of "early" and "late"
genes is evident (*111*).

In the poxvirus system five induced enzymes are early in the
sense that they appear soon after infection. However, there is not
quite the same distinct sequence of events as occurs in the T phage
system. Poxvirus DNA synthesis begins about the same time as induced enzymes are detectable; switch-off of enzyme synthesis occurs
about the time that viral DNA synthesis is completed (*75, 112*).
Wilcox and Cohen (*110*) have demonstrated that the LMW antigens,
described above, appear between 1- and 2-hr postinfection and their

production, like that of the enzymes, is switched off about 5-hr post-infection. Loh and Riggs (97) used fluorescent antibodies directed against NP, LS hemagglutinin antigens in an attempt to demonstrate the appearance of these entities during the course of infection. All these antigens did appear late in the course of infection and a sequence could be shown. LS appears at 4 hr, NP at 5–6 hr, and hemagglutinin at 10 hr after infection. Unfortunately, the sensitivity of the fluorescent antibody staining is low; using this technique one cannot be certain of the time at which antigen is first synthesized. The indirect immune precipitation technique following pulse labeling with radioactive amino acids (113) is by far the most sensitive method of measuring the appearance of classes of antigens.

Sequential development of precipitin lines with time of incubation of cells infected with rabbitpox has been observed by Appleyard et al. (95). Variable numbers of precipitin lines can be found in cells infected with host-range mutants of rabbitpox virus which do not fully replicate in them (114).

A more detailed study of the kinetics of formation of viral-coded proteins has been conducted by Salzman and Sebring (115). These investigators used antiserum from rabbits infected with live virus so that a full spectrum of antibodies to both early and late proteins could be obtained. Antigens were detected by a quantitative radio-chemical precipitin technique. Although there is a block in the normal transport of cellular RNA from nucleus to cytoplasm after vaccinia infection (116), one might expect host protein synthesis to continue until host mRNA, present at the start of infection, had decayed. However, 90% of cell protein synthesis is inhibited by 4-hr postinfection. The rapidity with which this occurs suggests that some mechanism other than transport blocks host protein synthesis. Since cell protein synthesis is also inhibited if infection takes place in actinomycin-treated cells (117), it follows that inhibition of host protein synthesis is controlled by the input virion. Salzman and Sebring (115) found that the absolute quantity of viral protein made between 2–3 hr postinfection was as great as the quantity synthesized in any subsequent 1-hr interval up to about 10-hr postinfection. Of the total quantity of viral directed protein, 65% was synthesized prior to 8-hr postinfection. Of those proteins incorporated into mature virus particles, 74% were made from 8–12-hr postinfection, a time in which 90% of infectious virus is formed. By combining autoradiography with agar diffusion of isotopically labeled viral antigen, Salzman and Sebring followed the

time of synthesis of a number of antigens, or more precisely, the time
at which viral-coded proteins became antigenic in the infection cycle.

Of particular interest is the large quantity of viral protein made
early in the infectious cycle. It is far in excess of the quantity required
to function enzymatically although some of the enzymes at least must
be included in this early class of proteins. Wilcox and Cohen (*110*)
found that the LMW early antigens did not represent thymidine
kinase.

B. DNA REPLICATION AND THE REGULATION OF
EARLY AND LATE EVENTS

1. *Regulation of Early Enzymes*

There is a clear correlation between the onset of DNA synthesis in
T4-infected *E. coli* and the time that early enzyme synthesis is
arrested.

In lambda-infected *E. coli,* exonuclease synthesis begins early and
is independent of formation of progeny DNA, while endolysin formation coincides with and is dependent on the appearance of progeny
DNA (*118*). Similar observations on the control of the early enzyme
deoxycytidylate deaminase and the late enzyme lysozyme in *B. subtilis* infected with phage 2C were made by Pene and Marmur (*119*).

If phage is irradiated prior to infection, or if certain mutants are
used, phage DNA synthesis does not occur and synthesis of early
enzymes continues well beyond the normal time (*111, 120, 121*). This
relationship between switch-off of early functions and DNA synthesis
holds for the poxvirus system (*112a*). Cells infected with uv-irradiated
poxvirus or infected in the presence of inhibitors of DNA synthesis
continue to make thymidine kinase (*66, 75*), alkaline deoxyribonuclease (*112*), neutral deoxyribonuclease, and DNA polymerase (*67*);
the normal switch-off of these enzymes fails to occur under such
conditions. The same effect is found if DNA synthesis is allowed to
proceed in the presence of bromodeoxyuridine (*75*). BUdR-substituted
DNA is, therefore, unable to effect certain functions even though
antigen detected by immunofluorescence is synthesized in the presence
of BUdR. These results indicate that progeny DNA carries out at
least one function not expressed by template DNA and that it is not
just the amount of DNA synthesized that is important. The messenger
for virus-induced thymidine kinase is remarkably stable; addition of
actinomycin D prior to the normal switch-off time leads to extended

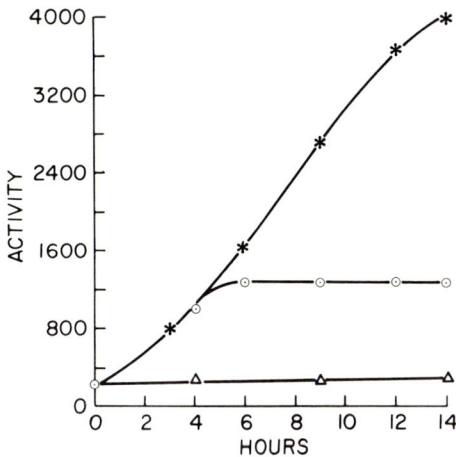

FIG. 9-6. Effect of actinomycin D on the switch-off of poxvirus-induced thymidine kinase. Increase in kinase activity in cultured cells infected with rabbitpox mutant U6/2 (⊙); increase in kinase activity in the same system when actinomycin D was added at 3-hr postinfection (*). [Reproduced from *Virology*, **26**, 738 (1965), with permission of Academic Press Inc., New York.]

synthesis of thymidine kinase (Fig. 9-6). This messenger stability was taken advantage of to show that switch-off of thymidine kinase synthesis required the synthesis of a protein or proteins which presumably acts by blocking messenger translation (*66*). The switch-off protein did not act catalytically in that a certain period of protein synthesis was necessary to achieve switch-off.

This experiment has been described by Spirin (*122*) as being the first to clearly demonstrate the content of gene expression at the translational level. In view of this, and because this experiment has been presented in recent textbooks (*74, 123*), it is important to ensure that only one interpretation of the experiment is possible. One concern is that puromycin which was used to arrest protein synthesis is not very specific for inhibition of protein synthesis since, as was subsequently shown (*65*), it can rapidly arrest viral DNA synthesis. In the original experiment (*66*) addition of puromycin must have arrested both viral DNA synthesis and enzyme synthesis. Upon removal of puromycin, the time required to establish switch-off could be a reflection of the time required to reinitiate viral DNA synthesis; the latter might have participated directly in enzyme switch-off. To cover this possibility the following experiment was carried out. Cells were in-

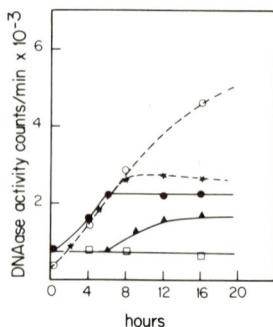

Fɪɢ. 9-7. Relationship between DNA synthesis and the synthesis of "acid" DNAase. Increase in acid DNAase in cowpox-infected cells (●); increase in acid DNAase (□) and alkaline DNAase (○) in cells infected in the presence of fluoro-deoxyuridine; increase in acid DNAase (▲) and alkaline DNAase (★) after infecting cells in the presence of fluorodeoxyuridine and then removing the in-hibitor 6-hr postinfection. [Reproduced from *Proc. Natl. Acad. Sci. U.S.,* **55,** 1581 (1966), with permission of the publishers.]

fected in the presence of fluorodeoxyuridine; extended thymidine kinase synthesis and the accumulation of replicator protein for DNA synthesis took place. After a time, streptovitacin was added to block protein synthesis and fluorodeoxyuridine was removed. Under such conditions viral DNA will replicate even though protein synthesis is blocked (*65*). Upon removal of streptovitacin, enzyme synthesis re-sumed. However, switch-off did not take place for about 4 hr after viral DNA synthesis was resumed. The original experiment is still valid, although the way in which the "repressor" protein terminates kinase synthesis remains to be elucidated.

It should be possible to add actinomycin D at times during the synthesis of DNA [occurring after the FUdR block, and hence not actinomycin sensitive (*65*)] and pinpoint the time at which the switch-off messenger is made and its relative stability.

2. A Late Enzyme

If cells are infected with poxvirus in the presence of FUdR and the inhibitor washed out some hours later, the timing of events is now seen to resemble that of the phage system (see Fig. 9-7). Early enzymes are synthesized soon after infection; after removal of the inhibitor, DNA synthesis then starts and soon after, the early enzymes are switched-off. An unexpected finding was that acid DNAase is not made in the presence of FUdR, but its synthesis does start up when viral DNA synthesis commences (*112*). That is to say, the acid DNAase is a "late" enzyme. Subsequent work (*83*) has shown that the amount of acid DNAase synthesized is directly proportional to the amount of viral DNA synthesized. Other factors operate to control the activity of this enzyme *in vivo;* the enzyme exists in a soluble and bound form and the switch-off of the increase in the soluble activity coincides with increase in the amount of enzyme in the bound state (Fig. 9-8). This acid DNAase is synthesized as late as 18-hr postinfection and represents the first late poxvirus function obtained in a highly purified state. The "neutral" DNAase and the acid DNAase must be distinct enzymes since the former is synthesized from input DNA templates (*67*).

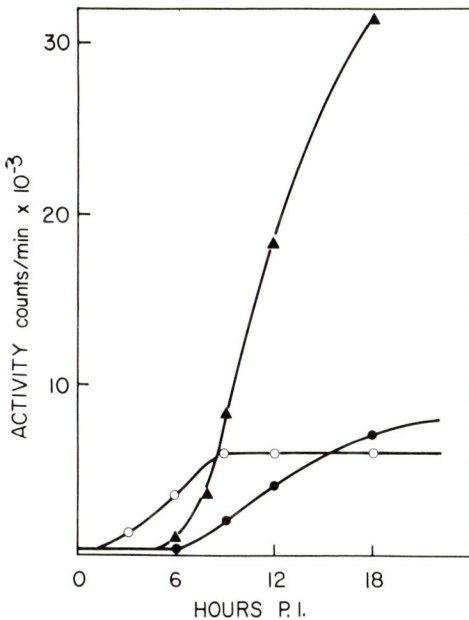

FIG. 9-8. Kinetics of synthesis of poxvirus-induced "acid" deoxyribonuclease. Increase in soluble activity (○); increase in activity in the particulate cell fraction (●); increase in activity of enzyme liberated from the particulate fraction with deoxycholate (▲). [Figure reproduced from *Virology,* **26,** 742 (1967), with permission of Academic Press, Inc., New York.]

3. *Viral Antigens*

The correlation between DNA replication and the timing of early and late events had also been shown by precipitation of ^{14}C-labeled viral protein by immunological techniques. Thus, Shatkin (*113*) showed that if viral DNA synthesis is inhibited, some viral proteins are synthesized, but the yield of protein is less than that in the uninhibited system and only a limited number, rather than a full spectrum of proteins is obtained. In a more detailed study Salzman and Sebring (*115*), using antibody against all classes of virus-coded proteins, demonstrated that in the normal system synthesis of early protein is blocked at about 6-hr postinfection and a new spectrum of proteins is made after this time. If DNA synthesis is prevented, synthesis of the spectrum of early proteins is continuous and the "late" spectrum of proteins fails to appear. Because of the immunological procedures employed, the date of Salzman and Sebring offers the most direct evidence that the "early" proteins observed are coded by the viral genome. It would be interesting to see if hemagglutinin is produced in the FUdR-inhibited system. Although Oda (*124*) reports that it is synthesized in mitomycin-blocked cells, it was not proved that viral DNA synthesis had been arrested. Holowczak and Joklik (*124a*) examined the structural polypeptides of the vaccinia virion by controlled degradation followed by gel electrophoresis. In this way they identified polypeptides associated with the viral core. Three to five of the structural proteins are coded by the parental genome and appear early. Two of these early structural polypeptides and one, designated VSP-4 which is synthesized at late times, are components of the core.

In summary, proteins synthesized during the course of poxvirus infection can be classified as "early" or "late" species with respect to the time that they appear in the replication cycle. The "early" species are those made before 4-hr postinfection which include thymidine kinase, DNA polymerase, alkaline and neutral deoxyribonucleases, LMW antigens of Wilcox and Cohen (*110*), and the "replicator protein" for DNA synthesis. Synthesis of the early protein is normally terminated about 5-hr postinfection and only a small proportion of this protein is incorporated into mature virus. Most of the protein incorporated into the virus particle is made "late." If DNA synthesis is inhibited, the synthesis of early proteins continues and the "late" proteins fail to appear. A possible exception to this last statement might be hemagglutinin synthesis. Induced acid deoxyribonuclease is unusual among the enzymes in that it continues to be synthesized as late as 18-hr postinfection. Even though it appears before 4–5-hr postinfection it is classified as "late" because its synthesis is directed by only progeny DNA, or to be precise, only after DNA has replicated.

A better nomenclature for proteins made during poxvirus replication might, therefore, be template DNA-directed and progeny DNA-directed proteins. Even this is not entirely satisfactory; if we assume that the induced enzymes are virus coded then the template-directed functions can be subdivided into those whose messenger is transcribed from virus in the "core" state [e.g., thymidine kinase (32)] and those whose messenger is synthesized after escape of viral DNA from the core. In the latter class is the DNA polymerase for which messenger is not formed when virus is arrested at the core stage (34).

Induced acid DNAase is the first poxvirus-induced function that has been obtained in a highly purified state and the availability of a microassay for this active protein makes it convenient to follow the synthesis of this late function (83). It is highly probable that, similiar to the acid DNAase, "late" proteins are synthesized as soon as DNA synthesis commences, although they become detectable immunologically only towards the end of DNA replication. This could be due to the fact that some time is necessary for the synthesized polypeptide to assume the correct antigenic configuration.

9-7. Regulation of Viral Messenger RNA Synthesis

Several hypotheses have been put forward to explain the control of synthesis of virus-induced proteins. These hypotheses include control at the level of (1) translation of specific cistrons, (2) regulation of total messenger RNA synthesis or destruction, and (3) transcription of specific messenger RNA species. With the introduction of the DNA–RNA hybridization technique (125a), it became possible to examine the general level at which the controls of viral gene expression are operative. Kano-Sueoka and Spiegelman (125) showed that during the course of phage T2 development differences were apparent in the mRNA synthesized during the course of infection. They suggested that these qualitative differences reflected functional differences in mRNA species.

Corresponding to the appearance of early and late proteins there appears early and late mRNA distinguishable by competition hybridization tests. This established that mRNA contained new types of molecules (126). Hall et al. (127) also showed a fraction of T2 mRNA synthesized at a late stage in late period was not synthesized at an early stage although messenger made early was still present at a late stage. The obvious implication of these findings is that control of the type of mRNA produced was responsible for the delay in the formation of late proteins and that regulation of early messenger can occur at this level of translation. However, the situation even in the phage system provides additional complexities in that some mRNA coding for late functions may be transcribed from parental genomes. Evidence for this has been advanced by Edlin, (128) who studied phenotypic reversion of T4 amber mutants by 5-fluorouracil. Furthermore, Bautz et al. (129) found that the lysozyme gene of T4 is transcribed to a limited extent immediately after infection, not transcribed for the remainder of the early period, and then extensively transcribed once progeny genomes are synthesized.

Unlike the rapidity with which T-even phages block the transcription of their host's genome, the inhibition of transcription of the poxvirus-infected HeLa cell genome takes place gradually from about 3-hr postinfection. By approximately 6–7 hr host genome transcription has ceased. Fortunately, there is an immediate block in the transport of host mRNA to the cytoplasm after infection (115). Moreover, by giving any short pulses (of the order of 15–20 min) of radioactive uridine to infected cells followed by their fractionation, the only newly synthesized RNA other than tRNA in the cytoplasmic fraction is viral mRNA.

A detailed examination of mRNA synthesis in poxvirus-infected cells has been carried out by Joklik and co-workers (62a, 130). They divided the cycle into early (preceding DNA synthesis) and late stages (from the start of DNA synthesis to maturation) and found that there is a progressive increase in the size of the mRNA synthesized. At early times the messenger RNA has a range of 8S–12S and with time this increases progressively to 16S–20S. All early mRNA sequences are transcribed at late times. The early messenger RNA is very stable in HeLa cells. This finding for early RNA's agrees with the particular observations on thymidine kinase and DNA polymerase messenger stability (66, 67). However, if we look at other functions we find that some early messenger is unstable, e.g., the messenger for alkaline DNAase (76). Oda and Joklik (130) find that late messenger is mostly unstable with a half-life of 1 hr or less but those sequences characteristic of early messenger are as stable as early messenger. The late messenger has sequences not found early. Soon after its formation, newly synthesized viral mRNA associates with ribosomes and as infection progresses the size of the polysome fraction also increases progressively.

In that early messenger differs from late messenger and that sequences are transcribed at late stages that are not transcribed early, the above results for poxvirus-infected HeLa cells parallel those described for T2-infected *E. coli*. A major objection to the above findings is that in the infected HeLa cell only about 50–60% of the input virions become uncoated (*23*). If the remaining fraction is arrested at the core stage, it will continue to produce "early" messenger throughout the entire replication period and this would account for the appearance of early sequences at late times. It has been clearly demonstrated that virus arrested at the core stage not only makes early messenger but makes even more than in the normal course of replication (*32*) (see Table 9-3). Competition hybridization experiments show that this very early messenger differs qualitatively from messenger made after uncoating but before DNA replication (*32*).

According to Becker and Joklik (*62a*) approximately 7 hr after infection of HeLa cells with the WR strain of vaccinia virus, the polysomes that were initially loaded with viral mRNA were extensively, if not completely, degraded and protein synthesis was arrested at the same time. This time of degradation is very close to the time at which switch-off of early enzyme synthesis takes place. On the other hand, in rabbitpox-infected HeLa cells, protein synthesis continues well beyond the time that enzyme switch-off occurs. The latter finding suggests that termination of enzyme synthesis was not a reflection of a general cessation of protein synthesis (*75*).

TABLE 9-3

Estimation of Relative Amounts of Viral RNA Produced by "Cores"

Virus	Treatment	CMP hybridized[a]	Relative amount of viral RNA
Normal	—	750	1
Normal	Protein synthesis blocked at infection	5000	6
UV-irradiated	—	1400	2
Heated virus	—	—	0

[a] Estimated from initial rate. See Ref. *32*.

Investigation of the polysome profile of rabbitpox-infected cells
(*112*) showed that even as late as 12-hr postinfection there was a reduction in the polysome peak but the polysome fraction was still high.
It is unlikely therefore, that random breakdown of polysomes could
account for the termination of enzyme synthesis from the stable messengers for thymidine kinase and DNA polymerase. Subsequent experiments refute the claim that polysomes loaded with viral messenger
are completely degraded at late times in the course of WR infection
(*130a*). The question remains of how the expression of stable messengers is prevented and whether the expression of unstable messengers
for early functions such as alkaline DNAase is blocked at the level
of transcription or translation.

There is some evidence that it is not the one process that is responsible for the arrest of both alkaline DNAase and thymidine kinase (*112*)
since in one poxvirus system alkaline DNAase synthesis is terminated
well in advance of the arrest of kinase synthesis. In the phage system
there are also examples of early enzymes with stable and unstable
messenger and again it raises the possibility of regulation at both
transcription and translation levels. For example, the messenger for
T4-induced thymidylate synthetase is unstable [of the order of 3 min
(*131*) while that of T4 and 122-induced deoxycytidylate hydroxymethylase is made during only the first 10 min after infection but has
a mean half-life of 11 min (*132*)].

The possibility that ribosomes fail to bind early messenger at some
time during the infectious cycle has been considered for both phage
and poxvirus systems. Friesen et al. (*133*) and Baldi and Haselkorn
(*134*) show that early messenger made at late times binds as well to
ribosomes as early messenger made early; early messenger made late
is associated with ribosomes. In contrast to these results, mRNA in
polysomes of poxvirus-infected cells is significantly depleted with
respect to sequences characteristic of early messenger (*130*). In both
the phage and poxvirus systems one can follow the over-all pattern
of messenger formation and fate by competition hybridization or one
can indirectly follow the messenger for a particular function. Both
approaches will be necessary for a detailed understanding of the
regulation of early functions.

A valuable tool for the study of poxvirus mRNA control in poxvirus
replication is the inhibitor isatin-β-thiosemicarbazone (IBT). For
some time isatin thiosemicarbazones have been used in the chemo-
therapy of smallpox infection (*135*). Until recently the mode of action

of this compound was quite unknown and all that could be said was that it inhibited viral maturation. At concentrations that inhibit virus replication by over 95% it causes no inhibition of uninfected cell metabolism. In infected cells it does not affect early mRNA synthesis, early enzyme synthesis, or viral DNA synthesis. Woodson and Joklik (136) have shown that with the onset of late functions in the presence of IBT, polysomes break down very rapidly, protein synthesis is depressed and there is a marked reduction in encapsidation of virions. The reason for the disappearance of polysomes in the presence of IBT is not inhibition of transcription of late messenger or inhibition of messenger attachment to ribosomes. The functional half-life of late messenger is reduced by the drug from 30 min to 5 min. It is postulated that IBT action is mediated through a protein synthesized under the direction of progeny DNA. This protein normally effects the switch-off of only some viral messengers, but IBT modifies this protein so that it now causes switch-off of all viral messengers. Thus the inhibitor is a useful probe to elucidate the role of protein in the switch-off phenomenon and to see if viral mRNA is prevented from functioning by introduction of chain breaks into the polymer or by dissociation from ribosomes.

Given that different messengers are transcribed at different times and that early messengers may either fail to be transcribed, or be transcribed but blocked somehow at the level of translation, perhaps directly, the problem remains of the signal for the transcription of late function. Or in other words, the key problem is the actual process of transcription of any portion of the genome. How does the polymerase recognize the start and finish of an operon and how does the replication of DNA facilitate the reading of late functions?

A full discussion of current information on gene transcription is beyond the scope of this article but several studies with phage systems warrant mention for they indicate important approaches to which the poxvirus system might be profitably applied or which provide suggestions to explain the sequence of events in the pox system.

Two conditional lethal mutants of phage T4 are able to synthesize DNA but unable to produce late protein in the restricted host (111). Green (137) examined one such mutant (Am N134) for its ability to synthesize late mRNA and found that although DNA was synthesized, DNA replication per se is insufficient to enable late mRNA synthesis to take place. In this case, however, failure to synthesize mRNA could have been due to failure of the DNA to replicate "normally" or to a change in the properties of the polymerase so that it could not recognize starting sites in the late region of the genome. When phage DNA or DNA extracted from infected cells at times during the infection was used as the primer for *E. coli* polymerase, only early mRNA synthesis was noted *in vivo* with all DNA's tested, even for DNA taken late. Heat denatured or renatured DNA did not yield late messenger; therefore, if synthesis of RNA is regulated by changes in the physical state of DNA the changes are more subtle than those tested.

Rapid advances in our understanding of gene transcription are coming from studies with lamboid phages, for the DNA of these can be separated into light and heavy strands and further into left- and right-hand halves. Hurwitz and co-workers (*138, 139*) studied the transcription of λ DNA with *E. coli* polymerase. These investigators showed that transcription of native templates was not random; nucleotide sequences located predominantly in the AT-rich half of λ DNA are preferentially copied throughout the duration of *in vitro* RNA synthesis. Synthesis begins at a site on the AT-rich half of the light strand and although it continues to be made late, the majority of late λ in RNA is transcribed from the heavy strand, predominantly from sites located on the GC-rich half, which contains late cistrons. In contrast to results for other bacteriophage systems, both strands of λ DNA are copied *in vivo* and *in vitro*. Of particular interest is the observation (*139*) that when gentle lysis was used to disrupt bacteria, transcription of endogenous DNA *in vitro* by early and late extracts of infected cells showed a temporal change in orientation resembling the strand-copying switch observed during *in vivo* RNA synthesis. The conclusion is that a fragile complex of the DNA template with a cellular component in a specific structural form of λ DNA is required for specificity. It is to be expected that studies on λ transcription will provide some understanding of why DNA replication is necessary for late messenger transcription in the poxvirus system, however possible advantages of the poxvirus system in the study of the transcription process needs to be more fully explored.

Perhaps certain initiating regions for operons become available at stages of infection when the virus is reduced to the core stage, when DNA is freed from the restriction of core coats, and when DNA begins to replicate, i.e., is in the appropriate physical state. It is also conceivable that different forms of poxvirus transcriptase are synthesized and that only in the appropriate form does the transcriptase copy late regions of the genome.

9-8. Discussion and Suggestions for Future Work

A simplified scheme of the replication cycle of poxvirus, in terms of work presented above, is this. (1) The outer coat of the virus is degraded by constitutive enzymes to the core stage. Within the core, RNA polymerase is closely associated with viral DNA. (2) The cores synthesize poxvirus mRNA as soon as they enter the cytoplasm. This mRNA represents several early functions such as thymidine kinase and uncoating requirements. (3) After DNA is uncoated, but before it replicates, other genes are transcribed. These include DNA polymerase, replicator protein, alkaline and neutral DNAases, and some low molecular weight antigens that become internal virus structural components. (4) DNA replicates and late genes are transcribed. These include the switch-off protein influencing early enzyme synthesis, acid DNAase, and high molecular weight antigens that become external virus structural components. (5) Assembly of the virus takes place.

Considering these steps it will be obvious that many questions remain to be answered. Starting with the virion itself, there is a need for fractionation and characterization of structural proteins by more sophisticated techniques of protein chemistry. We need to know the number of different proteins that make up the virus and their organization within the particle. Detailed studies on the synthesis of each protein will help to understand the process of viral assembly. There have been few studies where controlled degradation of the virus has been combined with electron microscopy and biochemical studies. Easterbrook (15) has made a valuable start in this direction showing that the *in vivo* stages of uncoating can be mimicked *in vitro* by detergent treatment of virions. It should be possible to couple this approach with biochemical and biological investigation of the products and their properties.

It would be interesting to try to isolate and study the lateral bodies. At present no one has suggested a function for these entities. The proteins split from the virion by constitutive enzymes within the phagosome must profoundly affect cell metabolism. An interesting problem is the control of particle digestion within the phagosome. A first question might be whether just one enzyme or a series of enzymes are involved in the breakdown of the outer coat. Within the cell there must be a control over virion degradation. Among the many possibilities, one can imagine that this results from inhibition of enzymatic degradation by a product of the outer coat, that conditions within the phagosome are such that core proteins are resistant to degradation, or simply that the core escapes digestion because the phagosome disrupts before core digestion takes place. If the latter is true, then what causes the phagosome to break down? Could one inhibit virus replication by chemically stabilizing the phagosome membrane?

The question of the uncoating protein is still in limbo and now resolves itself into whether there is a specific uncoating protein or whether it is just the act of protein synthesis that is essential to release DNA from the core. Transcription of viral mRNA by a polymerase that appears to be a component of the virion was unexpected until recently. Although one could argue that viruses replicating in the nucleus do not need to carry their own RNA polymerase, it might be worthwhile to examine the adenoviruses and herpesviruses for such an enzyme. Recent work has shown that reovirus contains a polymerase within the virion (139a). This is not unexpected considering that the RNA within the virus particle is double stranded.

RNA polymerase that is incorporated within mature virions must be synthesized during the infection cycle. The appearance of such an activity is demonstrable by using detergents. Since the polymerase activity of purified virions can be activated simply by mercaptoethanol treatment, it is not clear what structure in cytoplasmic extracts must be disrupted by detergent to show polymerase activity. Just how this transcriptase monitors the genome within the core needs to be studied. We know that some poxvirus genes are not transcribed at the core stage. Perhaps some proteins of the core act as specific repressors and their removal by the uncoating process then permits transcription of the remaining early genes by the input RNA polymerase. In contradiction to this idea is Woodson's (33) observation that the initial rate of viral mRNA synthesis drops rapidly under conditions where uncoating is allowed to proceed but where DNA replication is prevented. It suggests that the release of viral DNA from the core prevents reinitiation by the input transcriptase. On the other hand, as discussed above, poxvirus-induced DNA polymerase is synthesized under such conditions.

A more difficult question concerns the transcription of "late" genes after DNA replication. Perhaps DNA replication without concomitant synthesis of some repressor for "late" genes dilutes out this repressor so that these genes can then be transcribed. This postulated repressor would thus be a late function to be itself synthesized later during infection. In addition, a soluble transcriptase capable of initiating transcription of late genes might be synthesized. This newly synthesized transcriptase might initiate at late genes because of the dilution of repressor or because the replicating DNA is now in a favorable configuration for it to do so. The role of the dAT primed polymerase has yet to be elucidated. So far we have not been able to demonstrate transcription of any primer other than dAT. Even poxvirus DNA, although it binds to the enzyme, is not transcribed by this polymerase *in vitro*. Poxvirus DNA in the physical form of the replicating state, or enzyme in an appropriate configuration may be a prerequisite for DNA transcription. Mild methods for disrupting the core and solubilizing the polymerase associated with it need to be developed. Once this is achieved, we hope to be able to recombine the components and to study the influence of various components on transcription *in vitro*.

Study of the synthesis and release of mRNA by virus cores has scarcely begun. The percentage of the genome that is available for transcription could be estimated from the amount of RNA, of known specific activity, that can hybridize with a known amount of viral DNA. If it is possible to separate the two strands of poxvirus DNA as it is for λ DNA, one could see if a change in strand selection is involved in transcription at different stages of pox replication. Knowledge of the direction of release—i.e., whether the RNA is released with the 5′ or 3′ end first, or released randomly—would be useful in understanding messenger production by cores. So far, phage systems have not provided an explanation of the relation between DNA replication and late gene expression. If *in vitro* DNA synthesis could be achieved with isolated viral "factories," then perhaps one could examine the conditions necessary to transcribe late genes.

Investigations of enzyme regulation and messenger RNA synthesis have complemented each other as far as indicating sequential changes in translation and transcription. Studies on enzyme regulation led to the discovery of the extreme stability of some early messenger RNA and to the differing stabilities of early mRNA for different early functions. Syntheses of different early enzymes are not terminated simultaneously [cf. the regulation of DNAases compared to thymidine kinase in some systems (*112*) and the termination of thymidine kinase compared to that of DNA polymerase (*67*)]. This indicates that different early functions might be arrested by different mechanisms. Unfortunately we do not really know just how synthesis of any early enzyme is switched off but perhaps the isatin thiosemicarbazone studies will soon shed some light on this.

The induced enzymes have been poorly characterized. This is in part due to the fact that no enzymologist has studied them and to the fact that infected cultural animal cells are not a convenient source for enzyme purification. Nevertheless, a careful characterization of induced thymidine kinase is sadly overdue and would help to elucidate the significance of this enzyme in poxvirus replication. Whether or not there are any enzymatic steps in the assembly of viral components to form mature virions is not known.

In the work presented above there has been negligible reference to poxvirus genetics. The reader is referred to a review covering most of the work of Fenner and colleagues (*142*). Suffice it to say that the goal of mapping the entire poxvirus genome is a long way off. Genetic recombination between rabbitpox and other vaccinia viruses occurs with high frequency. Temperature-sensitive mutants and host-dependent conditional-lethal mutants are available. Mutants of rabbitpox which fail to grow in pig kidney cells vary greatly in their capacity to produce different viral antigens and to synthesize viral DNA (*140*). These mutants have not yet been fully characterized and the potential value of these mutants is as a tool for the study of particular viral functions rather than in mapping the viral genome by recombination.

The rationale behind the experiments used to study T-even phages has been applied frequently to study the sequence of events in poxvirus replication. In fact, the timing of events that does occur in poxvirus replication (for example, the induction of enzymes, the synthesis of DNA, and the arrest of early functions) resembles that in T phage development. Despite the similarities, there are aspects of poxvirus replication that are peculiar to it and these aspects are sufficiently intriguing to warrant investigation. It will be interesting to compare the biochemistry of poxvirus replication with that of the other cytoplasmic DNA viruses. Poxvirus has been of scientific interest from the time of its employment in Jenner's vaccination studies to its use as a model for epidemiological studies and as an agent for biological pet control (*141*). Now that we can carry out *in vitro* messenger RNA synthesis with the virus itself, molecular biologists should seriously consider this animal virus as a useful system to unravel some of the important problems in gene transcription.

REFERENCES

1. E. Jenner, 1798. Available as pamphlet, Vol. 4232, 1896, Army Medical Library, Washington, D.C.

2. S. S. Breese and C. J. DeBoer, *Virology*, **28**, 420 (1966).

3. A. Granoff, P. E. Came, and K. A. Rafferty, Jr., *Ann. N.Y. Acad. Sci.,* **126,** 237 (1965).

4. R. Leutenegger, *Virology,* **24,** 200 (1964).

5. W. K. Joklik, *Bacteriol. Rev.,* **30,** 33 (1966).

6. G. M. Woodroofe and F. Fenner, *Virology,* **16,** 334 (1962).

7. F. Fenner, "The Poxviruses" in *Viral and Rickettsial Diseases of Animals* Betts and York, eds.), Academic Press, Chap. XIII, in press.

7a. F. Fenner, personal communication.

8. F. Fenner and G. M. Woodroofe, *Virology,* **11,** 185 (1960).

9. H. T. Zwartouw, *J. Gen. Microbiol.,* **34,** 115 (1964).

10. W. K. Joklik, *Virology,* **18,** 9 (1962).

11. D. N. Planterose, C. Nishimura, and N. P. Salzman, *Virology,* **18,** 294 (1962).

12. J. Marquardt, S. E. Holm, and E. Lyeke, *Proc. Soc. Exptl. Biol. Med.,* **116,** 112 (1964).

12a. P. Gold and S. Dalbs, *Proc. Natl. Acad. Sci. U.S.,* **60,** 845 (1968).

13. D. Peters and D. Buttner, *European Regional Conference on Electron-microscopy, 3rd, Prague, 1964,* Vol. B, p. 377.

14. J. C. N. Westwood, W. J. Harris, H. T. Zwartouw, D. H. J. Titmuss, and G. Appleyard, *J. Gen. Microbiol.,* **34,** 67 (1964).

15. K. B. Easterbrook, *J. Ultrastructure Res.,* **14,** 484 (1966).

16. Y. Becker and Y. Chen, *Israel J. Med. Sci.,* **2,** 417 (1966).

17. J. Nagington and R. W. Horne, *Virology,* **16,** 248 (1962).

18. S. Dales, *Progr. Med. Virol.,* **7,** 1 (1965).

19. S. Dales, *J. Cell. Biol.,* **18,** 51 (1963).

20. R. Kajioka, L. Siminovitch, and S. Dales, *Virology,* **24,** 295 (1964).

21. S. Dales and R. Kajioka, *Virology,* **44,** 278 (1964).

22. S. Dales, *Proc. Natl. Acad. Sci. U.S.,* **54,** 462 (1965).

23. W. K. Joklik, *Cold Spring Harbor Symp. Quant. Biol.,* **27,** 199 (1962).

24. W. K. Joklik, *J. Mol. Biol.,* **8,** 263 (1964).

25. W. K. Joklik, *J. Mol. Biol.,* **8,** 277 (1964).

26. F. Fenner, *Brit. Med. J.,* 1962-2, 135.

27. G. P. Berry and H. M. Dedrick, *J. Bacteriol.,* **31,** 50 (1936).

28. P. Abel, *Z. Vererbungslehre,* **95,** 66 (1964).

29. P. Abel, *Z. Vererbungslehre,* **94,** 249 (1963).

29a. J. Degnen and B. R. McAuslan, unpublished.

30. S. Kit, L. J. Piekarski, and D. R. Dubbs, *J. Mol. Biol.,* **7,** 497 (1963).

31. W. Munyon and S. Kit, *Virology,* **29,** 303 (1966).

32. J. R. Kates and B. R. McAuslan, *Proc. Natl. Acad. Sci. U.S.,* **57,** 314 (1967).

33. B. Woodson, *Biochem. Biophys. Res. Commun.,* **27,** 169 (1967).

34. J. R. Kates and B. R. McAuslan, *Proc. Natl. Acad. Sci. U.S.,* **58,** 134 (1967).

35. C. A. Thomas and L. A. MacHattie, *Biochem. Rev.,* **36,** 485 (1967).

36. W. K. Joklik, *J. Mol. Biol.,* **5,** 265 (1962).

37. C. C. Randall, L. G. Gafford, and R. W. Darlington, *J. Bacteriol.,* **83,** 1037 (1962).

38. C. J. Pfau and J. F. McCrea, *Virology,* **21,** 425 (1963).

39. C. J. Pfau and J. F. McCrea, *Nature,* **194,** 894 (1962).

40. R. Langridge, *J. Mol. Biol.,* **2,** 192 (1960).

41. K. B. Easterbrook, *J. Virol.,* **1,** 643 (1967).

42. J. M. Hyde and C. C. Randall, *J. Bacteriol.,* **91,** 1363 (1966).

43. L. G. Gafford and C. C. Randall, *J. Mol. Biol.,* **26,** 303 (1967).

44. F. Frankel, *Proc. Natl. Acad. Sci. U.S.,* **49,** 366 (1963).

45. S. Kato, S. Kameyama, and J. Kamahora, *Biken's J.,* **3,** 135 (1960).

46. H. J. R. Cairns, *Virology,* **11,** 603 (1960).

47. S. Kit, D. R. Dubbs, and T. C. Hsu, *Virology,* **19,** 13 (1963).

48. N. P. Salzman, A. J. Shatkin, E. D. Sebring, and W. Munyon, *Cold Spring Harbor Symp. Quant. Biol.,* **27,** 237 (1962).

49. N. P. Salzman, *Virology,* **10,** 150 (1960).

50. S. Penman, K. Scherrer, Y. Becker, and J. E. Darnell, *Proc. Natl. Acad. Sci. U.S.,* **49,** 654 (1963).

50a. C. Jungwirth and I. B. Dawid, *Arch Ges. Virusforsch.,* **20,** 464 (1967).

51. K. B. Easterbrook and C. Davern, *Virology,* **19,** 509 (1963).

52. J. F. McCrea and M. B. Lipman, *J. Virol.,* **1,** 1037 (1967).

53. W. K. Joklik, *Virology,* **22,** 620 (1964).

54. V. N. Iyer and W. Szybalski, *Proc. Natl. Acad. Sci. U.S.,* **50,** 355 (1963).

55. M. Sekiguchi, A. Terawaki, T. Taguchi, and J. Kamawata, *Nature,* **183,** 1056 (1959).

56. N. Otsuji, M. Sekiguchi, T. Iijima, and Y. Takagi, *Nature,* **184,** 1079 (1959).

57. T. Ben-Porat, M. Reissig, and A. S. Kaplan, *Nature,* **190,** 33 (1961).

58. W. R. Smith and B. R. McAuslan, *J. Virol.,* **2,** 100b (1968).

59. A. Weissbach and D. Korn, *J. Biol. Chem.,* **237,** 3312 (1962).

60. E. Reich and R. M. Franklin, *Proc. Natl. Acad. Sci. U.S.,* **47,** 1212 (1961).

61. S. Kit, L. J. Piekarski, and D. R. Dubbs, *J. Mol. Biol.,* **7,** 497 (1963).

62. W. K. Joklik and Y. Becker, *J. Mol. Biol.,* **10,** 452 (1964).

62a. Y. Becker and W. K. Joklik, *Proc. Natl. Acad. Sci. U.S.,* **51,** 577 (1964).

63. G. C. Mueller and K. Kajiwara, in *Developmental and Metabolic Control Mechanisms and Neoplasia,* Williams and Wilkins, Baltimore, 1965, p. 452.

64. K. Lark, *Bacteriol. Rev.,* **30,** 3 (1966).

65. J. R. Kates and B. R. McAuslan, *J. Virol.,* **1,** 110 (1967).

66. B. R. McAuslan, *Virology,* **21,** 383 (1963b).

67. C. Jungwirth and W. K. Joklik, *Virology,* **27,** 80 (1965).

68. T. Ben-Porat and A. S. Kaplan, *International Congress on Microbiology, 9th, Moscow, 1966,* p. 463.

68a. A. S. Kaplan, personal communication.

69. L. Kucera and A. Granoff, *Bacteriol. Proc., Abstracts,* 158 (1967).

70. S. Kato, M. Takahashi, S. Kameyama, and J. Kamahora, *Biken's J.,* **2,** 353 (1959).

71. R. B. Arhelger, R. W. Darlington, L. G. Gafford, and C. C. Randall, *Lab. Invest.,* **11,** 814 (1962).

72. C. C. Randall and L. G. Gafford, *Am. J. Pathol.,* **49,** 51 (1962).

73. Y. Ichihashi and S. Matsumoto, *Virology,* **29,** 264 (1966).

74. S. E. Luria and J. E. Darnell, *General Virology,* 2nd ed., Academic Press, New York, 1967.

75. B. R. McAuslan, *Virology,* **20,** 162 (1963).

76. B. R. McAuslan, *Biochem. Biophys. Res. Commun.,* **19,** 15 (1965).

77. B. R. McAuslan, P. Herde, D. Pett, and J. Ross, *Biochem. Biophys. Res. Commun.,* **20,** 586 (1965).

78. S. Kit and D. R. Dubbs, *Virology*, **26**, 16 (1965).

78a. B. R. McAuslan, unpublished.

78b. B. R. McAuslan, in *Symposium on Enzyme Regulation and Metabolic Control, Mexico City, 1966*, National Cancer Institute Monograph 27, 1967.

79. D. R. Dubbs and S. Kit, *Virology*, **22**, 214 (1964).

80. W. Munyon and S. Kit, *Virology*, **26**, 374 (1965).

81. A. J. Shatkin and N. P. Salzman, *Virology*, **19**, 551 (1963).

82. J. R. Kates and B. R. McAuslan, *Biochim. Biophys. Acta*, **132**, 419 (1967).

82a. L. Eron and B. R. McAuslan, *Biochem. Biophys. Res. Commun.*, **22**, 578 (1966).

83. B. R. McAuslan and J. R. Kates, *Virology*, **26**, 742 (1967).

84. A. E. Oleson and J. F. Koerner, *J. Biol. Chem.*, **239**, 2935 (1964).

85. E. C. Short and J. F. Koerner, *Proc. Natl. Acad. Sci. U.S.*, **54**, 595 (1965).

86. S. Bose and N. Nossal, *Federation Proc.*, **23**, 272 (1964).

87. A. V. Paul and I. R. Lehman, *J. Biol. Chem.*, **241**, 3441 (1966).

88. H. V. Aposhian and D. M. Trilling, *Proc. Natl. Acad. Sci. U.S.*, **54**, 622 (1965).

89. K. Yamafuji and F. Yoshihara, *Nature*, **196**, 1340 (1962).

90. J. Mukai and K. Yamafuji, *Enzymologia*, **23**, 214 (1961).

91. I. R. Lehman, *Ann. Rev. Biochem.*, **36**, 645 (1967).

91a. A. Pitkanen, B. R. McAuslan, J. Hedgpeth, and B. Woodson, *J. Virol.*, **2**, 1363 (1968).

92. W. E. Magee, *Virology*, **17**, 604 (1962).

93. L. M. S. Chang and M. E. Hodes, *Virology*, **32**, 258 (1967).

94. H. V. Aposhian and A. Kornberg, *J. Biol. Chem.*, **237**, 519 (1962).

95. G. Appleyard, J. C. N. Westwood, and H. T. Zwartouw, *Virology*, **18**, 159 (1962).

96. G. Appleyard and J. C. N. Westwood, *J. Gen. Microbiol.*, **37**, 391 (1964).

97. P. C. Loh and J. L. Riggs, *J. Exptl. Med.*, **114**, 149 (1961).

98. F. Fenner, *Virology*, **5**, 502 (1958).

99. M. Oda, *Virology*, **25**, 664 (1965).

100. B. J. Neff, W. W. Ackermann, and R. E. Preston, *Proc. Soc. Exptl. Biol. Med.*, **118**, 664 (1965).

101. J. D. Stone, *Australian J. Exptl. Biol. Med. Sci.*, **24**, 191 (1945).

102. T. Shedlovsky and J. E. Smadel, *J. Exptl. Med.*, **75**, 165 (1942).

103. L. J. M. Randle and K. R. Dumbell, *J. Hyg.*, **60**, 41 (1962).

104. J. E. Smadel, in *Viral and Rickettsial Infections of Man*, 2nd ed. (Rivers and Horsfall, eds.), Lippincott, Philadelphia, 1952, p. 414.

105. G. Appleyard, H. T. Zwartouw, and J. C. N. Westwood, *Brit. J. Exptl. Pathol.*, **45**, 150 (1964).

106. G. Appleyard and J. C. N. Westwood, *Brit. J. Exptl. Pathol.*, **45**, 167 (1964).

107. H. T. Zwartouw, J. C. N. Westwood, and W. J. Harris, *J. Gen. Microbiol.*, **38**, 39 (1965).

108. J. Marquardt, S. E. Holm, and E. Lycke, *Virology*, **27**, 170 (1965).

109. G. H. Cohen and W. C. Wilcox, *J. Bacteriol.*, **92**, 676 (1966).

110. W. C. Wilcox and G. H. Cohen, *J. Virol.*, **1**, 500 (1967).

110a. J. G. Flaks, J. Lichtenstein, and S. S. Cohen, *J. Biol. Chem.*, **234**, 1507 (1959).

110b. J. S. Wiberg, M. L. Kirksen, R. H. Epstein, S. E. Luria, and J. M. Buchanan, *Proc. Natl. Acad. Sci. U.S.,* **48,** 293 (1962).

111. R. H. Epstein, A. Bolle, C. M. Steinberg, E. Kellenberger, E. Boy de la Tour, R. Chevally, R. S. Edgar, M. Susman, G. H. Denhardt, and A. Lelavsis, *Cold Spring Harbor Symp. Quant. Biol.,* **28,** 375 (1963).

112. B. R. McAuslan and J. R. Kates, *Proc. Natl. Acad. Sci. U.S.,* **55,** 1581 (1966).

112a. B. R. McAuslan, *International Congress on Microbiology, 9th, Moscow, 1966,* p. 501.

113. A. J. Shatkin, *Virology,* **29,** 292 (1963).

114. J. F. Sambrook, M. E. McClain, K. B. Easterbrook, and B. R. McAuslan, *Virology,* **26,** 738 (1965).

115. N. P. Salzman and E. D. Sebring, *J. Virol.,* **1,** 16 (1967).

116. N. P. Salzman, A. J. Shatkin, and E. D. Sebring, *J. Mol. Biol.,* **8,** 405 (1964).

117. A. J. Shatkin, *Nature,* **199,** 357 (1963).

118. J. J. Protass and D. Korn, *J. Biol. Chem.,* **241,** 475 (1966).

119. J. Pene and J. Marmur, *J. Virol.,* **1,** 86 (1967).

120. M. L. Dirksen, J. S. Wiberg, J. F. Koerner, and J. M. Buchanan, *Proc. Natl. Acad. Sci. U.S.,* **46,** 1425 (1960).

121. J. S. Wiberg, M. L. Kirksen, R. M. Epstein, S. E. Luria, and J. M. Buchanan, *Proc. Natl. Acad. Sci. U.S.,* **48,** 293 (1962).

122. A. S. Spirin, in *Current Topics in Development Biology* (A. A. Moscona and A. Monroy, eds.), Academic Press, New York, 1966, pp. 1–38.

123. B. David, R. Dulbecco, H. N. Eisen, H. S. Ginsberg, and W. B. Wood, *Microbiology,* Harper and Row, New York, 1967.

124. M. Oda, *Virology,* **21,** 533 (1963).

124a. J. A. Holowczail and W. K. Joklik, *Virology,* **33,** 717 (1967).

125. T. Kano-Sueoko and S. Spiegelman, *Proc. Natl. Acad. Sci. U.S.,* **48,** 1942 (1962).

125a. B. D. Hall and S. Spiegelman, *Proc. Natl. Acad. Sci. U.S.,* **47,** 137 (1961).

126. R. B. Khesin and M. F. Shemyakin, *Biokhimiya,* **27,** 761 (1962).

127. B. D. Hall, A. P. Nygaard, and M. H. Green, *J. Mol. Biol.,* **9,** 143 (1964).

128. G. Edlin, *J. Mol. Biol.,* **12,** 363 (1965).

129. E. K. F. Bautz, P. Kasai, E. Reilly, and E. A. Bautz, *Proc. Natl. Acad. Sci. U.S.,* **55,** 1081 (1966).

130. K. Oda and W. K. Joklik, *J. Mol. Biol.,* **27,** 395 (1967).

130a. W. K. Joklik, personal communication.

131. S. Bose and G. Warren, *Biochem. Biophys. Res. Commun.,* **26,** 285 (1967).

132. K. K. Mark and G. Streisinger, *International Congress on Biochemistry, 7th, Tokyo, 1967,* Vol. 5, Secretariat of the Seventh International Congress on Biochemistry, Tokyo, 1968. p. 677.

133. J. D. Friesen, B. Dale, and W. Bode, *J. Mol. Biol.,* **28,** 413 (1967).

134. M. Baldi and R. Haselkorn, *J. Mol. Biol.,* **27,** 193 (1967).

135. D. J. Bauer, *Brit. J. Exptl. Pathol.,* **44,** 233 (1963).

136. B. Woodson and W. K. Joklik, *Proc. Natl. Acad. Sci. U.S.,* **54,** 946 (1965).

137. M. H. Green, *International Congress on Microbiology, 9th, Moscow, 1966,* p. 509.

138. S. N. Cohen, U. Maitra, and J. Hurwitz, *J. Mol. Biol.,* **26,** 19 (1967).

139. S. N. Cohen and J. Hurwitz, *Proc. Natl. Acad. Sci. U.S.,* **57,** 1759 (1967).

139a. A. J. Shatkin and J. D. Sipe, *Proc. Natl. Acad. Sci. U.S.,* **61,** 1462 (1968).
140. F. Fenner and J. F. Sambrook, *Virology,* **28,** 600 (1966).
141. F. Fenner and F. N. Ratcliffe, *Myxomatosis,* Cambridge, New York, 1965.
142. F. Fenner and J. F. Sambrook, in *Ann. Revs. Microbiol.,* **18,** 47 (1964).

10

HERPESVIRUSES

BERNARD ROIZMAN

DEPARTMENT OF MICROBIOLOGY
THE UNIVERSITY OF CHICAGO
CHICAGO, ILLINOIS

10-1. Introduction

A. The Scope

Herpesviruses are not blessed with many reviews. The most recent and also the most extensive summation of the herpesvirus literature is that of Kaplan (1). The bibliography of the past three years hardly justifies a revision of this review. This chapter is different in scope, in that it is written for the prospective student of herpesviruses in search of a project and a perspective. It attempts to analyze the strength and weakness of data concerning the structure and replication of herpesviruses. By design, sections dealing with well-documented material are trimmed to the bone, whereas some obscure phenomena and controversial data are resurrected and discussed at length. This survey of the herpesvirus literature was completed in August, 1967 but a few, selected entries were made one year later. The bibliography, in keeping with the design, is selective but not balanced. Some omissions were inadvertent oversights. Others were prompted by charitable and humanitarian instincts.

B. Apologia

The first experimental transmission of a herpesvirus to a heterologous host was performed more than half a century ago (2). Since then, herpesviruses and diseases they cause have been the subject of several thousand articles. However, if the objective of much of the research effort in virology is a description of the information content of viruses and its expression in other biologic entities, the study of herpesviruses trail those of newer viruses boasting a much less illustrious ancestry. Possibly two factors account for this information gap. First, the molecular biology of animal viruses evolved from studies on pathogenesis and prevention of the disease they cause. While herpesvirus infections of man may be severe, they are rarely crippling, seldom lethal, and never cause epidemics of great military or economic importance. Their chief claim to infamy is recurrent infection of the cornea—a major cause of blindness in the United States. This, alas, is not an affliction common to legislators concerned with appropriation or to the trustees of foundations dedicated to the betterment of public health and thus it came about that the methodology for the studies of herpesviruses owes its existence largely to pressing needs of research on other viruses. The second and perhaps more important reason for the lag in studies of herpesviruses is that they are among the most complex if not also the most cantankerous viruses infecting animal cells. Nevertheless, herpesviruses and the diseases they cause share in common many unique and very fascinating features; some of which are dealt with in this chapter.

10-2. Chemistry and Architecture of the Herpesvirion

A. DEFINITION

1. *Derivation of the Name*

The word herpes is derived from ἑρπειν, meaning to creep. It has been used in medicine for at least twenty-five centuries. In the Hippocratic Corpus the term was applied to spreading cutaneous lesions of varied etiology, usually ulcerative, severe, and difficult to treat (*3*). In the intervening centuries the use of the term became more restricted: herpes zoster, derived from ζωστηρ, a girdle, alone retained the name herpes throughout the centuries. A clear account of herpes labialis as a distinct clinical entity appeared in 1694 (*3*). Herpes catarrhalis, pro genitalis, facialis, and simplex appear in the 18th and 19th centuries (*3*). Only after the discovery of the causative agent has it become clear that herpes genitalis, facialis, and febrilis are all different clinical manifestations of one disease, herpes simplex. The virus took on the name of the disease.

2. *Classification*

Herpesviruses are formally defined as large enveloped virions with a well-defined icosahedral capsid consisting of 162 capsomeres arranged around a DNA core. The viruses usually included in the herpesvirus group (*4*) are herpes simplex, B virus, marmoset virus, pseudo-rabies virus, equine herpesvirus, varicella-zoster virus, cytomegulo-viruses (of man, guinea pig, mouse, swine, etc.), a number of bovine (lumpy skin disease, infectious rhinotracheitis, and mammillitis viruses), canine, avian, rabbit, (virus III), and feline herpesviruses. Only common names are used in this chapter. The binomial name (*5*), which contains in addition to the generic "Herpesvirus" the name of the species in which the virus is found, generously allows only one herpesvirus per host species. The viruses meeting the structural and architectural criteria for inclusion into the herpesvirus group also share in common many unique features of their replicative processes. Among the most interesting aspirants for inclusion in the herpesvirus group are the viruses associated with Burkitt's lymphoma of children (*6–11*), Marek's disease of fowl (*12–15*), and Lucke's adenocarcinoma of frogs (*16–18*). These viruses superficially resemble in structure members of the herpesvirus group. However, the nucleic acid of these viruses has not been characterized as yet. These viruses are not considered in this review.

B. THE ARCHITECTURE

1. *General Description*

There is general agreement that the herpesvirion contains a core located in the center of the virion. The core is surrounded by a *capsid*. Recent studies suggest that the capsid may consist of three shells, an *inner, middle* and an *outer capsid*. The core and capsid form the nucleocapsid. The virion is defined in this chapter as a nucleocapsid surrounded by an envelope. In the same vein, the core surrounded by the hypothetical inner capsid and the nucleocapsid are collectively referred to as structural intermediates. The size and characteristics of the architectural components are best described in frame of reference of the methods which led to their discovery. The information concerning the structure of the herpesvirion is derived from studies on negatively stained lysates of infected cells and on thin sections of infected cells embedded in a suitable plastic.

2. *Negative Staining*

The most comprehensive discussion of the structure of herpesviruses as seen in negatively stained preparations is that by Wildy et al. (*19*). Briefly, the nucleocapsid and the virion are readily seen in lysates of infected cells stained with phosphotungstic acid (PTA) or with sodium silicotungstate. A third structure, smaller and possibly a core surrounded by an inner capsid, is also seen in freshly prepared lysates of infected cells and in material treated with formaldehyde immediately after homogenization of the cell (*20*). In negatively stained preparations the virion appears to be 150–180 mμ in diameter (Figs. 10-1a, 10-1b). Intact enveloped nucleocapsids are generally impervious to PTA. The stain readily penetrates virions modified by drying or by exposure to antibody reactive with the envelope (*21*). Virions appropriately treated and stained with PTA show a nucleocapsid approximately 100 mμ in diameter and an envelope 20 mμ thick (Fig. 10-1a). The envelope consists of two layers, each showing a repeating unit structure (Fig. 10-1b). Occasionally the outer layer of the envelope appears to consist of spikes; this happens particularly when excess negative stain fills the space between the repeating unit. The outer capsid (Fig. 10-1c) consists of 162 capsomeres arranged to form an icosahedron. The outer capsomeres appear to be elongated prisms (12 mμ), hexagonal in cross section; since they are partially filled with PTA it is inferred that the distal end is hollow (*19*). The third particle observed in PTA-stained preparations consists of a round body 20–25 mμ in diameter surrounded by a beaded structure 5–8 mμ in width (Figs. 10-1d, 10-1e). A hypothesis concerning the identity of this particle emerged from recent studies showing that both naked and enveloped nucleocapsids become disaggregated on prolonged isopycnic centrifugation in CsCl solutions (*22*). The effect of the CsCl appears to be much greater on nucleocapsids than on virions (Figs. 10-1f, 10-1g). Electron microscopy (*20*) of banded material dialyzed against distilled water and stained with PTA revealed the presence of numerous partially degraded nucleocapsids. These range from a small minority lacking a few outer capsomeres to numerous particles with just a few capsomeres projecting from the surface of a body approximately 75 mμ in diameter. The 75-mμ structure is very similar to the one described by Wildy (*19*) and might well be a middle capsid. Various considerations (*208*) have led to the conclusion that the 20–25-mμ particle surrounded by a beaded structure 5–8 mμ in width corresponds to the 50-mμ particle observed in preparations of bovine herpes (*23*) and consists of a core and inner capsid, respectively.

In evaluating the results obtained with negative staining two observations which detract from the value of the method should be noted. The first is that the process of staining and drying probably generates sufficient surface tension to cause considerable distortion of the material on the grid. This problem became of concern when it was observed that while the envelope is 20 mμ thick and the nucleocapsid has a diameter of 100 mμ the intact virion appears to have a diameter ranging from 150–180 mμ. The discrepancy between the calculated maximum diameter of 140 mμ and that actually observed might be due to collapse and flattening of the envelope due to surface tension. The nucleocapsids from fresh preparations appear to be more stable. However considerable flattening and distortion is frequently seen in nucleocapsids banded in sucrose or CsCl prior to negative staining

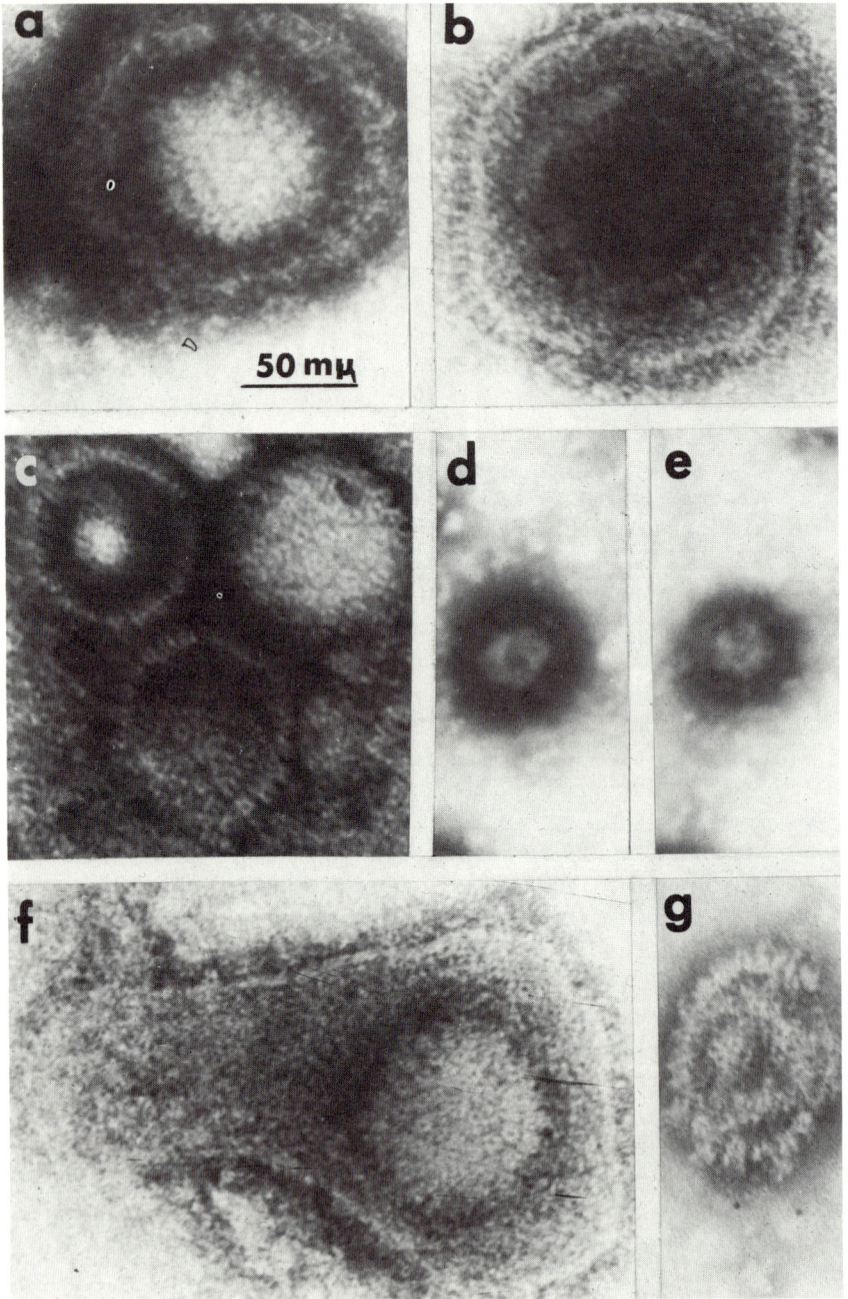

Fig. 10-1. Electron photomicrographs of negatively stained preparations of herpes simplex virus (20). (a) Virion showing detail of nucleocapsid surface,

(20). Such "collapsed" nucleocapsids would be expected to have a slightly larger diameter than undistorted ones. The distortion introduced by the method may well account for the variation in the size of the virion and nucleocapsid on record in the literature. The second point concerns variation in the penetrability of PTA. Early studies suggested that nucleocapsids impervious to PTA were "full," whereas those stained by PTA were devoid of cores. A recent report by Watson (21) dispels this myth.

3. Electron Microscopy of Thin Sections

Electron micrographs of thin sections stained with uranyl acetate and lead citrate or some other suitable stain usually show three different sized particles (20, 24–37). The smallest particle, 25–35 mμ in diameter, has been designated as the core and may be found in the nucleus of the infected cell (30, 32, 37). The presence of DNA in the core has been enzymatically shown by Epstein (32). The intermediate particle. 100 mμ in diameter, may be found in the nucleus (Figs. 10-2a, 10-2b), dispersed or in crystalline arrays (27, 30, 31, 33), and in the extracellular fluid. It consists of a nucleoid surrounded by two concentric shells. The inner shell is electron transparent, whereas the outer shell stains prominently (electron opaque). The largest particle is rarely found in the nucleus (26, 28, 30, 35). It is approximately 150 mμ in diameter and consists of a core and four concentric shells alternating in intensity of staining. It has been observed in this laboratory and by others (30, 34) that the core in the intermediate and large particles differ with respect to staining properties. This is not a consistent finding and its meaning remains obscure. If heavier staining is indicative of the presence of nucleic acid in the core, it could be argued that only nucleocapsids containing DNA become enveloped.

It has been generally assumed that (i) the intermediate and large particles seen in thin section correspond to the naked and enveloped nucleocapsid respectively, and (ii) electron-transparent shells are artifacts due to shrinkage of the nucleoid and nucleocapsid during fixation. This view leaves no structure corresponding to the inner nucleocapsid observed in negatively stained preparations and conveniently classifies

---

(b) virion showing detail of the structure of the envelope, (c) unenveloped nucleocapsids, (d, e) particle tentatively identified as a core surrounded by an inner capsid, (f) virion banded on isopycnic centrifugation in CsCl solution, (g) partially disrupted nucleocapsid recovered after isopycnic centrifugation in CsCl. Magnification b-g is the same as a.

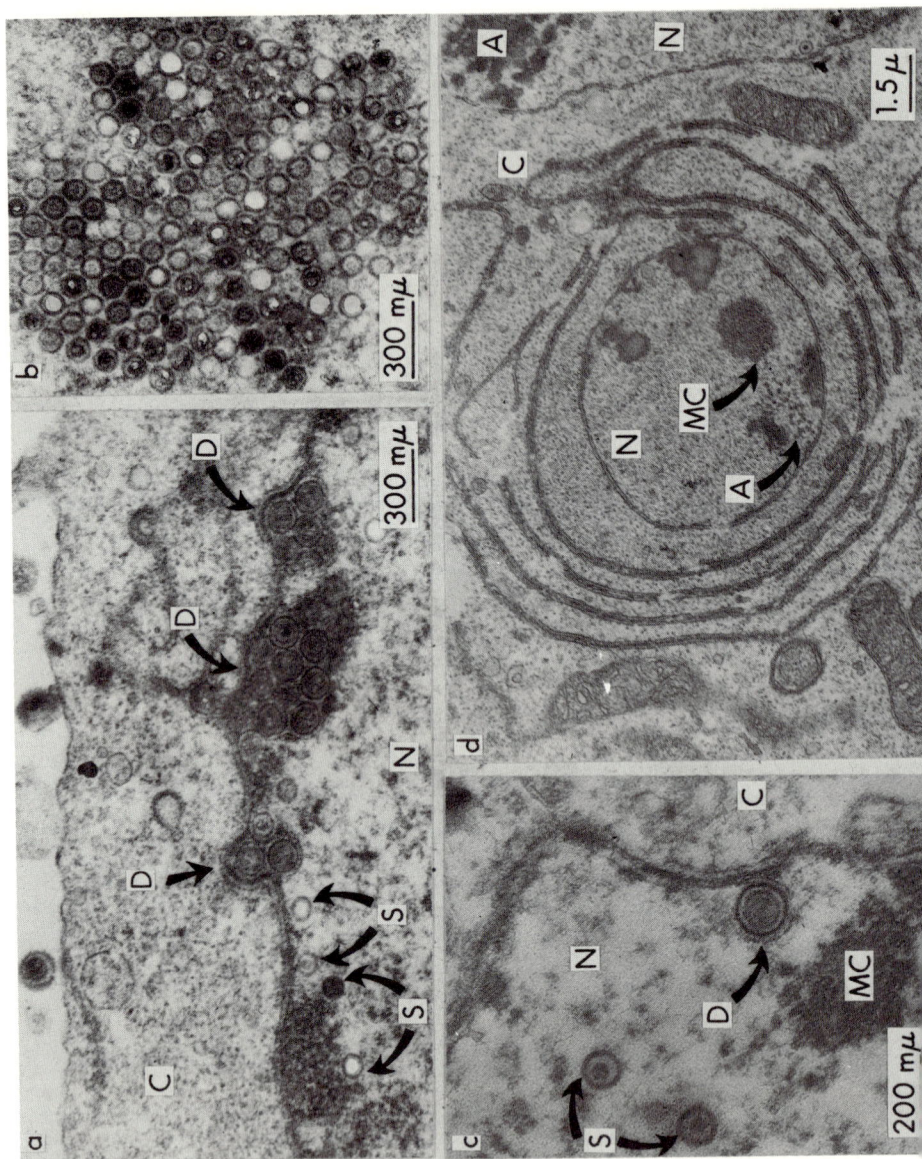

FIG. 10-2. Electron photomicrographs of thin sections of HEp-2 cells 18 hr after infection with herpes simplex virus (20). (a) Accumulation of viral particles with two electron-opaque shells between the layers of nuclear membrane, (b) intranuclear virus crystal, (c) intranuclear viral particle with two electron-opaque shells, (d) proliferation of membranes around the nucleus. Abbreviations: N, nucleus; C, cytoplasm; S, particle with single electron-opaque shell; D, particle with two electron opaque shells; MC, marginated chromatin; A, aggregates of

as artifacts any structure it cannot explain. It seems most unlikely that the nucleoid consists of DNA only, without some other component to ensure proper folding and to stabilize it; whether this hypothetical component accounts for the inner electron-transparent shell around the electron-opaque nucleoid is very uncertain. There is more substantial evidence that the second (outer) electron-transparent shell is not an artifact. Thus, throughout all stages of envelopment, the outer electron-opaque shell is invariably a fixed distance away from the nuclear membrane. These observations suggest that the space between the particles is not a void generated by shrinkage due to fixation but rather it is filled by a component (inner envelope) that stains poorly. It seems clear from the foregoing that other than the surface description furnished by Wildy et al. (19) and the strong likelihood that the DNA is in the center (32), the structure of the nucleocapsid is largely unknown.

C. STRUCTURAL COMPONENTS

1. Purification

Chemical analysis of viruses requires virus preparations of reasonable purity with respect to the specific component being analyzed. There have been numerous reports claiming preparations of virus free from certain host components. Thus consecutive differential (10,000 g × 30 min), rate (30,000 g × 90 min through 15–50% w/w sucrose), and isopycnic (105,000 g × 40 hr in CsCl initial density 1.28) centrifugations preceded by chromatography on brushite columns and followed by nuclease treatment will generally render virions free from detectable amounts of host nucleic acids at the cost of some 90% of the starting material (22, 39, 40). However, none of the published procedures, too numerous to list here, satisfactorily separated an artificial mixture of unlabeled virus and ^{14}C amino acid labeled debris from uninfected cells (40).

2. Viral DNA

Russell (41), and Ben-Porat and Kaplan (42) were, independently, the first to show by direct methods that herpesviruses contain DNA. To Ben-Porat and Kaplan must go the recognition for the demonstra-

---

electron-opaque bodies found frequently in association with particles containing a single electron-opaque shell.

tion that viral DNA differs significantly from host DNA with respect to its base composition—a finding of considerable usefullness in that it permits a simple and effective separation of the two. The base composition and size of the DNA of herpesviruses are shown in Tables 10-1 and 10-2, respectively. In general there is good agreement that (i) the DNA of all herpesviruses have a $G + C$ content greater than that of the animal cells in which they are grown (39, 42–49), (ii) the DNA is double stranded (41, 43), (iii) the molecular weight is in the range of $50–80 \times 10^6$ daltons (46–48). There is no evidence upon which to voice suspicion that herpesvirus DNA contains uncommon bases. As indicated earlier (Sec. 10-2, B, 3) the DNA is probably localized in the nucleoid (32).

3. Viral Proteins

It seems probable that proteins specified by the virus are present in the nucleoid, the capsid, and the envelope. Available information concerning the number and properties of the protein constituents of the virion is discussed in Sec. 10-3, D, 1.

4. Lipids

The presence of lipids is deduced primarily from the loss of biologic activity following exposure of the virus to lipid solvents (50) and to lipases (20). The nature of the lipids is unknown. It seems pertinent to mention here, however, that the source of genetic information for the synthesis of the lipid contained in the herpesvirion has not been unequivocably established. Virus grown in cells prelabeled with choline or in the presence of labeled choline became labeled and retained the label on isopycnic and rate centrifugations. Alas, virus was labeled not as well but with equal tenacity by artificially mixing unlabeled virus with labeled debris of uninfected cells (40).

D. Physical Properties of the Herpesvirion and of the Structural Intermediates

1. Buoyant Density of the Virion

On isopycnic centrifugation in CsCl solutions the herpesvirion bands at a density ranging from 1.255 to 1.280 depending on four factors, i.e., (i) length of centrifugation, (ii) prior treatment of the cell lysate, (iii) virus strain, and (iv) the cell in which the virus was produced.

In general, prolonged centrifugation in CsCl tends to increase the buoyant density. Thus the difference between the buoyant density obtained from preformed and self-forming gradients centrifuged at the same rate for 5 and 48 hr, respectively, may be as much as 0.02 g/cm³ (20, 22). Concomitant with the increase in buoyant density there is considerable disaggregation of virions (20, 22). Herpes simplex virus may be stabilized by pretreatment with formaldehyde (22). The formalinized virus bands at a density of 0.015 g/cm³ higher than the untreated one (22).

Mutants of herpes simplex have been shown to differ in buoyant density under specified conditions of centrifugation (52–55). In addition, the buoyant density of the virus may change if it is grown in a different host (56). However, the density of any one mutant produced in any one host is constant and reproducible (53, 56). The modification of herpes simplex virus induced by the host is likely to be a reflection of the structural components of the envelope contributed by the host. The evidence that host components are present is deduced from the nature of the envelopment process (Sec. 10-3, E, 2) and demonstrated more convincingly in agglutination tests with antibody against specific host-membrane antigens (57). It is perhaps not too surprising that the difference between the buoyant density of various herpesvirus mutants may also be a reflection of the composition and structure of the envelope. The conclusion is based on observations that mutants differing in buoyant density invariably differ with respect to surface characteristics such as elution from calcium phosphate gels and immunologic specificity and with respect to their effect on the social behavior of infected cells (50, 53–55, 58). The most plausible explanation for these covariant properties of the herpesvirus virion is that virus constituents of the envelope (i) confer immunologic specificity, (ii) determine the interaction with calcium phosphate gel and with CsCl ions, and (iii) are responsible for the modification of host membranes underlying the alteration in social behavior of infected cells.

2. Buoyant Density of the Structural Intermediates

The nucleocapsid (Fig. 10-1c) and particles (Figs. 10-1d, 10-1e) tentatively identified as cores are surrounded by an inner capsid band in CsCl solutions at a density of 1.305 (20, 22). Both of these particles are less stable in high salt solutions (Fig. 10-1g) and are less well stabilized by formaldehyde than virions (20, 22). A subviral particle containing viral DNA has also been found to band in

TABLE 10-1
Base Composition of Herpes Virus DNA

Virus	Authors	Method	C+G mole per cent
Herpes simplex	Ben-Porat and Kaplan 1962 (*42*)	Fractionation of ^{32}P-labeled deoxyribonucleotides	74
Herpes simplex		Spectrophotometrically (260 mμ)	
Herpes simplex	Russell and Crawford 1963 (*44*)	T_m	68
Herpes simplex		Buoyant density in CsCl	68
Herpes simplex		Spectrophotometrically (260/280 at pH 3)	67
Herpes simplex	Roizman et al. 1963 (*45*)	Buoyant density in CsCl	67
Herpes simplex	Russell and Crawford 1964 (*47*)	Buoyant density in CsCl	68
Herpes simplex	Lando et al. 1965 (*49*)	Fractionation of ^{32}P-labeled deoxyribonucleotides	65 ± 2.1
Pseudorabies	Ben-Porat and Kaplan 1962 (*42*)	Fractionation of ^{32}P-labeled deoxyribonucleotides	74
Pseudorabies	Russell and Crawford 1964 (*47*)	Buoyant density in CsCl	74
Pseudorabies	Kaplan and Ben-Porat 1964 (*46*)	Buoyant density in CsCl	73
Infectious bovine Rhinotracheitis	Russell and Crawford 1964 (*47*)	Buoyant density in CsCl	71
Equine herpes (equine abortion)	Darlington and Randall 1963 (*43*)	Fractionation of deoxyribonucleotides	56
(equine abortion)	Russell and Crawford 1964 (*47*)	Buoyant density in CsCl	55
(LK)	Russell and Crawford 1964 (*47*)	Buoyant density in CsCl	56
Human cytomegalo virus	Crawford and Lee 1964 (*48*)	Buoyant density in CsCl	58

TABLE 10-2

The Molecular Weight Estimations of Herpes Simplex

Virus	Authors	Method	Daltons
Herpes simplex	Ben-Porat and Kaplan 1962 (42)	Average DNA content per virion	4.5×10^6
	Russell and Crawford 1964 (47)	Sedimentation coefficient	68×10^6
Pseudorabies	Kaplan and Ben-Porat 1964 (46)	Band width in CsCl[a]	35×10^6
	Russell and Crawford 1964 (47)	Sedimentation coefficient	68×10^6
Infectious bovine Rhinotracheitis	Russell and Crawford 1964 (47)	Sedimentation coefficient	54×10^6
Equine herpes (LK)	Russell and Crawford 1964 (47)	Sedimentation coefficient	84×10^6
Human cytomegalo virus	Crawford and Lee 1964 (48)	Band width in CsCl[a]	32×10^6

[a] The values obtained by this method are approximately one-half the true value.

CsCl at a density of 1.410 g/cm^3 on isopycnic centrifugation of formalinized lysates of infected cells. The identity of this particle is unknown; it could not be detected in nonformalinized lysates centrifuged in CsCl.

3. Immunologic Specificity of Herpesvirions and of Structural Intermediates

Watson and Wildy (57) have shown in antibody agglutination tests that virions differ from nucleocapsids with respect to immunologic specificity. The envelope of the virion contains host-determinant antigens (57) in addition to viral antigens (55, 59). There is no information concerning the immunologic specificity of other structural intermediates.

4. Degradation of the Herpesvirions and of Structural Intermediates

It has been reported that ethyl ether disrupts the envelope without affecting the nucleocapsid and that the nucleocapsid, in contrast to the virion, disaggregates at pH 4 but remains stable following treatment with trypsin, ficin, papain, ethyl ether, and detergents (19). Studies on controlled degradation of herpesvirions are sorely lacking.

E. Architectural Components and Biologic Function

1. The Problems

The most important biologic function is the ability to reproduce on infection of a suitable host. This section therefore deals primarily with the features of the virion which determine the capacity to infect the cell. The experimental modifications of the virion which render it noninfectious are dealt with in part here, in part in Sec. 10-3, A, 1. The covariation between certain physical properties of the virion and its effect on the social behavior of cells is discussed in Secs. 10-2, D, 1 and 10-4, C, 2.

2. Infectivity: The Role of the Envelope

The discovery of enveloped and naked nucleocapsids (19) in herpesvirus preparations raised the sensible question whether either or both are infectious to the level of a transatlantic polemic. Holmes and Watson (60), corroborated by Siegert and Falke (37), reported that enveloped particles were more readily adsorbed to cells than naked ones, but they indicated that this may reflect only the size difference. Watson et al. (34) subsequently reported that in some preparations the number of plaque forming units exceeded the number of enveloped particles. This finding led to the conclusion that probably both kinds of particles were infectious albeit not necessarily with the same efficiency. This conclusion was challenged by Smith (61) who separated naked and enveloped particles on isopycnic centrifugation in CsCl density gradients and found infectivity associated only with enveloped particles. A priori it would seem that, assuming CsCl is not deleterious, separation of enveloped and nonenveloped nucleocapsids should be straightforward, whereas considerable error is inherent in the particle-counting technique. However, isopycnic centrifugation in CsCl is deleterious in that it causes disassembly of the herpesvirion (22). Moreover, electron microscopic comparison of the enveloped and non-enveloped forms recovered after centrifugation clearly indicates that naked nucleocapsids are more severely damaged by CsCl (20). Since Smith (61) does not furnish the amounts of naked and enveloped particles at the beginning and end of the centrifugation, his results cannot be used to differentiate between the two conflicting hypotheses.

It seems clear that the answer to the question whether the naked nucleocapsid is infectious requires unbiased separation of the two particles and this has not yet been done. Two points, however, can be made. First, there is ample evidence that enveloped virus is infectious. In addition to the evidence put forth by Holmes and Watson (60) and by Smith (61) it has been shown that lipid solvents (50) and lipases (51) inactivate the capacity to replicate. The implied conclusion that lipids are localized in the envelope is based on findings that (i) the envelope structurally resembles and is derived from host membranes (28–38), (ii) the envelope contains host antigenic determinants (57), and (iii) ether disrupts the envelope leaving the nucleocapsid intact (19). Moreover, the imperviousness of the enveloped nucleocapsid to PTA (21) and its relatively greater stability in salt solutions (51) suggest that the envelope should not be freely permeable to lipases. It would be expected therefore that inactivation of herpesvirus by lipases should follow second- or higher-order kinetics if the essential lipids were located in a structure other than the envelope. This is not the case (51), and it must be assumed that the lipid essential to infection forms the fabric of the envelope. The second point that should be emphasized is that an intact envelope is not a prerequisite for infection. Thus based on the observation that nucleocapsids are assembled in the nucleus but acquire the envelope while leaving it (Sec. 10-3, E, 2), the infectious virus present in extracts of nuclei and cytoplasm of infected cells was studied with respect to its physical properties (51). The data show that infectious nuclear virus sediments more slowly in sucrose density gradients and is considerably less stable in salt solutions than cytoplasmic virus. However, the infectivity of both nuclear and cytoplasmic preparations was inactivated by lipases. The data were not sufficient to determine whether the infectious nuclear virus is partially or completely unenveloped.

3. Infectivity—Inactivation

In addition to lipid solvent and lipases, herpesviruses are inactivated by trypsin (62), alkaline and acid phosphatases (63), urea (40), sodium deoxycholate (64), nitrous acid (65), ultraviolet light and X-ray (66, 67); and heat (68–73). The effects of all these agents are not in themselves unusual. Because many inactivation studies cannot be done with purified virus in a menstruum free from compounds protecting or competing with the virus for the inactivating agent, conclusions that one virus is more or less stable than another may well be unfounded. The critique is particularly applicable to any attempt to rationalize differences in the kinetics of heat inactivation reported from different laboratories. One method to control inactivation rate is to work with an artificial mixture of two viruses in which one serves as a standard for comparative purposes (53, 55).

The stability of herpesvirus has been of considerable concern in early studies of the kinetics of virus multiplications in cultured cells and in studies of virus preservation (68–75). Herpesviruses are clearly not very stable in cell culture medium at 37°C or, in fact, at any temperature above —20°C. In our experience herpes simplex is relatively stable for several weeks at 4°C as a 10% cell extract in distilled water (75, 76) and very stable for several years at —70°C as a 50% skim milk—50% cell extract in culture medium. Wallis and Melnick (77) reported that 1.0 M Na$_2$SO$_4$ or 1.0 M Na$_2$HPO$_4$ stabilize the virus remarkably well against heat inactivation at 50°C. In our experience 1.0 M Na$_2$SO$_4$ is of no particular advantage at 4°C or at —70°C. Alas, not too many experiments can be done at 50°C in 1.0 M Na$_2$SO$_4$.

10-3. Replication of Herpesviruses

A. INITIATION OF INFECTION

1. *Adsorption*

The rate of adsorption of herpesvirus like those of other viruses is dependent on the volume of virus inoculum, the presence of cations, the metabolic state of cells but not, within 4–37°C limits, on the temperature of incubation during adsorption. It takes approximately 2 hr to adsorb 200 PFU of herpes simplex virus to 10^6 cells in 1 ml of fluid (78, 79). The adsorption is even slower if the cells normally adhering to glass during growth are suspended prior to exposure to virus (80). Enveloped nucleocapsids adsorb more readily than naked nucleocapsids (37, 60). The nature of the receptor is unknown. Cells naturally lacking receptors for herpesvirus have not been described. However, HEp-2 cells exposed to parathyroid hormone (78) temporarily lost the capacity to absorb virus. Thyroid hormone in the same test accelerated absorption (81).

2. *Penetration*

Adsorbed virus may be prevented from infecting the cell by the timely addition of antibody to the extracellular fluid. Adsorbed virus inaccessible to antibody is defined as having penetrated into the cell. Penetration is temperature dependent and requires a cell capable of expending energy. Huang and Wagner (82) found that once the virus adsorbs to the cell, penetration, as defined, is relatively rapid. The virus is taken into the cell in a pinocytotic vesicle (37, 60), but very little is known of the fate of the virus afterwards.

3. *Interference with Adsorption to Cells*

Since 1962 several laboratories reported that sulfated polyanions, both natural (agar mucopolysaccharide, heparin) and synthetic, inhibit virus multiplication if present in the medium at the time of exposure of virus to cells (83–89). It has been reported that the inhibitory effect depended, within limits, on the degree of sulfation and the size of molecule; it was not dependent on the nature and degree of branching of the polysacchride backbone (87). The polyanions act primarily to prevent the adsorption of virus to cells (84, 86–88). The effect of sulfate groups in agar mucopolysaccharides may be neutralized by the addition of protamine (*85*). The effects of other polyanions are abolished by dilution.

B. General Survey of the Reproductive Cycle

1. *Information Content*

Ultimately an account of herpesvirus replication will contain a complete description of the physical properties, functions, amount, and time of synthesis of all the products specified by the virus in the infected cells. At present the number of products and their function is uncertain. According to current accounting practices, herpesvirus DNA carries information sufficient to make over 100 peptides each with an average molecular weight of 50,000 daltons. Herpesviruses are not as complex as T phages and moreover the number seems astronomical by comparison with the information content of papova viruses, which also multiply in the nucleus or with myxoviruses, which also have an envelope. What then is the need for so much genetic information?

A point should be made here, more for its heuristic value than to answer the question itself. The point is that all of the genetic information carried by the virus may not be expressed in cell cultures. The corollary is that analysis of the viral replication in cell culture may not reveal all of the genetic information carried by the virus. The argument is based on the fact that all laboratory strains were recovered from sick individuals at one time or another. Prior to that instant the information content of the virus was shaped and molded for many millenia for better survival in the complex multicellular organism it normally infects. Unlike the small DNA and RNA viruses, the herpesviruses have established in the course of evolution a unique relationship with the host they usually infect (*90*). The main feature of this relationship is that following primary infection, and in spite of the appearance of antibody, herpesviruses survive asymptomatically in some specific tissue for the lifespan of the host. The capacity to coexist is not an indication that the virus is incapable of inflicting injury: Herpesviruses indeed cause death or very severe illness in susceptible animals with which they do not normally come in contact in nature and moreover, they are uniformly destructive in cell cultures from those species. In the light of their natural history it seems reasonable to postulate that herpesviruses express their genetic potentialities more fully and effectively in the hosts with which they coexist than in the ones they destroy. Alas, it is difficult to carry out meaningful biochemical experiments in experimental animals infected with virus native to them. The cell culture is best suited for this purpose, but it is not the native habitat of herpesvirus in the evolutionary sense. For this reason alone, if one is needed, it seems unlikely that the entire information content of herpesviruses will ever be determined from studies on cultures of dispersed undifferentiated cells.

2. The Reproductive Cycle

Studies of the growth cycles of herpesviruses in more or less synchronously infected cells were reported by a number of investigators (68–70, 91–93). For the purpose of this discussion the reproductive cycle is best described in terms of the factors affecting four parameters, i.e., (i) the duration of the eclipse (time of appearance of new virus), (ii) duration of the reproductive cycle, (iii) the yield of virus per cell, (iv) virus release from cells.

(i) The duration of the eclipse varies from 3–8 hr for most herpesviruses. It is affected by the temperature (68, 70, 92) of incubation (increased below 34°C), by the multiplicity of infection, and by prior infection of cells with another mutant (94, 96) (decreased). Once a minimum eclipse period is attained (3 hr for pseudorabies in rabbit kidney cells, 5 hr for herpes simplex in HEp-2 cells) it cannot be shortened by increasing the multiplicity of infection (93, 94).

(ii) The duration of the reproductive cycle of herpesvirus varies considerably. The cycle of herpes simplex in HEp-2 cells lasts from 13–19 hr, depending on the multiplicity of infection and temperature of incubation; at 37°C and 50 plaque forming units per cell the cycle lasts 17 hr. The cycle of pseudorabies in rabbit kidney cells is somewhat shorter (93).

(iii) The virus yield from infected cells increases exponentially from the end of the eclipse phase until almost the end of the reproductive cycle. Under optimal conditions the yield of herpes simplex virus is 10,000 to 100,000 virions per HEp-2 cell. The best preparation of Watson et al. (34) contained about 10 virions per plaque forming unit; routine preparations of virus contain on the order of 100 to 1000 virions per plaque forming unit (92). The relative amounts of enveloped and naked nucleocapsids in herpesvirus-infected cells vary not only with the conditions of infection but also with the host species in which the virus is grown. The yield of virus is profoundly affected by the metabolic state of the cell (80), pH of the medium (95), and by the temperature of incubation (68, 70, 92), but not by the multiplicity of infection (80). Rapidly multiplying cells produce more virus than cells arrested by contact (density) inhibition (80). Cells grown and maintained in monolayer cultures usually yield more virus than those suspended after infection (80). The optimal temperature and pH for the reproduction of herpesviruses is strain specific (95, 96). Most mutants of herpes simplex virus multiply best at 34°C but some do better at 37°C (96).

(iv) Infectious herpesvirions first appear inside the infected cells. The release is generally slow and temperature dependent (70). The virus is released more readily from suspended cells than from those adhering to glass surfaces. At 34°C or below, herpes simplex virus release is slow and inefficient. The culture fluid may be discarded and the virus extracted by freezing and thawing or sonicating the infected cells in a small volume of fluid. Freezing and thawing releases numerous aggregates from infected cells; brief sonication to disperse the virus has been recommended (92).

3. *The Patterns of Macromolecular Synthesis*
   During the Reproductive Cycle

Information concerning DNA, RNA, and protein synthesis is available for BHK-21, HEp-2, and KB cells infected with herpes simplex (39, 97–102) and in part for rabbit kidney cells infected with pseudorabies virus (103). There have been no detailed studies of lipid or of carbohydrate synthesis in infected cells. The patterns for DNA, protein, and RNA synthesis in HEp-2 cells infected with herpes simplex are shown in Figs. 10-3–10-5. The main features of the data are an initial period of decline in the rate of synthesis of all three macromolecules (0–3 hr after infection), a period of leveling off or increase in macromolecular synthesis (3–9 hr after infection), and lastly, a period of gradual and irreversible decline. These three periods coincide approximately with (i) the inhibition of host macromolecular synthesis concomitant with the synthesis of nonstructural products specified by the virus, (ii) the synthesis of structural components of the virus, and (iii) assembly of the structural components into virions. The following sections deal with the synthesis of products specified by the virus and virus assembly. The inhibition of host macromolecular synthesis is discussed in Sec. 10-4, B, 2.

C. The Synthesis of Nonstructural Products Specified by the Virus

1. *mRNA*

By definition, viral RNA must be complementary to one of the strands of viral DNA. The operational definition of viral mRNA is that in addition to being complementary to viral DNA it should function by specifying the structure of viral proteins. To date evidence has been presented that (i) transcription of viral DNA is required for virus multiplication (*101, 104*), (ii) viral RNA is synthesized (*100, 101*), and (iii) new species of polyribosomes appear in the cytoplasm of infected cells (*98, 102*). The presence of viral RNA on the polyribosomes, although expected, has not yet been demonstrated.

According to Flanagan (*101*), in KB cells infected with herpes simplex virus, RNA annealable with viral DNA increases after infection, reaches peak levels 6–7 hr after infection, and declines slowly

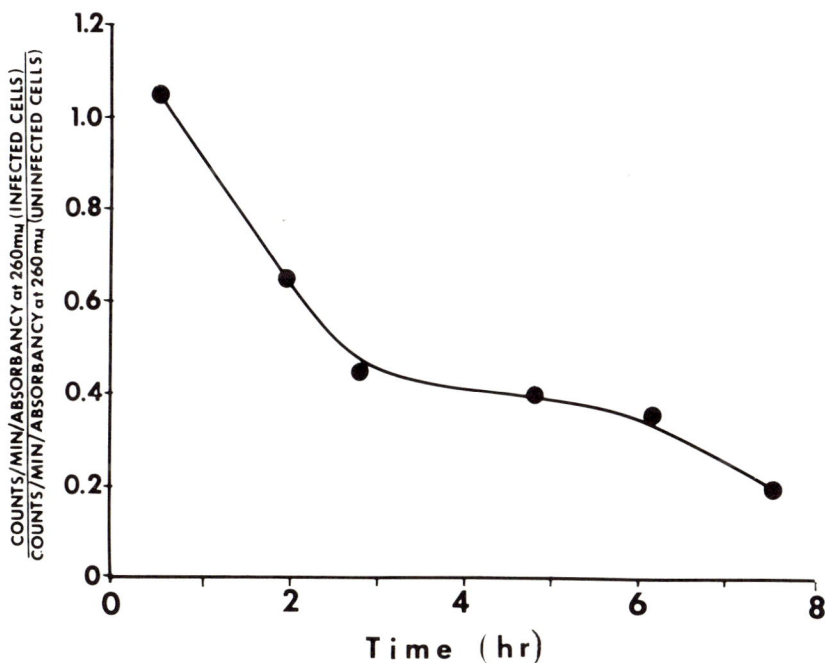

Fig. 10-3. The pattern of incorporation of ³H-uridine into RNA of HEp-2 cells infected with herpes simplex virus. The cells were pulse labeled for 30 min at various times after infection (*97, 99*).

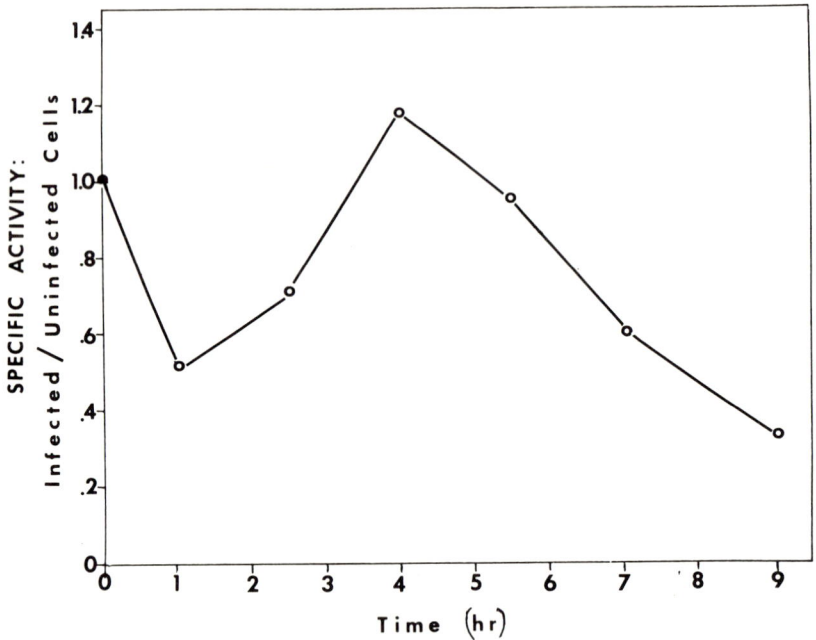

Fig. 10-4. The pattern of incorporation of ^{14}C amino acids into peptides of HEp-2 cells infected with herpes simplex virus. The cells were pulse labeled for 15 min at various times after infection (97, 98).

thereafter. The sedimentation coefficient of annealable RNA ranges between 12S and 32S but whereas longer (>32S) molecules anneal with viral DNA, those less than 12S do not. Flanagan noted that from 6–14 hr after infection the sedimentation patterns and presumably the size of the RNA molecules annealable to viral DNA remained constant. This finding would be expected if, 6 hr after infection the cell manufactured structural components of the virus exclusively. Hay et al. (100) published similar studies on the synthesis of herpes simplex virus RNA in BHK-21 cells. They agree with Flanagan on the general pattern of synthesis of viral RNA. They differ however in that they ascribe to viral RNA a sedimentation coefficient of approximately 20S and indicate moreover that a 4S RNA is also synthesized. The data of Hay et al. (100) suffer from the fact that in crucial experiments BHK-21 cells were not uniformly infected as evidenced by the synthesis of ribosomal RNA and by the annealing of a substantial proportion of labeled RNA with cellular DNA.

It seems reasonable that the viral RNA described by Flanagan and by Hay et al. functions as the mRNA backbone of cytoplasmic poly-ribosomes, but this remains to be proven. Nothing is known of (i) the site of synthesis, (ii) the source of the enzyme, (iii) the time of synthesis and the number of different kinds of viral RNA.

2. sRNA

In a series of papers published since 1965, Subak-Sharpe and co-workers (*100, 105–107*) presented data in support of the hypothesis that herpesviruses synthesize at least one but probably several species of sRNA. The importance of these experiments stems from two unique functions of sRNA corresponding to two sites on the molecule. On one site the sRNA accepts a specific activated amino acid to form amino-acyl–sRNA. On the other site it contains bases complimentary to a codon on the mRNA. When this site and the codon are aligned on the ribosome the amino-acyl SRNA condenses with the terminal amino-acid on the nascent peptide to form a peptide bond. Parenthetically, virologists have long sought to understand the mechanism by which viruses inhibit the synthesis of host proteins without affecting

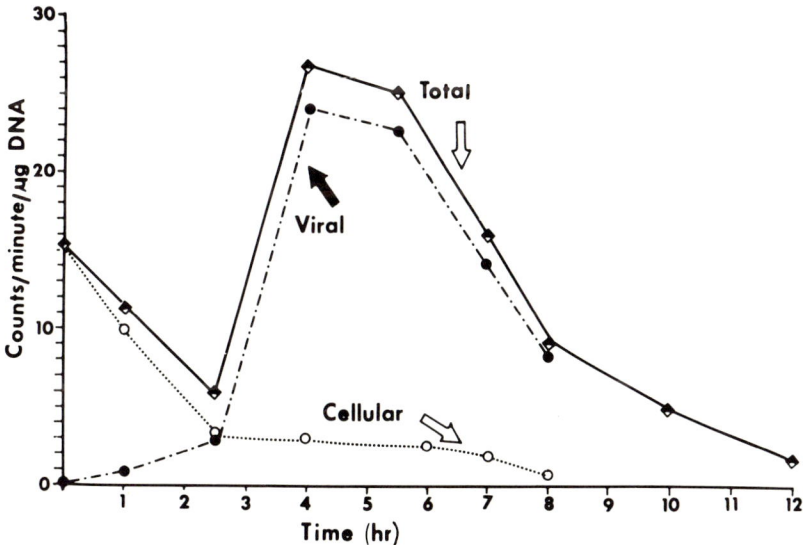

FIG. 10-5. The pattern of incorporation of ³H-thymidine into total, cellular, and viral DNA. The cells were pulse labeled for 15 min at different times after infection. Cellular and viral DNA were separated by isopycnic centrifugation in CsCl density gradients (*146*).

their own. It is easy to imagine the innumerable ways by which a virus could play havoc with host macromolecular synthesis were it but endowed with the information for the synthesis of its own sRNA and corresponding enzymes.

The justification for the initial experiments offered by Subak-Sharpe and Hay (*105*) was based on the assumption that in the course of evolution the amount of sRNA for each of the amino acids would become adjusted so that it is in direct proportion to the frequency of the codon in the mRNA. The argument was advanced that, since the G + C content of herpesvirus DNA is considerably greater than that of the mammalian cell DNA, it would be expected that the number of codons containing G, C, or both would be greater in viral mRNA than in cellular mRNA. Consequently, cells infected with DNA viruses rich in G + C should develop a shortage of sRNA with recognition sites complementary to codons containing G + C. The argument was refined on the basis of nearest-neighbor analyses and in a subsequent paper the particularly scarce sRNA species were identified as those containing complementary recognition sites for codons containing the doublet CpG (CpGpX and XpCpG). The authors then set forth to determine whether the virus overcomes the shortage in sRNA by directing the synthesis of its own sRNA or by directing the cell to make more of the scarce species of sRNA.

In the first series of experiments Subak-Sharpe and Hay (*105*) reported as evidence for the synthesis of new sRNA the coincidence of two activities in one fraction obtained by chromatography of RNA from infected cells on a methylated albumen-kieselguhr column. The fraction contained newly synthesized RNA complementary to viral DNA and RNA capable of accepting activated amino acids. The paper did not show whether the RNA complimentary to viral DNA was the one accepting the amino acid. It is conceivable that small fragments of labeled viral mRNA could have eluted together with unlabeled host sRNA synthesized before infection. Two recent experiments are more convincing. In the first, Subak-Sharpe, Shepherd and Hay (*106*) showed partial separation of arginyl–acyl–sRNA from infected and uninfected cells. In the second, they digested with ribonuclease T1 artificial mixtures of ^{3}H-labeled amino–acyl–sRNA from uninfected cells and ^{14}C-labeled amino–acyl–sRNA infected cells. The fragments were then chromatographed. Ribonuclease T1 hydrolyzes RNA at guanine–phosphate bonds exclusively. If the hypothetical sRNA specified by the virus differs from the corresponding host sRNA with respect to the position of the first guanine base following the terminal triplet pCpCpA, it would be expected that after digestion with T1 at least one set of fragments attached to the amino–acyl group of the sRNA from the infected cell should differ from those of uninfected cells. The data do, in fact, show a difference between arginyl–acyl–oligonucleotides from infected and uninfected cells but here again it could be argued that the separation is due to differences in methylation patterns of RNA synthesized before and after infection.

Pertinent to this section are two observations. First, in HEp-2 cells infected with herpes simplex virus the synthesis of 4S RNA and ribosomal RNA (as determined by electrophoresis in acrylamide gels) declines after infection. Ribosomal RNA synthesis however is inhibited more rapidly and more extensively than 4S RNA (*200*). Second, in recent studies (*201*) no differences have been found in the elution patterns of arginyl–acyl–sRNA from infected cells activated with enzymes from infected or from uninfected cells and chromatographed on reverse-phase freon-quaternary amine columns. The technique used in these experiments resolved two species of arginyl transfer RNA in both infected and uninfected cells. Among the many hypotheses that could account for this finding are (i) the virus does not direct the synthesis of a new transfer RNA in HEp-2 cells, (ii) the cell and viral transfer RNA are indistinguishable, and (iii) the chromatographic technique lacks resolving power.

3. *Enzymes*

This section is concerned with the time of synthesis and function of viral enzymes. In the past several years there have been reports on thymidine (TdR) kinase, thymidine monophosphate (TdR-MP) kinase, deoxyguanosine monophosphate (GdR-MP) kinase, deoxyadenosine monophosphate (AdR-MP) kinase, DNA nucleotidyltransferase (DNA polymerase), and deoxyribonuclease (DNAase). All of the enzymes studied to date concern themselves with the synthesis or with the breakdown of DNA, and very few have been studied in detail. Although a number are most probably synthesized according to information furnished by the virus, the evidence on this point is based on precedents established with bacteria infected with T phages and is not unequivocal. The problem stems from the fact that the functions performed by these enzymes are common to both infected and uninfected cells. In the order of preference, acceptable evidence that an enzyme is "viral" would be if (i) the enzyme were newly synthesized

in an infected host rendered totally incapable of expressing its own
genetic information and (ii) if it could be shown that the structure
of the enzyme is genetically determined by the virus. Because herpesviruses contain DNA replicating in the nucleus the evidence of choice
is difficult to obtain. The second type of evidence is unavailable possibly because techniques for large scale production of conditional
lethal mutants of animal viruses have not been developed as yet. The
evidence available at present consists largely of physical or immunological differences between the enzymes recovered from uninfected and
infected cells. In general, viral and cellular enzymes performing the
same function may differ with respect to immunologic specificity.
Immunologic differences alone, however, are insufficient evidence that
an enzyme in infected cells is "viral."

a. *TdR Kinase.* Thymidine kinase is a "scavenger" enzyme whose
function is to convert TdR to TdR-MP. The enzyme is present in
both "normal" and "neoplastic" cells grown *in vitro.* In BHK-21 cells
activity increased as much as 20-fold from 2–8 hr after infection with
herpes simplex virus (*108*). Thereafter, activity falls off.

Kit and Dubbs (*109, 110*) showed in an elegant series of studies that
TdR kinase activity was not essential for the growth of cells or of
herpes simplex virus in cell cultures. First, they obtained a BUdR
resistant strain of mouse fibroblasts lacking TdR kinase activity by
growing the cells in media containing the analog. TdR kinase activity
was induced in these cells by herpes simplex and vaccinia viruses, but
prevented by puromycin and actinomycin D. Subsequently they obtained TdR kinaseless mutants of herpes simplex virus by growing it
in BUdR resistant cells in the presence of the analog.

There are conflicting reports concerning the properties of the enzyme induced in infected cells. Compared with the enzyme from
uninfected cells, the TdR kinase from BHK-21 cells was reported to
have a low pH optimum and a low Km, to be relatively stable at 40°C,
and to be insensitive to inhibition by deoxythymidine triphosphate
(*108*). Moreover, the antigenic sepcificities of TdR kinase extracted
from BHK-21 cells infected with herpes simplex and rabbit kidney
cells infected with pseudorabies virus differed from those of uninfected
cells (*108, 111*). On the other hand, TdR kinase activity of African
green monkey kidney cells (BSC-1) could not be differentiated from
that of cells infected with herpes simplex virus with respect to thermal
stability and optimal temperature (*112*).

b. *TdR-MP Synthetase*. It was reported that TdR-MP synthetase activity did not increase in mouse fibroblasts, HeLa, and rabbit kidney cells infected with herpes simplex virus (*113*).

c. *TdR-MP, CdR-MP, AdR-MP, and GdR-MP Kinases*. Hamada et al. (*111*) reported that TdR-MP kinase activity increased in rabbit kidney cells infected with pseudorabies virus, whereas the activity of AdR-MP, GdR-MP, and CdR-MP kinases remained unaltered. The same laboratory previously reported (*114*) that TdR-MP kinase from pseudorabies-infected rabbit kidney cells was more stable at 37°C than the corresponding enzyme from uninfected cells. Prusoff et al. (*112*) observed a similar increase in TdR-MP kinase in African green monkey kidney cells infected with herpes simplex, but they were unable to differentiate between the properties of the enzyme in extracts of infected and uninfected cells. AdR-MP kinase from infected and uninfected cells could not be differentiated by antisera prepared against infected and uninfected cell extracts (*111*).

d. *DNA Polymerase*. Keir and Gold (*115*) reported that in BHK-21 cells infected with herpes simplex virus DNA polymerase activity of nuclei and of mitochondria–microsome fraction increased 2- to 6-fold between 2 and 5.5 hr after infection. The polymerases extracted from infected and from uninfected cells differed with respect to immunologic specificity, heat stability, primer, and cation requirements (*116*, *117*). However, Hamada et al. (*111*) could not differentiate in neutralization tests between DNA polymerase extracted from pseudorabies-infected and -uninfected rabbit kidney cells. The results of Hamada et al. are of interest because the rabbit is not a natural host for pseudorabies. The enzyme neutralization tests were done with rooster serum prepared against sonicates of infected stationary (arrested by contact or density inhibition) cultures of rabbit kidney cells. The serum prepared against the uninfected extract lacked neutralizing activity. The effect of the serum made against the infected cell extract on the primer itself was not tested.

e. *DNAase*. Increase in DNAase activity after infection with herpes simplex virus was reported by Keir and Gold (*115*) and by others (*118*, *119*). Morrison and Keir (*120*) reported that the enzyme extracted from infected BHK-21 cells is a DNA exonuclease capable of degrading both native and denatured DNA to deoxynucleoside 5′ monophosphates, whereas the enzyme extracted from uninfected cells is a DNA endonuclease effective against denatured DNA only. Rabbit antisera prepared against extracts of allotypic rabbit kidney cells infected with herpes simplex neutralized the DNA exonuclease from infected cells but not the endonuclease extracted from the uninfected cells.

D. The Synthesis of Structural Components

1. *Viral Proteins*

This section was intended to deal with the time, intracellular site, and general patterns of synthesis of structural proteins of herpesviruses. Two problems however necessitate that the topic be broadened to a consideration of the total protein and viral antigen synthesis in infected cells. First, the studies of the nature and number of herpesvirus structural proteins are still in their infancy. Second, is discussed in detail below, at the moment there is no reliable method for differentiating between structural and nonstructural herpesvirus proteins in extracts of infected cells. The dual emphasis on protein and on antigen synthesis is prompted by the fact that immunologic techniques are frequently used to measure viral proteins. While all proteins can be made antigenic, not all antigens are proteins. The work reviewed here followed three basic approaches. The first dealt with the general pattern and characteristics of protein synthesis in compartments of the infected cell. These data showed that the bulk of viral proteins was synthesized in the cytoplasm between 3 and 9 hr after infection. The second approach was to study the time of appearance and localization of viral antigens. The time of synthesis of viral antigens in general agrees very well with the time and pattern of synthesis of proteins measured directly. Lastly, labeled proteins extracted from cells and virus were solubilized and electrophoresed in acrylamide gels.

a. *Characteristics of Protein Synthesis in Infected Cells.* In HEp-2 cells infected with herpes simplex virus the rate of protein synthesis follows (Fig. 10-3) an initial period of decline (0–3 hr), a period of stimulation (3–8 hr) and, lastly, a period of gradual and irreversible decline (*97, 98*). Analysis of cells at different times after infection revealed close agreement between the patterns of (i) specific activities of homogenates of pulse-labeled cells, (ii) the specific activities of peptides extracted from the cytoplasm of pulse-labeled cells, (iii) the amounts of cytoplasmic polyribosomes recovered, and (iv) the amounts of pulse-labeled nascent peptides bound to these poly-ribosomes at different times after infection (98). This led to the conclusion that (i) early and late inhibition and intervening stimulation of protein synthesis are due to the corresponding breakdowns and reformation of polyribosomes and (ii) the bulk of viral proteins is made on cytoplasmic polyribosomes.

The shift in the synthesis of proteins, from cellular to viral is reflected in the profiles of polyribosomes in sucrose density gradient (98, 102, 121). As shown in Fig. 10-6 before and immediately after infection, cytoplasmic polyribosomes are polydisperse, but with a peak fraction sedimenting at 170S. These polyribosomes disappear between 1 and 2 hr after infection and are replaced by polydisperse polyribosomes showing a different profile on sedimentation in sucrose density gradient, particularly with respect to the sedimentation coefficient of the peak fraction. In HEp-2 cells productively infected

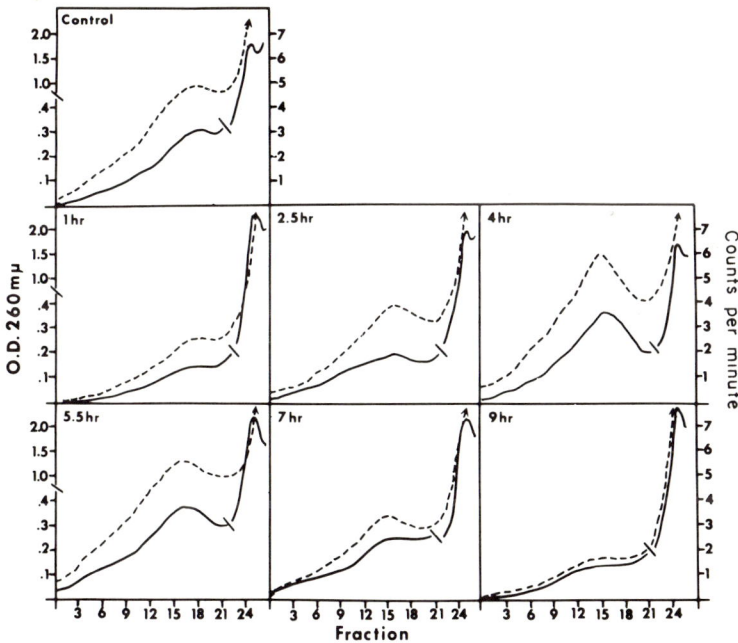

FIG. 10-6. The patterns of polyribosomes extracted from HEp-2 cells at various times after infection. The polyribosomes were extracted immediately after 15 min pulse labeling with ¹⁴Camino acids and centrifuged in 15–30% w/w sucrose density gradients (98, 102). Solid line, optical density at 260 mμ of polyribosomes banded in sucrose density gradient fractions; dashed line, counts per minute of ¹⁴C amino acids incorporated into nascent peptides (98, 102).

with the MP*dk⁻* strain of herpes simplex virus the sedimentation coefficient of the peak fraction is estimated at 270S (*98*). In DK cells
abortively infected with MP*dk⁻* virus, the peak polyribosome fraction
has a sedimentation coefficient of 220S. Finally, a mutant of MP*dk⁻*
designated MP*dk⁺sp* which grows, but poorly, in both HEp-2 and DK
cells produces equal amounts of 220S and 270S polyribosomes in the
two cell lines (*121*). It is of interest to note that HEp-2 cells infected with MP*dk⁻* produce predominantly enveloped nucleocapsids.
DK cells infected with the same virus yields nonenveloped nucleocapsids exclusively. The MP*dk⁺sp* virus produces in both cell lines
relatively more nonenveloped than enveloped nucleocapsids (*20*).
Nevertheless in considering the role of cytoplasmic polyribosomes
extracted 2.5 hr after infection and thereafter it should be pointed
out that (i) viral mRNA has not been demonstrated on these polyribosomes, (ii) nascent peptides on these polyribosomes have not been
identified with respect to structure or immunologic specificity as being
viral (largely because the bulk becomes insoluble on extraction),
and (iii) the correlation between the amount of 270S polyribosomes
and extent of envelopment will become significant if it is shown that
viral envelope proteins are made on these polyribosomes.

b. *Viral Antigens.* Immunologic analyses have been done to determine the time of synthesis, number (*122–128*), and site of localization (*129–138*) of viral antigens in infected cells. Interpretation of the
data hinges on evaluation of sera and the method of assay of the
antigen. Before considering the results in detail it is necessary to
discuss the source, preparation, and testing of antibody. Briefly there
are two sources of antibody, i.e., (i) human or animal convalescent
sera and (ii) hyperimmune sera prepared in a suitable animal. Each
class of antisera has advantages and disadvantages. Convalescent
sera are usually free from antibody reactive with uninfected cell
extracts, but the antibody titers to viral antigens are not nearly as
high as those obtained from properly immunized animals. Moreover,
there may be considerable variation in antibody titer to the various
viral antigens—a disadvantage in some respects which may nevertheless be very useful. Conversely, although hyperimmune sera can
be made very potent and reactive with a large number of viral antigens, such sera must be absorbed repeatedly with uninfected cell homo-
genate to ensure absence of antibody reactive with host antigens. It
should be noted in this connection that Watson et al. (*127*) reported an

ingenious procedure for obtaining hyperimmune sera free of antibody reactive against cellular antigens. They immunized rabbits with extracts of infected rabbit cells grown in rabbit serum. The rabbits, the rabbit cells, and the rabbit serum were all of the same allotype. The the antisera were reported to be free from antibody directed against rabbit cells grown in cultures. Two examples of differences between convalescent and hyperimmune rabbit sera are pertinent here. First, in immunofluorescence tests several batches of human, pooled γ globulin illuminated cytoplasmic viral antigens only (131, 137, 138). Rabbit hyperimmune sera, on the other hand, contained antibody against viral antigens localizing in both cytoplasm and nuclei of infected cells (138). Second, in the studies by Tokumaru (124, 125) the most potent human convalescent serum gave four distinct precipitin lines in gel diffusion tests when tested against an infected cell extract. Watson et al. (127) using their rabbit hyperimmune serum obtained 12 precipitin lines in gel diffusion tests. However, the flattering comparisons should not be taken to mean that hyperimmune sera will invariably have antibody in equal titers against all of the antigenic products specified by the virus in infected cells.

A point should also be made concerning preparation of antibody-reactive exclusively with specific antigens. Occasionally, most often by accident, procedures are found that yield antibody against specific classes of viral antigens. In this category falls the discovery that rabbit antibody made against boiled extracts of infected cells reacts exclusively with intranuclear viral antigens (137, 138). Most of the interest so far has been in the production of antibody to nonstructural proteins specified by the virus and conversely in the production of sera reacting with structural components of the virion only. Several comments concerning very obvious errors made in the past should be noted here. Firstly, the technique of immunizing rabbits with 3-hr infected cells in the expectation that such antisera will react with nonstructural components only or at least predominantly is probably not very useful for several reasons. (i) Not all of the virus adsorbed to the cell penetrates and becomes uncoated. The cell-associated virus is exceedingly difficult to remove. Once the sonicated extract is injected into the animal there is every probability that the cell-associated virus will infect the host. It is therefore most likely that the serum will contain antibody against virions and structural subunits. (ii) The fate of the coat protein of the infecting virus is unknown. It should be suspected until proven otherwise that at 3 hr after infection there is still enough coat protein in the infected cells to elicit antibody response. (iii) It could be that some structural components of the virus are synthesized early in the reproductive cycle in a manner analogous to those of vaccinia virus (139). Parenthetically, absorption of such sera with intact virions does not make them more acceptable since intact virions will remove antibody to surface antigens only. For this reason lack of neutralizing antibody is not an indication that a serum is free from antibody to structural components of the virus. Secondly, the technique of injecting into animals *live* concentrated virus for the purpose of obtaining antibody reacting exclusively with structural components is also not very useful. Pseudorabies, herpes simplex, herpes B, and other members of the herpesviruses have a very broad host range, and if the virus multiplies in the host, antibody against nonstructural viral proteins should also be expected. Even if infection aborts, as in the case of herpes simplex infection of DK cells (79, 99), structural and nonstructural proteins may be specified by the virus.

One and perhaps the unique virtue of the preceding discussion, is that it reduces considerably the number of reports contributing significantly to the problems concerning time and site of synthesis and the site of accumulation of herpesvirus antigens. Briefly, studies by Russell et al. (123) with a human convalescent serum and by Sabin (140) with both hyperimmune rabbit sera and convalescent human sera show that the formation of complement-fixing antigen in cells infected with herpes simplex virus takes place between 2 and 8 hr after infection. Thereafter, the complement-fixing titer of the antigen remains unchanged. There is as yet no immunologic verification of evidence presented in Sec. 10-3, D, 1 that herpesvirus proteins are synthesized in the cytoplasm (98).

The localization of herpesvirus antigen in infected cells has been studied with the aid of fluorescein-labeled antibody (129–138). The earliest report is that of Lebrun (129, 130) who found that in HEp-2 cells infected with herpes simplex virus the antigen was first localized in the nucleus (24 hr after infection) and subsequently (60–72 hr) appeared in the cytoplasm. Lebrun used a human serum. Other workers using human convalescent (131) and rabbit convalescent and hyperimmune sera (132) reported that viral antigen appeared first between 1 and 6 hr after infection in the cytoplasm at or near the nuclear membrane; nuclear fluorescence was not detected. Much later (138) it became apparent that the discrepancy between Lebrun's observation and those of subsequent investigators was due to differences in the specificity of antibody used in these studies. The HEp-2 cells infected with herpes simplex virus were found to contain five immunofluorescent elements. Three (small nuclear granules, large nuclear granules, and an amorphous mass filling the nucleus) shown in Figs. 10-7a, 10-7b contained antigen which reacted with a rabbit serum prepared against boiled, infected cell debris. A labeled pool of human antibody revealed antigens making up cytoplasmic granules and those responsible for a diffuse cytoplasmic fluorescence (Figs. 10-7c, 10-7d). All five immunofluorescent elements were demonstrable with a hyperimmune rabbit serum prepared against unheated infected cell debris. The antigens responsible for the diffuse cytoplasmic fluorescence and for the amorphous nuclear mass are synthesized early in infection and exist in a form which does not sediment on centrifugation at $79,000 \times g$ for 2 hr. The antigens comprising the nuclear and cytoplasmic granules arise relatively late in infection and are readily sedimented on centrifugation at $79,000 \times g$ for 2 hr. The identity of antigens responsible for the diffuse cytoplasmic fluorescence and amorphous nuclear mass are unknown. The cytoplasmic granule resembles in size accumulations of virions between the inner and outer layers of the nuclear membrane. Electron microscopy of nuclei shows several kinds of bodies and aggregations of subviral particles in the nucleus. The identity of the large and small granules is at the moment uncertain.

If viral proteins are made in the cytoplasm, as is indicated earlier in this section, and if the virus is assembled in the nucleus, as is shown in a later section, it would be expected that at least some antigens would be found in both cytoplasm and nucleus. The most significant and also the most puzzling finding of the immunofluorescence studies cited above is that viral antigens were segregated in the nucleus or in the cytoplasm; within the limits of detection each antigen accumulated in one compartment only. One hypothesis that could explain the findings is (i) that not all of the proteins made in the cytoplasm are transferred into the nucleus (202) and (ii) that viral subunits made in the cytoplasm aggregate and acquire a new antigenic specificity on entering the nucleus. A necessary corollary to this hypothesis is that the specific state of the subunits or the millieu of the cytoplasm prevents subunits from aggregating before they reach the nucleus. A priori, it would seem that aggregation of structural subunits in the cytoplasm would be most undesirable for the virus in that such

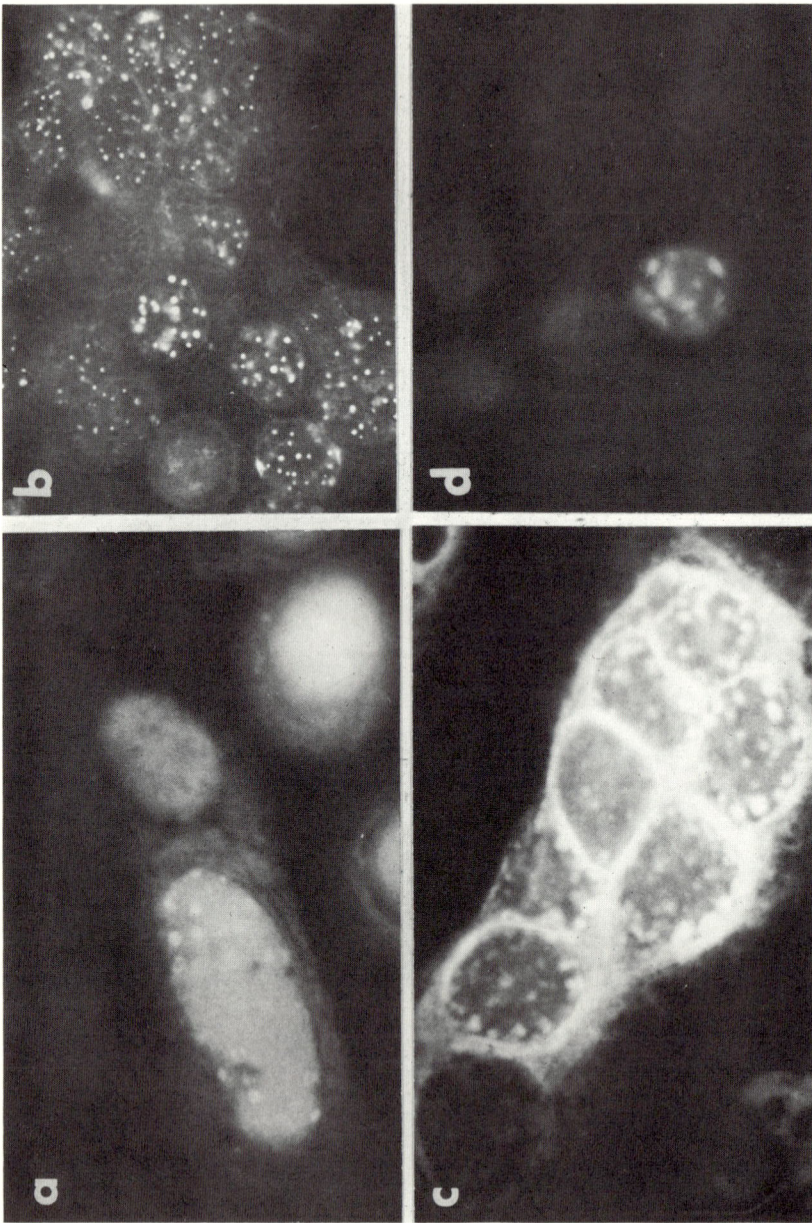

Fig. 10-7. Localization by immunofluorescence of herpes simplex virus antigens in HEp-2 cells 6 hr after infection (138). (a) Amorphous mass and large granules in nucleus of infected HEp-2 cell stained with labeled rabbit anti-boiled infected cell antibody, (b) small intranuclear granules in polykaryocyte stained with the same labeled antibody but after absorption with the supernatant obtained after centrifugation of infected cell extracts at 79,000 g for 2 hr, (c) diffuse cytoplasmic fluorescence and perinuclear granules in polykaryocyte stained with conjugated human antibody, (d) perinuclear granules in polykaryocyte stained with conjugated human antibody absorbed with 79,000 g supernatant.

aggregates would diffuse or be transported into the nucleus very slowly and inefficiently. Additional evidence to support or reject the hypothesis would have to come from studies with antibody against specific structural subunits of the virus.

c. *Analysis of Proteins by Electrophoresis in Acrylamide Gels.* As noted earlier, studies on the number and properties of herpesvirus proteins are in their infancy. A preliminary survey (*202*) revealed at least 25 proteins were synthesized at one time or another during infection of HEp-2 cells with herpes simplex virus. The proteins were made in the cytoplasm. The transfer of proteins from cytoplasm into the nucleus was slow and selective; less than half of the proteins synthesized during a short pulse appeared in the nucleus after a 3-hr chase. Moreover, proteins in three bands appeared to be restricted to the cytoplasm. Most of the proteins chased into the nucleus late in infection were identified on the basis of electrophoretic mobility as being structural viral proteins. All but a few proteins made late in infection could be chased into naked and enveloped nucleocapsids. The size of herpesvirus proteins, estimated by co-electrophoresis with poliovirus proteins of known molecular weight, range to about 125,000 daltons.

2. DNA Synthesis

In the past several years a considerable amount of information became available on DNA synthesis in pseudorabies- and herpes simplex-infected cells. The progress is due largely to the fact that the DNA's of the two herpesviruses have high $G + C$ molar base ratios and are readily separated from cellular DNA by isopycnic centrifugation in CsCl density gradients (Table 10-1). It is perhaps relevant to note that very little is known concerning the synthesis of viral DNAs more closely approximating host DNA with respect to base ratios.

The most extensive studies of herpesvirus DNA synthesis were done by Kaplan and Ben-Porat (*103, 141–143*). Their system consisted of pseudorabies and rabbit kidney cells. The system is particularly advantageous because only viral DNA is synthesized in rabbit kidney cells arrested prior to infection by contact (density) inhibition; host DNA synthesis along with cell division remain inhibited throughout the cycle (*145*). The other system that has been used is the HEp-2 cell infected with herpes simplex virus (*39, 146*). Alas, optimal virus yields are obtained from young, rapidly growing cells (*80*). In HEp-2 cells infected during rapid growth period, host DNA synthesis persists for several hours (Fig. 10-4) and for accurate determination of viral DNA it is necessary to centrifuge the extract containing DNA in CsCl solution (*39*).

a. *Pattern of Synthesis of Viral DNA.* Two basic methods have been used to estimate the pattern of synthesis of viral DNA. The first involved direct determination of specific activity of viral DNA in cells pulse labeled with radioactive thymidine at intervals after infection (*146*). The pattern obtained for herpes simplex virus DNA synthesis in HEp-2 cells is shown in Fig. 10-4. The salient features are that within a few hours after infection host DNA synthesis is

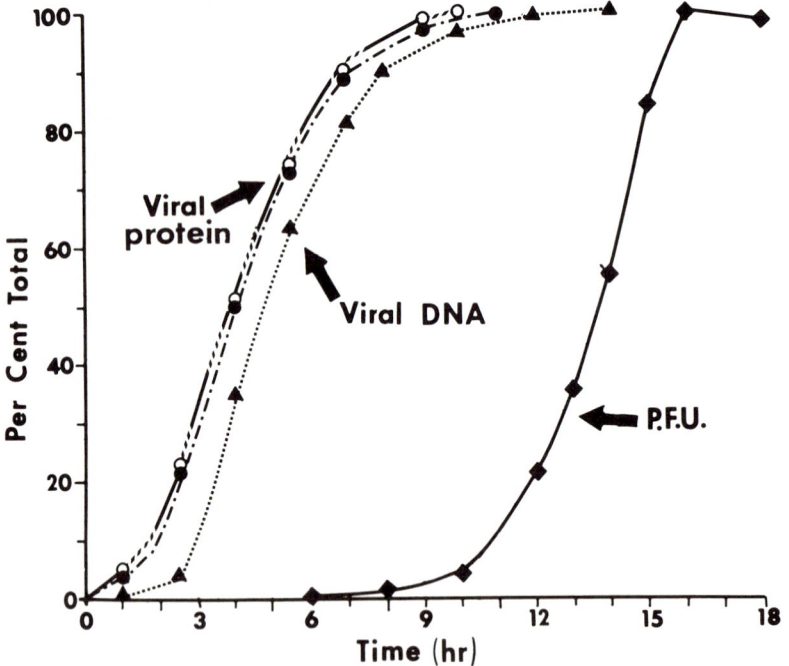

Fig. 10-8. The synthesis of viral protein, viral DNA, and assembly of herpes simplex virus in infected HEp-2 cells (*146*). Protein synthesis shown by filled circles was estimated from ^{14}C amino acid incorporation into peptides extracted from whole cell homogenates. The curve drawn through open circles was calculated from the ^{14}C amino acids incorporated in nascent peptides bound to cytoplasmic polyribosomes. The points shown for viral protein synthesis at 1 hr after infection are crude estimates based on extrapolation of measurements made at 2.5 hr after infection.

arrested and is replaced by viral DNA synthesis. The bulk of viral DNA is synthesized between 4 and 7 hr after infection; thereafter viral DNA synthesis declines slowly and irreversibly. On the basis of data shown in Fig. 10-4 it has been calculated (Fig. 10-8) that 6–9 hr elapse between the synthesis of DNA and formation of infectious virus.

The second method is based on a report by Salzman (147) that fluorodeoxyuridine (FUdR) blocked the synthesis of vaccinia DNA. When FUdR was added to infected cultures during virus multiplication, DNA synthesis was arrested but virus assembly continued for several hours. Salzman concluded that infectious virus made in the presence of FUdR contained DNA made prior to the addition of the

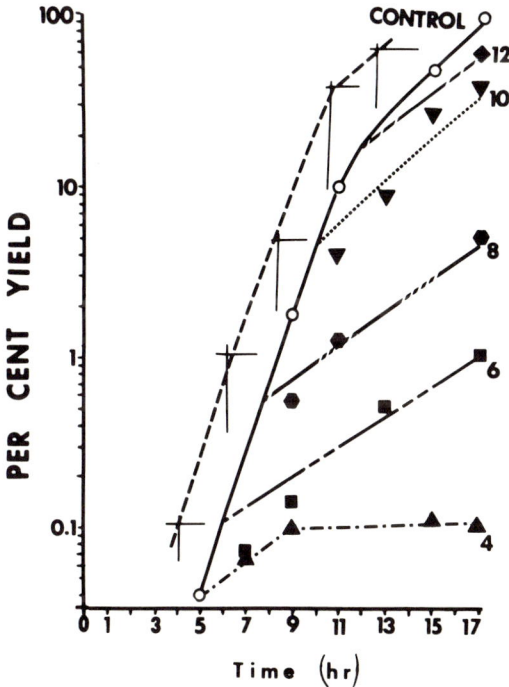

Fig. 10-9. The accumulation of viral DNA in HEp-2 cells infected with herpes simplex virus as estimated by a variation of Salzman's method (147). Approximately 20 μg/ml of IUdR of uridine were added at times shown to the right of each curve. A plot of the amount of infectious virus obtained 17 hr after infection against the time of addition of the drug yields a pattern presumed to show the accumulation of viral DNA which becomes incorporated into infectious virus (45).

drug. Accordingly, a plot of virus obtained in the presence of the drug against the time of addition of the drug should yield the pattern of accumulation of viral DNA which is encapsidated and forms virions. FUdR does not inhibit herpesvirus DNA synthesis in some established cell lines and therefore experiments similar to those reported by Salzman were done with iodouracildeoxyriboside (IUdR) (45) and bromuracildeoxyriboside (148). The pattern of the accumulation of herpes simplex virus DNA in HEp-2 cells is shown in Fig. 10-9. According to this method the DNA incorporated into virions is synthesized only 2 hr earlier. The results obtained by the two methods differ, but they are not truly comparable. The direct method measures all viral DNA and makes a few assumptions. However, only 10–15% of the total DNA becomes incorporated into virions and the pattern of total DNA synthesis may not be representative of the synthesis of DNA destined to become incorporated into virions (Sec. 10–3, E, 1). The indirect technique purports to measure the DNA incorporated into virions but it is subject to two classes of errors with opposite net effects (45). First, the patterns obtained with these drugs may falsely indicate that DNA is made sooner than it is actually made. This could be if (i) penetration of the drug is delayed, (ii) the phosphorylation of the drug is delayed, or (iii) incorporation of small amounts of the drug into DNA does not affect the function of viral DNA. Second, the pattern obtained with the drugs may falsely indicate that DNA is made later than it is actually made. This could be if (i) the drug affects the ability of viral DNA to function in protein synthesis required for the formation of virions or (ii) the drug alters the ability of the cell to support virus maturation. Since the work was done, the contribution of several class 1 errors was evaluated and found trivial (112). However, even though incorporation of IUdR into viral DNA could not be demonstrated in HEp-2 cells infected with herpes simplex virus (45), Smith and Dukes (149) reported the synthesis of defective virions, and Kaplan and Ben-Porat (150) were able to show that IUdR and BUdR become incorporated into DNA of pseudorabies virus and that the inhibition of virus assembly is due to the synthesis of fraudulent products. It may be concluded on that basis that the pattern of viral DNA accumulation obtained with halogenated pyrimidine nucleosides falsely indicates that the DNA is made later than it actually is. The magnitude of the error is not known. The question of whether the pattern of synthesis of total viral DNA differs from the pattern of synthesis of DNA withdrawn into virions is dealt with in Sec. 10-3, E, 1.

b. *Requirements for Viral DNA Synthesis.* Two series of experiments have been reported showing that protein synthesis after infection is required to initiate the synthesis of viral DNA. In the first series involving HEp-2 cells and herpes simplex virus (*39*), the exposure of infected cells to puromycin at any time between 0 and 3–4 hr after infection blocked the synthesis of viral DNA. However, cells exposed to puromycin between 4 and 6 hr after infection, i.e., after the onset of viral DNA synthesis, continued to incorporate radioactive thymidine into viral DNA albeit at a reduced rate. In view of the difference in size, base composition, and general structure of viral and cellular chromosomes, the observation that puromycin treatment immediately after infection blocks the onset of synthesis of viral DNA seems entirely reasonable. It indicates that viral DNA synthesis requires the participation of new enzymes made after infection. However, there is no simple, straightforward explanation for the fact that in cells treated with puromycin after the onset of viral DNA synthesis, the rate of incorporation of thymidine into viral DNA is reduced. The greatest reduction in rate of incorporation of thymidine into viral DNA occurred within a short time after addition of puromycin and could reflect a sudden change in the pool size of thymidine or a real change in the rate of synthesis of viral DNA. The first explanation is probably trivial but it must be considered, particularly in view of the possibility raised by Newton et al. (*151*) that the size of the thymidine pool in the infected cells may change during the reproductive cycle. The second explanation if true, would indicate that puromycin interfered with the availability of a necessary rate-limiting constituent such as an enzyme or possibly the primer itself. It is perhaps pertinent to note that viral DNA extracted from puromycin treated cells behaved on isopycnic centrifugation in CsCl as if it were highly fragmented (*39*).

Similar results were obtained in another series of experiments with rabbit kidney cells infected with pseudorabies virus (*141*). During the first hours after exposure to puromycin the rate of incorporation of thymidine into viral DNA dropped to one-half that observed in untreated cells. Thereafter the rate decreased but not very appreciably. In this instance the authors concluded that changes in the rate of thymidine incorporation reflect changes in DNA synthesis and that pseudorabies virus DNA synthesis requires concomitant synthesis of one or more proteins. The authors speculate that the function of the short-lived protein which must be synthesized continuously is to "prime" or initiate the synthesis of DNA. The conclusion is based on the fact that extracts from cells exposed to puromycin appear to contain all of the enzymes necessary to maintain DNA synthesis in a cell-free system consisting of cell extract, added deoxynucleoside triphosphates, and heat-denatured primer.

To sum up, it seems clear that viral DNA synthesis requires *de novo* protein synthesis. Once initiated, viral DNA synthesis continues in the absence of concomitant protein synthesis but probably at a reduced rate. The reduction in rate, if true, is not readily interpretable for lack of information concerning (i) thymidine pool size before and after addition to puromycin, (ii) integrity of replicating DNA in puromycin treated cells, and (iii) secondary effects of puromycin particularly on the intranuclear environment of the cell.

c. *Characteristics of Replication of Viral DNA.* Kaplan and Ben-Porat showed in studies reported between 1963 and 1967 (*141, 142*) that (i) pseudorabies virus DNA replicates in a semiconservative fashion and (ii) less than one-half of the DNA not integrated into virions and presumed available to function as templates is actually replicating.

d. *Site of Synthesis of Viral DNA.* Viral DNA synthesis takes place in the nucleus. For many years the conclusion was based on histochemical evidence (*152, 153*). Biochemical evidence was obtained in HEp-2 cells infected with herpes simplex virus. In cells fractionated immediately after a 5-min pulse labeling with ^{3}H-thymidine, labeled DNA was found in the nuclear fraction (*146*).

E. Virus Assembly

Information concerning assembly of herpesviruses is derived from two sources, i.e., studies of the fate of viral DNA synthesized in infected cells and electron microscopic studies of thin sections of infected cells.

The emerging picture is that viral DNA is synthesized and accumulates in intranuclear bodies similar to the factories of vaccinia. DNA in the periphery of the bodies is withdrawn and encapsidated by protein subunits entering from the cytoplasm. The nucleocapsid migrates to the nuclear membrane and becomes enveloped before, during, or after (?) passage through the nuclear membrane.

1. *Encapsidation of Viral DNA.*

We are concerned here primarily with the place, time, and mechanics of the process. Intranuclear viral particles consisting of a nucleoid, an inner electron-transparent shell, and an outer electron-opaque shell are seen as early as 4 hr after infection (*37, 123*). Between 16 and 24 hr after infection the particles aggregate at the nuclear membrane and around an electron-opaque nuclear body resembling in appearance the DNA factories of vaccinia (*30, 37, 123*). It seems likely, although evidence is lacking, that the body is the DNA factory and that DNA in the periphery of the body is encapsidized by protein subunits entering from the cytoplasm.

The mechanics of encapsidation are unknown. The withdrawal of DNA from the DNA pool into virions has been studied and, as expected, viral DNA becomes first incorporated into a structure (the nucleocapsid ?) impermeable to DNAse but different from the virion (*141, 143*). The process is slow (Fig. 10-8) and inefficient. The question has arisen whether viral DNA made late in the reproductive cycle has a greater probability of becoming encapsidized. Ideally, random withdrawal would be expected if DNA and protein synthesis were completely and randomly dispersed throughout the cell. Some deviation from randomness should be expected if proteins and DNA are synthesized in separate compartments and must diffuse toward each other in order to aggregate because the DNA at the periphery of the compartment would have a greater probability of being encapsidized than the DNA near the center. The expected deviation from randomness, however, may be too small to be readily demonstrable in a system in which only 15–20% of the DNA is incorporated into virions. Ben-Porat and Kaplan (*143*) concluded that envelopment is random. In their experiments (i) a number of stationary rabbit kidney cultures were infected with 10 plaque forming units of pseudorabies virus per cell and incubated in a medium containing 25 μg of fluorouracil and 0.3 μg of thymidine per ml; (ii) at 3, 5, and 7 hr after infection the medium from various cultures was replaced with a medium containing ^{14}C-thymidine; (iii) 1-hr afterwards the ^{14}C-thymidine was diluted 1000-fold with ^{12}C-thymidine; (iv) at intervals thereafter the medium was replaced with prewarmed fresh medium containing ^{12}C-thymidine. The media removed at various intervals were assayed with respect to infectivity and with respect to the total amount of DNA contained in the virus released from the infected cells. The findings were that although the total amount of label incorporated in 1 hr beginning with 3, 5, and 7 hr after infection differed, the proportion of labeled DNA incorporated into virus was independent of the time of labeling. However, it seems appropriate to ask whether one or more variables in the experiments could have obscured a small deviation from randomness. First, the question arises whether fluorouracil could have minimized the effects of the difference in the time the cells were pulse labeled. The question is pertinent particularly since it has been shown by Reissig and Kaplan (154) that the drug causes the synthesis of nonfunctional proteins. Second, the question arises whether virus released from cells is a representative sample of the total virus made in the infected cell. The argument stems from the fact that only a small fraction of total virus is released.

2. Envelopment

In the past fifteen years numerous papers have dealt with the envelopment of herpesviruses in various cells (28–38). There seems to be relatively good agreement that all herpesviruses are enveloped in a similar fashion. Most investigators see the same structures and probably the only thing that varies is the emphasis placed on one or another observation. Yet there is no unreserved agreement as to how herpesviruses are enveloped. Morgan et al. (30) admirably state the difficulty encountered in translating the static image obtained from a few thin sections which may or may not be representative of the whole cell into a comprehensive picture of rapidly moving events in viral replication. The difficulties were compounded with respect to envelopment of herpesviruses for the following reasons.

(i) Partially enveloped particles appear most frequently in apposition to the inner lamellae of the nuclear membrane. Enveloped particles frequently accumulate in the space between the inner and outer lamella of the nuclear membrane. It is generally accepted that the nuclear membrane is the site of envelopment (Fig. 10-2a).

(ii) Partially enveloped particles appear less frequently in the cytoplasm in apposition to undefined membranes and these observations led to suggestions that the virus can be enveloped by any "cytoplasmic" membrane. The photomicrographs do not make clear, however, whether the particles are being enveloped or unenveloped.

(iii) Enveloped particles are on rare occasion seen in the nucleus (Fig. 10-2c). Most frequently they are present in the cytoplasm surrounded by a membrane. For many years it has been claimed that the enveloped virus in the cytoplasm is contained in a vesicle or vacuole which serves to transport the virus from the nucleus to the extracellular fluid by a process of "reverse phagocytosis" (30, 204).

(iv) A characteristic feature of herpesvirus-infected cell is the formation of rings of double and sometimes triple layers of membrane around the nucleus (Fig. 10-2d). The function of these membranes have not been readily apparent. It has been suggested that the rings consist of reduplicated nuclear membrane deposited behind the virus in such a manner as to prevent rupture of the nucleus (30).

The solution to many of these puzzling observations may emerge from the finding (207) that in HEp-2 cells infected with herpes simplex virus, the reduplicated membranes form highly banded tubules originating at the outer lamallae of the nuclear membrane and opening into the extracellular fluid. The tubules contain enveloped virus. In cross section, the tubules appears as "vacuoles" containing enveloped virus. These observations suggest that tubules serve as a means of egress of the enveloped virus from the cell. The mechanism by which the tubules are formed and the source of genetic information are unknown.

F. REGULATION OF THE REPRODUCTIVE CYCLE

The keynote of life on Earth is reproducibility. Reproducibility inherently implies some forms of regulation. Life as we know it is the result of exceedingly rare errors compounded innumerable times—a true catastrophy in regulation. Viruses never cease to astonish us (considering the size and genetic potential) with the reproducibility of the growth cycle and with the fact that every one of the myriad progeny are an exact replica of the parent. This bespeaks of a very high order of regulation, but the existence of groups with many mutants and strains differing from each other is mute evidence that this regulation like that of their hosts is not foolproof. The question before us is just which aspects of the reproductive cycle are regulated, how, and, equally important, why. It is not profitable to consider the regulation of every event but some peculiar and unique problems associated with some groups and sequences of events should be noted. Before considering these, two points should be made. First, it seems desirable to stress that every succession of events by the virtue that it is a succession is, in fact, regulated. The second point is that in this discussion we should differentiate two kinds of regulation. It seems self-evident that some events are "committed" by a product of a viral gene, whereas others, possibly of no consequence to the virus, occur simply because there is no viral product to interfere with the event.

1. The Synthesis of Nonstructural Components

Accustomed as we are to finding a rich and varied storehouse of enzymes involved in DNA synthesis, transfer RNA, etc., in cultured animal cells, it seems puzzling that DNA viruses carry such information and expend the host's energy to synthesize more molecules concerned with the same function. The question is particularly appropriate if one considers that, while the doubling time of most animal cells is nearly equal to the reproductive cycle of DNA viruses, the amount of cellular DNA and protein synthesized by the cell greatly exceeds the amount of viral DNA (141, 145) and protein made during the same period. Some explanations to this puzzle seem trivial: for example, that the nonstructural components are evolutionary appendages dating back to a free living ancestor. Two other explanations seem far more probable. The first is that the cell grown in culture may bear little resemblance to its ancestors growing in a multicellular organism. Not all of the cells in the multicellular organism divide frequently and they would not therefore be expected to possess a full complement of enzymes required by the virus. One function of the nonstructural components would thus be to supplement the host and this is in accord with the observation that virus grows to higher titer in growing cells. Thymidine kinase is an example of a nonstructural component which is clearly in this category. The second hypothetical explanation is that some of the nonstructural products specified by the virus are endowed with specificity either with respect to the host or to the virus. As considered in more detail in Sec. 10-4, B, 2, one obvious prerequisite of the inhibitors of macromolecular synthesis induced by virus is that they differentiate between viral and cellular structures performing a similar function. Because of differences in size and structure of genetic material, it seems probable that in the environment of the cell viral polymerase may be uniquely suited to duplicate viral DNA just as host enzymes may be needed to duplicate host genetic material. Parenthetically, if this were true, measurements of the enzyme in tests characterized by a lack of specificity have a limited value.

Assuming that the function of nonstructural products is in part to supplement host functions and in part to differentiate between host and viral functions, the question arises why so few of the functions of the host are supplemented. It seems puzzling for example that supplemented viral enzymes synthesized in the host are usually those concerned with thymidine utilization and not, for example, those concerned with the deoxypurine utilization. Experimental data bearing on this question are totally lacking. It could be that thymidine and its derivatives have a regulatory function in uninfected animal cells and their availability in the non-dividing cell may therefore be controlled.

2. Regulation of the Synthesis of Structural and Nonstructural Components

The most persuasive argument for considering this topic is that, if viral DNA contains only one, nonrepeating codeword for each of the proteins specified by the virus, and if the DNA is transcribed in toto without interruptions, there would be as many enzyme molecules of each kind as there would be protein subunits. For each herpesvirus containing $162N$ subunits (where N equals the number of subunits per capsomere) there would be $162N$ molecules of each enzyme. This enormous excess of enzymes is not found. Another hypothesis is that the DNA contains only one, nonrepeating codeword for each enzyme and $162N$ repetitions of the codeword for the structural protein. This hypothesis, alas, postulates far more codewords than there is space in viral DNA. Elimination of the two extreme hypotheses leaves only one alternative: regulation of the amount and time of synthesis. Phage work has led to the hypothesis that viral proteins are synthesized sequentially; parental DNA serves as template for nonstructural and a few structural proteins synthesized early in infection, whereas progeny DNA serves as template for the bulk of the structural protein. This hypothesis raises questions concerning (i) how the synthesis of the late proteins is contained early in infection (if at all) and (ii) how the synthesis of "early" proteins is restricted late in infection.

Several interesting and very important experiments attempting to deal with the second of the two problems have been reported (155–157) and merit discussion. In the first of the papers, Kamiya et al. (155) reported an *in vitro* system measuring collectively the enzymes involved in the incorporation of deoxynucleotides into DNA. The authors showed that the limiting enzyme increased in activity during the first 6 hr, leveled off between 6 and 10 hr, and subsequently decreased. However, no leveling off or decrease in enzyme activity was observed in extracts of infected cells grown in a medium containing BUdR. This led to the conclusion that (i) the enzymes are regulated, (ii) substitution of BUdR for thymidine interferes with the regulation, and (iii) regulation was dependent not on presence of the DNA *per se* but upon the presence of newly synthesized competent DNA. The questions thus arise as to what accounts for (i) the leveling off of enzyme activity between 6 and 10 hr after infection and (ii) the decrease in activity of the enzymes thereafter. Subsequent papers, one by Kamiya et al. (*156*) and one by Zemla et al. (*157*), deal with these questions and conclude that between 6 and 10 hr after infection, enzyme synthesis is arrested, whereas after 10 hr the decrease in enzyme activity is due to leakage of intracellular protein into extracellular fluid as a consequence of viral release.

The loss of enzyme into extracellular fluid appears to be "regulation" by omission. It could be argued that if retention of enzymes were a selective advantage, mutants would have been found which did not cause enzyme leakage. In the light of the finding that at least 80% of total viral DNA is not encapsidized, it does not seem likely that retention of enzymes would be advantageous. In a different category is the conclusion that between 7 and 10 hr after infection enzyme synthesis is inhibited. This implies that regulation is "committed" by a product of a viral gene. For this reason, the evidence is worth examining in some detail.

Briefly, Kamiya et al. (*156*) showed that significant leakage of protein, begins 5–8 hr after infection. However, the protein responsible for the leakage is synthesized beginning approximately 4 hr after infection. This is shown by measuring loss of enzymes and leakage of proteins from cells exposed to puromycin at different times after infection. Thus there was no leakage at 15 hr from cells treated at 4 hr. Cells treated at 7 hr or later leaked as badly as untreated cells. Between 4 and 7 hr loss of enzyme and total protein was dependent on the time of exposure to the drug. With this experimental background the authors asked themselves whether the leveling off of enzyme activity between 6 and 10 hr was due to (i) inhibition of enzyme synthesis of (ii) concomitant synthesis and loss of enzyme. To answer this question two sets of cultures were infected and incubated in medium free of serum. At 7 hr after infection one set received puromycin. The treated and untreated cells were then assayed for loss of enzyme activity from cells and leakage of proteins into the extracellular fluid. The argument was that if there is concomitant synthesis and leakage of enzyme, it would be expected that the enzyme loss should be greater in puromycin treated cultures. The results show that treated and untreated cultures cannot be differentiated with respect to (i) the rate of enzyme loss and (ii) the rate of protein leakage. However, the interpretation of this experiment hinges on whether, at 7 hr after infection, there was any protein synthesis for puromycin to inhibit. The basis for this question is 2-fold. First, 7 hr in pseudorabies-infected cells correspond roughly to 9 hr in herpes simplex virus-infected cells. By 9 hr after infection 90% of the proteins synthesized during infection have already been made (Fig. 10-8). Second, the authors' argument is not supported by the evidence they present. Thus if late proteins were synthesized at 7 hr and if puromycin were to inhibit the synthesis of late proteins it would be expected that protein leakage from puromycin treated and untreated cells would differ. As indicated above, the leakage of protein from treated and untreated cells was exactly the same. It must be concluded therefore that no evidence emerged from this experiment that there was protein synthesis for the puromycin to inhibit between 7 and 10 hr after infection.

3. *The Rate of Virus Assembly*

Several of the puzzling observations that mesmerized this laboratory for an extended period concerned the rate of appearance of infectious virions in the infected cells. In one study (*94*), it was found repeatedly that in cells infected with mutant mP and superinfected with mutant MP 3 hr later, the MP progeny of the doubly infected cells appeared 1.4 hr earlier than in singly infected controls. More interesting, however, the rate of maturation of the MP mutant followed characteristically the rate of maturation of the mP mutant rather than its own. Thus began a series of studies (*45, 104*) whose objectives were to determine what factors affect the rate of virus multiplication. The most interesting observations were that cells exposed for brief intervals before 6 hr after infection to puromycin or to IUdR resumed the assembly of virus at the same rate as controls but only after a delay no less than the duration of the exposure to the drug (*45, 104*). Cells exposed briefly to the drug 6 hr or later after infection also recovered afterwards, but the rate of assembly of virus differed from that of controls. On the assumption that viral DNA is made between 4 and 6 hr after infection it was speculated that initiation of DNA synthesis required a critical concentration of enzymes and precursors (45). In retrospect the argument is irrefutable since the critical concentration could consist of a single molecule of DNA polymerase. Additional data to explain these observations have not been forthcoming.

4. *Compartmentalization of Herpesvirus Multiplication.*

As indicated earlier in the text, viral proteins are made in the cytoplasm, viral DNA is synthesized in the nucleus, and viral assembly begins in the nucleus and ends as the virus enters the cytoplasm. On the basis of these facts, we may speculate that (i) following uncoating viral DNA is transported (?) into the nucleus, (ii) the DNA is transcribed by cellular DNA-dependent RNA polymerase, (iii) the mRNA finds its way into the cytoplasm and directs the synthesis of viral proteins, (iv) the proteins enter the nucleus and begin replication of viral DNA, (v) step ii is repeated with viral enzymes(?), (vi) step iii is repeated, (vii) viral coat proteins enter the nucleus and begin assembly and, finally, (viii) the assembled virion accumulates in the cytoplasm. Two comments should be made concerning the scheme outlined above.

First, were all DNA viruses of the animal kingdom replicating in a similar fashion it could be argued that compartmentalization of viral multiplication is necessary because of the nature of the eukaryotic cell. It seems reasonable to expect that DNA viruses would be forced to utilize the nucleus—a cellular compartment apparently designed for optimal replication, transcription, and "packaging" of DNA. This however is not the case since vaccinia virus multiplies exclusively in the cytoplasm. It should be emphasized that it is not at all clear what factors determine whether a virus will multiply in the cytoplasm or in the nucleus, or how the virus reaches efficiently the specific site in which it will multiply.

Second, the scheme predicts enormous traffic of informational and structural macromolecules and of virus at the same time that portions of the nuclear membrane are being utilized to envelope the nucleocapsids. It is not at all clear how this traffic is regulated and what mechanisms determine the destination and transport of mRNA, protein subunits, nucleocapsids, etc., to and across the nuclear membrane. It is possible that elucidation of the structure and function of the double and triple membrane shells surrounding the nucleus (Fig. 10-7c) may shed some light on this problem.

10-4. The Infected Host

A. GENERAL SUMMARY

It is convenient to consider the interaction of herpesviruses with their hosts at three levels. The first and most important is the interaction with the cell in which the virus is multiplying. The most common effects of infection on cells grown in cultures are (i) changes in shape and structure of the nucleus, (ii) inhibition of host macromolecular synthesis and capacity to divide, and (iii) alteration in the structure of the cellular membrane and of immunologic specificity of the infected cells.

At the second level is the interaction of adjacent infected and uninfected cells. The infected cells exhibit new patterns of social behavior based primarily on changes in adhesiveness. The alteration in adhesiveness of cells infected with some virus mutants is manifested by the formation of loose aggregates of infected cells. At the other extreme are virus mutants which cause infected cells to fuse.

The most puzzling is the interaction between herpesviruses and the multicellular organisms they normally infect. Man, for example, is infected with herpes simplex virus between 6 months and 5 years of age. About 1% of those infected suffer mild or severe illness which runs its course in 1–3 weeks. For the rest the initial infection is inapparent or cannot be differentiated from other infectious episodes of infancy and childhood. However, as many as 75% of those who have contracted primary infection (as evidenced by the presence of antiviral antibody in their blood) are afflicted at some time during their lives with recurrent herpetic eruptions. The unique and puzzling feature of herpesvirus infection of man is that individuals subject to recurrent herpetic episodes can often predict the recrudescences accurately; the lesions appear following a specific physical or emotional provocation. It is now generally accepted that (i) following primary infection the virus is harbored in an inapparent form at some particular site and that (ii) specific stimuli associated with physical and emotional provocations of the host cause the virus to manifest itself in the form of typical herpetic lesions (90). The mechanisms by which the virus survives in the immune host and is inhibited from expressing its potentialities in the interim between recrudescences are largely obscure.

This section concerns the effects of herpesviruses on the cells in which they multiply and on the social behavior of cells in infected cultures. The interaction of herpesviruses with multicellular organisms falls outside the scope of this chapter and has been dealt with elsewhere (*90*).

B. The Cell

1. *Changes in Morphology*

The earliest changes following infection are seen in the nucleus. These are (i) margination of the chromatin, (ii) formation of an inclusion, (iii) swelling and distortion of the nucleus. The intranuclear inclusion described by Lipschutz (*158*) and named after him is granular and measures from 2 μ up to the diameter of the nucleus itself. Early in infection the inclusion is Feulgen positive (*159, 160*). Late in infection it turns Feulgen negative and becomes eosinophilic. Two other inclusions have also been described, one by Nicolau (*161*) in both nucleus and cytoplasm and one by Da Fano (*162*) in the cytoplasm.

Between 6 hr after infection and the end of the reproductive cycle the cell undergoes additional alterations in general appearance. In sparsely populated cultures of cells growing on glass surfaces, individual infected cells round up and ultimately detach from glass. In densely populated cultures, infected cells may adhere or fuse with adjoining infected or uninfected cells (*58*). Stoker and others (*163–165*) showed that infection can both prevent and abort mitosis. With few exceptions, considered in Sec. 10-5, the consequence of infection is cell death characterized by (i) inability to regain the capacity to synthesize its own macromolecules and (ii) loss of the capacity to multiply. Production of infectious progeny is not a prerequisite for cell death (*165*).

A number of investigators reported (i) amitotic nuclear division (*167–171*) and (ii) chromosome breakage (*172–177*) following infection with members of the herpesvirus group. These reports aroused interest particularly in view of the justifiable suspicion that all viruses may be potentially oncogenic. The published reports and photomicrographs suggest that following infection with herpes simplex or with pseudorabies the nuclei become distorted in shape and tend to fragment. The fragments, frequently unequal in size, remain attached

to each other. The mechanisms underlying the breakage of chromosomes remain obscure; the site and nature of aberrations induced by
herpesviruses cannot be readily differentiated from those occurring
spontaneously or those induced by a variety of mutagenic agents.
Evidence that herpesviruses cause a unique aberration at a specific
site is lacking.

2. Alteration in Host Macromolecular Metabolism

Cells replicating herpesviruses lose the capacity to synthesize their
own protein, DNA, and RNA (39, 97–102, 178). There is only scant
evidence that the inhibition is selective rather than absolute and it
is based on the finding that 4s RNA is not inhibited in HEp-2 cells
infected with herpes simplex as extensively as ribosomal RNA (200).
As shown in Fig. 10-3–10-5 the decrease in the synthesis of host
macromolecules begins immediately after exposure of cells to virus
and is essentially complete by 2–4 hr after infection. It is convenient
to designate the one or more molecules responsible for the inhibition
of the host as the inhibitor. In this section we are concerned with
(i) the time of synthesis of the inhibitor of host macromolecular synthesis, (ii) the nature of the inhibitor, (iii) the mechanism by which
the host is inhibited, and (iv) the need, in the evolutionary sense, for
the virus to inhibit the host.

Investigations conducted by two laboratories favor the hypothesis
that the inhibitor of DNA and protein synthesis is made early after
infection. The conclusion is based on two lines of evidence, i.e., (i)
puromycin arrests the inhibition of host DNA synthesis in rabbit
kidney cells infected with pseudorabies virus (178) and (ii) as shown
in Fig. 10-10, actinomycin D, 6-azauridine, and p-fluorophenyl-
alanine are each capable of preventing disaggregation of host polyribosomes in DK cells infected with herpes simplex virus (102). The
same experiments led to the tentative conclusion that the inhibition
of host macromolecular synthesis is mediated by a protein. Definitive evidence requires (i) an in vitro test system which differentiates
between host and viral macromolecular synthesis and (ii) biochemical
characterization of the inhibitor. On the other hand, studies by Newton (203) indicate that the inhibition of DNA and RNA synthesis is
due to a structural component of the virus.

The mechanism of inhibition of host macromolecular synthesis is
unknown. It it is not clear whether inhibition of protein synthesis, for
example, is the cause or consequence of the inhibition of cell nucleic

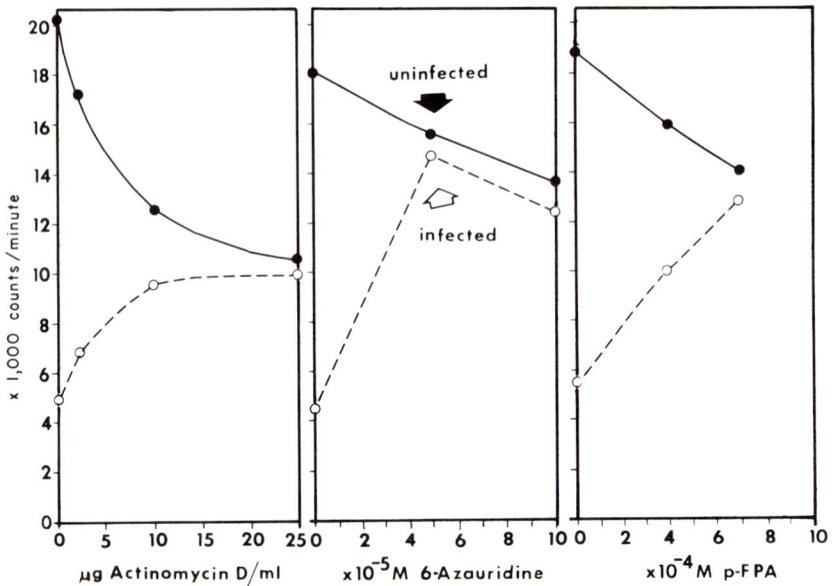

Fig. 10-10. The effect of actinomycin D, 6-azauridine, and *p*-fluorophenylalanine on the inhibition of host protein synthesis induced in DK cells by herpes simplex virus (*102*). The cells were treated with various concentrations of drugs prior to, during and after infection and pulse labeled with ^{14}C amino acids for 5 min prior to extraction and sedimentation of the cytoplasmic fractions in sucrose density gradients (15–30% w/w). The DK cells were extracted 2 hr after infection. The figures show the amount of ^{14}C-labeled nascent peptides recovered in the polyribosome region of the gradient. Solid line—infected cells; dashed line—uninfected cells.

acid synthesis. For heuristic if not factual reasons, it seems profitable to consider that the synthesis of each macromolecule is inhibited independently of the other two. The observation that herpes simplex virus inhibits protein synthesis more rapidly than actinomycin D does not detract from this view (*102*).

The brief account of the inhibition of host functions reflects the amount of available data but does little justice to the subject. The significance of the phenomenon stems from two considerations. First, the inhibitor must be able to differentiate between cellular and viral polyribosomes and between viral and cellular DNA with respect to both replication and transcription. There is at the moment no evidence that infected cells contain new or different nucleotides and sRNA differing in genetic coding properties from those of uninfected cells. If further studies bear this out, it would necessarily follow that the inhibitor functions by interacting directly with host templates. It is entirely conceivable that a virus could develop in the course of evolution an enzyme which has a greater affinity for viral nucleic acids than for cellular nucleic acids. It is far more difficult to visualize the evolutionary formulation of a macromolecule with greater affinity for host templates than its own particularly in view of the fact that herpesviruses multiply readily and effectively in a wide variety of species throughout the animal kingdom.

The second consideration relates not so much to the nature of the inhibitor as to the fact that herpesviruses contain genetic information for the inhibition of the host. It is not very clear why herpesviruses have acquired, retained, and express the capacity to inhibit the host. Recent studies with other viruses suggest that inhibition of host functions is not a prerequisite for virus multiplication and may in fact be deleterious. Thus SV_5 multiplies for many weeks in monkey cells apparently without harming them. In hamster cells, SV_5 completes one reproductive cycle and concomitantly destroys the host (179). With respect to SV_5, it is the cell that carries the seed for its own destruction. Similar observations on cells infected with herpesviruses have not been reported. On the contrary, the only available data pertinent to the problem raised here would seem to indicate that inhibition of host macromolecular synthesis is a prerequisite for virus multiplication. This conclusion is based on the observation that herpes simplex virus strain MPdk^- multiplies and effectively inhibits cells of human derivation but does not produce infectious progeny and does not effectively inhibit DK cells largely because the proteins specified in DK cells malfunction (99, 102). The significant finding is that in DK cells, viral DNA and proteins are synthesized only in cells infected at a multiplicity sufficiently high to inhibit the host. At low multiplicities of infection the cell makes interferon only (Fig. 10-11). The data would seem to indicate that (i) host response to infection and inhibition of host macromolecular synthesis are competing processes initiated simultaneously on infection and that (ii) at low multiplicities of infection the cells attain the upper hand only because the amount of effective inhibitor specified by the virus is insufficient to inhibit the host in time to prevent it from inhibiting the virus (99, 102). If the competing processes uncovered in this investigation are a general property of all cells infected with herpesvirus, one hypothesis that may account for the difference in be-

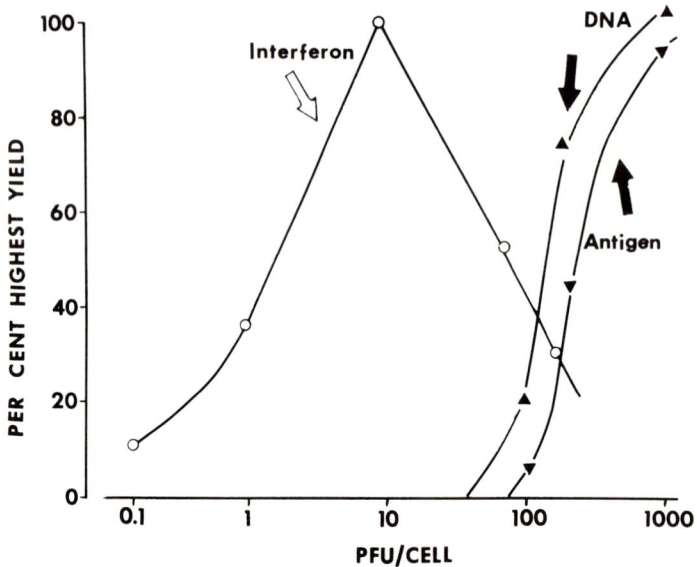

Fig. 10-11. The effects of multiplicity of infection of DK cells with MP*dk*⁻ strains of herpes simplex virus on its outcome (*99, 200*).

havior of herpesviruses and of SV₅ is (i) not all viruses elicit the synthesis of interferon and (ii) only viruses that elicit a host response have, in the course of evolution, acquired the capacity to inhibit the host.

3. *Alteration in Immunologic Specificity*

The studies on the immunologic specificity of cells infected with herpesviruses were prompted by the observation that viruses frequently alter the social behavior of cells (*58*). In general the shape, adhesiveness, and social behavior of the cell are inherited properties determined by the structure of the cellular membrane. The observation that viruses modify these properties implies that the structure of the cytoplasmic membrane has become altered. The evidence that infected cells acquire a new immunologic specificity was obtained with the aid of a test based on the observation that viruses fail to multiply in somatic cells injured by antibody and complement (*80*). In practice, cells are infected with herpes simplex virus, suspended, washed, and incubated at 37°C with appropriate amounts of antibody and complement. After 1 hr the cells are diluted in an appropriate medium and seeded on monolayer cultures of HEp-2 cells. The survivors produce plaques, whereas injured cells do not. Both antibody and complement are required for immune injury; antibody alone or complement alone is ineffective. The sole function of infected cells is to provide a measurement of the fraction of the test population that remains viable after exposure to antibody and complement. The sensitivity of the test stems from the fact that very few cells are needed since nearly every infected cell produces a plaque. The assay was initially standardized with 2-hr infected cells and antibody against uninfected cells. Pertinent here are the findings that the immunologic specificity of 2-hr infected cells could not be differentiated from that of uninfected cells and that the differentiation between "viable" and "killed" cells was not affected by the type of monolayer culture used for enumeration of infective centers.

The alteration of immunologic specificity after infection was demonstrated in tests employing 20–24 hr infected cells and rabbit sera prepared against infected cells (181). The tests showed that complement and unadsorbed anti-infected cell serum precluded the formation of plaques by 2-, 24-, and 48-hr infected cells. However, following absorption with uninfected cells the serum and complement precluded the formation of plaques by 24 and 48 hr infected cells only; the absorbed serum was not effective against 2-hr infected cells. Clearly, the 24- and 48-hr infected cells contain on their surface one or more antigens absent in uninfected cells. The conclusion that the membranes of infected cells become altered with respect to structure and immunologic specificity was corroborated in a study by Watkins (182) showing that HeLa cells infected with HFEM strain of herpes simplex acquire "stickiness" for sheep erythrocytes sensitized with rabbit antisheep erythrocyte serum. The adhesion of sensitized erythrocytes to the infected cells could be abolished by exposing the infected HeLa cells to antiviral serum. Normal rabbit serum and rabbit antisheep erythrocyte serum failed to prevent the adhesion of the sensitized erythrocytes to infected HeLa cells.

The weight of the evidence favors the hypothesis that the new antigen is a structural component of the viral envelope. The evidence consists of the following findings. (i) Absorption of serum with partially "purified" virus removes both neutralizing and cytolytic antibody (181). However, the weight of this evidence is relatively low since the most purified virus preparations were not free of host antigens.

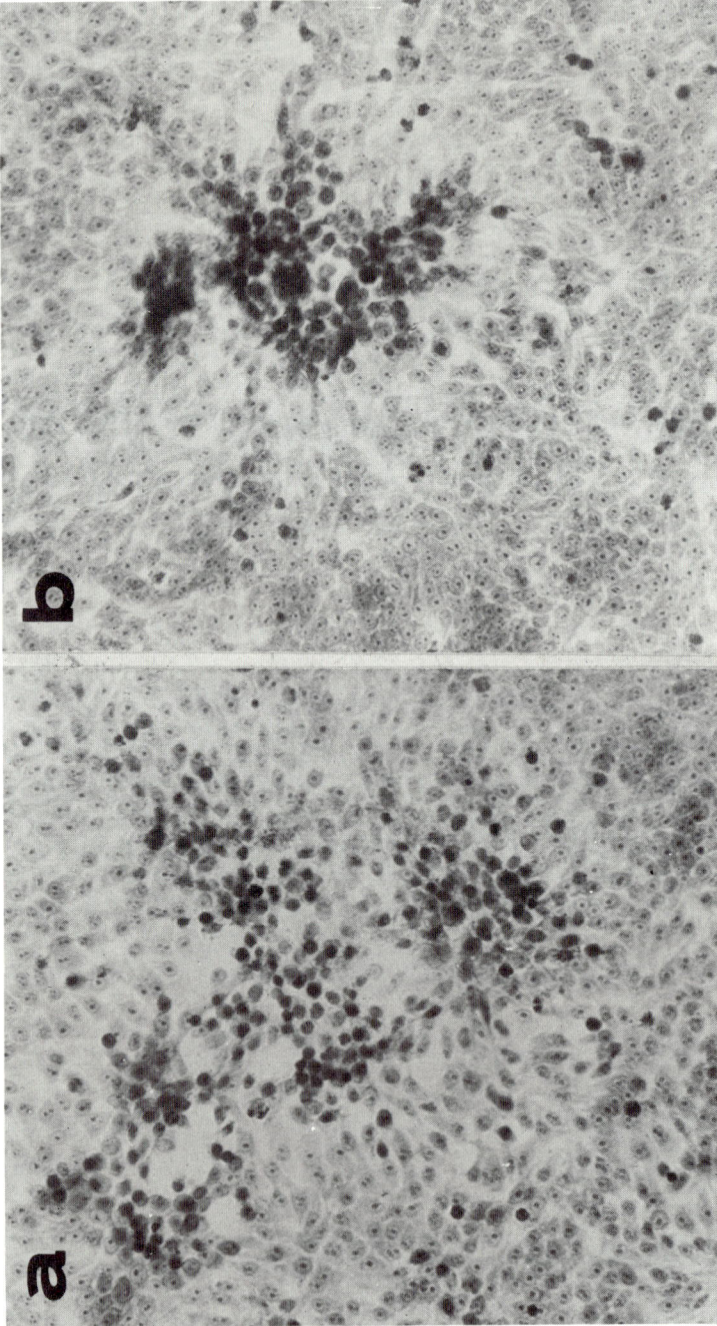

Fig. 10-12. The social behavior of cells infected with herpes simplex virus. HEp-2 cells grown on coverslips in Leighton tubes were infected at a multiplicity of 0.002 plaque forming unit per cell and incubated for 40 hr at 37°C before staining with Giemsa (59). Magnification ×135. (a) Strain VR3 obtained from Dr. Walter R. Dowdle, Communicable Disease Center, Atlanta, Georgia, (b) genital strain isolated from patient at the University of Chicago Hospitals, (c) facial strain isolated from patient of the University of Chicago Hospitals. (d) strain MP (185).

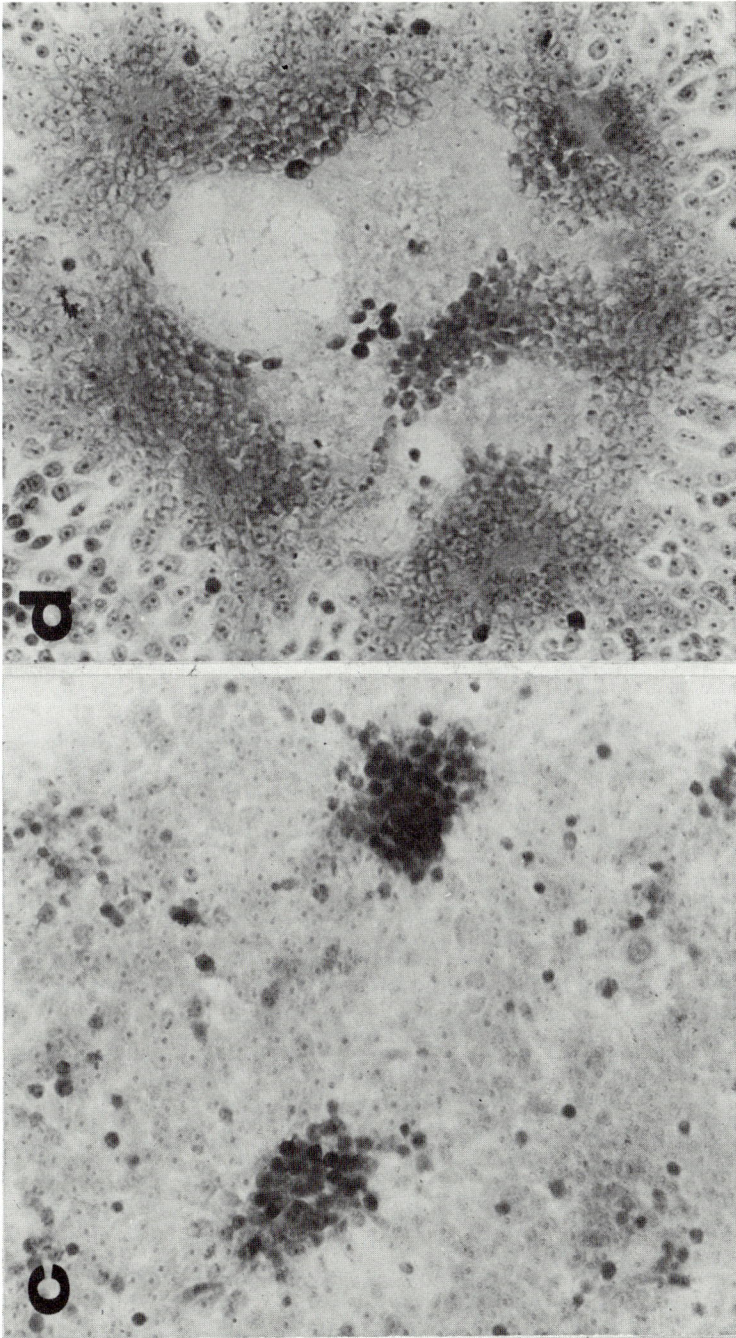

FIG. 10-12. (continued)

(ii) Assays of hyperimmune sera prepared against both infectious and noninfectious antigens fractionated from infected cells showed an excellent correlation between neutralizing and cytolytic titers (59). (iii) DK cells abortively infected with MP*dk*⁻ virus produce naked nucleocapsids only; envelopment does not take place (20, 22). Rabbit hyperimmune sera produced against extracts of abortively infected DK cells lack both neutralizing and cytolytic antibody (59).

C. The Social Behavior of Cells

1. *Description and Applications*

Numerous laboratories have reported the isolation of herpes simplex, pseudorabies, and herpes B virus strains differing with respect to their effects on cells (53, 55, 170, 183–191). Herpes simplex virus strains were recently (55) classified into four groups (Fig. 10-12), i.e., (i) strains causing rounding of cells but no adhesion or fusion, (ii) strains causing loose aggregation of rounded cells, (iii) strains causing very tight adhesion of rounded infected cells, and (iv) strains causing polykaryocytosis. The viruses comprising each group may differ in fine detail with respect to their effects on cells. Thus, polykaryocytes induced by various strains of herpes simplex virus differ in size and morphology (53, 54, 191).

A plaque assay of herpes simplex virus based on the alteration in the social behavior of infected cells has been described (52, 53, 166). The assay was prompted by the observation of Black and Melnick (192) that the spread of herpes B virus from cell to cell by direct extension is not blocked by neutralizing antibody. In practice HEp-2 cells grown in monolayer cultures are exposed to appropriate dilutions of virus, then overlayed with a medium containing human, pooled γ globulin as the source of antibody. Plaques develop after 40–48 hr of incubation at 37°C. To simplify the enumeration of plaques, the liquid overlay is removed, the cells are fixed with methanol, stained with Giemsa, then air dried. The plaque morphology and cell interactions caused by some strains are so different as to permit simultaneous assays of artificial mixtures of two or more mutants. Provided the number of plaques is not excessive, there is no reciprocal interference between the strains and the plaque counts are proportional to the concentration of virus in the artificial mixture. Studies based on the use of artificial mixtures have revealed that viruses differing with respect to their effects on the behavior of cells invariably differ also with respect to (i) buoyant density in CsCl, (ii) stability at 40°C, (iii) patterns of elution from brushite columns, and (iv) immunologic specificity (50–54).

2. The Basis for the Alteration in the Social Behavior of Infected Cells

The mechanism by which herpesviruses alter the social behavior of infected cells is largely unknown. The little that is known is based on a few experiments and much deductive reasoning and covers four points as follows.

(i) The structure of the cell which interacts with other cells and determines the adhesiveness is the cytoplasmic membrane. It follows that changes in the social behavior of cells invariably reflect corresponding changes in function, structure, and chemical composition of cellular membranes.

(ii) The alterations in social behavior induced by herpesviruses are also induced by other viruses and by chemicals such as certain lipids, lipophilic substances, and parathyroid hormone (58). Moreover, the alterations induced by the same virus in different cell lines may vary considerably (53, 58). On the basis of these observations it must be concluded that the cell responds in a similar fashion to a variety of dissimilar agents in a specific and reproducible manner.

(iii) As indicated in Sec. 10-2, D, 1, the hypothesis which best fits the available data is that cellular membranes must be modified by products specified by the virus in order for the virion to become enveloped. The hypothesis envisions that the product inducing the modification in the membranes is also responsible for the immunologic specificity of the virus and of the infected cells, and for the alteration in the cytoplasmic membranes resulting in an altered social behavior of the infected cells. One prediction of the hypothesis has been fulfilled. Thus mutations in the virus resulting in an altered structure of the envelope invariably conferred a new immunologic specificity to the virus and induced in infected cultures a new pattern of social behavior (53, 55). Ultimately, however, it will be necessary to show that a defined structural component of the envelope is, in fact, responsible for the alteration of the social behavior of infected cells.

(iv) The conclusion that the cell responds in a similar fashion to the action of numerous agents raises the question as to what selective processes in evolution have led to this particular cellular response. The answer is obviously unknown, but the point should be stressed that the function of animal cells in the artificial *in vitro* environment of the cell culture and in the whole animal differs considerably. In culture, cells act as single entities competing independently for survival. In the animal they are dependent components of a complex multicellular organism readily expendable if they constitute a threat to the life of the animal. In evolution, the selective pressure operates only at the level of the entire animal. From this point of view we may speculate that the ease with which viruses and other agents appear to modify cellular membranes and hence behavior, while undesirable and probably lethal for the cell in culture, may be very desirable for the organism as a whole. From the point of view of the host, prolonged association of the virus with the host is undesirable. It is very unlikely, for example, that cells bearing herpesvirus antigens on their surfaces would survive very long in an immune host. The modification of the cell and its subsequent removal by an immune mechanism is not always beneficial to the host. Two examples may be cited of injury to the host resulting from immune reaction with cells altered by viruses. Thus choriomeningitis virus is asymptomatic in nonimmune (tolerant) mice; development of immunity to the virus results in the death of animals (*193, 194*). Recently it has been shown that infants with maternal antibody to respiratory syncytial virus develop a more serious illness than infants lacking antibody to the virus (*195*).

Alteration in the function and structure of cellular membranes of infected cells is not the only cellular response to viral infection which appears to favor the organism rather than the infected cell. Thus the function of interferon appears to be to protect uninfected cells from becoming infected; it does not usually protect the infected cells synthesizing it. The logical consequences of this hypothesis is that (i) the selective pressure should favor the organism whose cells undergo the most extensive alterations as a result of viral infection and (ii) the "successful" virus is one that multiplies without altering the cell surface and without killing the host. There are no compelling examples that fulfill the first prediction. With respect to the second, it should be reiterated that SV_5 multiplies for many weeks without symptoms in cells derived from monkeys—its natural host—but multiplies only briefly and causes polykaryocytosis in cultures of hamster cells in which it is not naturally found (*179*).

10-5. Nonpermissiveness

A. DEFINITION

In *productive* infection dealt with in the preceding sections, the virus multiplies and the cell is defined as *permissive*. In *abortive* infection, infectious progeny is not made and the cell is defined as *nonpermissive*. Interest in nonpermissiveness stems from the expectation that any virus would replicate in any multiplying cell provided (i) the virus penetrates and is uncoated, (ii) the virus carries all of the genetic information necessary for its multiplication, and (iii) the genetic code is universal. It seems rather clear that whereas the genetic code may be universal, the host range of most viruses is not. Nonpermissive cells are invaluable for the study of cellular determinants of infection.

B. GENERAL SURVEY

Information concerning nonpermissiveness with respect to herpesviruses is very fragmentary and may be summarized as follows:

(i) Mitotic HEp-2 cells will absorb but not replicate herpes simplex virus and are not recruited into viral polykaryocytes. The cells regain the capacity to support viral multiplication after mitosis (*58, 131*). The refractiveness of dividing cells may be related to the observation that protein synthesis is inhibited during mitosis (*196*).

(ii) Treatment of cells with interferon (dealt with in another chapter) ultraviolet light (1000 ergs/mm² at 2537 Å), or heat (15 min at 45°C) before infection greatly diminish the yield of herpesviruses. The effectiveness of ultraviolet light pretreatment (*198*) is of interest particularly in view of the reports (*66, 69*) that the capacity of cells to support the multiplication of herpesviruses is highly resistant to X-irradiation. The refractiveness of heated cells appears to be specific for DNA viruses; the multiplication of RNA viruses is unaffected (*199*). Arginine deprivation (*197*) is particularly effective late in infection (*22*). It has been reported that in arginine-deprived cells the virus uncoated (*205*) and viral DNA is made (*206*). Recent experiments revealed that arginine-deprived cells failed to synthesize several proteins specified by the virus (*209*). The function of these proteins is unknown.

(iii) Cells derived from different species vary greatly in their capacity to support viral multiplication. Hamster (BHK-21), and mouse (L, mouse embryo) cells for example, support the multiplication of herpes simplex virus but not nearly as well as HEp-2 or HeLa cells (*21, 78*). The factors limiting virus yield from these cells are unknown. DK cells, a continuous line derived from dog kidney, are completely refractory to wild strains of herpes simplex virus (*79*). Current studies indicate that infection aborts because virus is unable to specify the synthesis of functional proteins. The conclusion is based on two lines of evidence. First, some viral functions are expressed, but not very effectively. Thus inhibition of host macromolecular synthesis is very slow and ineffective (*99, 102*), the yields of viral DNA are only 5% of the yield obtained from HEp-2 cells, and, lastly, the nucleocapsid made in DK cells is less stable on centrifugation in CsCl solution than that produced in HEp-2 cells (*20, 22*). Second, some functions are missing entirely. Thymidine kinase is not induced (*144*) and nucleocapsids are not enveloped (*20, 21*). The reason why wild strains of herpes simplex are unable to specify functional proteins remains obscure.

ACKNOWLEDGMENTS

I should like to acknowledge my indebtedness to Mrs. Norma Coleman and Mrs. Deirdre Kondor for assistance in the preparation of the manuscript, to another B. Roizman for keeping the references straight, and to numerous colleagues for many useful discussions. The published and unpublished data from our laboratory cited here were accumulated with the aid of grants from the American Cancer Society (E 314D), the National Science Foundation (GB 4555), the U.S.P.H.S. (CA 08494), and the Whitehall Foundation.

I should also like to express my thanks to Academic Press and to the American Society of Microbiology for permission to republish figures which appeared originally in *Virology* and the *Journal of Virology*, respectively.

REFERENCES

1. A. S. Kaplan, in *Basic Medical Virology* (James E. Prier, ed.), Williams and Wilkins, Baltimore, Md., 1966, Chap. 16.

2. C. E. Van Rooyen and A. J. Rhodes, *Virus Diseases of Man*, Thomas Nelson & Sons, New York, 1948, pp. 155–168 and pp. 169–261.

3. T. L. Beswick, *Med. Hist.*, **6**, 214 (1962).

4. I. B. Wilner, *A Classification of the Major Groups of Human and Other Animal Viruses,* 3rd ed., Burgess Publishing Co., Minneapolis, Minn., 1966, p. 78.

5. C. H. Andrewes, *Advan. Virus Res.,* **9,** 271 (1962).

6. M. A. Epstein, B. G. Achong, and Y. M. Barr, *Lancet,* **1964-II,** 702.

7. M. A. Epstein and Y. M. Barr, *J. Natl. Cancer Inst.,* **34,** 231 (1965).

8. G. T. O'Connor and A. S. Rabson, *J. Natl. Cancer Inst.,* **35,** 899 (1965).

9. S. E. Stewart, E. Lovelace, J. J. Whang, and V. Anomah Ngu, *J. Natl. Cancer Inst.,* **34,** 319 (1965).

10. K. Hummeler, G. Henle, and W. Henle, *J. Bacteriol.,* **91,** 1366 (1966).

11. J. Yamaguchi, Y. Hinuma, and J. T. Grace, Jr., *J. Virol.,* **1,** 640 (1967).

12. P. M. Biggs, H. G. Purchase, B. R. Bee, and P. J. Dalton, *Vet. Res.,* **77,** 1339 (1965).

13. P. A. L. Wight, J. E. Wilson, J. G. Campbell, and E. Fraser, *Nature,* **216,** 804 (1967).

14. M. A. Epstein, B. G. Achong, A. E. Churchill, and P. M. Biggs, *J. Natl. Cancer Inst.,* **41,** 805 (1968).

15. H. G. Purchase and P. M. Biggs, *Res. Vet. Sci.,* **8,** 440 (1967).

16. O. W. Fowcell, *J. Biophys. Biochem. Cytol.,* **2,** 725 (1956).

17. P. D. Lunger, *Virology,* **24,** 138 (1964).

18. C. W. Stackpole and M. Mizell, *Virology,* **36,** 63 (1968).

19. P. Wildy, W. C. Russell, and R. W. Horne, *Virology,* **12,** 204 (1960).

20. S. Spring, B. Roizman, and J. Schwartz, *J. Virol.,* **2,** 384 (1968).

21. D. H. Watson, *Symp. Gen. Soc. Microbiol.,* in press, 1968.

22. S. B. Spring and B. Roizman, *J. Virol.,* **1,** 294 (1967).

23. D. S. Bocciarelli, Z. Orfei, G. Mondino, and A. Persechino, *Virology,* **30,** 58 (1966).

24. C. Morgan, S. A. Ellison, H. M. Rose, and D. H. Moore, *Proc. Soc. Exptl. Biol. Med.,* **82,** 454 (1953).

25. C. Morgan, S. A. Ellison, H. M. Rose, and D. H. Moore, *J. Exptl. Med.,* **100,** 195 (1954).

26. M. Reissig and J. L. Melnick, *J. Exptl. Med.,* **101,** 341 (1955).

27. C. Morgan, E. P. Jones, M. Holden, and H. M. Rose, *Virology,* **5,** 568 (1958).

28. M. G. P. Stoker, K. M. Smith, and R. W. Ross, *J. Gen. Microbiol.,* **19,** 244 (1958).

29. D. Falke, R. Siegert, and W. Vogell, *Arch. Ges. Virusforsch.,* **9,** 484 (1959).

30. C. Morgan, H. M. Rose, M. Holden, and E. P. Jones, *J. Exptl. Med.,* **110,** 643 (1959).

31. C. Morgan and H. M. Rose, *International Conference on Electron Microscopy, 4th, Berlin, 1958,* Vol. 5, Springer, Berlin, 1960, Chap. 2, p. 590.

32. M. A. Epstein, *J. Exptl. Med.,* **115,** 1 (1962).

33. M. H. McGavran and M. C. Smith, *Exptl. Mol. Pathol.,* **4,** 1 (1965).

34. D. H. Watson, P. Wildy, and W. C. Russell, *Virology,* **24,** 523 (1964).

35. M. A. Epstein, *J. Cell Biol.,* **12,** 589 (1962).

36. P. Becker, J. L. Melnick, and H. D. Mayor, *Exptl. Mol. Pathol.,* **4,** 11 (1965).

37. R. S. Siegert and D. Falke, *Arch. Ges. Virusforsch.,* **19,** 230 (1966).

38. F. H. Shipkey, R. A. Erlandson, R. B. Bailey, V. I. Babcock, and C. M. Southam, *Exptl. Mol. Pathol.,* **6,** 39 (1967).

39. B. Roizman and P. R. Roane, Jr., *Virology,* **22,** 262 (1964).

40. B. Roizman, unpublished work, 1965.

41. W. C. Russell, *Virology,* **16,** 355 (1962).

42. T. Ben-Porat and A. S. Kaplan, *Virology,* **16,** 261 (1962).

43. R. W. Darlington and C. C. Randall, *Virology,* **19,** 322 (1963).

44. W. C. Russell and L. V. Crawford, *Virology,* **21,** 353 (1963).

45. B. Roizman, L. Aurelian, and P. R. Roane, Jr., *Virology,* **21,** 482 (1963).

46. A. S. Kaplan and T. Ben-Porat, *Virology,* **23,** 90 (1964).

47. W. C. Russell and L. V. Crawford, *Virology,* **22,** 288 (1964).

48. L. V. Crawford and A. J. Lee, *Virology,* **23,** 105 (1964).

49. D. Lando, J. De Rudder, and M. P. De Garilhe, *Bull. Soc. Chim. Biol.,* **47,** 1033 (1965).

50. B. Roizman and P. R. Roane, Jr., *Virology,* **19,** 198 (1963).

51. S. Spring and B. Roizman, *J. Virol.,* **2,** 979 (1968).

52. B. Roizman and P. R. Roane, Jr., *Virology,* **15,** 75 (1961).

53. B. Roizman and L. Aurelian, *J. Mol. Biol.,* **11,** 528 (1965).

54. H. Kohlhage, *Arch. Ges. Virusforsch.,* **14,** 358 (1964).

55. P. M. Ejercito, E. D. Kieff, and B. Roizman, *J. Gen. Virol.,* **2,** 357 (1968).

56. P. G. Spear and B. Roizman, *Nature,* **214,** 713 (1967).

57. D. H. Watson and P. Wildy, *Virology,* **21,** 100 (1963).

58. B. Roizman, *Cold Spring Harbor Symp. Quant. Biol.,* **27,** 327 (1962).

59. B. Roizman and S. Spring, *Proceedings of the Conference on Cross Reacting Antigens and Neoantigens,* William & Wilkins, Baltimore, Md., 1967, pp. 85–96.

60. I. H. Holmes and D. H. Watson, *Virology,* **21,** 112 (1961).

61. K. O. Smith, *Proc. Soc. Exptl. Biol. Med.,* **115,** 814 (1964).

62. I. Gresser and J. F. Enders, *Virology,* **13,** 420 (1961).

63. H. Amos, *J. Exptl. Med.,* **98,** 365 (1953).

64. S. P. Bedson and J. V. T. Gostling, *Brit. J. Exptl. Pathol.,* **39,** 502 (1958).

65. S. Ivanicova, R. Skoda, V. Mayer, and F. Sokol, *Acta Virol.,* **7,** 7 (1963).

66. W. F. Powell, *Virology,* **9,** 1 (1959).

67. P. R. Roane, Jr., and B. Roizman, *Biochem. Biophys. Acta,* **91,** 168 (1964).

68. A. E. Farnham and A. A. Newton, *Virology,* **7,** 449 (1959).

69. A. S. Kaplan, *Virology,* **4,** 435 (1957).

70. M. D. Hoggan and B. Roizman, *Virology,* **8,** 508 (1959).

71. J. G. Stevens and N. B. Groman, *Am. J. Vet. Res.,* **24,** 1158 (1963).

72. T. F. McN. Scott, D. L. McLeod, and T. Tokumaru, *J. Immunol.,* **86,** 1 (1961).

73. G. Plummer and B. Lewis, *J. Bacteriol.,* **89,** 671 (1965).

74. C. Wallis, C. S. Yang, and J. L. Melnick, *J. Immunol.,* **89,** 41 (1962).

75. K. Munk and W. W. Ackerman, *J. Immunol.,* **71,** 426 (1953).

76. B. Roizman, unpublished work, 1967.

77. C. Wallis and J. L. Melnick, *J. Bacteriol.,* **90,** 1632 (1965).

78. B. Roizman, *Proc. Natl. Acad. Sci. U.S.,* **48,** 795 (1962).

79. L. Aurelian and B. Roizman, *Virology,* **22,** 262 (1964).

80. B. Roizman, and P. G. Spear, *J. Virol.,* **2,** 83 (1968).

81. B. Roizman, *Proc. Natl. Acad. Sci. U.S.,* **48,** 973 (1962).

82. A. S. Huang and R. R. Wagner, *Proc. Soc. Exptl. Biol. Med.,* **116,** 863 (1964).

83. A. Vaheri and K. Penttinen, *Ann. Med. Exptl. Biol. Fenniae (Helsinki),* **40,** 334 (1962).

84. A. Vaheri and K. Cantell, *Virology,* **21,** 661 (1963).

85. A. A. Tytell and R. E. Neuman, *Proc. Soc. Exptl. Biol. Med.,* **113,** 343 (1963).

86. A. J. Nahmias and S. Kibrick, *J. Bacteriol.,* **87,** 1060 (1964).

87. A. J. Nahmias, S. Kibrick, and P. Bernfeld, *Proc. Soc. Exptl. Biol. Med.,* **115,** 993 (1964).

88. R. Benda, *Acta Virol.,* **10,** 376 (1966).

89. Gy. Hadhazy, F. Lehel, and L. Gergely, *Acta Microbiol. Acad. Sci. Hung.,* **13,** 145 (1966).

90. B. Roizman, in *Perspectives in Viology,* Vol. IV (M. Pollard, ed.), Harper & Row, New York, 1965, pp. 283–304.

91. M. G. P. Stoker and R. W. Ross, *J. Gen. Microbiol.,* **19,** 250 (1958).

92. K. O. Smith, *J. Bacteriol.,* **86,** 99 (1963).

93. A. S. Kaplan and A. E. Vatter, *Virology,* **7,** 394 (1959).

94. B. Roizman, *Proc. Natl. Acad. Sci. U.S.,* **49,** 165 (1963).

95. B. Roizman, *Proc. Soc. Exptl. Biol. Med.,* **119,** 1021 (1965).

96. B. Roizman, *Virology,* **27,** 113 (1965).

97. B. Roizman, G. S. Borman, and M. Kamali-Rousta, *Nature,* **206,** 1374 (1965).

98. R. J. Sydiskis and B. Roizman, *Science,* **153,** 76 (1966).

99. L. Aurelian and B. Roizman, *J. Mol. Biol.,* **11,** 539 (1965).

100. J. Hay, G. J. Koteles, H. M. Keir, and H. Subak-Sharpe, *Nature,* **210,** 387 (1966).

101. J. F. Flanagan, *J. Virol.,* **1,** 583 (1967).

102. R. J. Sydiskis and B. Roizman, *Virology,* **32,** 678 (1967).

103. A. S. Kaplan and T. Ben-Porat, *Virology,* **19,** 205 (1963).

104. B. Roizman, in *Proceedings of the 17th Annual Symposium, M. D. Anderson Hospital and Tumor Institute,* William & Wilkins, Baltimore, Md., 1963, pp. 205–223.

105. H. Subak-Sharpe and J. Hay, *J. Mol. Biol.,* **12,** 924 (1965).

106. H. Subak-Sharpe, W. M. Shepherd, and J. Hay, *Cold Spring Harbor Symp. Quant. Biol.,* **31,** 583 (1966).

107. J. Hay, H. Subak-Sharpe, and W. M. Shepherd, *Biochem. J.,* **103,** 69 (1967).

108. H. G. Klemperer, G. R. Haynes, W. I. H. Shedden, and D. H. Watson, *Virology,* **31,** 120 (1967).

109. S. Kit and D. R. Dubbs, *Biochem. Biophys. Res. Commun.,* **11,** 55 (1963).

110. S. Kit and D. R. Dubbs, *Biochem. Biophys. Res. Commun.,* **13,** 500 (1963).

111. C. Hamada, T. Kamiya, and A. S. Kaplan, *Virology,* **28,** 271 (1966).

112. W. H. Prusoff, Y. S. Bakhle, and L. Sekely, *Ann. N.Y. Acad. Sci.,* **130,** 135 (1965).

113. P. M. Frearson, S. Kit, and D. R. Dubbs, *Cancer Res.,* **25,** 737 (1965).

114. M. Nohara and A. S. Kaplan, *Biochem. Biophys. Res. Commun.,* **12,** 189 (1963).

115. H. M. Keir and E. Gold, *Biochim. Biophys. Acta,* **72,** 263 (1963).

116. H. M. Keir, J. Hay, J. M. Morrison, and H. Subak-Sharpe, *Nature,* **210,** 369 (1966).

117. H. M. Keir, H. Subak-Sharpe, W. I. H. Shedden, D. H. Watson, and P. Wildy, *Virology,* **30,** 154 (1966).

118. B. R. McAuslan, P. Herde, D. Pett, and J. Ross, *Biochem. Biophys. Res. Commun.,* **20,** 586 (1965).

119. J. Flanagan, *J. Bacteriol.,* **91,** 789 (1966).

120. J. M. Morrison and H. M. Keir, *Biochem. J.,* **103,** 70 (1967).

121. R. J. Sydiskis and B. Roizman, *Virology,* 34, 562 (1968).

122. E. Gold, P. Wildy, and D. H. Watson, *J. Immunol.,* **91,** 666 (1963).

123. W. C. Russell, E. Gold, H. M. Keir, H. Omura, D. H. Watson, and P. Wildy, *Virology,* **22,** 103 (1964).

124. T. Tokumaru, *J. Immunol.,* **95,** 181 (1965).

125. T. Tokumaru, *J. Immunol.,* **95,** 189 (1965).

126. C. Hamada and A. S. Kaplan, *J. Bacteriol.,* **89,** 1328 (1965).

127. D. H. Watson, W. I. Shedden, A. Elliot, T. Tetsuka, P. Wildy, D. Bourgaux-Ramoisy, and E. Gold, *Immunology,* **11,** 399 (1966).

128. S. Fujiwara and A. S. Kaplan, *Virology,* **32,** 60 (1967).

129. J. Lebrun, *Extra. Bull. Microscop. Appl.,* **2,** 94 (1956).

130. J. Lebrun, *Virology,* **2,** 496 (1956) .

131. B. Roizman, *Virology,* **13,** 387 (1961).

132. S. Nii and J. Kamahora, *Bikens J.,* **6,** 145 (1963).

133. K. Munk and H. Fischer, *Arch. Ges. Virusforsch.,* **15,** 539 (1965).

134. J. F. O'Dea and J. K. Dineen, *J. Gen. Microbiol.,* **17,** 19 (1957).

135. R. Benda, O. Prochazka, L. Cerva, H. Rehn, and V. Hronovsky, *Acta Virol.,* **10,** 149 (1966).

136. F. Rapp, L. E. Rasmussen, and M. Benyesh-Melnick, *J. Immunol.,* **91,** 709 (1963).

137. P. R. Roane, Jr., and B. Roizman, *Virology,* **29,** 668 (1966).

138. B. Roizman, S. B. Spring, and P. R. Roane, Jr., *J. Virol.,* **1,** 181 (1967).

139. N. P. Salzman and E. D. Sebring, *J. Virol.,* **1,** 16 (1967).

140. A. B. Sabin, private communication, 1966.

141. A. S. Kaplan and T. Ben-Porat, *Symposium, International Congress of Microbiology,* Moscow, 1966.

142. A. S. Kaplan, *Virology,* **24,** 19 (1964).

143. T. Ben-Porat and A. S. Kaplan, *Virology,* **20,** 310 (1963).

144. B. Roizman, *Proceedings of the International Wenner-Gren Symposium,* Pergamon, London, 1960, p. 73.

145. A. S. Kaplan and T. Ben-Porat, *Virology,* **11,** 12 (1960).

146. R. Sydiskis and B. Roizman, unpublished work, 1966.

147. N. P. Salzman, *Virology,* **10,** 150 (1960).

148. P. Siminoff, *Virology,* **24,** 1 (1964).

149. K. O. Smith and C. D. Dukes, *J. Immunol.,* **92,** 550 (1964).

150. A. S. Kaplan and T. Ben-Porat, *J. Mol. Biol.,* **19,** 320 (1966).

151. A. Newton, P. P. Dendy, C. L. Smith, and P. Wildy, *Nature,* **194,** 886 (1962).

152. A. Newton and M. G. P. Stoker, *Virology,* **5,** 560 (1958).

153. K. Munk and G. Sauer, *Virology,* **22,** 153 (1964).

154. M. Reissig and A. S. Kaplan, *Virology,* **16,** 1 (1962).

155. T. Kamiya, T. Ben-Porat, and A. S. Kaplan, *Biochem., Biophys. Res. Commun.,* **16,** 410 (1964).

156. T. Kamiya, T. Ben-Porat, and A. S. Kaplan, *Virology,* **26,** 577 (1965).

157. J. Zemla, C. Coto, and A. S. Kaplan, *Virology,* **31,** 736 (1967).

158. B. Lipschutz, *Wien. Med. Wochshr.,* **71,** 231 (1921).
159. H. V. Crouse, L. L. Coriell, H. Blank, and T. F. McN. Scott, *J. Immunol.,* **65,** 119 (1950).
160. W. Wolman and A. Behar, *J. Infect. Diseases,* **91,** 63 (1952).
161. S. Nicolau, *Compte. Rend. Soc. Biol. Paris,* **126,** 326 (1937).
162. C. Da Fano, *J. Pathol. Bacteriol.,* **26,** 85 (1923).
163. M. G. P. Stoker and A. A. Newton, *Ann. N.Y. Acad. Sci.,* **81,** 129 (1959).
164. M. G. P. Stoker, *Symp. Soc. Gen. Microbiol.,* **9,** 142 (1959).
165. J. T. Vantis and P. Wildy, *Virology,* **17,** 225 (1962).
166. M. D. Hoggan, B. Roizman, and T. B. Turner, *J. Immunol.,* **84,** 152 (1960).
167. T. F. McN. Scott, C. F. Burgoon, L. L. Coriell, and H. Blank, *J. Immunol.,* **71,** 385 (1953).
168. A. S. Kaplan and T. Ben-Porat, *Virology,* **8,** 352 (1959).
169. M. Reissig and A. S. Kaplan, *Virology,* **11,** 1 (1960).
170. D. Falke, *Virology,* **14,** 492 (1961).
171. S. Nii and J. Kamahora, *Biken's J.,* **6,** 33 (1963).
172. B. Hampar and S. A. Ellison, *Nature,* **192,** 145 (1961).
173. B. Hampar and S. A. Ellison, *Proc. Natl. Acad. Sci. U.S.,* **49,** 474 (1963).
174. H. F. Stich, T. C. Hsu, and F. Rapp, *Virology,* **22,** 439 (1965).
175. M. Benyesh-Melnick, H. F. Stich, F. Rapp, and T. C. Hsu, *Proc. Soc. Exptl. Biol. Med.,* **117,** 546 (1964).
176. F. Rapp and T. C. Hsu, *Virology,* **25,** 401 (1965).
177. M. Boiron, J. Tanzer, M. Thomas, and A. Hampe, *Nature,* **209,** 737 (1966).
178. T. Ben-Porat and A. S. Kaplan, *Virology,* **25,** 22 (1965).
179. K. V. Holmes and P. W. Choppin, *J. Exptl. Med.,* **124,** 501 (1966).
180. B. Roizman and P. R. Roane, Jr., *J. Immunol.,* **87,** 714 (1961).
181. P. R. Roane, Jr., and B. Roizman, *Virology,* **22,** 1 (1964).
182. J. F. Watkins, *Nature,* **202,** 1364 (1964).
183. T. Tokumaru, *Proc. Soc. Exptl. Biol. Med.,* **99,** 55 (1957).
184. A. Gray, T. Tokumaru, and T. F. McN. Scott, *Arch. Ges. Virusforsch.,* **8,** 60 (1958).
185. M. D. Hoggan and B. Roizman, *Am. J. Hyg.,* **70,** 208 (1959).
186. H. C. Hinze and D. L. Walker, *J. Bacteriol.* **82,** 498 (1961).
187. S. Nii and J. Kamahora, *Biken's J.,* **4,** 255 (1961).
188. H. Kohlhage and R. Siegert, *Arch. Ges. Virusforsch.,* **12,** 273 (1962).
189. K. E. Schneweiss, *Zentr. Bakteriol. Parasitenk., Abt. I, Orig.,* **186,** 467 (1962).
190. K. Munk and D. Donner, *Arch. Ges. Virusforsch.,* **8,** 529 (1963).
191. C. E. Wheeler, *J. Immunol.,* **93,** 749 (1964).
192. F. L. Black and J. L. Melnick, *J. Immunol.,* **74,** 236 (1955).
193. E. Traub, *Perspectives Virol. Symp.,* **1959,** 160.
194. M. Volkert, *in Perspectives in Virology,* Vol. IV, Harper & Row, New York, 1965, p. 269.
195. R. M. Chanock, C. B. Smith, W. T. Friedewald, R. M. Parrot, B. R. Forsyth, H. V. Coates, A. Z. Kapikian, and M. A. Gharpure, *in First International Conference on Vaccines against Viral and Rickettsial Diseases of Man,* Pan American Health Organization, International Conference on Virus and Rickettsial Vaccines, Washington, D. C., Scientific Publications 147, 1967, pp. 53–61.

196. J. M. Salb and P. I. Marcus, *Proc. Natl. Acad. Sci. U.S.,* **54,** 1353 (1965).

197. R. W. Tankersley, Jr., *J. Bacteriol,* **87,** 609 (1964).

198. B. Roizman, unpublished data, 1965.

199. M. A. Gharpure, *Virology,* **27,** 308 (1965).

200. E. K. Wagner and B. Roizman, unpublished data, 1968.

201. V. L. Morris, E. K. Wagner, and B. Roizman, unpublished data, 1968.

202. P. G. Spear and B. Roizman, *Virology,* **36,** 545 (1968).

203. A. Newton, *Proceedings First International Congress for Virology,* Helsinki, *Finland, 1968,* in press, 1969.

204. S. Nii, C. Morgan, and H. M. Rose, *J. Virol.,* **2,** 517 (1968).

205. V. R. M. Inglis, *J. Gen. Virol.,* **3,** 9 (1968).

206. Y. Becker, V. Olshevsky, and J. Levitt, *J. Gen. Virol.,* **1,** 111 (1967).

207. J. Schwartz and B. Roizman, *Virology,* in press, 1969.

208. B. Roizman, S. B. Spring, and J. Schwartz, *Federation Proc.,* in press, 1969.

209. S. B. Spring, B. Roizman, and P. G. Spear, in preparation.

11

DNA-CONTAINING BACTERIOPHAGE

CHRISTOPHER K. MATHEWS

DEPARTMENT OF BIOCHEMISTRY
UNIVERSITY OF ARIZONA COLLEGE OF MEDICINE
TUCSON, ARIZONA

11-1. Introduction

The study of bacteriophages, or bacterial viruses, has played an enormous role in establishing our present ideas on the transmission of genetic messages and on general problems of virus replication. While great diversity exists among the viruses, and even among different bacteriophages which attack the same host, certain properties of bacteriophages are common to all viruses. These include (1) the presence of only one type of nucleic acid (DNA or RNA) in a mature virus particle, (2) a single molecule of nucleic acid per virus particle, insofar as is presently known, (3) a virtual absence of metabolic activity in the absence of a susceptible host cell, and (4) a carefully timed replication cycle within the infected cell.

In order to study virus multiplication biochemically, one must be able to initiate the growth cycle simultaneously in a large, genetically homogeneous cell population. This is easily done with bacteriophages, since one can isolate the host in pure culture and grow large populations rapidly in chemically defined media. In addition, one can subject either the phage or its host to genetic manipulation, such that, by exploring the biochemical consequences of an infection in which either the virus or the host bacterium bears a genetically defined lesion, one can gain insight into the normal infective cycle. These advantages enabled S. S. Cohen and others to carry out definitive biochemical investigations on phage growth as early as the mid-1940's, while comparable experiments with animal viruses could not be approached until a decade later when suitable tissue culture techniques had been developed.

The original, and still most fundamental, questions of bacteriophage reproduction could be phrased in the context of the results of Ellis and Delbrück (1), who in 1939 published what is now generally accepted as the first quantitative study of this process. These workers devised a procedure known as the one-step growth experiment. A growing bacterial culture is infected with phage and shortly thereafter is diluted by a large factor to prevent readsorption of progeny phage on uninfected or unlysed cells. Samples are removed at various times and assayed for infectious titer by virtue of their ability to form plaques when plated on a lawn of susceptible bacteria. Ellis and Delbrück found that at 37°C the concentration of infectious particles remained constant until about 25 min. At this time the titer increased rapidly by some 100-fold, and did not subsequently change. A typical one-step growth curve is shown in Fig. 11-1. Analysis of similar curves led Ellis and Delbrück to conclude that there are three major periods in the life cycle of a virulent bacteriophage: (1) adsorption onto a bacterial cell; (2) multiplication of the phage within the bacterium (latent period); and (3) cell lysis, with concomitant release of the newly formed phage into the medium. During the latent period each infected cell forms only a single plaque on a Petri plate seeded with bacteria no matter how many completed phage particles it might contain, for any phage released subsequent to plating are trapped in the solid agar medium. Lysis, however, disperses the cells throughout the medium, and it was possible to show that a single phage particle can initiate the formation of a plaque.

Another important early study of phage growth was made by Doermann (2), who devised techniques for artificial lysis of cells before the end of the normal growth cycle. This allowed him to investigate the intracellular development of phage by removing samples during the latent period, lysing them, and plating. Results of this type of growth curve, also shown in Fig. 11-1, demonstrated that during the first 12 min after infection by phage T2 no infective particles are detectable inside the bacterial cell. Such particles begin to be formed after this time until, at about 15 min, there is an average of 1 plaque-forming unit per infected cell. The period during which infected cells contain no plaque-forming particles is known as the eclipse. A central problem

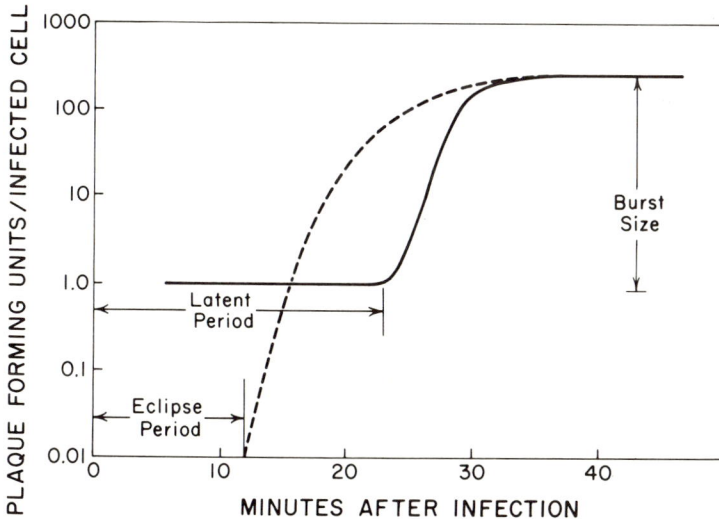

FIG. 11-1. Typical one-step growth curve for a T-even bacteriophage. Cells were infected as described in the text. Samples were removed and plated either with premature lysis (dashed line) for assay of total phage or without lysis (solid line) for assay of plaque-forming centers (phage plus infected cells).

of bacterial virology, and the one which receives the greatest emphasis in this chapter, is the question of how the viral genetic material inter-acts with metabolic machinery of the host cell during the eclipse in such a way that mature particles are formed subsequently; in other words, how is infectivity lost and then regained?

Fortunately for the field of bacterial virology, an informal agree-ment was reached in the mid-1940's among most of the workers in the then-fledgling field to confine their studies to the seven members of the T series of bacteriophages, whose common host is *Escherichia coli*. This compact was reached to counteract the early tendency of each researcher to isolate and study his own phage–host system, a practice which made extremely difficult a meaningful comparison of data obtained in different laboratories. Within the group of T phages, the T-even phages, T2, T4, and T6, comprised the most popular experi-mental material. Limiting the energies of early phage workers to these three closely related viruses without question contributed substantially to the spectacular progress of the field of bacterial virology. See Ref. *3* for a collection of delightful essays describing both the scientific and personal aspects of the development of this area.

In spite of the great amount of attention which the T-even phages have received, a surprising number of fundamental questions about these viruses remain unanswered. This chapter attempts to describe the status of this as-yet uncharted territory, including such questions as: How is the host cell killed? What are the mechanisms of DNA replication and genetic recombination? What is the function of the "internal protein" which enters the infected cell along with the viral DNA? How is the DNA "packaged" to give a mature virus particle? What are the timing mechanisms which control the synthesis of virus-specific proteins? What factors determine the time of lysis of the infected cell?

Bacteriophages have formed the subject matter of two excellent textbooks (*4, 5*). In addition, bacteriophage reproduction was com-prehensively reviewed in 1963 by Champe (*6*) and by Cohen (*7*). These and other reviews lay by far the heaviest emphasis among the bacteriophages upon the widely studied T-even phages. However, many other phage systems present research opportunities no less challenging than those offered by the T-even phages, and just as relevant to problems of virology and of biology in general. These include questions such as the nature of the lysogenic state, in which the chromosome of a temperate bacteriophage becomes integrated into that of its bacterial host; the mode of replication of single-stranded

DNA, found in certain small bacteriophages; and the mechanism of replication of a relatively recently discovered class of phages which contain RNA as its genetic material (the last-mentioned topic is treated elsewhere in this volume). These non-T-even systems have received greatly increased attention within the past four years, as illustrated in Fig. 11-2. This figure shows the number of papers appearing in the *Journal of Molecular Biology* in each of the past six years which deal primarily with T-even phages and those dealing with all other phages. This chapter attempts to discuss most of the current research topics of greatest interest in bacterial virology, drawing examples from many phage systems. However, because of the historical and conceptual framework provided by work on the T-even phages, these viruses are discussed first and most extensively.

11-2. T-Even Bacteriophages

A. STRUCTURAL FEATURES

Like all viruses, T-even bacteriophages contain a core of nucleic acid, in this case DNA, surrounded by a protein coat. Unlike most viruses,

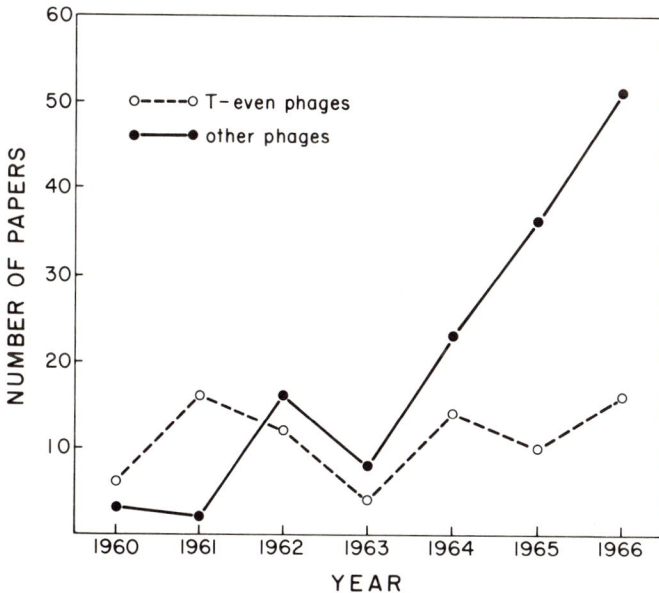

FIG. 11-2. Papers published in the *Journal of Molecular Biology* dealing primarily with T-even phages (open circles) and all other phages (closed circles).

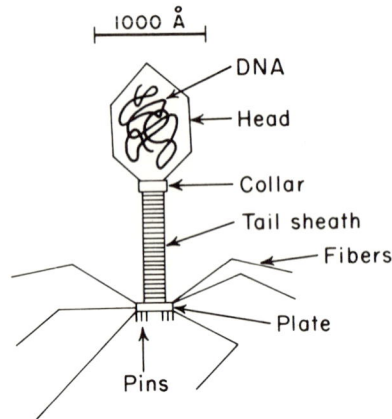

Fig. 11-3. Schematic representation of the structure of a T-even bacteriophage particle.

the external structure is complex, consisting of several different types of protein molecules, or aggregations of molecules. This was evident from the earliest electron micrographs taken in the early 1940's (8, 9) which revealed the now-familiar tadpole-like structure of these phages. Subsequent work, using refined electron microscopic techniques and various chemical methods for dissociation of the virus particle into its several components (cf. 10), has revealed the structure diagrammed in Fig. 11-3.

Most of the protein forms the hexagonal-shaped head membrane which surrounds the DNA. The head is composed of about 1000 identical protein subunits of molecular weight 80,000 per subunit (10, 11). Attached to one of the vertices of the head is a tail of remarkably complex design. The tail consists of two coaxial cylindrical structures, an outer sheath and an inner core. At the proximal end, near the point of attachment to the head, is a collar structure (12). At the distal end is a hexagonal structure, the baseplate, to which are attached six tail fibers (13). Several shorter structures, the tail pins, are also attached to the baseplate (14).

A good deal of information about the tail structure has come from studies on isolated tail components. The tail sheath contains 144 subunits of molecular weight 55,500 each (15). The sheath is a remarkable structure, capable of contracting during the early stages of infection from a length of 800 Å to 350 Å, with a concomitant thickening (10). This contraction appears to result at least partially from hydrolysis of bound ATP. For some time it had been known that T2 particles possess ATPase activity (16). This activity has now been shown to reside in isolated sheaths (17). In addition, T2 particles bind nucleoside triphosphates, mostly ATP, to the extent of 140 molecules per particle, or about one molecule per subunit (18). Studies with isolated tail sheaths have revealed that removal of bound ATP by prolonged dialysis shortens and thickens the sheath but that the original structure is regained upon readdition of ATP (15).

Tail cores have also been isolated in pure form (19). The T2 core has a molecular weight of 487,000 and consists of five identical subunits. These are entwined about each other to form a hollow, cable-like structure, some 1000 Å in length, with an internal diameter of about 25 Å. The tail fibers, the only other tail components yet to be isolated, consist of four subunits per fiber, each subunit having a molecular weight of about 100,000 (10).

A lysozyme activity is associated with the distal end of the tail. This probably participates in the early stages of infection by dissolving a portion of the bacterial cell wall. The protein responsible for this activity has a molecular weight of about 15,000 (20) and is similar to the lysozyme which is formed in infected cells and which participates in lysis of the cell. The lysozyme in intact phage may be associated with, or identical to, a distal tail component such as the baseplate.†

Another protein component of whole T-even phage does not form part of the external structure. This is the so-called internal protein, a component which has been estimated to represent between 3 (21) and 7% (22) of the total protein. Of unknown function, the internal protein appears to be complexed with the DNA inside the phage head. It is injected into the infected cell along with the DNA.

A number of small molecules are structural components of intact T-even phage. In addition to the internal protein, the interior of the head contains a polypeptide which is relatively rich in aspartic acid, glutamic acid, and lysine and which forms about 1% of the carbon of the phage (23). Small amounts of the polyamines putrescine and spermidine are also found complexed with the DNA (24). The tail sheath contains, in addition to bound nucleoside triphosphates, calcium ions in the amount of about one per sheath subunit (25). Finally, a recent report from Kozloff's laboratory, which has contributed heavily

---

† Two recent studies (435, 436) establish that the lysozyme associated with phage particles does not participate in the initiation of infection.

to our understanding of phage tail structure and function, demonstrates the existence of pteridine components in phage tails (26). Phage T4B contains 6 molecules per particle of a substance identified as pteroyl pentaglutamic acid, a folic acid-like molecule. This pteridine is not found in uninfected cells, which contain pteroyl triglutamate as the most abundant folic acid compound. Phage T4D contains a pteridine component which is similar to, but not identical with, the pteridine found in T4B. The presumed function of these intriguing components is discussed in Sec. 12-2, B, in connection with the mechanism of viral invasion.

In addition to the protein and low molecular weight components discussed above, the T-even phage particle contains DNA to the extent of about 60% of the weight of the phage, or some 200,000 nucleotide pairs. This has been shown to be a single molecule, 50 μ in length (27, 28), with a molecular weight of about 1.3×10^8. T-even phage DNA molecules—at least those from T2 and T4—have two other remarkable properties which were originally suggested on the basis of genetic experiments. In 1964 Streisinger, Edgar, and Denhardt (29) showed, by extensive recombinational analysis, that the genetic map of T4 was topologically equivalent to a circle. Since there was no physical evidence for the circularity of the molecule, it was suggested that the molecule, while linear, is circularly permuted with respect to base sequence, i.e., that a population of DNA molecules, while having the same linear sequence of genes, begin and end at different points. This was confirmed in the laboratory of Thomas (30, 31). In one experiment (31), DNA was denatured and then reannealed. Examination of the product in the electron miscroscope showed circular structures. This is expected if random reassociation leads to pairing of strands whose base sequences are complementary but which begin and end at different points, a prediction of the circular permutation model.

A second property of T-even DNA molecules is terminal redundancy, i.e., some of the genetic information at one end of the molecule (about 1%) is repeated at the other end. This was originally suggested from studies on an unusual class of heterozygotes in T4 (29, 32). The model has been confirmed by MacHattie, Ritchie, Thomas, and Richardson (33). These authors treated T2 DNA with an exonuclease which removes nucleotides stepwise from the 3' ends of polynucleotide chains in duplex molecules. After limited digestion the molecules were subjected to annealing conditions. Electron microscopic examination showed circular structures, which are expected if the exonuclease treatment exposes 5′-terminated single-stranded regions of complementary sequence at either end of the molecule. The experiments illustrating terminal redundancy and circular permutation are illustrated schematically in Fig. 11-4.

Not only is T-even DNA terminally redundant but the extent of redundancy appears to depend upon the over-all length of the DNA molecule. Streisinger, Emrich, and Stahl (34) found that crosses of phage carrying long deletions yield a greater frequency of terminal redundancy heterozygotes than crosses in which the phage carried short deletions or point mutations. This startling discovery suggests that each T4 particle contains, on the average, a "headful" of DNA. The mechanism by which the length of the DNA molecule is controlled during phage maturation is completely unknown at present.

T-even phage DNA has an additional chemical characteristic which

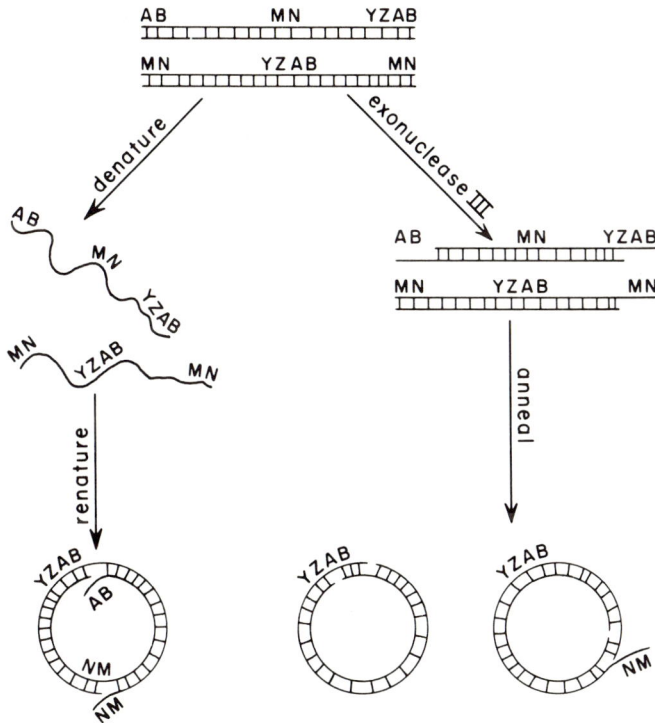

Fig. 11-4. Diagram of experiments proving that T-even phage DNA is circularly permuted (left-half) and terminally redundant (right-half) (31, 33).

TABLE 11-1

Glucosylated Hydroxymethylcytosine Content of
T-Even Phage DNA's[a]

Phage	HMC	α-glucosyl HMC	β-glucosyl HMC	β(1,6)-glucosyl-α-glucosyl HMC
T2	25%	70%	0	5%
T4	0	70%	30%	0
T6	25%	3%	0	72%

[a] Data from Ref. 40.

renders it unique. Instead of the normal base, cytosine, these DNA molecules contain 5-hydroxymethylcytosine (35). In addition, the hydroxymethyl groups are glucosylated, with the type of bond formed and the extent of glucosylation a function of the phage strain (36–40). This is illustrated in Table 11-1.

B. EARLY STEPS IN INFECTION

The first step in infection of E. coli by T-even bacteriophages is adsorption of the virus to the bacterial surface. In 1953 Anderson (41) showed, by electron microscopy, that phage adsorb to bacteria tail first. At about the time Hershey and Chase were performing their classical blendor experiment (42; cf. below), which showed that DNA is the only virus component, aside from the other minor head constituents (21, 23), to enter the infected cell. The two major questions relative to understanding the early stages of T-even phage infection are: (1) What is the mechanism of adsorption? and (2) How does the viral DNA enter the infected cell? Since it is clear that the DNA must pass through the tail during injection, the latter question is intimately tied up with the problem of the function of the various components of the phage tail.

It was clear from the early days of phage research that bacterial cell walls must contain receptor sites for adsorption of specific bacteriophages. The fact that only a few bacterial species are susceptible to infection by any one bacteriophage provided strong circumstantial evidence for this idea. More convincing was the fact that a sensitive bacterium could mutate to a form which resisted infection by a particular phage or group of phages. Thus, a single-step mutation in E. coli confers resistance simultaneously to T3, T4, and T7, suggest-

ing that the three phages share a common receptor site. The receptor
site is believed to undergo a spontaneous chemical reaction with a
component of the phage tail which results in adsorption. Bacterial
metabolism does not appear to be necessary for this step, as shown
by adsorption of phages to purified cell-wall material (*43, 44*).

Biochemical characterization of phage receptor sites was not possible until further information had been gained about the complex chemistry of bacterial cell walls. Weidel and his colleagues (*45*) showed
that the *E. coli* cell wall contains two outer layers—a lipoprotein and
a lipopolysaccharide. Studies on the ability of the purified cell-wall
components to inactive phage (by adsorbing them and causing them
to inject their DNA) showed that the lipoprotein layer contains the
receptors for T2 and T6, while the lipopolysaccharide adsorbs T3, T4,
and T7. The receptor for T5 is borne in a small lipopolysaccharide–
lipoprotein structure which is easily removed from the bacterial surface by dilute alkali (*46*).

The components of the phage tail which are responsible for recognition of the bacterial receptor site are the tail fibers. This was shown
by the fact that isolated fibers bind to bacteria and with the same
specificity as the phages from which they are derived (*47*). In certain
strains of phage the fibers must be in a particular orientation in order
for adsorption to occur. It has been known for some years that
certain strains of T4 and T6 require the presence of L-tryptophan
as an activating cofactor for adsorption (*48, 49*). Electron-microscope
studies by Cummings (*50*) showed that in T4B, one such strain,
tryptophan causes the conversion of the phage from an inactive form,
in which the fibers are oriented toward the head, to an active form,
with the fibers extended distally from the end of the tail. It has been
known for some time that indole prevents activation and competes
with tryptophan to bring about this effect (*51*). Interestingly enough,
indole also inhibits adsorption by a T2 strain, T2H, even though this
phage does not require tryptophan for adsorption (*52*). In the presence
of indole the tail fibers of T2H are oriented in the nonextended form
(*52*). These observations and others led Kanner and Kozloff (*52*) to
suggest that indole and tryptophan are able to act as π-electron donors
in a charge transfer complex with some component of the phage tail,
and that formation of such a complex is involved in adsorption. A
systematic search for some tail component which could act as electron
acceptor in such a complex led Kozloff and Lute (*26*) to the discovery
of folic acid-like compounds in phage tails (see Sec. 11-2, A). Although

the precise role of these compounds in the tail fiber orientation reaction is not known, the presence of these compounds does not appear to result from experimental artifact. First, the number of pteridine molecules per phage particle is six, equal to the number of tail fibers. Secondly, the pteridine component of phage T4B, pentaglutamylpteroylglutamic acid, is different from any pteridine known to exist in uninfected *E. coli,* and third, different phages have bound pteridines which are different from each other. Further elucidation of the origin, site of attachment to phage, and biochemical role of these pteridines should be one of the fascinating chapters in phage biochemistry.

The precise steps following adsorption have remained unclear in spite of a great amount of effort at both the biochemical and morphological levels aimed at clarifying this process. It was known that after adsorption the lysozyme located in the phage tail digests a portion of the bacterial cell wall (*43, 53*). Shortly thereafter, the tail sheath contracts and the DNA is forced through the tail core, past the cell membrane, into the interior of the host bacterium. A number of unresolved questions regarding this process have recently been clarified by Simon and Anderson (*14, 54*). In a spectacular series of electron micrographs, these authors have demonstrated the existence of six short tail fibers attached to each tail baseplate, in addition to the already-known long fibers. These short fibers, which are probably the same as another structure, the tail pins, appear to play a role in positioning the baseplate above the bacterial surface and in attracting the particle close to the surface. Simon and Anderson have also shown that the baseplate undergoes an extensive conformational change during contraction of the tail sheath; in fact, they believe that this change, from a hexagon to a six-pointed star, is the event which triggers contraction of the sheath. At the same time a plug appears to be removed from the hollow tail in the region of the baseplate. These events combine to force the hollow core away from the baseplate and through the bacterial cell wall. Previously, it had not been clear whether the core actually penetrated the wall. Following this, the DNA passes through the core. Interestingly, the inner diameter of the core, some 25 Å, is only slightly larger than the diameter of a double-stranded DNA molecule. Thus, the DNA must pass through in "single file," and jamming may be prevented.

A schematic picture of early events in phage infection, drawn in light of the results of Simon and Anderson, is presented in Fig. 11-5. A number of important questions about this process are still unanswered, including the following: (1) What is the mechanism of recognition of receptor sites? (2) What is the mechanism of contraction of the tail sheath? The possible role of ATP hydrolysis in this process has been mentioned above. Recently Moody has performed a careful electron microscopic study of this process (55, 56) and has proposed a model for contraction based on a rearrangement of sheath subunits. (3) How is the DNA injected and how does it traverse the cell's lipid membrane? This process is completed within 1 min. Expenditure of metabolic energy is apparently not required, for phage will inject DNA into heat-killed or cyanide-treated cells or, indeed, into isolated cell-wall fragments. Undoubtedly, a complete answer will require a clearer picture than we now have of the mode of packing of the 50-μ-long DNA molecule into the phage head, which is less than 0.1 μ in length. It is believed that the polyamines and basic polypeptide which enter the cell along with the DNA play a role in the injection process by neutralizing part of the negative charge on the DNA. It has been thought by some that the internal protein, which is also basic, plays a similar role. However, although it can bind to DNA, the physiological role of the internal protein appears more complex. The evidence for this statement is largely circumstantial, viz., the fact that internal proteins from phages T2 and T4 do not cross react with one another immunologically (57), even though the external antigens do. This demonstrates an element of species specificity which would not be expected if the sole function were neutralization of electric charge. The exact role of the internal protein remains one of the more important unsolved problems in phage biology. A number of functions have been suggested, e.g., that the protein acts as a

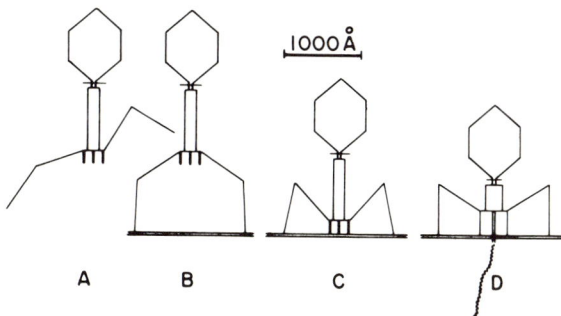

FIG. 11-5. Schematic representation of major steps in adsorption by phage T4, as described by Simon and Anderson (14).

"condensation principle" around which DNA can condense during viral maturation (58), that it acts as a regulator of genetic transcription during the infective cycle (59). Thus far, however, no firm evidence has been obtained for either of these proposals.

Although there are many gaps in our knowledge of the individual steps in phage adsorption and penetration, the over-all picture helps us to understand the phenomenon of the eclipse, that period during which infective phage cannot be liberated from infected cells. Once the DNA is injected into the cell, it is physically removed from its "adsorption organ," namely the phage tail assembly. Thus, if DNA is now released by mechanical rupture of an infected cell, it has no way in which to attach to and penetrate another cell. However, the DNA, once injected into a cell, is infective, as shown by the fact that removal of the head and tail adsorbed to the outside of the cell by mechanical shearing does not affect the ability of the cell to produce progeny phage (42). Thus, it now becomes appropriate to ask what are the metabolic consequences of the presence of phage DNA inside the cell.

C. Metabolism of T-Even Phage-Infected Bacteria

1. Macromolecular Metabolism

Cohen's pioneering experiments on macromolecular metabolism in T2 phage-infected cells (60) raised a number of important questions which were to occupy the attention of phage workers for many years thereafter. The basic experiment of this study was the following: Phage T2 was added to a growing culture of *E. coli* B at a sufficiently high multiplicity so that essentially the entire cell population was infected simultaneously. Samples were removed at various times and assayed chemically for DNA, RNA, and protein. DNA synthesis ceased immediately after infection, but at about 8 min after infection it commenced again at a rate some 5-fold higher than the preinfection rate. Protein synthesis continued at the same rate as that existing just before infection. RNA synthesis, as measured chemically, was negligible.

2. Arrest of Host-Cell Genetic Functions

The discovery of Wyatt and Cohen that T-even phage DNA contains 5-hydroxymethylcytosine (HMC) instead of cytosine (35) made it possible to follow the synthesis of phage-specific DNA. Hershey, Dixon, and Chase (61) found that only HMC-containing DNA is synthesized in infected cells. Thus, there appeared to be some mechanism for the arrest of bacterial DNA synthesis. Cohen's experiments on RNA synthesis had already suggested that the formation of bacterial RNA molecules ceases after infection. Examination of the proteins synthesized after infection led to the conclusion that infection also blocks the synthesis of bacterial proteins. The synthesis of bacterial enzymes and the formation of induced enzymes were both blocked after infection (62, 63). It therefore appears that one of the earliest metabolic events in T-even phage infection is an inhibition of all nucleic acid and protein synthesis of the host. The mechanism of this event, and its relation to the primary lethal effect on the infected cell, are unknown, in spite of a great deal of research activity in this area. An early theory which seemed plausible was that the abolition of host genetic activity is due to the physical destruction of the host chromosome. It was known that bacterial DNA breaks down after infection, presumably due to the action of phage-specific nucleases (61; also cf. below). However, Nomura et al., in a careful series of experiments, showed this hypothesis to be untenable (64). Physical studies of bacterial DNA removed from cells 5 min after infection with T4, a time at which host-cell macromolecular synthesis had ceased, showed no apparent degradation. Not only did the bacterial DNA maintain its structural integrity, it maintained its functional integrity as well. At 5 min after infection the chromosomes of infected male bacteria were transferred to phage-resistant, β-galactosidase-negative female cells by genetic mating. That the transferred chromosomes maintained their genetic capabilities was proven by the induction of β-galactosidase in the recipient cells.

Other work in Nomura's laboratory has implicated the action of a phage gene in the abolition of host-cell genetic activity. In the presence of chloramphenicol, an inhibitor of protein synthesis, the formation of bacterial RNA—both ribosomal and transfer—was not blocked (65). A similar conclusion was reached by Hayward and Green (66), who were studying the inhibition of development of the temperate phage λ by infection with T4. The major site of inhibition was the formation of λ-specific messenger RNA (cf. below), and T4-specific protein synthesis was required in order for complete inhibition to occur.

It seems unlikely that the action of a phage gene product is solely responsible for inhibition of bacterial macromolecular synthesis, since it has been known for some time that phage ghosts, prepared by osmotic shock, can also adsorb to sensitive bacteria, kill them (in terms of their ability to reproduce), and block the synthesis of bacterial protein and nucleic acid (66, 67). Such ghosts, of course, have no DNA to inject, and, therefore, phage genes are not expressed. Nomura et al. (68) have recently reinvestigated this apparent anomaly, by studying host-cell shut-off at various multiplicities of infection with T4 in the presence of chloramphenicol. At relatively high multiplicities, e.g., 15 to 18 phage per cell, significant inhibition (70–95%) of cellular DNA and RNA synthesis was observed. In light of this Nomura et al. conclude that there are two separate mechanisms which act to bring about the arrest of host-cell genetic activity: first, a chloramphenicol-sensitive effect, which is probably mediated through the action of a phage gene product, and a chloramphenicol-insensitive effect which is dependent upon the multiplicity of infection. The nature of the first effect is completely unknown at present. The second effect is probably mediated through the action of a phage structural protein at the bacterial surface, since, as noted above, phage ghosts can evoke host-cell shut-off.

The role of phage genes in effecting arrest of the synthesis of both inducible and constitutive β-galactosidase has recently been reinvestigated by Kaempfer and Magasanik (68a, 68b) and Kaempfer and Sarkar (68c). These authors studied the kinetics of cessation of enzyme synthesis following T2 phage infection or inducer removal. They conclude that the primary effect of infection is to turn off the synthesis of host messenger RNA (cf. Sec. 12-2, C, 3). Because the kinetics of shut-off of either induced or constitutive enzyme formation are identical, they conclude that the effect is on the synthesis of mRNA, not its metabolic stability. Under the conditions used by these authors, protein synthesis is not required for this shut-off. Unfortunately, the authors did not discuss their results in light of the work of Nomura et al. (68), and, because different experimental conditions and measurements were used in the two laboratories, it is difficult at this time to state confidently whether a phage gene is involved in the arrest of bacterial genetic activity. Kaempfer and Magasanik did, however, provide some intriguing data which implicated a bacterial gene in the process (68a). In host bacteria carrying an F episome, kinetics of shut-off of enzyme synthesis after infection are identical with those following removal of inducer. However, when the host cells bear no sex factor, phage infection causes more rapid shut-off. Evidence is presented that host mRNA is degraded twice as rapidly after infection of F⁻ cells than F⁺ cells. The apparent role of the sex factor in preventing mRNA degradation is unknown.

Kaempfer and Magasanik (68a) discuss the possibility that the inhibition of mRNA synthesis after infection is caused by changes in the cell surface. That dramatic changes in the properties of the membrane occur after infection has been well documented (68d).† Such changes could disrupt a membrane attachment site for bacterial DNA (68e) and render it unable to be transcribed. Further insight into this possibility may be gained by using colicins as model systems. Colicins are bacteriocidal agents, protein in nature, which are synthesized in certain bacterial strains under the genetic control of episomes called colicinogenic factors and which are released to the medium after a process which is at least superficially similar to the induction of temperate phage (for review see Ref. 69). Different colicins have different, but characteristic, effects on bacterial metabolism. Nomura (70) has extensively studied the mode of action of one such agent, colicin K, whose over-all effect is similar to that of T-even phage infection. Adsorption of a particle of colicin K to a bacterial surface results in inhibition of the syntheses of RNA, DNA, and protein in the bacterium. The effect is reversible, for brief treatment of an inhibited cell population with trypsin restores the ability of the population to synthesize macromolecules. Thus, a particle of colicin K, while remaining bound at the bacterial surface, can effect internal metabolic changes of the most profound nature. Elucidation of the action of colicins, as well as the inhibitory effects of phage infection, should provide challenging and fundamental problems for biologists and biochemists in the next few years.

3. Messenger RNA

The role played by T-even phages in the development of the concept of messenger RNA is perhaps too well known to be discussed at length here, but a brief reiteration is useful for understanding some later sections of the chapter. As stated above, Cohen (60) could detect

---

† Silver, Levine, and Spielman (411) have recently published an exhaustive study of cation fluxes and permeability changes accompanying T2 infection. Alterations in membrane structure have been probed by Buller and Astrachan (412) and Furrow and Pizer (413), who have detected changes in the pattern of phospholipid synthesis occurring after T4 infection.

no *net* RNA synthesis in T2-infected cells. ^{32}P-phosphate incorporation experiments did show a small amount of activity in an RNA fraction but this was not rigorously characterized as radioactive RNA. The problem was taken up by Volkin and Astrachan some 10 years later (*71, 72*). They established that a small amount of ^{32}P is incorporated into RNA after phage infection, and, further, that the base composition of the newly synthesized (labeled) RNA matches that of the DNA of the infecting phage. Because of this, Volkin and Astrachan suggested that the labeled RNA is an intermediate in the synthesis of phage DNA. It must be recalled, however, that this work was done before anyone had a clear idea of the various roles of RNA in the transmission of genetic informatiton, indeed, before the development of *in vitro* systems capable of carrying out incorporation of amino acids into acid-insoluble material. During the period of 1956 to 1961 evidence was accumulating from many systems that a species of RNA acts as an intermediate in the transfer of information from the gene to the protein-synthesizing apparatus. Jacob and Monod (*73*) summarized the evidence available in 1961 and predicted the properties of a hypothetical "messenger RNA." This theoretical molecule was remarkably similar to the Volkin–Astrachan RNA, viz., in mimicking the base composition of phage DNA and in being metabolically unstable (rapidly synthesized and degraded). Nomura, Hall, and Spiegelman (*74*) presented the first experimental evidence that the rapidly labeled T2-specific RNA might be a genetic messenger. Rapidly labeled RNA in T2-infected cells could be physically separated from RNA present in the cells before infection. Moreover, extraction of labeled RNA from cells found most of the radioactivity in a ribosomal fraction. This point was followed up by Brenner, Jacob, and Meselson (*75*), who used a combination of density- and radioisotope-labeling experiments in conjunction with density-gradient centrifugation techniques to establish the following points: (1) No new ribosomes are made after infection by a virulent phage; (2) RNA synthesized after infection becomes attached to ribosomes present in the cells before infection; (3) these pre-existing ribosomes are the site of postinfection protein synthesis. Thus, this classic series of experiments established our present conception of the ribosome as a nonspecific "workbench" of protein synthesis, which follows the instructions provided by messenger RNA, an informed genetic intermediary. It also explains why Cohen (*60*) was unable to demonstrate net synthesis of RNA after T2 infection. If no species of cellular RNA are formed after infection, and if the phage mRNA is degraded about as rapidly as it is formed, as expected from the instability of this molecule (73), then a chemical assay for RNA would fail to record an increase in total intracellular concentration with time after infection.

Further properties of phage messenger RNA are discussed below, in connection with the problem of control of macromolecular syntheses during a phage replication cycle.

4. Early and Late Proteins

It is stated above that infection with T-even phages does not alter the over-all rate of protein synthesis in infected cells, yet the formation of bacterial proteins ceases. One must conclude, therefore, that proteins made after infection are virus specific. However, very little of the protein synthesized early after infection is incorporated in mature phage particles (76). This was shown by feeding ^{35}S-sulfate as an isotopic protein precursor to infected cells. When the isotope was present only during the first 5 min after infection, less than 10% of the label incorporated into protein was eventually found in mature virus. When the isotope was added at later times, however, this value rose to about 60%. Although the precise function of the early proteins which were not incorporated into mature phage was not immediately known, early evidence suggested that they play a role in DNA synthesis. In 1947, Cohen and Fowler (77) treated T2-infected cultures with 5-methyltryptophan. Not only was protein synthesis inhibited, as expected, but so was DNA synthesis. Subsequent studies with chloramphenicol, another inhibitor of protein synthesis, showed that the time at which protein synthesis is blocked is critical in terms of effecting a simultaneous inhibition of DNA synthesis (78). When the inhibitor was present from the time of infection, DNA synthesis was essentially completely blocked, but when it was added after 5 min significant DNA synthesis was observed, and when added after 10 min, virtually no inhibition of DNA synthesis occurred. These experiments, which were performed in 1956, confirmed that the early proteins participate in DNA replication. Once a supply of these proteins is present in the cell, about 5 min after infection, further protein synthesis is not required for initiation and maintenance of DNA replication. Incidentally, it should be pointed out that internal protein, a structural component of phage, is synthesized throughout the entire course of infection (79). This probably accounts for the small amount of early-synthesized protein which is incorporated into whole phage.

At any rate, phage workers in the mid-1950's were faced with the problem of identifying the precise biochemical roles of the other early proteins in DNA synthesis.

The first insight into the above question came from Cohen's laboratory. In 1954, Barner and Cohen (*80*) discovered that infection of a thymine-requiring auxotroph with T2 overcame the nutritional deficiency, i.e., thymine was not required for phage production. This suggested that the phage somehow introduced into the infected cell a metabolic system for thymine synthesis, a phenomenon which Cohen referred to as "virus-induced acquisition of metabolic function" (*81*). At the same time Cohen's laboratory was investigating the biosynthesis of 5-hydroxymethylcytosine, the unique pyrimidine component of T-even phage DNA (*35*). Both of these lines of investigation resulted in the discovery of two new enzyme activities in phage-infected bacteria. In 1957, Flaks and Cohen (*82*) reported on the properties of these enzymes. Deoxycytidylate hydroxymethylase catalyzes the tetrahydrofolate-dependent conversion of dCMP to 5-hydroxymethyl-deoxcytidylate (dHMP). Thus, the biosynthesis of the virus-specific pyrimidine occurs at the mononucleotide level. Similarly, large amounts of thymidylate synthetase were found in cells infected by T-even phages or T5. This enzyme carries out a methylation, also tetrahydrofolate dependent, of deoxyuridylate to thymidylate. Although the enzyme is found in uninfected *E. coli* (*83*), its activity increases severalfold after infection. Moreover, the thymine-requiring *E. coli* strains which Barner and Cohen had been studying (*80, 84*) lack dTMP synthetase, but the enzyme is still found in these auxotrophs after infection with T-even phages or T5. This is consistent with the idea that the phage contains the genetic information for production of its own dTMP synthetase, an idea which is discussed in more detail below.

At about the same time as the above work was being done, Kornberg and his associates were beginning to achieve success in their investigations of the biosynthesis of DNA (cf. *85*). Turning their attention to T2-infected bacteria (*86*), they were able to demonstrate several more new enzyme activities which, like dCMP hydroxymethylase, are not found in uninfected cells and some activities which, like dTMP synthetase, are present in uninfected cells but whose activity increases severalfold after infection. A number of other laboratories became involved in this field and more phage-induced enzyme activities were discovered, such that there are now about 19 new enzyme activities known to be associated with T-even phage infection. These are discussed in greater detail below. The phenomenon of virus-specific enzyme synthesis now appears to be quite general. As we can see in subsequent sections, most phages which have been studied can induce new enzyme activities in infected cells. Animal viruses contain similar capabilities, as discussed elsewhere in this volume.

5. Origin of Phage-Induced Enzymes

The origin of the new enzyme activities found in phage-infected cells was not immediately clear. The most plausible theory was that new enzyme molecules were synthesized under the genetic control of the infecting virus. However, other possibilities could be entertained. For example, phage infection might somehow derepress a bacterial gene to produce a protein not present before infection, or, the phage-induced enzymes might exist as inactive enzyme precursors or "zymogens" in the uninfected cell. The latter hypotheses had some appeal in light of the fact that some of the phage-induced enzyme activities are also found, albeit at lower levels, in uninfected cells. However, Greenberg, Somerville, and DeWolf (87) found that T2-induced dTMP synthetase could be chromatographically separated from the corresponding preinfection enzyme and that the two enzymes differed in a number of physical and kinetic properties. Similar data have been published for several of the other of the "early enzymes" (cf. below). This type of evidence, however, does not prove that the enzyme is not of bacterial origin. In order to fully exploit the potentially enormous advantages of the phage-induced enzymes for studying gene–protein relationships and for probing the metabolic roles played by these enzymes, it was necessary to prove unambiguously that these enzymes are viral gene products, synthesized de novo after infection.

An essentially biochemical approach to this problem was followed in Cohen's laboratory. Flaks, Lichtenstein, and Cohen (88) showed that inhibitors of protein synthesis blocked the appearance of dCMP hydroxymethylase in T6-infected cells, suggesting that appearance of the activity results from de novo protein synthesis. Pizer and Cohen (quoted in Ref. 89) increased the sensitivity of the hydroxymethylase assay to the point where they could state that there was less than one molecule of active dCMP hydroxymethylase per cell of uninfected E. coli, as compared with about 8000 molecules per cell 15 min after infection. Mathews, Brown, and Cohen carried out an in vivo labeling experiment which clearly established that synthesis of this enzyme after infection is entirely *de novo* (*90*). *E. coli* B$_{45}$, a methionine auxotroph, was grown in the presence of radioactive methionine in order to label bacterial proteins. The cells were placed in nonradioactive medium and infected with T6. Enzyme was isolated from the cells and purified by a method previously shown to remove all contaminating bacterial protein. The purified enzyme was nonradioactive, indicating that it must have been formed entirely from material present in the cell after infection.

Although the above experiments clearly ruled out any theory of phage-induced enzyme formation which depended upon the existence of preformed enzyme precursors, it did not rule out the possibility that these enzymes are the products of derepressed bacterial genes. Some experiments on the properties of phage-induced dTMP synthetases strongly suggested that this enzyme is a viral gene product (*91*). The enzyme induced by T2 was found to differ in kinetic properties and in inhibition by 5-fluorodeoxyuridylate from that induced in the same *E. coli* strain by T6. This would not be expected if both phage were derepressing the same gene. However, the most direct evidence on the viral origin of phage-induced enzymes came from studies with conditional lethal phage mutants. The development of these mutants constitutes one of the most important recent advances in virology and molecular biology. Because of the widespread applications of these mutants to problems of phage development, a brief digression into some relevant phage genetics appears appropriate at this point.

6. Conditional Lethal Mutants

Early studies on the genetics of T-even bacteriophages utilized markers which, while easily detectable by choice of appropriate plating conditions, were not readily identifiable in terms of biochemical function. The reason for this is quite obvious; mutations which affected known metabolic reactions were quite apt to be lethal, and viruses bearing such mutations could not be scored as mutant or wild type, since they would not produce plaques under any conditions. Consequently, markers such as host-range specificity, resistance to acridine dyes, and aberrant plaque morphology were used for genetic investigation with little knowledge of the biochemical bases for the observed phenotypes. In order to correlate genetic and biochemical functions, what was needed was a set of conditional lethal mutations, i.e., mutations which would exhibit the mutant phenotype under one set of conditions (restrictive) but behave like wild-type under another set of conditions (permissive). The latter property would enable one to propagate a phage bearing a mutation, while the former would allow one to identify such a mutation and study its effects in biochemical terms. Two types of conditional lethal mutations were developed for phage T4 in the early 1960's, primarily by Edgar and Epstein and their collaborators (92, 93). Edgar's laboratory isolated a large number of temperature-sensitive (ts) mutants, i.e., mutants which could grow and produce plaques at one temperature (30°C) but were unable to multiply at a higher temperature (42°). (Wild-type T4 can grow normally at 42°.) As one might expect, failure of growth at 42° results from the fact that the protein specified by the mutant locus is unable to function at that temperature. This was first shown, for phage systems, by Wiberg and Buchanan (94), who demonstrated that certain ts mutants of phage T4 which were unable to synthesize DNA in infected cells at 42° produced a dCMP hydroxymethylase which was inactivated at 40°. Moreover, the hydroxymethylating enzymes made by two such mutants differed from each other in several properties, including temperature sensitivity. This established that (1) the phage does carry the genetic information for virus-induced enzymes and (2) the mutant gene in the phages under study is the structural gene for the hydroxymethylase.

The other set of conditional lethal mutations, developed by Epstein and his colleagues, goes by the trivial name of amber (am; see Ref. 93) but can be more accurately called suppressor sensitive, in light of knowledge which did not exist when these mutants first became available. The mutants were originally isolated as strains which could grow on a particular strain of E. coli K12, called CR63, but could not propagate on E. coli B. They were used for biochemical studies before the nature of the mutation was recognized (95). The work of Dirksen, Hutson, and Buchanan (96) provided an early clue to the biochemistry of the amber mutation. Several mutants, which mapped in the same gene when tested by recombinational analysis, were unable to induce dCMP hydroxymethylase in E. coli B but could form the enzyme in E. coli CR63. Three permissive strains of E. coli K12, including CR63, were used. For each mutant the enzyme induced in any of the permissive strains differed in temperature sensitivity from the enzyme induced by the same phage mutant in the other bacterial strains. This suggested a role of the host bacterium in the translation of the genetic message of the phage, or, to quote Dirksen et al., "the mutation in the phage genome gives rise to nonsense or unacceptable missense in E. coli B, and to acceptable missense in the permissive hosts" (96).

The above suggestion was clarified and extended in a series of brilliant experiments performed in the laboratories of Brenner and Garen. A detailed discussion of this work is beyond the scope of this chapter, but an excellent review is contained in a recent article by Lengyel (*97*). Sarabhai, Stretton, Brenner, and Bolle (*98*) showed that under restrictive conditions the amber mutation terminates translation of a polypeptide chain. A series of T4 mutants were studied which were amber with respect to the ability to synthesize the phage head protein. Each mutant formed a fragment of the head protein under restrictive conditions. The length of each fragment, relative to the length of the entire head protein, depended upon the position of the mutation within the gene, as determined by genetic mapping. This not only proved that the amber mutation causes chain termination under restrictive conditions but that a gene is co-linear with the amino acid sequence of the polypeptide chain which it specifies.

Expression of the amber mutation under permissive conditions involves the action of suppressor genes which allow insertion of specific amino acids at the point in translation where the amber mutation occurs. The coding triplet giving rise to the amber phenotype has been shown to be AUG (*99, 100;* cf. *97*). In at least one case, suppression involves mutation in a gene for transfer RNA, which gives rise to an altered form which can recognize the amber triplet (*101*). The biochemical bases for all forms of suppression have not yet been elucidated.

As mentioned above, Edgar and Epstein and their co-workers isolated a large number of *ts* and *am* mutants of T4 and mapped the position of each mutation by means of complementation and recombination tests. About **70** different genes have been identified by this procedure. All of the genes lie on a circular map, as mentioned in Sec. 11-2, A. Mutants in each gene have been studied under restrictive condition for identification of the gross biochemical defect, e.g., failure of DNA synthesis, inability to synthesize a certain structural component, etc. Some of the mutants have been investigated more intensely than others, such that it has been possible to identify the specific gene product affected by the mutation. However, a large number of the genes have not been characterized biochemically. A number of these genes are known to affect steps in DNA replication or in the process of maturation (formation of mature particles). Obviously, the nature of these processes would be greatly clarified if the precise function of all genes affecting each process were known.

Figure 11-6 presents a genetic map for T4 based upon the work alluded to above and more recent work from the same laboratory (102). It also includes the position of several other genes which have not yet been mentioned in this chapter. Where possible the protein specified by each gene is indicated, along with the gross function affected. Details of the identification of some of these gene products are referred to in subsequent pages of the text.

7. Properties of the Phage-Induced Enzymes

Figure 11-7 presents a picture of our present knowledge of new and increased enzyme activities in T-even phage-infected bacteria. The properties and metabolic role of each are discussed briefly below.

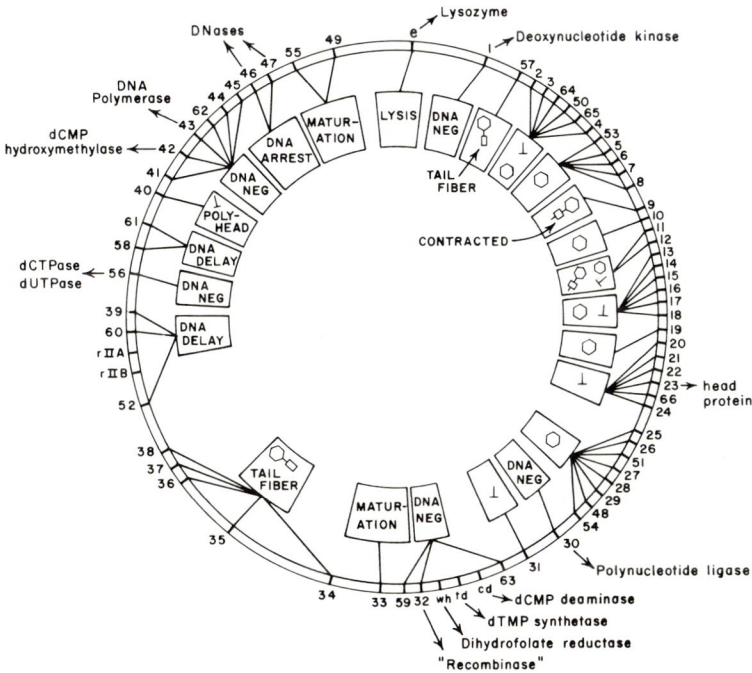

Fig. 11-6. Approximate location of markers on the T4 genetic map. Map distances not given. The inner symbols indicate the defective phenotype of conditional lethal mutants: either particles seen by electron microscopic examination of defective lysates or the name of the function affected. Presumed or demonstrated gene products are indicated outside the circle. Modified from Refs. 92, 93, 102, and 123.

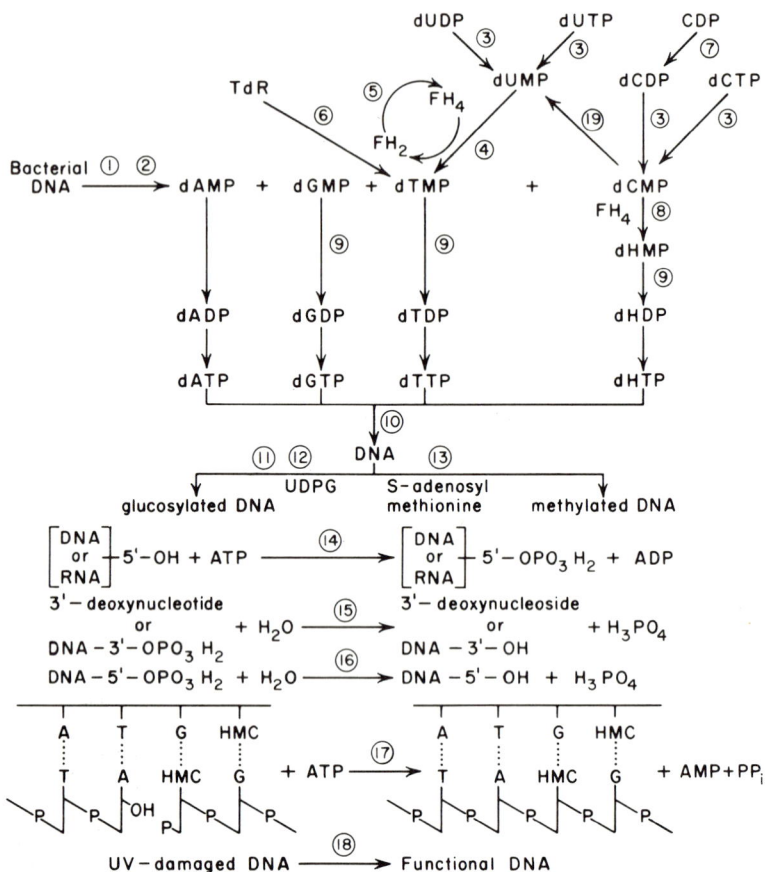

Fig. 11-7. Enzymes of DNA metabolism which are induced in T4 infection
1 and 2, DNase; 3, deoxyuridine and deoxycytidine di- and triphosphatase; 4,
dTMP synthetase; 5, dihydrofolate reductase; 6, thymidine kinase; 7, CDP reductase; 8, dCMP hydroxymethylase; 9, deoxynucleotide kinase; 10, DNA polymerase; 11 and 12, glucosyl transferases; 13, DNA methylase; 14, polynucleotide
kinase; 15 and 16, polynucleotide phosphatases; 17, polynucleotide ligase; 18, uv
repair enzyme; 19, dCMP deaminase.

Deoxyribonucleases. At least three new deoxyribonuclease activities
are found in T2-infected cells (*103–105*). One of these activities is
closely associated with phage-specific DNA polymerase (cf. below;
105, 106). Actually, DNase was the first activity found to be specifically associated with phage infection. In 1952, Pardee and Williams
(*107*) reported increased levels of DNase in T2-infected cells. How-

ever, this was thought to result from destruction of an endogenous inhibitor of a bacterial enzyme (108). Short and Koerner (105) have now separated the new DNase activities from the various *E. coli* DNases, showing each to be a distinct entity. Wiberg (109) has shown that T4 amber mutants mapping in gene 46 or 47 are unable to bring about the breakdown of bacterial DNA after infection, although it is not yet clear that these are structural genes for DNases. The precise functions of the phage-specific DNases are not known. Presumably one function is to increase the size of the deoxyribonucleotide pool for phage DNA synthesis. However, there is also presumptive evidence suggesting a direct role for the DNases in DNA synthesis: (1) the association of a DNase activity with DNA polymerase throughout extensive purification (105, 106); (2) the fact that amber mutants in gene 46 or 47 are DA or "DNA arrest" mutants (92). This means that DNA synthesis comes to a halt shortly after it has begun. Phage yields of these mutants are about 7% those of wild-type T4 under the same conditions. This would not be expected to occur solely as a consequence of the blockage of deoxynucleotide release from bacterial DNA, for two-thirds of the DNA phosphorus of T2 comes from materials not in the cell at the time of infection (110).

Deoxynucleoside Di- and Triphosphatase. The absence of cytosine from T-even phage DNA led Kornberg and his associates (86) to look for metabolic routes whereby dCTP, the immediate precursor to DNA-cytosine, could be removed in the infected cell. They discovered a phage-induced activity which splits dCTP to dCMP and pyrophosphate. Later Zimmerman and Kornberg (111) found that a purified dCTPase fraction also split dCDP to dCMP and inorganic phosphate. At about the same time, Greenberg and Somerville (112) were studying an enzyme in uninfected *E. coli* which splits dUTP to dUMP and pyrophosphate. Recently Greenberg (113) showed that the activity of this enzyme increases up to tenfold after infection by T2. This seemed to result from synthesis of a new enzyme, for the phage-induced dUTPase could also split dUDP, whereas the host enzyme could not. Greenberg observed several other differences in the properties of the two preparations. Simultaneously, Warner and Barnes (114) published evidence that in T4-infected cells dCTP, dCDP, dUTP, and dUDP are all cleaved by the same enzyme. dCTPase and dUTPase activities remained together throughout extensive purification. Moreover, extracts of cells infected with T4 amber mutants defective in gene 56 were identical with uninfected cell extracts with respect to the ability to cleave dCTP, dCDP, dUTP, or dUDP (114, 115). This suggests that gene 56, which is classified as DO, or DNA-negative (102), is a structural gene for a polyfunctional deoxyribonucleoside di- and triphosphatase. Such an enzyme would play a dual role in the metabolism of the infected cell: (1) removing deoxycytidine polyphosphates from the pool of DNA precursors and (2) providing a new pathway for synthesis of dUMP for thymidylate synthesis. Actually, gene 56 mutants are not completely DNA-negative (109). A small amount of DNA is synthesized by such mutants, but it is rapidly degraded, probably because it contains cytosine and as such is susceptible to the action of phage-specific nucleases.

Thymidylate Synthetase. The discovery of this enzyme was mentioned in Sec. 11-2, C, 4. That the phage-induced activity results from formation of a new enzyme was indicated by Simon and Tessman (116), who isolated "thymidine-requiring" (td) mutants of T4. Shapiro, Eigner, and Greenberg (117) showed that such mutants could not induce dTMP synthetase. The metabolic role of this enzyme in infected cells was of some interest, for it duplicates an activity already present in the cell before infection. Mathews (118) showed that under some conditions growth rates and phage yields of a td mutant were reduced 2- to 3-fold relative to wild type. Thus, the enzyme is useful, but not essential, for phage multiplication.

Dihydrofolate Reductase. During the transfer and reduction of the one-carbon group of methylenetetrahydrofolate in thymidylate synthesis, the reduced pyrazine ring of THF is oxidized to give dihydrofolate as a reaction product (119). Consequently, the action of dihydrofolate reductase is necessary to regenerate THF so that dTMP synthesis can continue. Like dTMP synthetase, DHF reductase is found in uninfected cells, but its activity rises up to 20-fold following infection by T-even phages or T5 (120). This is due to synthesis of an enzyme molecule which differs from the preexisting enzyme in several properties, including substrate specificity, response to inhibitors, stability, and sedimentation coefficient (121). Moreover, the reductases induced by T2, T4, T5, and T6 are different from each other, indicating that the enzyme is a viral gene product (122). Definitive genetic evidence on this point has been obtained. Hall, Tessman, and Karlstrom (123) devised a selective plating procedure for mutants suspected to be deficient in ability to induce DHF reductase. Under plating conditions of limiting deoxyuridine, presumed DHF reductase mutants gave plaques surrounded by a small halo, possibly caused by excretion of accumulated deoxyuridine compounds to surrounding cells, whose growth was consequently more extensive than elsewhere on the plate. Because of this, these mutants were termed "white" (*wh*). As expected, the mutants are indeed unable to induce DHF reductase (*122, 124*). Hall (*124*) and Mathews (*122*) have provided evidence that the *wh* gene is the structural gene for the reductase. Like dTMP synthetase, the metabolic requirement for the phage-induced reductase appears to be quantitative, not qualitative, for *wh* mutants can multiply on wild-type strains, albeit at reduced rates (*125*). Under these conditions the *wh* gene leads to a partial thymidylate deficiency.

Deoxycytidylate Deaminase. This enzyme, absent from uninfected *E. coli,* is found after infection by T-even phages (*126*). Thus, it appears to provide a new pathway for dUMP synthesis. Again, the function of this enzyme appears to be quantitative, i.e., it acts to increase the pool of pyrimidine nucleotide DNA precursors. dCMP deaminaseless mutants (*cd*) of T4 can multiply in wild-type *E. coli,* although at a somewhat lower rate than wild-type (*126*). In light of the apparent dispensability of the deaminase, it is interesting that the viral enzyme is subject to complex regulatory effects by pyrimidine nucleotides, notably inhibition by dTTP and activation by dCTP or hydroxymethyl-dCTP (*127–129*). These effects are similar to those previously observed with dCMP deaminase from animal tissues (*130, 131*). Whether these apparent regulatory effects observed *in vitro* actually operate *in vivo* has not yet been determined. However, it is remarkable that evolution has maintained, in the relatively simple T-even phage genome, information for the synthesis of complex feed-back binding sites.

Hall, Tessman, and Karlstrom have mapped the *td, wh,* and *cd* genes (*123*). They are located quite close together on the T4 chromosome, distinct from most of the other genes controlling steps in DNA synthesis. These authors have suggested the existence of a coordinate unit of expression governing the formation of the synthetase, reductase, and deaminase, but studies on the kinetics of induction of these enzymes (*132, 133*) would appear to contradict this interpretation.

Ribonucleotide Reductase. Cohen, Barner, and Lichtenstein (*134*) showed that T6 phage-specific RNA is rapidly degraded and, in a crude cell-free system, the resultant ribonucleotides are converted to the corresponding deoxy compounds. Later Cohen and Barner (*135*) showed that the activity of a reductive system for cytidine nucleotides increases some 15-fold after infection with T6; they also reported that this increase in activity is not detectable earlier than 5 min after infection. Beck, however, has reported that a CDP reductase activity is induced in T2-infected cells with the same kinetics as those of the other early enzymes (*136*). Further studies on this interesting enzyme have not been reported, at least from phage systems.

Thymidine Kinase. A recent report by Hiraga et al. presents evidence that a new species of thymidine kinase is induced in T2 and T4 infection (*137*). The virus-induced enzyme is distinguished from the bacterial kinase by its greater heat lability. However, Okazaki and Kornberg (*138*) had earlier reported that T2 infection has no effect on the intracellular activity of this enzyme. This interesting question obviously deserves further study.

Deoxycytidylate Hydroxymethylase. The discovery of this enzyme and proof of its *de novo* origin are discussed in Secs. 11-2, C, 4 and 11-2, C, 5 (*82, 90*). Conditional lethal mutants of T4 which are defective in gene 42 cannot induce this enzyme under restrictive conditions (*95*). Evidence that this gene is the structural gene for the hydroxymethylase has been cited above (*94, 96*). As one might expect, gene 42 mutants are virtually completely blocked in DNA synthesis (*132, 139*).

Deoxynucleotide Kinase. Kornberg et al. (*86*) and Somerville, Ebisuzaki, and Greenberg (*140*) both described a new activity in T2-infected cells, namely the phosphorylation of hydroxymethyldeoxycytidylate to the di- and triphosphate levels. Kornberg et al. observed, as did Bessman (*141*), that the activities of dTMP kinase and dGMP kinase, both detectable before infection, rise steeply after infection. The level of dAMP kinase, which is much more active before infection than the other deoxynucleotide kinases, does not increase. Bello and Bessman (*142, 143*) presented persuasive biochemical evidence that the three activities, namely phosphorylation of dHMP, dTMP, and dGMP, are catalyzed by a single phage-specific enzyme in T2-infected cells. This is in distinct contrast to uninfected cells, where a separate enzyme catalyzes the phosphorylation of each nucleotide. The products of the multifunctional T2 kinase are the corresponding deoxynucleotide diphosphates. These are converted to the triphosphates by an already very active, nonspecific enzyme present in uninfected cells (*144*). The low specificity is inferred from the fact that dHDP, certainly not a normal metabolite in growing *E. coli*, is readily phosphorylated by this system.

Recently Duckworth and Bessman (*145*) have presented more extensive biochemical evidence, and convincing genetic evidence, that a single, trifunctional deoxynucleotide kinase is induced in T4 infections. T4 amber mutants defective in gene 1 [a DO mutation (102)] are unable to induce any of the three activities under restrictive conditions (109, 145). Moreover, revertants of these mutants regain all three activities simultaneously. Finally, the kinase induced by a gene 1 mutant in a permissive E. coli strain differs in properties from the wild-type enzyme, establishing that gene 1 is indeed the structural gene for deoxynucleotide kinase.

DNA Polymerase. The activity of this enzyme, as measured in crude extracts, rises some 12-fold in T2 infections (86). Aposhian and Kornberg (146) showed that the increase results from synthesis of a new polymerase which can be separated from the host enzyme and which differs from it primarily in requiring a single-stranded DNA primer, while the bacterial polymerase exhibits a slight preference for native DNA. Goulian (147) has discussed the action of the T4-specific DNA polymerase in vitro. The enzyme requires the presence of a free 3'-hydroxyl group. It appears to form a strand complementary to that of the single-stranded primer. However, the two strands of the resultant duplex molecule fail to separate in vitro. The relationship between these observations and the role of phage-induced DNA polymerase in DNA replication in vivo awaits clarification.

de Waard, Paul, and Lehman (148) and Warner and Barnes (149) showed that T4 am and ts mutants mapping in gene 43 cannot induce DNA polymerase under restrictive conditions. The polymerase isolated from a culture infected with a gene 43 ts mutant was more temperature sensitive than the wild-type enzyme, suggesting that gene 43 is the structural gene for the T4 DNA polymerase. It is hoped that further studies on this system will help to clarify the precise metabolic role of DNA polymerase, a question about which there is still considerable disagreement. A provocative study has been published by Speyer and his colleagues (150, 151). A ts mutation was bred into a number of rII mutants of T4 (for a discussion of the rII mutation see Sec. 12-2, G). In each case reversion frequencies of the rII mutation to wild type were greatly increased by the presence of the ts mutation in the same chromosome. This suggests that the phage polymerase plays a role in base selection during DNA replication, and that a faulty polymerase makes mistakes more often than the normal enzyme.

Hydroxymethylcytosine Glucosyl Transferases. As discussed previously, some of the hydroxymethyl groups of T-even phage DNA-HMC are glucosylated. Kornberg et al. (*86*) discovered in T2-infected cells a new enzyme which could transfer glucose from uridine diphosphate glucose to DNA. Since the patterns of glucosylation differ in T2, T4, and T6 DNA's, one might expect different types of glucosylating enzymes to be induced by each phage. Kornberg and his associates purified the glucosylases induced by each phage and showed that the enzyme activities corresponded to the glucosylation patterns observed in isolated DNA's (*152–154*). All three phages induce a transferase which transfers glucose to DNA-HMC in an α linkage. However, the three enzymes are not identical; they show a definite species specificity. T2 DNA, which presumably has been glucosylated by the T2 glucosylase *in vivo*, will not serve as a substrate for the T2 enzyme *in vitro*. However, it can serve as a substrate for further glucosylation by the T4 and T6 enzymes.

It can be recalled from the data of Table 11-1, that T4 DNA contains monoglucosyl residues in both α- and β-linkages, whereas T6 DNA contains β, α-diglucosyl residues (*155*). T4-infected cells contain a β-glucosyl transferase, similar to the α-glucosylating enzyme, while T6 induces a β-glucosyl transferase which transfers glucose to monoglucosylated HMC residues. The species specificity of these enzymes has not yet been explained, e.g., the fact that only 70% of the HMC residues in T2 are glucosylated. The enzymes appear to recognize some aspects of DNA secondary structure, for native DNA's are far better substrates than heat-denatured material (*154*).

The discovery of DNA glucosylating enzymes allowed an explanation to be made, in biochemical terms, for at least one case of host-induced modification, originally described by Luria and Human (*156*). Growth of T2 and T6 on certain *E. coli* strains gave rise to progeny which could not grow in *E. coli* but could give plaques when plated on *Shigella dysenteriae*. It was found that the *E. coli* strains exhibiting this phenomenon were deficient in the synthesis of UDPG (*157–159*). Thus, phage DNA formed in such cells is under-glucosylated and, consequently, is more susceptible to nucleolytic attack (*160*) by *E. coli* nucleases. However, such molecules are apparently metabolically stable in *Shigella*. This phenomenon has formed the basis for an isolation of T2 and T6 mutants lacking the capacity to induce the α-glucosyltransferase (*161*). These were selected as strains unable to grow on *E. coli* B but able to multiply in *Shigella*. These strains apparently produce a nonglucosylated DNA. A T4 mutant defective in α-glucosyl transferase has also been isolated by Hosoda (162). This mutant, however, was obtained by accident. A T4 amber mutant mapping in gene 30 (DNA negative) was found to carry the glucosyl transferase mutation as a second defect. The transferase mutation is not the mutation mapping in gene 30, for other gene 30 mutations are able to form the α-glucosyl transferase. In contrast to the T2 and T6 glucosyl transferase mutants, the T4 mutant can plate on *E. coli* B as well as on *Shigella*. Apparently the DNA is glucosylated by the β-glucosyl transferase to an extent sufficient to preclude nucleolytic degradation.

Using Hosoda's α-glucosyl transferase mutant, Georgopoulos (163) has isolated mutants which are unable to plate on *E. coli* B but can grow on *Shigella*. These strains cannot induce either α- or β-transferase. Revertants were isolated by plating on *E. coli* B. These could induce either the α- or the β-enzyme but not both. Revertants gaining the α-activity plated equally well on *E. coli* B or *Shigella*. However, revertants with the β-activity had relative plating efficiencies on B vs. *Shigella* of less than 0.5, even though their DNA's were fully glucosylated. It would appear, therefore, that β-glucosylation of all of the HMC does not by itself confer full ability to overcome the restriction of growth in *E. coli* B.

DNA Methylase. Hausmann and Gold (164) have recently shown that infection of *E. coli* B with T1, T2, or T4 leads to increases in the DNA methylase activity of crude extracts of up to 30-fold. Although the new activity has not been purified to the point where one can state confidently that new enzymes are being synthesized, the data thus far obtained are in harmony with this conclusion. Infection with λ or T7 had no effect on the activity of this enzyme, while infection with T3, T5, or T6 led to decreases in methylase activity. The precipitous decline in activity after T3 infection is due to the appearance in these cells of an active enzyme which cleaves the methyl group donor, S-adenosylmethionine, to thiomethyladenosine and homoserine (165).

In phage, as in other systems, the biological significance of nucleic acid methylation is unknown. A role in host-induced modification has been suggested (166, 167) but no definitive evidence obtained. In this context it should be noted that T2 containing unmethylated DNA (produced by co-infection of *E. coli* B with T2 and ultraviolet light-inactivated T3) is biologically indistinguishable from normal T2, which contains methylated DNA (165). Finally, it should be observed that of the T series of bacteriophages the pattern of DNA methylation correlates roughly with ability to induce methylase activity. T1, T2, and T4, which can induce the activity, contain methylated bases in their DNA, as does T7, which does not affect the activity of the preexisting enzyme. T3, T5, and T6 DNA's are not methylated. This presents an interesting difference, and one of the few major ones, between T6 and the other T-even phages.

Polynucleotide Kinase. Richardson (*168*) described an activity in T4-infected cells which transfers orthophosphate from ATP to 5'-hydroxyl termini of a wide variety of nucleic acid compounds, including DNA, RNA, and even nucleoside 3'-monophosphates. The enzyme, named polynucleotide kinase, was highly purified. It is undetectable in host cells. Novogrodsky and Hurwitz (*169*) found that the same enzyme, also induced by T2, is undetectable in T1- or T5-infected cells. The enzyme shows low specificity, not only for phosphate acceptor, but also for phosphate donor. ATP, UTP, CTP, and GTP are all able to act in this latter role (*170*). Because of the low specificity of the enzyme *in vitro*, its role *in vivo* remains to be determined.† As of this writing, no phage mutants have been isolated which lack the capacity to induce the kinase. Such mutants would help greatly in elucidating the role of the enzyme, particularly the question of its possible involvement in DNA synthesis.

3'-Deoxynucleotidase and 5'-Polynucleotide Phosphatase. In a recent report Becker and Hurwitz (*171*) have described two new activities associated with T-even phage infection: (1) an enzyme which removed phosphate groups from 3'-mononucleotides and from 3'-phosphoryl termini in DNA and (2) an enzyme which removes terminal 5'-phosphate groups from DNA and RNA. The 3-deoxynucleotidase will not attack RNA, 5-phosphate esters, or ribonucleotides. In this respect it is much more specific than a similar enzyme isolated from some strains of *E. coli*. The 5' polynucleotide phosphatase, active with both DNA and RNA, will not cleave mononucleotides. So far, the physiological roles of these two enzymes are completely unknown. It is possible to propose a role for the 3'-deoxynucleotidase in regulation of DNA polymerase activity, for polynucleotides containing 3'-phosphate termini are potent inhibitors of the latter enzyme (*172*). However, as Becker and Hurwitz pointed out, there is no known mechanism for the generation of 3'-phosphate ends of DNA in *E. coli*.

† Because of the low specificity, it should also be recognized that this enzyme might well be responsible for the increased thymidine kinase activity observed by Hiraga et al. (*137*).

Polynucleotide Ligase. The middle months of 1967 witnessed a period of active research on the enzymatic breakage and joining of DNA strands, work primarily carried forth in the laboratories of Gellert, Richardson, Hurwitz, and Lehman. This research appears to hold considerable promise for an eventual understanding of genetic recombination at the molecular level. Weiss and Richardson (*173*) described an enzyme in T4-infected cells which catalyzes the covalent joining of two segments of an interrupted strand in a DNA duplex (see reaction, Fig. 11-7). The reaction requires ATP, with the products being AMP and PP_i. The enzyme is routinely assayed with "nicked" T7 DNA, obtained by treating native T7 DNA successively with pancreatic DNase for brief periods, phosphatase to remove phosphate groups in the resultant "nicks," and T4 polynucleotide kinase with γ-labeled ATP to introduce radioactive phosphate termini into the "nicks." The action of polynucleotide ligase then fixes the labeled phosphate in phosphodiester linkages, which are stable to further phosphatase treatment and can then be assayed as acid-insoluble radioactivity. Although Weiss and Richardson were unable to detect significant activity with this assay in uninfected *E. coli*, other investigators (*174–176*), using different assay conditions, were able to detect a similar enzyme. This was originally discovered as an enzyme which could convert DNA of phage to covalent circles (*174*). A novel feature of the bacterial enzyme is that it requires as a cofactor not ATP but DPN (*177, 178*). The products of the reaction, in addition to uninterrupted DNA, are AMP and nicotine mononucleotide (*178*).

In their original oral report, Weiss, Live, and Richardson (*179*) mentioned that T4 amber mutants in gene 30 were unable to induce polynucleotide ligase. Fareed and Richardson have now carried this work further to identify gene 30 as the structural gene for this enzyme (*180*).

In 1964 Mead (*181*) described an enzyme activity in T2-infected cells which incorporates oligodeoxyribonucleotides into polydeoxyribonucleotides in the presence of a DNA acceptor. Mead suggested that the reaction, which required ATP, involved joining of polynucleotide fragments and that it might play a role in genetic recombination. The possible relation of this enzyme activity to the more recently characterized polynucleotide ligase is unclear. From the fragmentary details given in Mead's report, it appears possible that his activity is the same as polynucleotide ligase. However, the more recent workers have not discussed their work with reference to Mead's report. Moreover, Mead did not present evidence that the enzyme activity was specifically associated with phage infection.

Ultraviolet Repair Enzyme. T4 phage is more sensitive to uv light than T2 (*182*). T4 carries a gene, called the v gene, which confers radiation resistance (*183*). Haber (*184*) has presented preliminary evidence that the v gene controls the formation of a phage-specific uv repair enzyme. As does the corresponding bacterial enzyme, the phage enzyme probably acts to excise thymine dimers, a major lethal uv photo-product (*185, 186*). Ultraviolet survival curves were determined, using bacterial plating hosts in which the dimer-excision enzyme was either present or absent. The curves were identical, suggesting that the phage induces its own dimer-excision enzyme and that this is used, rather than the bacterial enzyme, to repair uv-damaged phage DNA.

In light of the interspecific differences existing among patterns of viral enzyme synthesis directed by different T-even phages, e.g., in methylation and glucosylation, it is of some interest to know whether a genome of one phage can replicate using the early protein of a heterologous phage. Eckart (*187*) has performed a careful study of this question. Bacteria were infected with a T2 strain bearing a long deletion in the rII region (cf. Sec. 11-2, G), such that reversion to wild type was not possible. At 6 min, chloramphenicol was added to inhibit further protein synthesis and the cells were superinfected with a bromodeoxyuridine-revertible rII mutant of T4 in either the presence or absence of BUdR. Chromosome replication was determined from the extent by which BUdR caused reversion, a process requiring DNA replication. Eckart found that T4 genomes do replicate under these conditions, although to only about 10% of the level observed when the primary infection is by the homologous phage, T4.

D. DNA Replication in T-Even Phage-Infected Bacteria

The experiments of Hershey, Dixon, and Chase (*61*), referred to in Sec. 11-2, C, 1, showed that the synthesis of hydroxymethylcytosine-containing DNA commences between 6 and 7 min after infection by T2. By the time the first infective progeny are detectable, phage DNA is present to the extent of 40–80 phage-equivalent units per cell. Thereafter, production of DNA and infectious units proceed at equal rates such that there is a constant excess of 50–100 phage units of DNA per cell. This "incomplete phage DNA" is distinguished from the DNA of mature phage by its noninfectivity, its sensitivity to the action of deoxyribonuclease, and its failure to sediment in centrifugal gradients which sediment mature particles. Stent and Maaloe (*188*) and Hershey (*189*) provided evidence that incomplete phage DNA was actually precursor DNA, i.e., it is subsequently incorporated into mature viral progeny. These experiments, as well as parallel genetic studies (cf. Ref. *190*) presented an over-all picture of DNA metabolism in which infected cells accumulate a pool of phage precursor DNA. Molecules are withdrawn at random from this pool, for packaging into virus particles. This process, while probably not unique, is certainly different from any known mode of cellular DNA replication. A number of models have been put forward to explain T-even phage replication, and many ingenious experiments have been performed. However, the exact nature of the process is still unknown.

Following the success of Meselson and Stahl (*191*) in demonstrating semiconservative replication of bacterial DNA, several investigators attempted similar experiments with phage systems. Such studies, however, were complicated by the fact that parental T-even DNA becomes extensively fragmented during a cycle of growth in a bacterium. When parental phage DNA is labeled with ^{32}P, label is found in many of the progeny phage (*192, 193*). This is probably a reflection of the high rate of genetic recombination characteristic of these phages. Successful performance of a Meselson–Stahl experiment requires that one be able to demonstrate a "hybrid" DNA molecule, composed of one strand of parental material and one strand of newly synthesized material. Fragmentation of parental DNA would preclude the existence of structures containing whole strands of this material. However, if parental label is incorporated as relatively large parts of new DNA molecules, then it should be possible to isolate fragments after mechanical rupture of replicating DNA. Such an experiment was successfully performed by Kozinski (*194*). ^{32}P-labeled T4 was used to infect *E. coli* growing in the presence of the thymidine analog, 5-bromodeoxyuridine. Under these conditions, DNA which is synthesized entirely after infection is nonradioactive and "heavy," i.e., it has a characteristic high density resulting from substitution of bromouracil for thymine. Parental DNA is radioactive and "light." Any truly semiconservative replicating structures should have half the specific radioactivity of the parental molecules and a density halfway between "heavy" and "light." When Kozinski examined the DNA isolated from the infected culture described above, he found that radioactivity sedimented in a CsCl gradient in the position of "light" den-

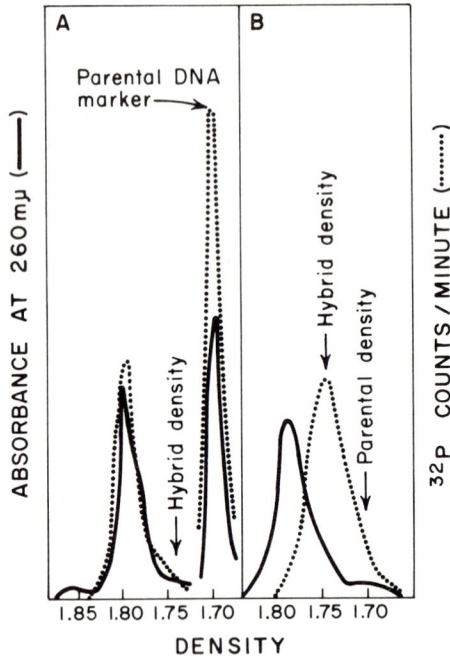

FIG. 11-8. Demonstration of semiconservative replication of T4 DNA replication, as performed by Kozinski (194). Cells are infected with ³²P-labeled phage in the presence of 5-bromouracil. Thus, DNA which is newly synthesized is denser than parental DNA, while any replicative hybrid should have an intermediate density and be radioactive. Phage are purified from the resultant lysate, and DNA is extracted and centrifuged in a CsCl gradient. In A, radioactive parental DNA is added as a marker. In B, the isolated DNA is fragmented by sonication prior to centrifugation. For further explanation, see text.

sity. This is expected if the parental DNA is subdivided into fragments none of which is large enough to appreciably affect the density of the resultant molecule. However, when Kozinski subjected DNA molecules to sonic oscillation, a treatment which reduces the length of DNA molecules without affecting their density, he found that all of the label sedimented in the hybrid density region, indicating that semiconservative replication was taking place. From the extent of fragmentation brought about by the sonic treatment, Kozinski could estimate that parental label is incorporated into new DNA molecules in fragments amounting to between 5 and 10% of the length of the genome. A schematic picture of his experiment is shown in Fig. 11-8. A similar experiment was performed by Roller (195), who showed that 80–85% of parental T4 DNA is transferred to progeny and that most of the transferred material is in pieces of the order of 8–10% of the length of a whole T4 DNA molecule. Shahn and Kozinski (*196*) have recently presented evidence confirming that this 8–10% is transferred to each new molecule as a single piece.

What molecular species are formed during the replication of T-even phage DNA? The principal approach to this question has been to label the input DNA and/or the replicating DNA with a suitable isotopic precursor and examine the distribution of label in a centrifugal density gradient. Frankel (*197*) and Kozinski and Kozinski (*198*) have described a form of replicating T4 DNA which sediments much more rapidly in sucrose gradients than DNA from mature T4. The structure appears to be linear (*197*), not circular. Frankel estimated that each molecule of this species contains between 15 anad 20 phage equivalents of DNA. The molecule contains regions of single strandedness, as indicated by the decrease in sedimentation coefficient accompanying brief treatment with an appropriate endonuclease. Although it is difficult *a priori* to see how such a structure might be formed, its existence suggests a possible explanation for the generation of terminally redundant, circularly permuted molecules in mature phage. The model, proposed by Frankel, is illustrated in Fig. 11-9. An unknown mechanism, the "headful cutting device," removes equal lengths of DNA from the replicating structure. Since, as discussed in Sec. 11-2, A, the headful contains more than a single complement of genetic information, the DNA molecules resulting from headful cutting are not unique with respect to beginning and end.

Kozinski, Kozinski, and James (*199*) have recently questioned Frankel's conclusion that replicating DNA is a linear molecule containing many phage equivalents of DNA. A replicative hybrid was isolated from a CsCl gradient following infection with a light, radioactive phage in heavy, nonradioactive medium. Although the material

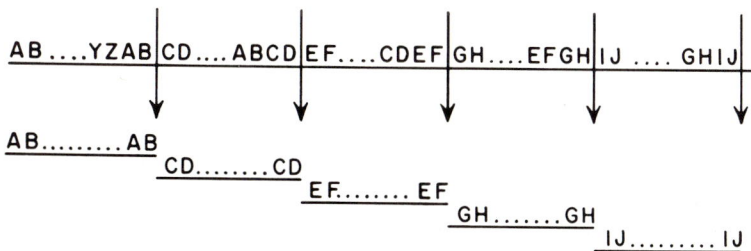

AB....YZAB CD....ABCD EF....CDEF GH....EFGH IJ ....GHIJ

AB.........AB
    CD........CD
        EF........EF
           GH.......GH
              IJ.........IJ

FIG. 11-9. A scheme for generating a population of circularly permuted, terminally redundant molecules (*197*).

sedimented very rapidly in sucrose gradients, electron microscopy showed that it is only about twice as long as a DNA molecule isolated from mature phage. However, the structure was highly coiled and twisted. Kozinski et al. concluded that the resultant compactness of the structure could at least partially explain its rapid sedimentation properties. In another experiment these investigators showed that, under conditions permitting replication but not recombination, the density of the hybrid molecule does not increase with time after infection, as would be expected if replication involves the covalent attachment of new material to a hybrid structure containing equal amounts of old and new material. Kozinski et al. do not dispute the possibility that longer structures might be formed through recombination or that circularly permuted molecules could be generated by the process described above. As one might expect, studies on DNA replication in phage-infected cells have been closely connected with studies on recombination, studied at both the genetic and biochemical levels. It is beyond the scope of this chapter to discuss in detail how genetic investigations of recombination have contributed to our understanding of the nature of replicating viral DNA molecules and their interaction. For an excellent discussion, see Chap. 10 of Ref. *190*. Recent investigations, notably in the laboratories of Kozinski and of Tomizawa, have attempted to explain recombination in terms of observed molecular events. Anraku and Tomizawa (*200*) infected bacteria mixedly with radioactive (^{32}P-labeled) and heavy (bromouracil-labeled) T4 in the presence of 5-fluorodeoxyuridine (FUdR), an inhibitor of DNA synthesis (*201*). DNA isolated 15 to 20 min after infection showed some "joint molecules," which contained both labels, held together by hydrogen bonds. At later times (30–45 min) DNA molecules could be extracted in which heavy and radioactive portions were linked covalently. Anraku and Tomizawa state that DNA replication is not required for formation of either molecular species, even though replication was not completely blocked in their experiments (cf. *132*). At any rate their data indicate that recombination in T4 takes place by a process of breakage and joining of DNA strands, rather than a copy–choice process. Similar conclusions have been reached from genetic experiments performed by Simon (*202*). The studies of Anraku and Tomizawa, moreover, show that any model advanced to explain recombination must include the formation first of hydrogen-bonded joint molecules, followed by a covalent linkage process.

In a more recent communication, Tomizawa, Anraku, and Iwama (*203*) have presented evidence suggesting an essential molecular event as a prerequisite to both replication and recombination. An amber mutant, TAamA453, which had previously shown to be essentially totally blocked in DNA synthesis (*204*), was found to be unable to form the joint molecules described above. The mutant is defective in gene **32**. Although the function of this gene is unknown, it is apparently involved in both replication and recombination. This study, and a similar study by Frankel (*204*), point up the great advantages of using DNA-defective amber mutants as tools for examining phage DNA replication. As mentioned earlier, the biochemical functions of a number of these mutants are known. However, an equal number remain unexplored. It would appear that studies on the molecular events associated with infection by some of these mutants under nonpermissive conditions represents one of the most fruitful approaches for probing phage DNA replication.

Another area deserving of further exploration relates to very early events in DNA replication. Because net formation of DNA commences at about **7** min after infection, earlier periods have been relatively ignored. However, Frankel (*197*) showed that a small amount of newly formed DNA is present as early as 5 min after infection by T4. This DNA sediments in sucrose gradients at the same rate as DNA extracted from mature phage. Murray (*205*) has estimated the amount of this early DNA to be about 0.5 phage equivalent per cell at 6 min. Neither protein synthesis nor the action of phage-specific DNA polymerase seems to be required for formation of this material. The function of this early DNA in the over-all replication cycle is unknown at present. However, its existence is of some interest in light of earlier studies on the kinetics of stabilization of infected complexes both to ultraviolet light inactivation (*206*) and to the lethal effects of the decay of ^{32}P contained in the infecting particles (*207*). Both of these stabilizing effects, either of which would seem to require some transfer of genetic information to a new structure, are discernible by 5 min, a time at which no significant DNA replication has occurred (*207*). It will be of some interest to see whether these stabilization effects are related to the formation of early DNA.†

---

† Further studies on early DNA (*437*) indicate that it results from the addition of nucleotides to parental DNA molecules, to the extent of about 0.06 phage equivalent unit of new material per parental molecule.

A final area of active research interest in the topic of phage DNA replication relates to the nature of the cellular site for DNA replication. Kozinski and Lin (*208*) showed that replicating T4 DNA enters a proteinaceous complex within the infected cell. Apparently the proteins involved in complexation are bacterial in origin, for chloramphenicol does not inhibit the process. The question of the number of sites per cell for DNA replication has been approached by Snustad in an elegant series of genetic experiments (*209, 210*). Snustad found that infection of cells under restrictive conditions with a constant, low multiplicity of wild-type phage and increasing multiplicities of a conditional lethal mutant leads to a sharp decrease in the number of infected cells which yield progeny. This is interpreted as competition between infecting genomes for sites of heterocatalytic expression. At high multiplicities of mutant infection, no wild-type genomes are expressed. Snustad believes that genomes which can express themselves are drawn at random from the pool of infecting genomes. Further, he has analyzed his data to allow a determination of the number of genomic expression sites per cell. This number is 4 for one of his bacterial strains, 6 for another. The questions of the nature of these sites and the factors governing the number of sites per cell should be of considerable future interest, particularly in view of current work on the control of replication of bacterial chromosomes (*211*).†

E. TEMPORAL CONTROL OF GENE EXPRESSION

The existence of at least two classes of phage-specific proteins in T-even-infected cells, namely the "late" proteins which are incorpo-

---

† Two important papers on T4 DNA replication have appeared recently. Werner (*414*) has analyzed the distribution of growing points in replicating T4 DNA by means of 5-bromouracil density-labeling techniques. Early during the period of DNA replication there are between 1.0 and 1.5 growing points per molecular equivalent of DNA. However, these are not distributed uniformly, there being but 0.15–0.19 phage equivalent lengths of DNA between growing points. This suggests that several successive rounds of replication are initiated from a single site, which is not itself duplicated.

Sugimoto, Okazaki, and Okazaki (*415*) have examined the role of polynucleotide ligase in T4 DNA synthesis. Cells infected with temperature-sensitive gene 30 mutants of T4 (ligase-defective) were transferred to 43° during a period of active DNA synthesis. Large amounts of short DNA fragments accumulated. This supports a model of discontinuous DNA replication which involves synthesis and subsequent joining of short segments, with the latter reaction catalyzed by polynucleotide ligase (*416*).

rated into viral progeny, and the "early" proteins, most of which are not (see Sec. 11-2, C, 4), suggested that there are control mechanisms which direct the syntheses of specific proteins at particular times in the replication cycle. In general, it appears that the early enzymes of DNA metabolism are formed within the first 10 to 15 min after infection, while the syntheses of structural proteins and lysozyme commence at about 8 to 10 min. The mechanism by which the infected cell arrests the expression of early genetic functions and activates the expression of late ones has absorbed the energies of many workers in the past several years.

In 1960 Dirksen et al. (212) showed that when *E. coli* is infected with ultraviolet light-inactivated T2, the normal control mechanism is inoperative. The levels of early enzymes continued to increase for up to an hour, instead of leveling off at about 15 min as in normal infection. The final specific activities reached were increased some 4-fold relative to normal infection. Subsequent work in other laboratories showed that uv irradiation of phage also blocks the expression of late genetic functions, including the syntheses of structural proteins (213) and lysozyme (214).

In their original report Dirksen et al. discussed a number of possible factors which could be responsible for the loss of control in uv-phage-infected cells, including the failure of such cultures to synthesize DNA. This suggestion was put into sharper focus by Wiberg et al. (95), who studied the control of early enzyme synthesis in cells infected with a variety of T4 amber mutants. These authors found that mutants which were essentially incapable of inducing DNA synthesis behaved similarly to uv-irradiated phage, i.e., early enzyme formation was exaggerated and extended. Other amber mutants which were not completely blocked in DNA synthesis exhibited near-normal control of early enzyme synthesis. Subsequent workers (132, 215) found that amber mutants which could not shut off the formation of early enzymes were unable to direct late protein synthesis. Using a fluorescent antibody technique for visualizing viral antigens in single infected cells, Sercarz (215) showed that the block in late protein formation is complete. A small amount of DNA synthesis, such as that occurring in wild-type infection in the presence of antimetabolites such as fluorodeoxyuridine, does permit late protein synthesis (132, 215, 216). Mathews (132) was able to estimate that between 5 and 10 phage equivalent units of DNA must be synthesized per infected cell in order for lysozyme to be synthesized at a normal rate. Apparently this must be newly synthesized DNA, for Mathews and Kessin (*139*) observed no lysozyme in *E. coli* B infected with DNA-blocked *am* or *ts* mutants, even when the multiplicity of infection was as high as 20.

A number of workers have asked whether regulation of phage gene expression is at the level of transcription (i.e., different mRNA's made at different times) or translation (i.e., different mRNA's utilized for protein synthesis at different times). Although the answer to this question is still far from clear, it appears that both types of regulation are involved. Kano-Sueoka and Spiegelman (*217*) approached this question as follows: One T2-infected culture was pulse labeled with a ^{14}C-containing RNA precursor at an early time in infection, while another culture was pulsed late with an ^{3}H-containing RNA precursor. The two cultures were mixed, and RNA was extracted and subjected to column chromatography on methylated albumin-kieselguhr. Elution patterns for the two isotopes were different, indicating that different mRNA species are made at different times in infection. Hall, Nygaard, and Green (*218*), using newly developed assays for DNA-RNA hybridization, examined RNA preparations which were pulse labeled early or late in infection. They found that all of the RNA species formed early in infection are also synthesized at later times but that new forms of RNA, not present at early times, are synthesized late in infection.

Bautz and his collaborators (*219–221*) have devised techniques for studying the synthesis of gene-specific mRNA molecules. Pulse-labeled RNA from wild-type T4-infected cells was hybridized with DNA from mutants containing deletions of entire genes—either the rII gene or the *e* gene, which is the structural gene for lysozyme. The RNA remaining unhybridized in each case is considered to be specific to the deleted gene. Appropriate hybridization techniques were used to follow the formation of each of these messengers during the infective cycle. The rII gene is transcribed throughout the course of infection. However, the lysozyme gene is transcribed once very early in infection and then is not transcribed again until after 6 min. The early burst of lysozyme mRNA synthesis is also observed with T4 *am* N82, an amber mutant unable to induce lysozyme formation. Kasai and Bautz (*221*) concluded that failure of the lysozyme mRNA to be translated early acts to repress further synthesis of this molecule.

Using indirect evidence, Edlin (*222*) has proposed a model for regulation of T4 protein synthesis which operates primarily at the level of translation. Edlin found that 5-fluorouracil causes phenotypic reversion of amber mutants, with the maximum effect being exerted at that time in the infective cycle when the mutant function is expressed. Presumably, fluorouracil acts by being incorporated into mRNA as uracil and occasionally being read by the protein-synthesizing apparatus as cytosine (223). Edlin's model proposes that mRNA for late functions is more stable metabolically, and more tightly bound to ribosomes, than early-function mRNA. This model is based, at least in part, upon experiments suggesting that late function mRNA can be transcribed from input DNA. A late-function *am* mutation was bred into a genome containing two *ts* mutations. Cells were infected and incubated in the presence of FU at 42°, a nonpermissive temperature for DNA synthesis. At various times the cells were diluted away from FU, plated, and incubated at 30°, a permissive temperature. Under these conditions, functional late mRNA can only be made during the early period of incubation since FU is not present later, yet most of the infected cells were able to produce plaques. Edlin suggested from this experiment that late-function mRNA was being transcribed from nonreplicating DNA at 42°. However, he did not check to see whether DNA synthesis was completely blocked under his conditions. Moreover, he does not appear to have considered the possibility that FU might cause phenotypic reversion of the *ts* mutations, a situation that would lead to replication at 42°.

Baldi and Haselkorn (224) have recently performed an experiment showing that late T4 mRNA is bound to ribosomes no more strongly than early mRNA.

Several laboratories have studied phage-specific RNA synthesis *in vitro* as an approach to the question of selective transcription as a regulatory mechanism. Geiduschek et al. (225) used competitive DNA-RNA hybridization to characterize the RNA formed by *E. coli* RNA polymerase in the presence of T2 or T4 DNA. In both cases all of the polynucleotide sequences formed early after infection *in vivo* were represented in the *in vitro* synthesized product, while late-type messenger was undetectable. This result is in agreement with the idea, stated above, that input DNA cannot transcribe late-function genes. Geiduschek et al. also found that their *in vitro* product contained very little RNA complementary to early mRNA synthesized *in vivo*. This suggests that the same mechanism governing strand selection for transcription in the infected cell is operative *in vitro*, in agreement with the results of others (226, 227).

A possible mechanism for translation-level control of phage protein synthesis was suggested by Sueoka and Kano-Sueoka (228). These workers discovered a structural change in a leucine-specific transfer RNA fraction, as reflected in an altered elution pattern of leucyl-tRNA's from a methylated albumin column. In their original report they stated that the structural modification occurred at about 8 min after infection by T-even phages, and they suggested that synthesis of an altered tRNA might be a necessary prelude to translation of late-function mRNA. Closer attention to the conditions of infection (229) indicated that the structural modification is actually one of the earliest events in infection, being complete by 1 or 2 min after infection. This suggests that the modification might actually be involved in the shutoff of bacterial protein synthesis. The modification appears to be specific, in that tRNA's for 16 other amino acids besides leucine show no detectable change in chromatographic behavior after phage infection. Moreover, the modification appears to be directed by a phage gene, since it is evoked by all T-even phages, but not by T-odd phages, phage lambda, T2 ghosts, or colicins. These later results of Kano-Sueoka and Sueoka (229) have been confirmed by Waters and Novelli (230), who have discovered a second modification in leucyl-tRNA, occurring quite late in infection by T2. It should be emphasized that no regulatory significance has yet been directly ascribed to either of these modifications.

Another explanation for regulation of T4-specific protein synthesis has been put forth by Shalitin and Sarid (231, 232). These investigators observed that polyamines leak out of infected cells soon after infection, as might be expected from earlier-observed changes in the permeability of infected cells (68d). Moreover, putrescine appears to selectively inhibit the transcription of certain regions of the T4 DNA molecule in vitro. Shalitin and Sarid proposed that putrescine specifically blocks transcription of late-function genes in vivo and that time-dependent leakage of putrescine from the cell allows these genes to be transcribed. This theory is difficult to reconcile with the fact that certain DNA-negative amber mutants are completely blocked in expression of late functions. Presumably, infection by such mutants would lead to leakage of polyamines also, since other early functions are expressed. Furthermore, one must account for the facts that late-function messengers are not transcribed in vitro, even in the absence of putrescine (225) and that the intracellular level of putrescine actually increases after infection (233).

In many ways the T-even phage-infected bacterium presents an attractive model system for studying cellular differentiation. However, any comprehensive explanation of regulation in this system must take into account its ever-increasing complexity. Early investigations in this area were oriented toward the idea that there are but two classes of phage-specific proteins: "early" and "late" (cf. *234*). Recent studies, using pulse-labeling of proteins in combination with disk–gel electrophoresis and radioautography (*235, 236*), have indicated that there are at least four classes of phage-specific proteins, as judged from their timing of synthesis. Moreover, it has been known for some time that at least one protein, the internal protein, is synthesized throughout the entire course of infection. Finally, it has been demonstrated that "turn off" of an early function does not necessarily lead to "turn on" of a late one (*132, 139*).

F. Maturation and Virus Assembly

As stated in Sec. 11-1, no infective particles can be demonstrated in T-even phage-infected cells until about 12 min after infection. This is 4 to 5 min after the appearance in the infected cell of newly synthesized DNA and viral antigens. How are these building materials brought together and assembled into the complex structure associated with the T-even phage particle? For many of the simpler viruses one can invoke a spontaneous self-assembly process, akin to crystallization, as the major agent in effecting virus maturation. However, the T-even phages seemed too complex for maturation to occur solely by self-assembly, and it seemed reasonable *a priori* to predict that certain enzymatic steps might be involved in maturation. Three experimental approaches have been most useful in studying this process: first, electron microscopic examination of ultrathin sections of infected bacteria, as pioneered by Kellenberger and associates (*237–239*); second, use of various inhibitors of maturation; third, and perhaps potentially most informative, studies on viral morphogenesis *in vitro*, as developed by Edgar and Wood (*102, 240*).

Electron microscopic examination of infected cells reveals that DNA, which is formed first as long fibrils, condenses toward the end of the eclipse period into hexagonal-shaped packets about the size of a complete phage particle. At first these packets are not surrounded by a head membrane, for disruption of the cell yields more packets than head membranes. The mechanism of this condensation process remains one of the major unsolved problems of viral maturation, as does the nature of the previously mentioned "headful cutting device," which determines the length of each DNA molecule. Apparently, certain "late" proteins are required for condensation to occur, for when chloramphenicol is added at the eighth minute, condensation does not take place, and a giant pool of unmatured DNA is formed in the cell. However, Piechowski and Susman (241) have shown that, once certain late proteins have formed, maturation can occur in the absence of protein synthesis.

Once the DNA has condensed, each packet becomes "packaged," or surrounded by head protein. Undoubtedly, this process has some aspects of spontaneity since it involves the joining together of approximately 1000 identical subunits of head protein. Tails, and then tail fibers, are added to the already completed heads.

Electron microscopy has also been used to characterize the large number of amber and temperature-sensitive mutants which are defective in late steps of virus development (92). For example, artificial lysates of cells infected with gene 23 mutants contain morphologically distinct tails, complete with tail fibers, but no heads. Evidence from other lines (98, 242) has confirmed that gene 23 is the structural gene for the head protein. About 40 genes dealing with the synthesis of late proteins and the maturation process have now been discovered, and most of these have been characterized electron microscopically. Some of this information has been presented diagrammatically in Fig. 11-6.

A number of inhibitors of maturation have been reported. 9-Aminoacridine blocks both maturation and lysis (243), without affecting over-all protein or DNA synthesis. Both the protein and the DNA formed in the presence of the 9-aminoacridine are functional, as shown by their incorporation into mature phage once the inhibitor has been removed (244). At present the mode of action of 9-aminoacridine is unknown. Actinomycin D, already well known as an inhibitor of DNA-dependent RNA synthesis, has been shown to block maturation specifically, when applied at concentrations too low to affect over-all polymer synthesis (245). Korn (246) has shown that, in the presence of the drug, replicating DNA of the type described by Frankel (197) accumulates and there is little DNA of the type extracted from mature phage. In this respect the action of actinomycin is similar to that of chloramphenicol. Korn has suggested that actinomycin blocks a step in the condensation of DNA.

Edgar and Wood have recently developed a novel and informative approach to the problem of T4 morphogenesis (*102, 240*). These workers have devised a method for observing certain steps of maturation *in vitro*. An example of this approach is as follows. Cells were infected with a virus bearing mutations in genes 34, 35, 37, and 38, genes previously shown to be necessary for tail fiber synthesis. Such a phage mutant directs the synthesis of noninfective progeny which appear complete in the electron microscope except for the absence of tail fibers. These particles were isolated by breaking open the infected cells and purifying by differential centrifugation. They were then added to an extract of cells infected with a gene 23 mutant, which can make tails and fibers, but no heads. Infectivity of this mixture increased 1000-fold, and electron microscopy confirmed that complete phage particles were indeed being assembled from the components of the cell-free mixture. This *"in vitro complementation"* approach, has now been used to test, pairwise, all of the 40 genes known to be involved in T4 morphogenesis. By this approach, the 40 genes have been divided into 13 complementation groups, i.e., groups within which complementation has been shown not to occur. Moreover, the sequence of action of the genes has been at least partially determined. The experiments reported so far show that there are three independent lines of assembly, leading, respectively, to heads, tails, and tail fibers. Sixteen gene products have been shown to participate in head formation, although the *in vitro* experiments have so far demonstrated only three separate steps. Similarly, 19 gene products take part in tail formation, although only 5 distinct steps have been recognized. Joining of completed heads to completed tails is a spontaneous process, although another gene product (gene 9) is required for the attachment to become irreversible. Four distinct steps, involving the action of five gene products, lead to the formation of complete tail fibers. Attachment of the fibers to the head–tail complex appears to be catalyzed by a factor which has some of the properties of an enzyme. This factor is presumed to be a viral gene product, since it is not found in extracts of uninfected bacteria. However, mutants unable to synthesize the factor have not yet been isolated.

G. Lysis, Lysis Inhibition, and the rII Function

As with most virulent phages, the final event in the replicative cycle of T-even bacteriophages is rupture, or lysis, of the infected cell, with progeny virus being released to the medium. The immediate causative

agent is a phage-specific lysozyme, which is the product of the viral
e gene (*247*). However, the timing of lysis is subject to other, more
complex control factors which are still but poorly understood. Under
the usual conditions of a one-step growth experiment—i.e., a brief
adsorption period at high cell density, followed by dilution of the
culture to prevent readsorption of newly liberated phage—the latent
period is of the order of 25 min. However, when adsorption and virus
growth are both carried out at relatively high cell concentrations, the
latent period is greatly lengthened, sometimes by several hours. Virus
production continues during this extended period, such that final burst
sizes are as high as 1000 or more. Yet, as discussed Sec. 11-2, E,
lysozyme is detectable in infected cells from about 10 min onward.
How is its lytic action prevented during the elongated latent period?

The earliest studies on the timing of lysis were reported in 1948 by
Doermann (*248*), who showed that secondary infection of an infected
cell by a new virus particle is responsible for the extended latent
period, a phenomenon which he termed lysis inhibition. In concentrated cultures, such reinfection occurs when those few earliest-lysing
cells release viral progeny which reinfect the remainder of the cell
population. Although the mechanism by which reinfection delays lysis
is unknown, it appears to involve the energy metabolism of the infected cell. Thus cyanide, an agent known to block terminal respiration, causes premature lysis. One could reasonably infer that blockage
of energy metabolism halts the synthesis of new cell-wall material,
necessary to counteract the destruction of such material by lysozyme,
with the result being the observed dissolution of the cell.

Further light on the above supposition was provided by studies on
cell respiration in bacteria infected with an *e* mutant of T4 unable to
induce lysozyme formation (*249*). Such bacteria fail to lyse naturally,
so reinfection does not occur. However, both phage production and
oxygen uptake in these cells ceased at about 25 or 30 min after infection. On the other hand, when superinfection is carried out by addition
of phage to the medium, phage production and oxygen consumption
both continue for an extended period. It would appear, therefore, that
the ultimate effect of superinfection is to prolong the metabolic life-
time of the infected cell. The mechanism of this effect is completely
unknown, as is the timing mechanism leading to metabolic death of the
nonsuperinfected cell at about 25 min.

The T4 *e* gene and its product, lysozyme, have proven to be a fruitful system for experimental verification of some theoretical postulates
concerning the nature of the genetic code. Lysozyme is a small protein,

containing but 160 amino acids, and mutants in the e gene are easily recognized by appropriate plating techniques. Thus, structural analysis of lysozyme molecules produced by certain proflavine-induced mutants (250, 251) has confirmed an earlier prediction (252) that the acridine dyes exert their mutagenic effects by causing addition or deletion of bases to or from a polynucleotide chain. Moreover, such experiments have verified that translation of messenger RNA occurs in a 5' to 3' direction.†

A somewhat related genetic system which has formed a cornerstone of molecular genetics, particularly in the hands of Benzer and his colleagues (253), is the rII system. The r, or rapid lysis, mutants are characterized by loss of lysis inhibition. Thus, burst sizes in concentrated cultures tend to be lower than in similar cultures infected with r^+, or wild-type, strains. However, when plated on a wild-type strain such as $E.$ $coli$ B, r mutants give larger plaques than r^+, because all of the infected cells lyse. In phage T4 there are at least three genes giving the r phenotype (254), the rI, rII, and rIII genes. The rII system is particularly favorable for study because it is conditionally lethal, i.e., it fails to produce plaques when plated on certain $E.$ $coli$ strains which are lysogenic for phage lambda. Thus, one can easily identify revertants, even when they occur at very low frequency, by plating large numbers of phage on $E.$ $coli$ K12 (λ).

The rII gene contains two cistrons, termed A and B, as defined by cis–$trans$ complementation tests. Obviously, some knowledge of the nature of the products of these cistrons would be important for understanding both the metabolic block in rII-infected K12 (λ) and the mechanism of lysis inhibition. However, this seemingly modest goal has eluded the grasp of some of the most skilled experimentalists in bacterial virology. Many workers have tried to isolate an rII protein. However, not until late 1967 was even partial success recorded (255), and there is still virtually no clue to the function of this material.

An alternate approach, which to date has been only slightly more successful, is an examination of the physiological properties of $E.$ $coli$ K12 (λ) infected with rII mutants. Garen (256) found that metabolism of such cells proceeds normally for the first 10 min, with DNA synthesis, protein synthesis, and respiration proceeding on schedule. Some critical event occurs at about 10 min, and all of the above processes

---

† The complete amino acid sequences of T2 and T4 lysozymes have now been published (438, 439). Interestingly, the sequences of the two proteins are identical except for three positions. These differences could easily have arisen through alterations each involving a single base.

are affected. Garen found that this event could be prevented, and phage development would occur, if relatively high concentrations of magnesium were present in the medium. Garen suggested that magnesium ions, needed for essential metabolic processes, might leak out of rII-infected cells to a greater extent than r+-infected cells, due to defective repair of the bacterial membrane after infection (68d). This possibility was tested further by FerroLuzzi-Ames and Ames (257), who found that polyamines also effect phenotypic repair of rII mutants, as does magnesium. These workers found that putrescine leakage from infected cells is more extensive in rII-infected cells than in r+-infected bacteria. At about the same time Brock (258) showed that there are no generalized permeability differences between cells of K12 (λ) infected by rII or r+. At this stage, however, the data could not show whether the polyvalent cations, such as polyamines and magnesium, were participating directly in essential metabolic processes or whether they were necessary for stabilization of the membrane to prevent loss of some other essential metabolite.

Sekiguchi (259) has recently published an extensive and careful study of the physiology of rII-infected cells. He showed that the apparent stimulatory effect of magnesium is actually due to counteraction of an inhibition by sodium of the uptake of polyvalent cations. Thus, rII mutants can multiply in low-sodium medium in the absence of added magnesium. Addition of sodium to rII-infected cultures inhibits magnesium uptake. Sekiguchi also observed a sharp decrease in oxidative phosphorylation in rII-infected bacteria relative to cells infected by wild type. This decrease begins immediately after infection, and Sekiguchi suggested that the resultant depletion of ATP is responsible for the generalized metabolic failure which occurs at 10 min. It is possible that more ATP is required for ion transport in rII-infected cells than in those infected by r+ and that this leads to the observed depletion. However, although these experiments do help to pinpoint the specific metabolic defect associated with rII infection, they have not yet led to a recognition of the nature and role of the rII gene product nor of the critical metabolic change associated with lysogenization of phage λ.

11-3. T-Odd Bacteriophages

The T-odd phages are not nearly as closely related to each other as are the T-even phages. In general, the T-odd phages have smaller heads, containing less DNA, than the T-evens. In addition, T-odd DNA's are "normal" with respect to base composition in that they contain adenine, guanine, cytosine, and thymine. These viruses have not been studied as systematically as the T-even phages. This section does not try to review all of the literature in this area, but mentions only relatively recent studies of special interest.

Of the T-odd phages, T5 is most closely related to the T-even phages, with respect to physical dimensions, amount of DNA per phage, and metabolic events in infected cells. T1, T3, and T7 do not stimulate a net DNA synthesis after infection, apparently deriving most of the deoxynucleotides needed for virus synthesis from breakdown of the host-cell DNA (7). However, T5 does elicit considerable net DNA synthesis and induces the formation of several early enzymes. These include thymidylate synthetase (84), dihydrofolate reductase (120, 122), DNA polymerase (260), deoxyribonuclease (261), and a polyfunctional deoxynucleotide kinase capable of phosphorylating dAMP, dGMP, dCMP, and dTMP (262). Interestingly, dCMP deaminase is not induced (263), nor are polynucleotide kinase (169) or 3'-deoxynucleotidase (171). Other possible enzymes, such as dUTPase or ribonucleotide reductase, have not, to this writer's knowledge, been examined. As mentioned in Sec. 11-2, C, 7, one new enzyme activity has been found to be associated with infection by T3, namely the cleavage of S-adenosylmethionine to thiomethyladenosine and homoserine (165). Apparently the activity is not essential to the infection process, since mutants have been isolated which lack ability to induce the enzyme (264). Such mutants appear to multiply normally. Hausmann and Gomez (265) have isolated a number of amber mutants of T3 and of T7 which are defective in DNA synthesis, but to date none of the mutants has been identified in terms of a specific metabolic defect.

Thomas and his collaborators have studied the structures of T3, T5, and T7 DNA's, using the methods described in Sec. 11-2, A (30, 266–268). All three DNA's contain unique base sequences, i.e., they are not circularly permuted (266). T3 and T7 DNA's contain regions of terminal redundancy (267) while T5 DNA does not seem to have been examined for this property. However, T5 DNA has the unusual feature of containing single-strand interruptions at defined points in both strands of the duplex molecule (268). Preliminary evidence suggests that these interruptions are repaired in vivo after infection by T5. In this context it would be of interest to know whether a T5-specific polynucleotide ligase (173) might exist and play a role in this repair.

The metabolism of T5-infected cells has been studied over the past several years by McCorquodale, the Lannis, and their co-workers. The process of injection of T5 DNA is somewhat slower than that of the T-even DNA's. Y. T. Lanni (*269*) showed that the process could be disrupted by mechanical shearing of infected phage-bacterium complexes in a Waring blendor. Lanni, McCorquodale, and Wilson (*270*) showed that a fragment of DNA consisting of about 8% of the total viral DNA complement is introduced first and that shearing prevents transfer of the remaining DNA. Lanni (*271*) demonstrated a requirement for phage-specific protein synthesis, directed by the first step transfer material (FST-DNA) before the remaining 92% of the DNA could enter the cell. McCorquodale, Oleson, and Buchanan (*236*) have used the combined gel–electrophoresis–autoradiography technique of Levinthal et al. (*235*) to study the classes of proteins synthesized in T5-infected cells. There appear to be three major classes. Class I proteins are synthesized by FST-DNA, as shown by the fact that they are formed after blending of infected complexes. These proteins appear to play roles in the cessation of bacterial protein synthesis, breakdown of bacterial DNA (*272*), and eventual shut-off of their own synthesis. Class II proteins apparently include most of the "early enzymes," such as thymidylate synthetase, while the Class III proteins represent structural proteins and other factors involved in maturation.

11-4. Temperate Bacteriophages

A. DEVELOPMENT OF THE CONCEPT OF LYSOGENY

The T-phages are typical virulent bacteriophages, i.e., viruses in which infection invariably leads to cell death and virus production. There is another major class of phages, the so-called *temperate* bacteriophages, in which infection need not always be lethal. A cell infected by a temperate bacteriophage can, depending upon environmental conditions, either produce a crop of viral progeny and undergo cell death by lysis (lytic response) or it can incorporate the genome of the virus into its own genetic apparatus (lysogenic response). Under the latter conditions the bacterium remains viable, but the latent capacity to undergo a lytic cycle is passed on to each daughter upon cell division, and is retained through many subsequent generations.

Although it had been known since the early 1920's that certain bacterial strains are lysogenic, or capable of producing lysis, it was thought by many that this was a reflection of contamination of bacterial cultures by phage. The latter hypothesis was made untenable by McKinley's demonstration (273) that lysogeny was maintained even when bacteria were grown in the presence of a specific phage antiserum, ruling out the existence of phage as an extracellular contaminant. Moreover, Burnet and McKie (274) demonstrated that artificial lysis of lysogenic bacteria did not release infective phage particles; thus, intracellular contamination by phage in lysogeny was also ruled out. However, the early literature on temperate bacteriophages and lysogeny is marked by lively controversy. The present conception of lysogeny did not receive general acceptance until the work of Lwoff in the late 1940's and early 1950's (275).

Lwoff's first experiment was to examine single lysogenic cells of *Bacillus megaterium* through many generations (276). A cell was cultivated in a drop of medium, on a microscope stage. After each division, one of the daughter cells was plated on a nonlysogenic indicator strain, to see whether it was capable of producing virus. This capability was maintained through at least 19 generations. It was also observed that, although the culture fluid generally contained no phage, occasionally a single cell in the microscopic field would disappear. When this happened, about 100 infective virus units could be found in the culture fluid. Thus, the lysogenic cell appeared to maintain virus in a noninfective state, a state which Lwoff termed prophage. The prophage is reproduced at cell division and transmitted to each daughter cell. Occasionally, the prophage becomes *induced* to form viral progeny, the result being lysis and concomitant cell death. Lwoff felt that the frequency with which lysogenic cells become induced is a function of some environmental condition. Accordingly, Lwoff, Siminovitch, and Kjeldgaard (277) searched for some way to induce a lytic response in a large fraction of a lysogenic population. After a great deal of effort, they found that irradiation of a lysogenic strain of *B. megaterium* with small doses of ultraviolet light led to lysis of 99.9% of the bacteria, with concomitant virus production. A large number of agents is now known to function as inducers, including X-rays, nitrogen mustards, Mitomycin C, and thymine starvation. These agents all appear to act by virtue of their ability to interfere with nucleic acid metabolism.

Biochemical research on temperate phage replication and the nature of lysogeny began in earnest somewhat later than corresponding work on the virulent phages, due at least partially to the emphasis placed upon the T-even phages by the early post-World War II school of phage workers in the U.S. However, this area has fluorished in recent years and has presented many opportunities completely unavailable to workers on the T phages. Work on temperate phages has become all the more relevant to general biological problems in light of an increasing body of evidence, both presumptive and direct, which suggests that cells of higher organisms can harbor viruses in an inactive state and produce infective virus when exposed to appropriate environmental stimuli (cf. Ref. *275;* also other chapters, this volume).

The virus of choice for most research on lysogeny and temperate phage replication has been the coliphage lambda (λ), and most of what follows is drawn from studies on this virus. However, where appropriate, examples from other temperate phage systems are cited.

B. STRUCTURAL FEATURES OF PHAGE λ

Like the T-phages, λ consists of a hexagonal head attached to a tail (*278*). The tail is thinner and about half again as long as a T-even phage tail. It appears to be less complex morphologically than its T-even counterpart. The head is somewhat smaller than the head of T-even phage, which is not surprising in view of the fact that it contains considerably less DNA. The λ DNA molecule is some 17 μ in length and has a molecular weight of about 33×10^6 (*279, 280*). It is a double-stranded molecule, containing no polynucleotide interruptions. The amount of DNA in a λ virion has been estimated as sufficient to code for some 40 to 45 proteins (*281*). To date some 25 genes have been identified and mapped.

A great deal of information has been gained about the structure of λ DNA, and its relation to the genetic map, through the use of a biological test devised by Kaiser and Hogness (*282*). Since the test has also been of great value in studies on the intracellular fate of λ DNA during lysogenization and vegetative growth, it is described briefly here. DNA is isolated from λ bearing a given genetic marker. The DNA is incubated with a bacterial culture which has been infected with a strain of λ not bearing the marker in question. The latter acts as a "helper phage," allowing plaque formation to take place at high efficiency. The added DNA is taken up into the cells, and incubation is continued. During this period some of the cells initiate virus production, and some of the added DNA becomes incorporated into vegetative, or replicating, genomes by recombination. After lysis of the phage-producing cells, a sample of the culture is plated under conditions allowing recognition of the marker initially present on the added DNA. The frequency with which the marker appears in the viral progeny is a measure of the biological activity of the added DNA.

The above test has been used to prove that λ DNA is not circularly permuted with respect to base sequence (283). DNA containing two distant markers was subjected to shear, a treatment which breaks the molecules in the middle to yield two duplex molecules of half the original length (284). The two markers were transferred independently to recipient cells, showing that they were on different portions of the fragmented molecule, a situation which could not obtain if the sequence were permuted.

Like the T-even phages, λ DNA contains a small amount of information repeated at both ends of the molecule. However, the redundant information is in the form of short single-stranded regions of DNA (285), some ten nucleotides long. The single-stranded regions at either end of the molecule are mutually complementary, such that circular molecules or concatemers (end-to-end polymers) are formed when the DNA is subjected to annealing conditions (280). The single-stranded regions, sometimes termed "cohesive ends" are necessary for biological activity of λ DNA or its fragments in the Kaiser–Hogness transformation test. Treatment of λ DNA with DNA polymerase, which renders the single-stranded ends double stranded, destroys the activity of the DNA (286). Moreover, artificially formed circles are inactive (287).

The λ system is the first for which it was possible to correlate the position of a gene on a genetic map and on a DNA molecule. Meselson and Weigle (288, 289) carried out genetic crosses with λ in which one of the parents contained a density label. The amount of label found in the recombinant progeny was that predicted on the assumption of colinearity between the genetic map and the DNA molecule. This has been carried further by an approach which utilizes fragmentation of λ DNA molecules and biological assay of the fragments (281), such that the physical positions of a considerable number of λ genes on the DNA molecule are now known.

C. Metabolism of λ-Infected Bacteria

1. *Over-all Aspects*

Infection of *E. coli* with λ or another temperate phage can lead either to a lytic cycle of growth, similar to that already discussed for the virulent phages, or to lysogenization of the infecting genome. The relative frequency with which an infected cell becomes lysogenic depends both upon the nature of the virus and environmental conditions. For instance, lowering the temperature generally increases the frequency of lysogenization. The specific steps in the establishment and maintenance of lysogeny are discussed below. In general it appears that the first step is insertion of the phage chromosome at a specific location on the bacterial chromosome. Once the phage genome is inserted, or integrated, it is maintained in a largely inactive state, replicating in synchrony with the bacterial chromosome. The phage genome directs the synthesis of an immunity substance, or repressor, which prevents expression of most of the phage genes, whether they be on the lysogenized phage genome or on the chromosome of a superinfecting phage. Thus, the lysogenic cell is immune to superinfection by the same, but not by different, phages. Immunity, the failure of an injected chromosome to replicate, is to be distinguished from resistance, which is the failure of a given phage to adsorb to a bacterial surface. Mutations in the gene specifying the immunity substance give rise to phage which are unable to establish lysogeny. Such mutants give clear plaques when plated on sensitive bacteria, in contrast to the turbid plaques given by wild-type phage. In the latter case, turbidity results from lysogenization and growth of some of the cells in the area of the plaque.

Induction occurs under conditions where the level of immunity substance in the cell is insufficient to maintain the other phage genes in the repressed state. As mentioned previously, most of the common inducing agents appear to act by virtue of their ability to disrupt nucleic acid metabolism, which, in turn, can block either the formation or the activity of the repressor.

2. Transduction

An aspect of the biology of temperate bacteriophage which has been extremely useful to bacterial geneticists is the phenomenon of transduction, originally discovered in 1952 by Zinder and Lederberg (290). Because it is not directly germane to the topic of this chapter, transduction is treated only briefly here. Excellent discussions may be found in the textbooks of Hayes (291), Stent (5), and Luria and Darnell (190).

Transduction is the transfer of genetic information from one bacterium to another by a temperate bacteriophage. There are two types of transduction, general and specialized. In general transduction, any genetic marker can be transferred, albeit at very low frequency. A lysate of a generalized transducing phage, such as P1 or P22, is able to transfer a given genetic character at a frequency of the order of 10^{-5}–10^{-7} per infectious particle. The best explanation of this phenomenon is that during phage maturation, segments of bacterial DNA are occasionally packaged into virus heads. Transducing phages are equal in density to nontransducing phages of the same type, indicating that the same amount of DNA is included in each phage head. Since it is possible for several genes, representing some 2% or more of the *E. coli* genome, to be cotransduced, it is probable that at least some transducing phages contain exclusively bacterial DNA wrapped in a phage coat (*292*).

In order for a transduced gene to be hereditarily stable, it must become integrated into the chromosome of the recipient cell, by a process akin to recombination. This phenomenon is known as complete transduction. More often, the transduced marker does not become integrated but does express itself in the recipient. However, in this instance, known as abortive transduction, the transferred material fails to replicate and is transmitted to only one daughter when the cell divides.

In specialized transduction, the other major type, only certain genetic markers are transferred by a given phage. Specialized transduction is exemplified by the case of phage λ, which can bear the bacterial genes necessary for galactose utilization (*293*) or those governing the synthesis of biotin (*294*) but no others. This appears to result from the fact that the λ prophage is inserted in the *E. coli* K12 chromosome at a site between, and adjacent to, the *gal* and *bio* markers. Transduction of the *gal* genes has received the most attention to date. This results from an occasional faulty replication of a λ chromosome, such that a portion of the bacterial chromosome containing the *gal* region is replicated along with the prophage. When this occurs, a corresponding region of the phage genome is deleted. The viral progeny resulting from such an event, therefore, carry the *gal* genes but lack some essential phage functions and are unable to replicate independently. Such particles are called λ *dg* (defective-galactose). Unlike generalized transducing phage, different λ *dg* preparations have different densities, indicating that the amount of λ DNA deleted is subject to variation.

Transduction, in particular general transduction, has been of great utility for genetic mapping where no suitable system is available for bacterial mating or transformation. In addition, the existence of the phenomenon suggests a teleological reason for lysogeny. Presumably the long-range beneficial effects of the existence of bacterial viruses which can effect the interchange of genetic material outweigh the damaging short-term effects of occasional killing of a bacterial cell by a temperate phage.

3. *Viral Genes and Functions*

To an even greater extent than with the T phages, identification of biochemical function in temperate phages has been preceded by isolation of phage mutants showing over-all abnormalities in development. Physiological studies on these mutants have permitted characterization of specific metabolic functions. As with the T phages, most of the known genes have been discovered by isolation of conditional lethal mutants (*295*). These include temperature-sensitive and amber (or, to use the terminology of λ workers, suppressor-sensitive) mutations. Campbell (*295*) isolated a large number of suppressor-sensitive (*sus*) mutants and, by use of complementation tests, assigned each mutation to one of 18 cistrons, termed A, B, C, . . . R. Mutants in gene R were unable to form the virus-specific lysozyme, and this gene has been subsequently shown to be the structural gene for lysozyme (*296*). Brooks (*297*) studied the ability of various *sus* mutants of λ to lysogenize a nonpermissive host and to form a vegetative gene pool following uv induction. Her data led her to conclude that genes N, O, and P control essential early steps in DNA replication, while genes Q, R, and B acted later in the vegetative cycle. Q, like R, appears to be concerned with lysis of the host cell, while B performs some essential role in virus assembly.

Dove (*298*) and Joyner et al. (*299*) have carried out biochemical studies on cells infected with Campbell's *sus* mutants. In accord with the genetic studies of Brooks, mutants in genes N, O, and P are blocked in DNA replication. As was found for T4, these DNA-defective mutants are blocked in late functions, including formation of lysozyme, tail antigen, and mRNA late in infection. In contrast to the situation observed with amber mutants of T4, the block in late gene function is not complete. Mutants in gene Q synthesize DNA with normal kinetics but are also blocked in late functions, including mRNA synthesis. This tends to rule out a gene dosage effect as the sole causative agent leading to failure of late gene function in DNA-defective mutants, and suggests, in analogy with T4, that a structural modification of DNA might be necessary for transcription of late-acting genes.

All of the late functioning genes A–M allow DNA synthesis, but the DNA formed by mutants in genes A–F is noninfective when assayed in the Kaiser–Hogness transformation test. Dove has concluded, therefore, that genes A–F are structural genes for enzymes involved in steps of DNA maturation, such as the production of cohesive ends. Preliminary evidence suggests that J is the structural protein for phage antigen, presumably a tail protein.

Weigle (*300*) has devised an *in vitro* complementation test for formation of infectious λ particles, similar to that described for T4 by Edgar and Wood (*102*). Mutants in genes A–F are "tail donors" in this system, i.e., they produce tails active in the complementation test but not heads, while mutants in genes G–M produce heads but no tails. This would imply that one of the genes A–F is the structural gene for the head protein, which is of interest in light of the idea that in T4 the head protein might play a role in condensation of DNA for packaging. As with T4, the assembly of fully formed heads and tails appears to be a spontaneous process.

Another type of mutation, one which cannot be studied in virulent phages, leads to fully (as opposed to conditionally) defective function (*301*). Such mutations are obtained with temperate phages by appropriate mutagenic treatment of a lysogenic bacterial strain. Thus, the prophage bears a mutation and is unable to direct a cycle of virus growth after induction. Since most phage genes are not required for maintenance of the lysogenic state, the defective prophage can continue to replicate in synchrony with the bacterial chromosome. The presence of the prophage in the bacterium can be detected in various ways, perhaps most readily by the fact that bacteria lysogenic for a defective prophage maintain the immunity pattern characteristic of the infecting virus. Moreover, the defective prophage can be shown to replicate under conditions where the missing function is supplied by superinfection with a related phage.

One defective lysogen which has received considerable attention is called λ_{T11}. Radding (*302*) showed that a lysogenic strain bearing a mutation in the T11 gene, while unable to produce phage after uv induction, produced abnormally high levels of a phage-specific deoxyribonuclease which had previously been shown to be associated with vegetative development of phage λ (*303*). The T11 mutant is also blocked in DNA replication and in late functions (*298, 299*). Current evidence suggests that the T11 gene is a regulatory gene, necessary for "turning off" the synthesis of early protein and allowing subsequent developmental steps to be initiated (*304, 305*).

The temperate phage-associated DNase is of some interest, inasmuch as it is the only "early" enzyme thus far discovered (*306, 307*). Apparently the functioning of the DNase is not essential for phage multiplication. Radding, Szpirer, and Thomas (*308*) have mapped the genetic region responsible for synthesis of this enzyme. The region does not fall in any of the cistrons previously shown to be essential for virus growth. However, the enzyme may participate in some other function, such as prophage integration. At any rate, the products of other early genes necessary for phage DNA replication have not yet been characterized.

As with T-even phages, plaque-type mutations have been useful in the elucidation of the life cycle of temperate bacteriophages. The most commonly used markers are mi (minute plaque), h (host range), and c (clear plaque). The latter marker has been of great value in studies on maintenance of the lysogenic state. Clear plaques result when all of the infecting phage fail to lysogenize their hosts, such that all infected cells lyse. Clear plaque mutants of λ fall into three cistrons: c_I, c_{II}, and c_{III} (*309, 310*). There is now excellent evidence, discussed below, that c_I is the structural gene for a repressor, which acts to prevent the other genes of the prophage from functioning. Genes c_{II} and c_{III} probably act transiently to prevent bacterial nucleic acid synthesis during the establishment of lysogeny. Failure of any of these genes to function results in the absence of a lysogenic response to infection.

A final type of mutation affecting temperate phages, and one which has proven especially useful in studies on the establishment of lysogeny, gives a phenotype of decreased buoyant density relative to wild-type phage, and is called b. The most widely studied representative of this class of mutants is $\lambda b2$ discovered by Kellenberger, Zichichi, and Weigle (*311*) during purification of λ by equilibrium density-gradient centrifugation. $\lambda b2$, a deletion mutant, contains 18% less DNA than wild-type λ, accounting for its altered density. Apparently the deleted portion includes a region which is responsible for recognition of a specific site on the bacterial chromosome during the establishment of lysogeny (cf. Sec. 12-4, C, 4). Accordingly, this mutant cannot lysogenize its host, and, if a lytic response does not occur, the $\lambda b2$ chromosome remains in the cell in a nondividing state, being transmitted to only one daughter at cell division.

The order and approximate location of the above genes, as determined by recombinational analysis of mixed infections with λ, are shown in Fig. 11-10. Like the λ DNA molecule itself, the map is linear.

4. Establishment and Maintenance of Lysogeny

The most widely accepted general picture of the establishment of lysogeny is as follows. DNA enters the cell after infection. Provided that sufficient repressor is made to inactivate most of the phage genes so that a lytic cycle does not ensue, phage DNA becomes inserted into the bacterial chromosome at a specific region and becomes a prophage. That the prophage is DNA is indicated by comparison of the rates of inactivation of prophage and homologous free phage by decay of incorporated ^{32}P (312). The rates are equal, suggesting that both represent DNA "targets" of equal size. That the prophage is associated with a specific region of the bacterial chromosome is shown by genetic mapping of the prophage by means of transduction or bacterial conjugation (313). Most temperate coliphages have a unique attachment site. These sites are distributed over the entire bacterial chromosome.

There are a number of ways in which one could visualize prophage attachment to the bacterial chromosome a priori. Most phage biologists believe that the phage chromosome is inserted linearly into the bacterial chromosome. Campbell (314) has proposed a mechanism for λ whereby this might be effected. After infection the λ DNA molecule forms a covalently bonded circular structure. This becomes associated with the host chromosome at a region of specific genetic homology. A single crossing over then results in linear insertion of the phage genome. Prophage detachment after induction could occur through a reversal of the above process. An occasional faulty circularization after induction could account for transduction of the closely linked

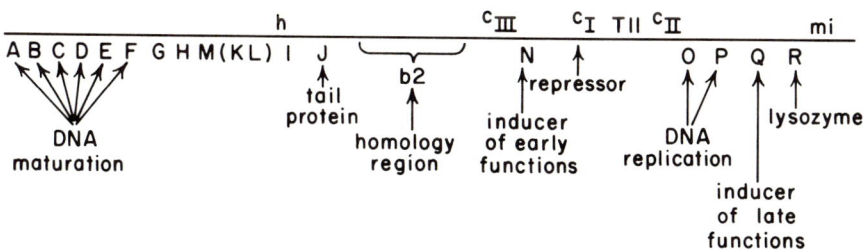

FIG. 11-10. Approximate location of markers on the vegetative map of λ. Gene products or functions are indicated below (modified from Dove, Ref. 298).

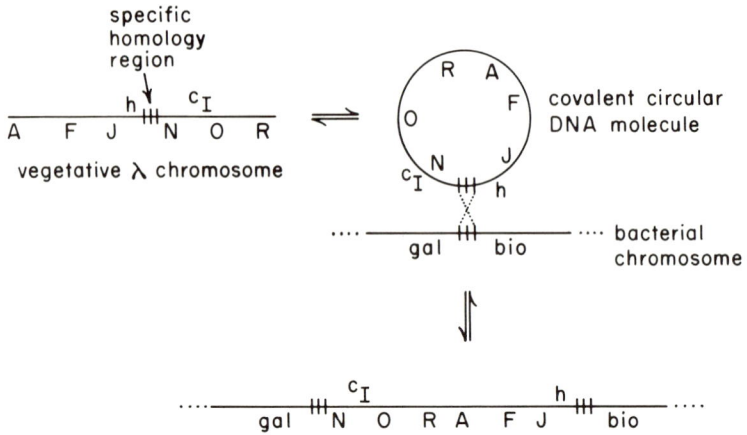

FIG. 11-11. The Campbell model (*314*) for establishment of lysogeny.

gal or *bio* genes. The basic postulates of the Campbell model are shown in Fig. 11-11.

Although direct evidence for Campbell's formulation model is still lacking, the model presents the best way of explaining a large number of experimental observations, including the following:

(1) The order of genetic markers in the λ prophage, as determined by interrupted mating experiments, is a circular permutation of the order as determined in vegetative phage (*315*). Whereas the vegetative map gives the order shown in Fig. 11-10, namely A to R, the prophage map starts at c_{III} and runs to J. Moreover, these markers are in the expected spatial relationship to the adjacent bacterial markers *gal* and *bio* (*316*).

(2) Superinfection of a lysogenic culture with a homologous phage occasionally results in formation of a double lysogen, or a bacterium containing two prophage at the same site. Genetic studies of such lysogens suggest that the superinfecting genome is inserted linearly into the already-present prophage (*317*).

(3) As mentioned above, the deletion mutant λ*b*2 behaves as though it had lost that segment of phage DNA which is homologous to a segment of bacterial DNA at the prophage attachment site. As such, the mutant can undergo vegetative growth but is unable to form a prophage.

(4) The model predicts that covalent circular molecules of λ DNA should be detectable under appropriate conditions of infection. Such

molecules have, in fact, been isolated from λ-infected sensitive (*318*) or immune (*319*) cells, and from lysogenic cells after induction (*320*). Presumably, formation of circles would involve first, complementary base pairing of the cohesive ends of the linear λ DNA molecule, followed by the formation of covalent phosphodiester bonds between the termini. As mentioned in Sec. 11-2, C, 7, an enzyme capable of carrying out the latter reaction has been described in uninfected *E. coli* (*174*, *175*). Circular DNA molecules were characterized first by observing that they sedimented more rapidly in sucrose gradients than linear molecules of the same size. They also are resistant to the action of exonucleases, and they show no strand separation upon treatment with heat or alkali. In general, it appears that circular molecules produced *in vivo* are more highly twisted than circles formed *in vitro,* as judged by electron microscopy.†

Although the precise mechanism responsible for the presumed crossover and integration is unknown, there is genetic evidence that a virus-directed enzyme is involved. Signer and Beckwith (*321*) proposed that the phage directs synthesis of a specific "recombinase." The existence of a virus-coded enzyme could explain why the formation of double lysogen of λ is a relatively rare event (*317*). Presumably the immunity substance of the first prophage would prevent the superinfecting genome from synthesizing its own recombinase. The rare instances of double lysogenization which do occur would result from nonspecific action of bacterial enzymes. In line with this reasoning, Fischer-Fantuzzi (*322*) has found that a nonimmune lysogen bearing a cryptic prophage can lysogenize another λ genome quite readily. Moreover, λb2 also lysogenizes quite readily in this system. Presumably the λb2 DNA can interact with λ DNA already present in the prophage, to account for its insertion adjacent to the first prophage. This suggests that λb2, while lacking a DNA region homologous to the bacterial attachment site, can synthesize its own recombinase in nonimmune bacteria.

† An important recent paper by Menninger et al. (*417*) provides further support for the Campbell model. λ phage with a density label and a host modification property were allowed to lysogenize a male strain of bacteria. The chromosome containing the prophage was transferred to a female strain, whereupon zygotic induction occurred, due to the absence of repressor in the recipient cells. A certain proportion of the progeny retained both the density label and the host modification property. This indicates that an entire phage genome can be attached to a bacterial chromosome during lysogenization and detached intact upon induction.

Also in line with the idea of a phage-controlled recombinase is the recent isolation of *int* mutants of λ (*323*) and the *Salmonella* phage P22 (*324*), which cannot become integrated as prophage. Unlike λ*b*2, which exhibits the same phenotype, the *int* mutants are revertible, indicating that they bear point mutations. Moreover, some of the mutants are temperature sensitive (*324*), suggesting that the gene product is a protein, presumably the recombinase. Finally, mixed infection with λ *int* and λ*b*2 leads to integration of the *int* genome, in line with the idea that λ*b*2 is able to synthesize the recombinase (*323*).

Once the prophage is established, how is lysogeny maintained? Obviously, this must involve repression of those genes whose action is necessary for vegetative phage growth. In 1961, Jacob and Monod (*73*) published their classic paper on genetic regulation of protein synthesis. In formulating a general theory of regulation, they took into account genetic data on the clear-plaque mutants of λ, which, as we have seen, are unable to lysogenize their hosts (*309*, *310*). Jacob and Monod proposed that the c_I gene is the structural gene for a repressor, or immunity substance. This was considered to be a protein which binds to a specific site on the prophage and prevents expression of most viral genes by blocking their transcription. The theory is formally the same as that proposed for the control of closely linked bacterial genes, such as the *lac* "operon," or the group of genes whose action is responsible for fermentation of lactose.

A large body of indirect evidence is in accord with the existence of a protein repressor in lysogenic cells. This includes the existence of temperature-sensitive c_I mutants (*325*) (presumably these synthesize a heat-labile repressor), and the fact that in mixed infections the clear-plaque phenotype is recessive to wild type, suggesting that the c_I gene product can diffuse and act on another genome. Ptashne has recently performed a brilliant experiment (*326*), which provides the first direct evidence for existence of a λ repressor. Bacteria bearing a prophage were infected with a λ strain bearing an amber mutation in gene c_I, such that no c_I gene product is synthesized. ^{14}C-Leucine was added to label any proteins formed after infection. A parallel culture was infected with an otherwise isogenic λ strain which bore a wild-type c_I gene. ^{3}H-Leucine was added to the latter culture. In both cultures the immunity provided by the already-lysogenized phage genome prevented expression of most of the superinfecting phage genes, with the exception of c_I. Moreover, both of the superinfecting phage strains bore mutations in gene N, a very early-acting gene whose action appears to be necessary for synthesis of most phage-specific proteins. Finally, bacterial protein synthesis in the two cultures was suppressed by massive uv irradiation of the bacteria before superinfection (because of this, a noninducible prophage was present in the bacterium). After a period of protein labeling, the two cultures were mixed, and the c_I gene product was isolated and fractionated on the basis of the assumption that it was the only protein labeled with ^{3}H but not with ^{14}C. It proved to be a protein, of molecular weight about 30,000.

Ptashne tested his repressor preparation for its ability to bind specifically to λ DNA, as measured by co-sedimentation of repressor and DNA in sucrose gradients (327). To make his test critical, he isolated DNA from wild-type λ and from λ imm^{434}, a related phage derived from recombination between λ and the related phage 434. This strain contains the immunity region of 434, but virtually all of its other genes are those of λ. Thus, λ imm^{434} is sensitive to the 434 repressor only. As expected, the repressor binds very tightly to λ DNA, but binding to λ imm^{434} DNA was undetectable. Although the precise mode of action of repressor is still not known, this system presents an unparalleled opportunity to clarify the general problem of regulation of gene activity.

As mentioned previously, there are three genes giving the c phenotype, with c_I being the structural gene for the repressor. Apparently the c_{II} and c_{III} genes function transiently during the establishment of lysogeny. Levine and Smith (328, 329) have studied the biochemistry of nonlysogenizing mutants in the *Salmonella* phage P22. Only two c genes appear to function in this system, as opposed to the three in λ. By use of temperature-sensitive c mutants and appropriate temperature-shift experiments, Levine and Smith could identify the time periods during which each gene must function during the establishment of lysogeny. One gene functions from the 7th to the 11th minute after infection, while the other, presumed to be the structural gene for repressor, begins to function at about 16 min, and its continued action is necessary for maintenance of lysogeny. Smith and Levine (329) have devised a technique for following syntheses of viral and bacterial DNA independently. In normal infection the rate of viral DNA synthesis increases until about 6 min and then decreases, reaching zero near 16 min. In infection by the early-acting c gene mutant, the inhibition of phage DNA synthesis is not observed at 6 min. DNA synthesis continues until lysis occurs, at about 25 min. Mutants in the later-acting gene show the early inhibition, but massive DNA synthesis starts at 16 min and continues throughout a 50-min latent period.

The c mutants of λ are to be distinguished from λ_{vir}, a strain which is also unable to establish lysogeny, and which gives the clear plaque phenotype. λ_{vir} is a rather complex multiple mutant whose phenotype, unlike that of c mutants, is dominant in mixed infections with wild-type λ. Lisio and Weissbach (330) have presented evidence that λ_{vir} functions are partially, but not completely, sensitive to λ repressor.†

From the above discussion, one could infer that a lysogenic cell becomes induced when the level of repressor is not sufficiently high to maintain lysogeny. This could occur a priori either through inactivation of repressor already present or through failure of the repressor to be synthesized. Ogawa and Tomizawa (331, 332) favor the former view. They have studied bacteria infected by $\lambda b2$, which, as stated earlier, cannot establish lysogeny. Cells which do not undergo a lytic response upon infection bear the $\lambda b2$ chromosome in a non-dividing state. As such, it is segregated out at cell division, being transmitted to only one cell. Immunity, however, persists for about four generations after segregation, indicating that the repressor is growth stable. Blockage of repressor formation, therefore, would not be expected to give the essentially immediate lytic response which is observed.

As a corollary to their work, Tomizawa and Ogawa (333) have studied the multiplication of rII mutants of T4 in nonlysogenic, immune cells obtained after segregation out of a $\lambda b2$ chromosome. They have concluded that the λ repressor, and not the presence of the prophage per se, is the agent responsible for inhibition of rII multiplication in cells lysogenic for λ.

Assuming that induction does act via inactivation of repressor, one is faced with the question of the nature of the inactivation. In the case of a heat-inducible prophage, the repressor, presumably a temperature-sensitive protein, is denatured at high temperature. However, as stated earlier, most of the general inducing agents appear to interfere with DNA metabolism. Goldthwait and Jacob (334) proposed that in

† Ptashne and Hopkins (418) have recently shown that λ_{vir} contains mutations in its operators which decrease the affinity of its DNA for the λ repressor in an in vitro system. This system has been further exploited to show that the repressor binds to at least two separate operators, which independently control separate operons.

the absence of DNA replication some DNA precursor builds up in the cell and binds to the repressor, causing inactivation. In this context it is of interest that λ lysogenized in a recombination-deficient mutant of *E. coli* is defective in its ability to be induced by uv light (*335*). Presumably the Rec- mutation in the bacterium involves an alteration of nucleic acid metabolism which also affects a vital step in the initiation of vegetative growth.

5. *DNA Synthesis in Vegetative Growth*

Most of what we know about λ DNA replication has been learned within the past two years. Dove and Weigle (*336*) found that DNA isolated from cells undergoing vegetative growth is inactive in the Kaiser–Hogness infectivity test, as was DNA isolated from lysogenic bacteria. This was true whether vegetative growth was initiated by induction or by infection with a c_I mutant. Conversion of infecting phage DNA to the inactive form took place immediately after infection and did not involve phage-specific proteins. Infectious DNA was formed late in infection with the aid of virus-specific proteins. These findings suggested that a series of structural modifications in DNA takes place during the infective process.

Weissbach and his co-workers and Smith and Skalka have examined the molecular species of DNA present at various times after induction of lysogenic bacteria. The basic approach is to label replicating DNA with ^{3}H-thymidine, isolate the DNA, and observe the sedimentation patterns in sucrose gradients. Circular λ DNA molecules have been detected after induction of a lysogen (*320*). However, only a small proportion of the total labeled DNA is found as circles. Moreover, the circles are metabolically stable; they are not packaged into mature phage particles. On the other hand, they are probably products of replication since *sus* mutants in genes N, O, and P appear unable to form them after induction of suitable lysogens (*337*).

Although circular DNA molecules do not appear to be direct intermediates in the formation of mature phage particles after induction, there remains the possibility, as discussed previously, that they play a direct role in lysogenization. Salzman and Weissbach (*338*) measured the fraction of labeled input DNA which becomes circularized after infection of sensitive or immune cells. The proportion was similar whether infection was carried out at high or low multiplicity, suggesting that circularization does not, as might be envisaged, represent a mechanism for disposal of excess phage genomes.

Both Smith and Skalka (*339*) and Salzman and Weissbach (*340*) have obtained evidence for the existence of concatemers (end-to-end DNA molcules several phage units in length) as replicative intermediates in vegetative growth. These are characterized by their high sedimentation rates, as compared with mature λ DNA, and by their extreme sensitivity to shear. This presents an interesting parallel to studies on T4 DNA replication, discussed in Sec. 11-2, D. Pulse-chase experiments indicate that the first replicative intermediate formed has the same sedimentation rate as mature λ DNA (*341*). This is then converted to a more rapidly sedimenting form and, shortly before lysis, is again converted to a form having the same sedimentation rate as mature λ DNA. The latter form is probably the infective DNA observed by Dove and Weigle to be formed toward the end of the latent period. λ *sus* mutants in genes A or D, which, as we have seen, are unable to effect DNA maturation, accumulate large amounts of the presumed concatemer, and are unable to convert them to mature phage DNA.

Weissbach and Salzman (*341*) have recently presented evidence that the early replicative intermediate—that which cosediments with DNA—does not contain the 5′-single-stranded, or cohesive, ends characteristic of mature phage DNA. Mature DNA bearing a ^{14}C label was mixed with ^3H-labeled early-replicating DNA, and the mixture was subjected to annealing conditions. The ^{14}C-labeled material sedimented more rapidly after annealing than before, while the sedimentation rate of the ^3H-labeled material did not change. Thus, it did not appear to undergo circularization, as would be expected by the presence of cohesive ends.

A tentative and partial scheme for λ DNA replication after induction, based upon the work of Weissbach and Salzman, is as follows:

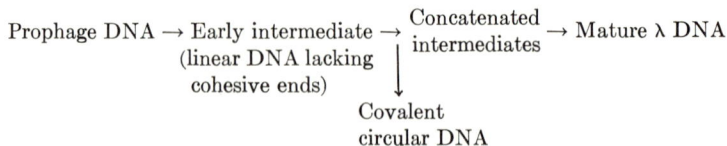

Prophage DNA → Early intermediate → Concatenated intermediates → Mature λ DNA
(linear DNA lacking cohesive ends) ↓
Covalent circular DNA

A number of aspects of the scheme remain unsettled,† particularly the mechanism of formation of the early intermediate. If the Camp-

---

† In fact, much of the scheme is contradicted by recent work of Young and Sinsheimer (*419*) and of Ogawa, Tomizawa, and Fuke (*440*), who have presented evidence that covalent circular forms of λ DNA are intermediates in replication.

bell model for lysogenization and prophage detachment is correct, then one would expect a covalently bonded circle to be at least a transient intermediate in the formation of early replicating DNA, yet those circles which have been detected do not appear to be replicative intermediates. Information on the order of genes in the early intermediate would obviously be highly informative, but the gathering of such information could present formidable technical difficulties.

6. *Regulation of Gene Expression*

Because the physical positions of genes on the λ DNA molecule have been shown to correspond with genetic map positions (*281, 289*), the λ system presents favorable material for studying the thesis that temporal control of gene expression results from selective transcription of those genes which are being expressed. One can further ask whether the entire complement of genetic information resides on one or both strands of the DNA molecule. Two properties of the λ DNA molecule greatly facilitate such studies. First, as stated previously, the molecule can be split into two halves by controlled shearing. One knows which genes reside on the right half of the molecule and which on the left. Because the base compositions of the two halves differ, they can be separated by density-gradient centrifugation (*284*). Secondly, one strand contains a number of deoxycytidine-rich clusters (*342*) and thus binds polyguanylate relatively tightly. This increases the density of the strand (the so-called C strand) to the point where it can be separated from the complementary W strand in a density gradient.

Examination of the genetic map of λ (see Fig. 11-10) reveals that, for the most part, the early-acting genes are clustered at the right-hand portion of the molecule, the late-acting genes are found to the left, and those genes governing lysogenization are found in the middle. One might expect, therefore, that mainly right-hand regions of the molecule would be transcribed early in vegetative growth, in correspondence with the genes being expressed at that time. A number of investigators (*343–346*) have tested this idea. The general approach is as follows. RNA is pulse labeled at various times during phage development. The product is tested for its ability to hybridize with W strands, C strands, right halves, or left halves. Taylor, Hradecna, and Szybalski (*346*) used a somewhat different approach to determine which half is transcribed: They prehybridized their mRNA to DNA from a λdg which bears only the left-hand λ genes. To date the results of Taylor et al. (*346*) are most definitive, but results from the other laboratories are in substantial agreement.

The major findings of Taylor et al. are as follows: In a noninduced lysogen, 0.03–0.6% of the total pulse-labeled RNA is phage-specific. Presumably this represents mRNA corresponding to the c_I region. It is transcribed mostly from the W strand. Early after induction the above figure rises to about 2%, and at later periods in induction 6–12% of the pulse-labeled RNA hybridizes with λ DNA. Thus, induction of prophage development represents a true gene derepression.

The earliest RNA formed after induction hybridizes with the W strand and comes mostly from the right half. At later periods in the growth cycle 85% of the mRNA is transcribed from the C strand, progressively more and more from the left arm. As expected, mutants defective in early functions O and P continue to synthesize RNA mainly from the W strand. Thus, the nature of the mRNA synthesized at different periods of phage growth appears to correlate quite well with the genes known to be functioning at each period. It is also of interest that both strands apparently contain meaningful genetic information, in contrast to the situation observed earlier with several smaller phages (*347–349*).

Cohen, Maitra, and Hurwitz (*350*) have used the above-described approach to study the RNA transcribed from λ DNA *in vitro* by purified RNA polymerase. The right half is transcribed predominantly *in vitro*, and no change in the pattern is observed with time. Apparently a "switch" which acts *in vivo* to direct transcription from one strand to its complement fails to act *in vitro*. The nature of this (or these) switch is a question of considerable interest to Szybalski and his associates (*346*).

Skalka, Butler, and Echols (*351*) have carried their studies on selective transcription to the point of examining the behavior of λ mutants which are defective in genes presumed to play a regulatory role. Failure of gene N to function results in extremely limited λ-specific mRNA synthesis. As had been previously proposed (*305*), the action of this gene appears to be required very early for the subsequent expression of almost all other genes. A gene Q mutant, which, it will be recalled, can synthesize DNA but not late proteins, is defective in its ability to transcribe the left-hand portion of the DNA molecule. Finally, a defective prophage bearing the T11 mutation synthesizes an abnormally high level of mRNA throughout the period studied. This RNA is transcribed almost exclusively from the right half. It will be recalled that this defective lysogen overproduces early proteins, including the λ-specific DNase, but DNA synthesis and late protein formation do not take place to a significant extent (305). Thus, gene transcription appear to correlate well with gene expression in these three regulatory mutants.

11-5. Single-Stranded DNA Bacteriophages

Two groups of small coliphages are attracting considerable attention because, being the smallest known DNA-containing biological systems, they present an excellent opportunity to effect a complete correlation of genetic structure with biological function. These include the two closely related tailless phages φX174 and S13, and the filamentous male-specific phages f1, fd, and m13. All of these phages are characterized by the presence of a single molecule of DNA which is single stranded (352) and circular (353). The DNA is in all cases quite small, having a molecular weight of about 1.7×10^6 in φX174 and S13, and 1.3×10^6 in the filamentous phages. This is sufficient to code for about seven to ten proteins of average size. Thus, it becomes feasible with these viruses to identify every gene and gene product. Such identification, of course, will not necessarily lead to complete elucidation of the viral life cycle, since these phages with their small complement of genetic information, undoubtedly rely more on bacterial functions for their replication than do the larger phages we have been discussing (354).

Phages f1, fd, and M13 comprise a relatively recently discovered class of viruses which exist as long filaments, some 8500 Å long by 60 Å in diameter (355, 356). These viruses will only infect male strains of E. coli, i.e, strains which can function as donors in bacterial mating. They have the unusual property of not killing the host cell when virus growth begins. The infected cell continues to grow and divide, with mature virus being released, not by lysis of the cell, but by extrusion through the cell wall.

By far the most extensively studied of these small DNA phages, in terms of biochemistry of replication, is φX174, which has been investigated primarily by Sinsheimer and his associates (354). The related phage S13 has been subjected to extensive genetic analysis by the Tessmans (357, 358). Because of the similarities between the two phages, the biochemical and genetic studies have been mutually reinforcing, and attainment of the goal of identification of all viral genes and functions appears a definite possibility. Seven genes have been identified in S13, on the basis of complementation studies with a large number of conditional lethal mutants (*358*). This appears to account for between 70 and 100% of the entire complement of genetic information. Of the seven genes, two are involved in determination of coat protein structure, one is concerned with the synthesis of a replicative form of DNA (see below), one is concerned with lysis, and three are of currently unknown function, possibly the formation of mature single-stranded DNA from the replicating form. Insofar as they have been studied, the genetics and the molecular details of viral replication appear to be quite similar for all of the small DNA phages.

Shortly after infection by ϕX174 the parental DNA is converted to a "replicative form" (RF). Like the single-stranded parental DNA, the RF is infective to spheroplasts of *E. coli* (*359*). Determinations of density, heat denaturation, and uv-sensitivity suggested that the RF is a double-stranded DNA molecule. This was confirmed by electron microscopic examination (*360*) and by the resistance of RF to attack by an exonuclease (*361*). Apparently one of the first functions of the parental DNA, therefore, is to prime the synthesis of a complementary strand to give a closed circular molecule. Most of the circles are twisted and contain no single strand breaks (*362*). These early-synthesized RF molecules serve as templates for synthesis of more RF molecules by semiconservative replication. The RF molecules also serve as templates for transcription, with the mRNA formed *in vivo* being complementary to that DNA strand of the RF which is complementary to parental DNA (*349*). Thus, ϕX mRNA is identical in base composition and sequence to mature ϕX single-stranded DNA.

At about 8 min after infection, single-stranded DNA begins to be formed. Little is known about the nature of this process, but it is known that RF DNA is not converted to single-stranded (SS) DNA. Rather, all of the SS DNA which appears in progeny phages arises from the medium (*363*). As is discussed below, only a limited number of RF molecules participate in SS DNA synthesis.

In light of the expectation that host-cell functions are involved to a greater extent in the replication of small viruses than in large ones, it is of some interest to know the effects of ϕX174 infection on bacterial metabolism. Rueckert and Zillig (*364*) reported that there was no effect of infection on the synthesis of host-cell DNA, RNA, or protein. Using a lysis-defective virus mutant, which prolongs the latent period, Lindqvist and Sinsheimer (*365*) found that, while bacterial RNA and protein syntheses are not affected, the formation of bacterial RNA and protein syntheses are not affected and the formation of bacterial DNA ceases at 12–14 min. This effect appears to be mediated by a phage-specific protein. It is of interest that this effect coincides with the shutoff of RF DNA synthesis and the initiation of SS DNA synthesis. Since the over-all rate of DNA synthesis increases markedly at this time, it appears that the phage-directed turn-off of bacterial DNA synthesis could allow the diversion of the bacterial replicative apparatus to viral needs at precisely the time when it can be most effectively utilized.

A body of evidence is accumulating in favor of the idea that ϕX174 replication occurs only at a limited number of sites in the infected bacterium. While the nature of these sites, and the enzymatic functions they perform, must be considered as matters for speculation, it would appear that enzymatic machinery for bacterial DNA repair and/or replication is involved (*354*). It has been known for some time that parental DNA is not transferred to progeny phages. That the RF DNA which contains parental material plays a special role in replication was suggested by an experiment of Denhardt and Sinsheimer (*366*). Cells were infected with ^{32}P-labeled phage of high specific activity, and incubation was continued for a short period in nonradioactive medium, such that several new RF molecules per cell were synthesized. The cells were stored frozen for a period sufficient to inactivate the parental DNA-containing RF's. Upon thawing the cells were found to be unable to produce viral progeny, even though they contained newly synthesized, unlabeled RF. Thus, the RF plays a special role, or occupies a special site, which cannot be taken over by new RF molecules.

Yarus and Sinsheimer (*367*) used a genetic approach to gain further evidence for existence of the essential site. Cells were infected simultaneously with four different mutants, and the number of genotypes occurring in single bursts was examined to get an idea of the number of genomes being replicated per cell under various conditions. The ^{32}P inactivation experiment of Denhardt and Sinsheimer had been performed with starved cells, in order to synchronize the infection. Yarus and Sinsheimer found that usually only one genome is replicated in such cells, suggesting the existence of only one site per cell or, at the most, two. Log phase cells supported the replication of up to four genomes. Those genomes not being replicated did appear to form RF molecules. Therefore, it would appear that the essential bacterial step occurs after RF formation. Denhardt, Dressler, and Hathaway (*368*) have presented evidence implicating a bacterial repair or replication enzyme at this step. They have isolated an *E. coli* mutant which can adsorb ϕX174 but cannot support its replication. Parental DNA is converted in the mutant to RF which is active in the spheroplast infectivity test. However, the RF does not replicate further. The replication-deficient mutant is also deficient in its ability to undergo recombination.

11-6. Bacteriophages Infecting *Bacillus subtilis*

The discovery of DNA-mediated transformation of biochemical characteristics in *Bacillus subtilis* (*369*) made this organism attractive as a system for genetic investigation. To facilitate such work, a number of phages active against *B. subtilis* have been isolated and characterized (*370*). For the most part, DNA from these phages is infective, i.e., it can be taken up by cells competent to undergo genetic transformation and direct the formation of infective virus units (*371*). This "transfection" process is to be distinguished from infectivity systems developed for coliphages, which either depend upon prior conversion of cells to spheroplasts or upon preinfection of recipient cells with helper phage. It has even been reported that competent *B. subtilis* cells can take up DNA from polyoma or vaccinia virus and produce infectious viruses (*372, 373*). These results, however, have not received universal acceptance (*190*).

Investigation of a number of transfection systems has revealed that some viral DNA's exhibit a first-order relationship between DNA concentration and infective centers formed in the test (*374*), while with some other DNA's, plaque-forming ability is proportional to the third or fourth power of DNA concentration (*375*). The latter result indicates that a number of DNA molecules must interact inside the infected cell in order to produce a plaque, even though viral DNA was isolated in an undegraded condition. Current evidence (*376, 377*) suggests that transfecting DNA is partially degraded and that fragments from several molecules must recombine in order to yield a single infective unit. The reason why some DNA's exhibit this response, while others apparently do not is not clear. As discussed below, a number of the *B. subtilis* phages have DNA of aberrant base composition, which may result in insensitivity to attack by bacterial nucleases. It can be recalled that T-even phage DNA is resistant to attack by *E. coli* nucleases. To the writer's knowledge, no correlation of DNA structure with infectivity in transfecting phage DNA's has been published.

The abovementioned discoveries of unusual base compositions of *B. subtilis* phage DNA's have opened up new areas for investigation of how a virus alters the metabolism of its host cell. Phage SP2 contains uracil in its DNA instead of thymine (*378*). Kahan (*379*) has reported briefly on new enzyme activities formed in cells infected by this virus. These include a specific phosphatase which cleaves thymidylic acid, and a kinase which phosphorylates dUMP to dUDP. Thus, the virus apparently contains its own enzymatic machinery for excluding thymine from its DNA and incorporating uracil, in distinct contrast to the situation which obtains in cellular organisms.

The DNA of the transducing phage PBS2 also contains uracil (*378*), and it has been reported to be glucosylated (*380*). Takahashi and Marmur believe that glucose is bound through guanine and cytosine residues, but little is known about the enzymology of glucosylation.

Another group of *B. subtilis* phages, typified by SP8, contains 5-hydroxymethyl uracil completely substituted for thymine (*381*). This is also glucosylated to the extent of about one mole per mole of phosphorus (*382*). The glucose is bound covalently in some unknown manner. Interestingly, a host-range and temperature-sensitive mutant of SP8 contains mannose instead of glucose. Kahan, Kahan, and Riddle (*383*) have, in a preliminary communication, reported on new enzyme activities in SP8-infected cells which can account for the replacement of thymine by hydroxymethyl uracil. These include a dCMP deaminase and an enhanced activity of dUTPase. A similar activity which cleaves dTTP to dTMP was also detected. Hydroxymethyl-dUMP was formed via a tetrahydrofolate-dependent dUMP hydroxymethylase, analogous to the dCMP hydroxymethylase characteristic of T-even phage infections. Roscoe and Tucker (*384*) have examined the enzymology of DNA synthesis in cells infected by φe, a phage related to SP8 in that its DNA contains hydroxymethyl uracil instead of thymine, but different in that its DNA is not glucosylated. Like SP8, this phage also induces a dCMP deaminase and a dUMP hydroxymethylase. However, φe manifests a different mechanism for exclusion of thymine from DNA. Infection by φe leads to formation of a protein molecule which acts as a specific and irreversible inhibitor of thymidylate synthetase (*385*). Further studies on the mode of interaction between these two proteins should be of considerable general interest, aside from considerations of viral metabolism.

Still another mechanism for excluding thymine from phage DNA has been put forth by Aposhian and Tremblay (*386*). These investigators have studied phage SP5C, another virus which contains hydroxymethyl uracil in its DNA. This phage induces a highly specific dTMP phosphatase, similar to that found in SP2-infected cells by Kahan. Since the three groups of workers who have studied the enzymology of thymine replacement by hydroxymethyl uracil have used three different phages, it will be of some interest to see how many of the three mechanisms are utilized by any one phage, and to see whether any of these mechanisms plays any role other than exclusion of thymine from DNA.

A final phage-specific enzyme has been found in *B. subtilis* infected with SP3, a phage with apparently "normal" base composition (*387*). This is a DNase which prefers heat-denatured DNA to native DNA as a substrate.† The role of the enzyme is unknown.

Pène and Marmur (*388*) have initiated studies on the control of gene functions in *B. subtilis* infected by phage 2C, another virus containing hydroxymethyl uracil instead of thymine. Studies on mRNA formation in this system are facilitated by the fact that *B. subtilis*, unlike *E. coli*, is permeable to actinomycin D. The results of Pène and Marmur are similar to those obtained with the T-even phages, i.e., DNA replication is required for expression of late functions, and mRNA for these late functions appears to be transcribed only from newly replicated DNA.

11-7. Concluding Remarks

The above discussions have been limited to phage–host systems which have been investigated from the standpoint of metabolism of the infected cell. The past few years have witnessed an awakening interest in isolation and characterization of new bacteriophages. It is hoped that these, as well as the better known phages, will receive increased attention from the standpoint of replication. Our present state of knowledge reveals remarkable diversity in the ways in which

---

† This enzyme also has the unusual property cleaving DNA sequentially to give a product composed almost entirely of dinucleotides (*422*).

a virus parasitizes its host. Further knowledge of this diversity should present more opportunities to learn about the metabolism of the host cell itself, as well as the nature of virus replication *per se*.

ADDENDUM

Since the initial draft of this chapter was completed, a number of important papers dealing with DNA phages have been published, and it seems appropriate to briefly describe some of the most interesting. The problem of regulation in T4-infected bacteria continues to be an object of intense study in several laboratories. Geiduschek and his collaborators (*389–391*) have used competitive DNA–RNA hybridization experiments to study the control of transcription *in vivo* and *in vitro*. An *in vitro* system capable of synthesizing late species of mRNA has been developed (*389, 390*), and it has been shown that transcription of these species requires the presence of an active product of gene 55. Thus, gene 55, which had previously been implicated in the maturation process (*92*), appears to direct the formation of a product which acts as a positive control element for transcription of late genes. At this writing, the precise nature of the gene 55 product is unknown.

Bolle, Epstein, Salser, and Geiduschek (*391*) have performed an exhaustive study of the synthesis and stabilities of T4mRNA species synthesized *in vivo*. The general findings of Hall, Nygaard, and Green (*218*) were confirmed, and it was further shown that there are no inherent differences in metabolic stability between early and late species [similar conclusions have been reached by Greene and Korn (*392*)]. No late mRNA species were formed early, in contrast to the observations of Bautz et al. (*220*) on the synthesis of lysozyme-specific mRNA. Finally, in studies not yet reported in detail, these workers have demonstrated the existence of at least three different classes of early RNA, whose syntheses begin at different times early in infection. This finding appears to complement the observations of Levinthal et al. (*235*) on the existence of several classes of early proteins.†

---

† Bolle et al. have extended their competitive hybridization studies (*420*) to show that T4 DNA replication is a prerequisite to transcription of late mRNA species. Thus, DNA-negative amber mutants cannot synthesize late classes of RNA. Similar results have been obtained by the author (*421*). However, although DNA synthesis is a necessary condition, it is not sufficient, for maturation-defective mutants (genes 33 and 55), while able to synthesize DNA, cannot form late RNA species.

Guha, Salser, and Szybalski (*393*) have accomplished the formidable technical feat of separation of the strands of T4 DNA, by taking advantage of the differential binding of polyuridylic acid by the two strands. DNA–RNA hybridization studies of pulse-labeled RNA and the isolated strands show that one strand is transcribed exclusively during the first 10 min after infection, while both strands are transcribed at later times. Competitive hybridization experiments indicate that all of the late mRNA species are transcribed exclusively from the strand which is *not* transcribed early. So far these findings are consistent with other data, primarily genetic, on the direction of transcription (i. e., clockwise or counterclockwise) of certain T4 genes.†

Another approach to regulation of phage protein synthesis involves development of appropriate *in vitro* protein-synthesizing systems. Salser, Gesteland, and Bolle (*394, 395*) have observed T4 lysozyme activity in an amino acid incorporating system programmed with RNA extracted from T4-infected cells. Moreover, when the RNA was extracted from cells infected with a T4 strain bearing an amber mutation in the structural gene for lysozyme, no lysozyme was produced unless the other components of the system were derived from a bacterial strain carrying a suppressor mutation. Therefore, this system provides a sensitive assay, both for *in vitro* expression of a viral gene and for suppression.

Celis and Conway (*396*) have described an *in vitro* amino acid incorporating system which is dependent upon added T2 DNA. Gel electropherograms and peptide maps of the products are similar in some respects to the corresponding patterns of proteins formed early after T2 infection *in vivo*. However, to date no specific early enzyme activities have been detected in the *in vitro* synthesized product.

Elements of the translation apparatus of T4-infected cells have been studied for further clues to the mechanism of the virulent parasitism of this virus. Neidhardt and Earhart have continued their study of amino acyl-tRNA synthetases after infection. These workers had previously demonstrated (*397*) an altered valine-activating enzyme which appears after infection. More recently (*398*), they have found

† Kasai, Bautz, Guha, and Szybalski (*423*) have extended these findings to show that the mRNA's specific for the rII function and the *e* gene (coding for lysozyme) are both transcribed from the same strand, namely the strand which is normally transcribed early in infection. Thus, although the *e* gene is classified as "late" in terms of its function, it does have some properties of early genes.

that no postinfection alteration occurs in the activating enzyme patterns for several other amino acids; moreover, they have demonstrated rigorously that an unmodified phenylalanine-activating enzyme of the host must be present for successful infection to occur.

Hsu, Foft, and Weiss (399, 400) have used radiosulfur labeling to study tRNA patterns before and after T-even phage infection. A new ³⁵S-labeled tRNA appears after infection with T4. This species hybridizes specifically with T4 or T2 DNA, indicating that it is a viral gene product. Present data suggest that this new species has acceptor activity for leucine. Thus, it is quite possible that it is the same as the new leucine-specific tRNA studied earlier in the laboratories of Sueoka (228, 229) and Novelli (230). The metabolic role of this new species is as yet unknown.

Frankel (401) has reported new data concerning the nature of replicating T4 DNA. In alkaline sucrose gradients, material is seen which sediments two to three times more rapidly than strands of denatured T4 DNA. Sensitivity of this material to shear suggests that it consists of concatenates of single T4 DNA strands.

Wood et al. (402) have used the in vitro complementation approach to study an enzyme-like factor whose action is necessary for the ultimate step in the assembly of individual T4 virions, namely, the joining of completed heads to completed tails. Mutants mapping in cistron 63 are unable to induce formation of the factor. Interestingly, although the action of the factor is definitely a late function, the factor itself is synthesized at a constant rate from the onset of infection.

Kozloff and his colleagues (403) have pursued the characterization of structural pteridine compounds in T4 to the point of identifying the T4D pteridine as dihydropteroylhexaglutamate. In a search for proteins which might bind this compound in intact virions, they have obtained preliminary evidence that dihydrofolate reductase is a structural component of T4D. This is somewhat surprising in light of the fact that the presence of the viral reductase is not absolutely required for phage growth (125).

Coliphage λ is becoming an increasingly popular experimental system, both for studies on regulation of gene expression, and for correlation of chromosome structure and gene function. The Campbell model of prophage integration has received support from the isolation, in at least three laboratories (323, 404, 405) of point mutants unable to integrate into the host chromosome. Gingery and Echols (404) have mapped several of these mutants and find that they occupy two different regions of the chromosome; one set of mutants maps immediately to the right of the λb2 region, believed to be the specific homology region necessary for recognition by λ of its attachment site. Presumably, at least one of the two integration genes governs the formation of a "recombinase," which functions both in prophage insertion and detachment after induction.[†]

Studies on the functions of genes c_I and N have provided further clues to the mechanism of maintenance of lysogeny and early events following the lifting of repression. Echols, Pilarski, and Cheng (406) have shown that partially purified λ repressor (326) inhibits transcription of λ DNA *in vitro*, a direct confirmation of the predictions of the Jacob–Monod model for regulation of gene expression. In addition, Green et al. (407) have provided indirect evidence that the λ repressor also blocks λ DNA replication.

Konrad (408) has analyzed the function of gene N and has presented evidence that a protein product of this gene must be formed during vegetative growth before any other λ genes can be transcribed.[‡] The protein is apparently unstable metabolically.

Howard (409) has identified a new λ gene, which is responsible for exclusion of rII mutants of T4 in lysogenic cells. It had previously been thought that the immunity substance, or c_I gene product, is responsible for rII exclusion. Now Howard has isolated exclusion-deficient mutants, which map between genes N and c_I. Since rII exclusion occurs in the lysogenic state, Howard's work shows that c_I is not the only gene to be expressed during lysogeny.

Doerfler and Hogness (428) have presented a new approach to the question of gene orientation in phage λ. Strands of λ DNA were separated (429) and heteroduplex molecules formed by renaturation with one strand from wild-type DNA and the complementary strand from a mutant bearing an amber mutation in gene N. Under experimental conditions which prevented excision and repair of the non-complementary region, the heteroduplex molecules were tested for N gene activity in the Kaiser–Hogness transformation system, and the

[†] Further studies (424–426) have shown that one class of mutants is defective in vegetative recombination but not prophage insertion and excision, while the other class is defective in integration but not in general recombination. Presumably, the defective gene product in the latter class of mutants is a site-specific enzyme which only functions in integrative recombination.

[‡] Radding and Echols (427) have presented genetic data that the N gene product acts to initiate and/or support the synthesis of two early λ proteins, the exonuclease (308) and the so-called β protein (305).

template strand for the N gene product was identified as that which is transcribed from right to left.

Davis and Davidson (*430*) have designed an approach to the physical mapping of deletion mutations in phage λ, an approach which should have general applicability. DNA from wild-type phage and from a mutant bearing a deletion are mixed, denatured, and renatured. Heteroduplex molecules, which contain one wild-type and one mutant strand, contain a "bush," or collapsed single-stranded region consisting of wild-type DNA which has no complement in the area of the mutation. The bushes can be visualized in the electron microscope. Thus, one can determine both the position and the extent of the deletion.

Meselson and Yuan (*431*) have begun to approach the problem of host-controlled restriction and modification at the molecular level. As an example of this phenomenon, λ which are grown on *E. coli* B are unable to grow on K strains of *E. coli;* DNA injected under these conditions is rapidly degraded. However, the DNA of λ grown in K is modified by the host in such a way that it resists degradation upon subsequent growth of K. Meselson and Yuan have isolated an unusual endonuclease from strain K, an enzyme which will degrade DNA from unmodified phage but fails to recognize modified DNA. Interestingly, this enzyme requires S-adenosylmethionine, ATP, and magnesium ions. The nature of the cofactor requirement is not known. A very similar enzyme has been isolated by Linn and Arber (*432*), in studies on a restriction system for the single-stranded DNA phage fd.

Marmur and his colleagues (*433, 434*) have reported on a fascinating temperate phage, PBSX, of *Bacillus subtilis*. The phage, which is induced following mitomycin treatment, is defective, being unable to replicate when added to a bacterial culture, and, in fact, not even being able to inject its DNA. Hybridization and transformation studies have shown that the DNA which is packaged into phage heads is almost entirely bacterial in origin. This is not "accidental" packaging of bacterial DNA, as seen in the generalized transducing phages, for induced cells have a mechanism which reduces bacterial DNA specifically to fragments of about 22S, the same size as that isolated from purified phage. This seems to be similar to the "headful cutting mechanism" postulated in T4-infected cells (*197*). Taxonomically, PBSX seems to be intermediate between bacteriocins and the better-known bacteriophage discussed elsewhere in this chapter.

One of the most interesting recent developments in phage biochemistry is the use of the φX174 system to demonstrate *in vitro* formation of biologically active DNA (*410*). Goulian, Kornberg, and Sinsheimer used a combination of DNA polymerase and polynucleotide ligase to form intact RF molecules from single-stranded ϕX174 DNA. Both strands of the enzymically synthesized product were infective to spheroplasts, and the newly formed strand could be used as template to direct the synthesis of further infective single-stranded molecules.

REFERENCES

1. E. L. Ellis and M. Delbrück, *J. Gen. Physiol.,* **22**, 365 (1939).

2. A. H. Doermann, *Carnegie Inst. Wash. Yearbook* **47**, 176 (1948).

3. J. Cairns, G. S. Stent, and J. D. Watson, (eds.), *Phage and the Origins of Molecular Biology* (Cold Spring Harbor Laboratory of Quantitative Biology, 1966).

4. M. H. Adams, *Bacteriophages,* Wiley (Interscience), New York, 1959.

5. G. S. Stent, *Molecular Biology of Bacterial Viruses,* Freeman, San Francisco, 1963.

6. S. P. Champe, *Ann. Rev. Microbiol.,* **17**, 87 (1963).

7. S. S. Cohen, *Ann. Rev. Biochem.,* **32**, 83 (1963).

8. H. Ruska, *Naturwiss.* **28**, 45 (1940).

9. S. E. Luria and T. F. Anderson, *Proc. Natl. Acad. Sci. U.S.,* **28**, 127 (1942).

10. S. Brenner, G. Streisinger, R. W. Horne, S. P. Champe, L. Barnett, S. Benzer, and M. W. Rees, *J. Mol. Biol.,* **1**, 281 (1959).

11. H. Van Vunakis, W. Baker, and R. Brown, *Virology,* **5**, 327 (1958).

12. T. F. Anderson, *Proc. European Regional Conf. Electron Microscopy, Delft 1960,* **II**, 1008 (1960).

13. H. Fernandez-Moran, *Symp. Intern. Soc. Cell Biol.,* **7**, 411 (1963).

14. L. Simon and T. F. Anderson, *Virology,* **32**, 279 (1967).

15. N. Sarkar, S. Sarkar, and L. M. Kozloff, *Biochemistry,* **3**, 511 (1964).

16. P. P. Dukes and L. M. Kozloff, *J. Biol. Chem.,* **234**, 534 (1959).

17. A. S. Tikhonenko and B. F. Polglazov, *Biokhimiya,* **28**, 340 (1963).

18. R. Wahl and L. M. Kozloff, *J. Biol. Chem.,* **237**, 1953 (1962).

19. S. Sarkar, N. Sarkar, and L. M. Kozloff, *Biochemistry,* **3**, 517 (1964).

20. G. Koch and W. J. Dreyer, *Virology,* **6**, 291 (1958).

21. A. D. Hershey, *Virology,* **1**, 108 (1955).

22. T. Minagawa, *Virology,* **13**, 515 (1961).

23. A. D. Hershey, *Virology,* **4**, 237 (1957).

24. B. N. Ames, D. T. Dubin, and S. M. Rosenthal, *Science,* **127**, 814 (1958).

25. L. M. Kozloff and M. Lute, *Biochim. Biophys. Acta,* **37**, 420 (1960).

26. L. M. Kozloff and M. Lute, *J. Mol. Biol.,* **12**, 780 (1965).

27. I. Rubenstein, C. A. Thomas, and A. D. Hershey, *Proc. Natl. Acad. Sci. U.S.,* **47**, 1113 (1961).

28. J. Cairns, *Cold Spring Harbor Symp. Quant. Biol.,* **28**, 43 (1963).

29. G. Streisinger, R. S. Edgar, and G. H. Denhardt, *Proc. Natl. Acad. Sci. U.S.,* **51**, 775 (1964).

30. C. A. Thomas and L. A. MacHattie, *Proc. Natl. Acad. Sci. U.S.,* **52,** 1297 (1964).

31. C. A. Thomas and I. Rubenstein, *Biophys. J.,* **4,** 93 (1964).

32. J. Séchaud, G. Streisinger, J. Emrich, J. Newton, H. Lanford, H. Reinhold, and M. M. Stahl, *Proc. Natl. Acad. Sci. U.S.,* **54,** 1333 (1965).

33. L. A. MacHattie, D. A. Ritchie, C. A. Thomas, and C. C. Richardson, *J. Mol. Biol.,* **23,** 355 (1967).

34. G. Streisinger, J. Emrich, and M. M. Stahl, *Proc. Natl. Acad. Sci. U.S.,* **57,** 292 (1967).

35. G. R. Wyatt and S. S. Cohen, *Biochem. J.,* **55,** 774 (1953).

36. R. L. Sinsheimer, *Science,* **120,** 551 (1954).

37. E. Volkin, *J. Am. Chem. Soc.,* **76,** 5892 (1954).

38. M. A. Jesaitis, *Nature,* **178,** 637 (1955).

39. J. Lichtenstein and S. S. Cohen, *J. Biol. Chem.,* **235,** 1134 (1960).

40. I. R. Lehman and E. A. Pratt, *J. Biol. Chem.,* **235,** 3254 (1960).

41. T. F. Anderson, *Cold Spring Harbor Symp. Quant. Biol.,* **18,** 197 (1953).

42. A. D. Hershey and M. Chase, *J. Gen. Physiol.,* **36,** 39 (1952).

43. W. Weidel, *Z. Naturforsch.,* **6b,** 251 (1951).

44. W. Weidel, *Ann. Inst. Pasteur,* **84,** 60 (1951).

45. W. Weidel and J. Primosigh, *J. Gen. Microbiol.,* **18,** 513 (1958).

46. W. Weidel and E. Kellenberger, *Biochim. Biophys. Acta,* **17,** 1 (1955).

47. R. C. Williams and D. Fraser, *Virology,* **2,** 289 (1956).

48. T. F. Anderson, *J. Cellular Comp. Physiol.,* **25,** 17 (1945).

49. T. F. Anderson, *J. Bacteriol.,* **55,** 637 (1948).

50. D. J. Cummings, *Virology,* **23,** 408 (1964).

51. M. Delbrück, *J. Bacteriol.,* **56,** 1 (1948).

52. L. C. Kanner and L. M. Kozloff, *Biochemistry,* **3,** 215 (1964).

53. G. Koch and W. Weidel, *Z. Naturforsch.,* **11b,** 345 (1956).

54. L. D. Simon and T. F. Anderson, *Virology,* **32,** 298 (1967).

55. M. F. Moody, *J. Mol. Biol.,* **25,** 167 (1967).

56. M. F. Moody, *J. Mol. Biol.,* **25,** 201 (1967).

57. L. Levine, J. L. Barlow, and H. Van Vunakis, *Virology,* **6,** 702 (1958).

58. E. Kellenberger, *Advan. Virus Res.,* **8,** 1 (1962).

59. D. M. Chaproniere-Rickenberg, H. R. Mahler, and D. Fraser, *Virology,* **23,** 96 (1964).

60. S. S. Cohen, *Cold Spring Harbor Symp. Quant. Biol.,* **12,** 35 (1947).

61. A. D. Hershey, J. Dixon, and M. Chase, *J. Gen. Physiol.,* **36,** 777 (1953).

62. S. Benzer, *Biochim. Biophys. Acta,* **11,** 383 (1953).

63. A. P. Levin and K. Burton, *J. Gen. Microbiol.,* **25,** 307 (1961).

64. M. Nomura, K. Matsubara, K. Okamoto, and R. Fujimura, *J. Mol. Biol.,* **5,** 535 (1962).

65. M. Nomura, K. Okamoto, and K. Asano, *J. Mol. Biol.,* **4,** 376 (1962).

66. W. S. Hayward and M. H. Green, *Proc. Natl. Acad. Sci. U.S.,* **54,** 1675 (1965).

67. R. C. French and L. Siminovitch, *Can. J. Microbiol.,* **1,** 757 (1955).

68. M. Nomura, C. Witten, N. Mantei, and H. Echols, *J. Mol. Biol.,* **17,** 273 (1966).

68a. R. O. R. Kaempfer and B. Magasanik, *J. Mol. Biol.,* **27,** 453 (1967).

68b. R. O. R. Kaempfer and B. Magasanik, *J. Mol. Biol.,* **27**, 475 (1967).

68c. R. O. R. Kaempfer and S. Sarkar, *J. Mol. Biol.,* **27**, 469 (1967).

68d. T. T. Puck and L. H. Lee, *J. Exptl. Med.,* **101**, 151 (1955).

68e. F. Jacob, S. Brenner, and F. Cuzin, *Cold Spring Harbor Symp. Quant. Biol.,* **28**, 329 (1963).

69. M. Nomura, *Cold Spring Harbor Symp. Quant. Biol.,* **28**, 315 (1963).

70. M. Nomura, *Proc. Natl. Acad. Sci. U.S.,* **52**, 1514 (1964).

71. E. Volkin and L. Astrachan, *Virology,* **2**, 149 (1956).

72. L. Astrachan and E. Volkin, *Biochim. Biophys. Acta,* **29**, 536 (1958).

73. F. Jacob and J. Monod, *J. Mol. Biol.,* **3**, 318 (1961).

74. M. Nomura, B. D. Hall, and S. Spiegelman, *J. Mol. Biol.,* **2**, 306 (1960).

75. S. Brenner, F. Jacob, and M. Meselson, *Nature,* **190**, 576 (1961).

76. A. D. Hershey, A. Garen, D. K. Fraser, and J. D. Hudis, *Carnegie Inst. Wash. Yearbook,* **53**, 210 (1954).

77. S. S. Cohen and C. B. Fowler, *J. Exptl. Med.,* **85**, 771 (1947).

78. J. Tomizawa and S. Sunakawa, *J. Gen. Physiol.,* **39**, 553 (1956).

79. T. Minagawa, *Virology,* **13**, 515 (1961).

80. H. D. Barner and S. S. Cohen, *J. Bacteriol.,* **68**, 80 (1954).

81. J. G. Flaks and S. S. Cohen, *J. Biol. Chem.,* **234**, 1501 (1959).

82. J. G. Flaks and S. S. Cohen, *Biochim. Biophys. Acta,* **25**, 667 (1957).

83. M. Friedkin and A. Kornberg, in *The Chemical Basis of Heredity* (W. D. McElroy and B. Glass, eds.), Johns Hopkins Press, Baltimore, Md., 1957, p. 609.

84. H. D. Barner and S. S. Cohen, *J. Biol. Chem.,* **234**, 1987 (1959).

85. A. Kornberg, *Enzymatic Synthesis of DNA,* Wiley, New York, 1961.

86. A. Kornberg, S. B. Zimmerman, S. R. Kornberg, and J. Josse, *Proc. Natl. Acad. Sci. U.S.,* **45**, 772 (1959).

87. G. R. Greenberg, R. Somerville, and S. DeWolf, *Proc. Natl. Acad. Sci. U.S.,* **48**, 242 (1962).

88. J. G. Flaks, J. Lichtenstein, and S. S. Cohen, *J. Biol. Chem.,* **234**, 1507 (1959).

89. S. S. Cohen, *Federation Proc.,* **20**, 641 (1961).

90. C. K. Mathews, F. Brown, and S. S. Cohen, *J. Biol. Chem.,* **239**, 2957 (1964).

91. C. K. Mathews and S. S. Cohen, *J. Biol. Chem.,* **238**, 367 (1963).

92. R. H. Epstein, A. Bolle, C. M. Steinberg, E. Kellenberger, E. Boy de la Tour, R. Chevalley, R. S. Edgar, M. Susman, G. H. Denhardt, and I. Lielausis, *Cold Spring Harbor Symp. Quant. Biol.,* **28**, 375 (1963).

93. R. S. Edgar and R. H. Epstein, *Sci. Am.,* **212**, 70 (1965).

94. J. S. Wiberg and J. M. Buchanan, *Proc. Natl. Acad. Sci. U.S.,* **51**, 421 (1964).

95. J. S. Wiberg, M.-L. Dirksen, R. H. Epstein, S. E. Luria, and J. M. Buchanan, *Proc. Natl. Acad. Sci. U.S.,* **48**, 293 (1962).

96. M.-L. Dirksen, J. C. Hutson, and J. M. Buchanan, *Proc. Natl. Acad. Sci. U.S.,* **50**, 507 (1963).

97. P. Lengyel, *J. Gen. Physiol.,* **49**, 305 (1966).

98. A. S. Sarabhai, A. O. W. Stretton, S. Brenner, and A. Bolle, *Nature,* **201**, 13 (1964).

99. S. Brenner, A. O. W. Stretton, and S. Kaplan, *Nature,* **206**, 994 (1965).

100. M. G. Weigert and A. Garen, *Nature,* **206**, 992 (1965).

101. M. R. Capecchi and G. N. Gussin, *Science,* **149**, 417 (1965).

102. R. S. Edgar and W. B. Wood, *Proc. Natl. Acad. Sci. U.S.,* **55,** 498 (1966).

103. A. B. Stone and K. Burton, *Biochem. J.,* **85,** 600 (1962).

104. A. E. Oleson and J. F. Koerner, *J. Biol. Chem.,* **239,** 2935 (1964).

105. E. C. Short and J. F. Koerner, *Proc. Natl. Acad. Sci. U.S.,* **54,** 595 (1965).

106. Z. J. Lucas, *Federation Proc.,* **24,** 286 (1965).

107. A. B. Pardee and I. Williams, *Arch. Biochem. Biophys.,* **40,** 222 (1952).

108. L. M. Kozloff, *Cold Spring Harbor Symp. Quant. Biol.,* **18,** 109 (1953).

109. J. S. Wiberg, *Proc. Natl. Acad. Sci. U.S.,* **55,** 614 (1966).

110. S. S. Cohen, *J. Biol. Chem.,* **174,** 281 (1948).

111. S. B. Zimmerman and A. Kornberg, *J. Biol. Chem.,* **236,** 1480 (1961).

112. G. R. Greenberg and R. Somerville, *Proc. Natl. Acad. Sci. U.S.,* **48,** 247 (1962).

113. G. R. Greenberg, *Proc. Natl. Acad. Sci. U.S.,* **56,** 1226 (1966).

114. H. R. Warner and J. E. Barnes, *Proc. Natl. Acad. Sci. U.S.,* **56,** 1233 (1966).

115. J. S. Wiberg, *Federation Proc.,* **25,** 785 (1966).

116. E. H. Simon and I. Tessman, *Proc. Natl. Acad. Sci. U.S.,* **50,** 526 (1963).

117. D. M. Shapiro, J. Eigner, and G. R. Greenberg, *Proc. Natl. Acad. Sci. U.S.,* **53,** 874 (1965).

118. C. K. Mathews, *J. Bacteriol.,* **90,** 648 (1965).

119. A. J. Wahba and M. Friedkin, *J. Biol. Chem.,* **236,** PC11 (1961).

120. C. K. Mathews and S. S. Cohen, *J. Biol. Chem.,* **238,** PC853 (1963).

121. C. K. Mathews and K. E. Sutherland, *J. Biol. Chem.,* **240,** 2142 (1965).

122. C. K. Mathews, *J. Biol. Chem.,* **242,** 4083 (1967).

123. D. H. Hall, I. Tessman, and O. Karlstrom, *Virology,* **31,** 442 (1967).

124. D. H. Hall, *Proc. Natl. Acad. Sci. U.S.,* **58,** 584 (1967).

125. C. K. Mathews, *J. Virol.,* **1,** 963 (1967).

126. K. Keck, H. R. Mahler, and D. Fraser, *Arch. Biochem. Biophys.,* **86,** 85 (1960).

127. D. H. Hall and I. Tessman, *Virology,* **29,** 339 (1966).

128. W. H. Fleming and M. J. Bessman, *J. Biol. Chem.,* **242,** 363 (1967).

129. G. F. Maley, D. U. Guarino, and F. Maley, *J. Biol. Chem.,* **242,** 3517 (1967).

130. G. Geraci, M. Rossi, and E. Scarano, *Biochemistry,* **6,** 183 (1967).

131. G. F. Maley and F. Maley, *J. Biol. Chem.,* **239,** 1168 (1964).

132. C. K. Mathews, *Biochemistry,* **5,** 2092 (1966).

133. H. R. Warner and N. Lewis, *Virology,* **29,** 172 (1966).

134. S. S. Cohen, H. D. Barner, and J. Lichtenstein, *J. Biol. Chem.,* **236,** 1448 (1961).

135. S. S. Cohen and H. D. Barner, *J. Biol. Chem.,* **237,** PC1376 (1962).

136. W. S. Beck, personal communication.

137. S. Hiraga, K. Igarashi, and T. Yura, *Biochim. Biophys. Acta,* **145,** 41 (1967).

138. R. Okazaki and A. Kornberg, *J. Biol. Chem.,* **239,** 269 (1964).

139. C. K. Mathews and R. H. Kessin, *J. Virol.,* **1,** 92 (1967).

140. R. Somerville, K. Ebisuzaki, and G. R. Greenberg, *Proc. Natl. Acad. Sci. U.S.,* **45,** 1240 (1959).

141. M. J. Bessman, *J. Biol. Chem.,* **234,** 2735 (1959).

142. M. J. Bessman and L. J. Bello, *J. Biol. Chem.,* **236,** PC72 (1961).

143. L. J. Bello and M. J. Bessman, *J. Biol. Chem.,* **238,** 1777 (1963).

144. L. J. Bello and M. J. Bessman, *Biochim. Biophys. Acta,* **72,** 647 (1963).

145. D. H. Duckworth and M. J. Bessman, *J. Biol. Chem.,* **242,** 2877 (1967).

146. H. V. Aposhian and A. Kornberg, *J. Biol. Chem.,* **237,** 519 (1962).

147. M. Goulian, *Federation Proc.,* **26,** 396 (1967).

148. A. de Waard, A. V. Paul, and I. R. Lehman, *Proc. Natl. Acad. Sci. U.S.,* **54,** 1241 (1965).

149. H. R. Warner and J. E. Barnes, *Virology,* **28,** 100 (1966).

150. J. F. Speyer, *Biochem. Biophys. Res. Commun.,* **21,** 6 (1965).

151. J. F. Speyer, J. D. Karam, and A. B. Lenny, *Cold Spring Harbor Symp. Quant. Biol.,* **31,** 693 (1966).

152. S. R. Kornberg, S. B. Zimmerman, and A. Kornberg, *J. Biol. Chem.,* **236,** 1487 (1961).

153. S. B. Zimmerman, S. R. Kornberg, and A. Kornberg, *J. Biol. Chem.,* **237,** 512 (1962).

154. J. Josse and A. Kornberg, *J. Biol. Chem.,* **237,** 1968 (1962).

155. S. Kuno and I. R. Lehman, *J. Biol. Chem.,* **237,** 1266 (1962).

156. S. E. Luria and M. L. Human, *J. Bacteriol.,* **64,** 557 (1952).

157. S. Hattman and T. Fukasawa, *Proc. Natl. Acad. Sci. U.S.,* **50,** 297 (1963).

158. A. Shedlovsky and S. Brenner, *Proc. Natl. Acad. Sci. U.S.,* **50,** 300 (1963).

159. N. Symonds, K. A. Stacey, S. W. Glover, J. Schell, and S. Silver, *Biochem. Biophys. Res. Commun.,* **12,** 220 (1963).

160. C. C. Richardson, *J. Biol. Chem.,* **241,** 2084 (1966).

161. H. R. Revel, S. Hattman, and S. E. Luria, *Biochem. Biophys. Res. Commun.,* **18,** 545 (1965).

162. J. Hosoda, *Biochem. Biophys. Res. Commun.,* **27,** 294 (1967).

163. C. P. Georgopoulos, *Biochem. Biophys. Res. Commun.,* **28,** 179 (1967).

164. R. Hausmann and M. Gold, *J. Biol. Chem.,* **241,** 1985 (1966).

165. M. Gefter, R. Hausmann, M. Gold, and J. Hurwitz, *J. Biol. Chem.,* **241,** 1995 (1966).

166. M. Gold and J. Hurwitz, *Cold Spring Harbor Symp. Quant. Biol.,* **28,** 149 (1963).

167. W. Arber, *J. Mol. Biol.,* **11,** 247 (1965).

168. C. C. Richardson, *Proc. Natl. Acad. Sci. U.S.,* **54,** 158 (1965).

169. A. Novogrodsky and J. Hurwitz, *J. Biol. Chem.,* **241,** 2923 (1966).

170. A. Novogrodsky, M. Tal, A. Traub, and J. Hurwitz, *J. Biol. Chem.,* **241,** 2933 (1966).

171. A. Becker and J. Hurwitz, *J. Biol. Chem.,* **242,** 936 (1967).

172. C. C. Richardson, C. L. Schildkraut, and A. Kornberg, *Cold Spring Harbor Symp. Quant. Biol.,* **28,** 9 (1963).

173. B. Weiss and C. C. Richardson, *Proc. Natl. Acad. Sci. U.S.,* **57,** 1021 (1967).

174. M. Gellert, *Proc. Natl. Acad. Sci. U.S.,* **57,** 148 (1967).

175. A. Becker, M. Gefter, and J. Hurwitz, *Proc. Natl. Acad. Sci. U.S.,* **58,** 240 (1967).

176. B. M. Olivera and I. R. Lehman, *Proc. Natl. Acad. Sci. U.S.,* **57,** 1426 (1967).

177. B. M. Olivera and I. R. Lehman, *Proc. Natl. Acad. Sci. U.S.,* **57,** 1700 (1967).

178. S. B. Zimmerman, J. W. Little, C. K. Oshinsky, and M. Gellert, *Proc. Natl. Acad. Sci. U.S.,* **57,** 1841 (1967).

179. B. Weiss, T. R. Live, and C. C. Richardson, *Federation Proc.,* **26,** 395 (1967).

180. G. C. Fareed and C. C. Richardson, *Proc. Natl. Acad. Sci. U.S.,* **58,** 665 (1967).

181. C. G. Mead, *Proc. Natl. Acad. Sci. U.S.,* **52,** 1482 (1964).

182. S. E. Luria, *Proc. Natl. Acad. Sci. U.S.,* **33,** 253 (1947).

183. W. Harm, *Virology,* **19,** 66 (1963).

184. K. Haber, *Biochem. Biophys. Res. Commun.,* **23,** 502 (1966).

185. R. B. Setlow and W. L. Carrier, *Proc. Natl. Acad. Sci. U.S.,* **51,** 226 (1964).

186. R. P. Boyce and P. Howard-Flanders, *Proc. Natl. Acad. Sci. U.S.,* **51,** 293 (1964).

187. W. Eckart, *J. Mol. Biol.,* **18,** 292 (1966).

188. G. S. Stent and O. Maaloe, *Biochim. Biophys. Acta,* **10,** 55 (1953).

189. A. D. Hershey, *J. Gen. Physiol.,* **37,** 1 (1953).

190. S. E. Luria and J. E. Darnell, *General Virology,* 2nd ed., Wiley, New York, 1967.

191. M. Meselson and F. W. Stahl, *Proc. Natl. Acad. Sci. U.S.,* **44,** 671 (1958).

192. C. Levinthal, *Proc. Natl. Acad. Sci. U.S.,* **42,** 394 (1956).

193. G. S. Stent and N. K. Jerne, *Proc. Natl. Acad. Sci. U.S.,* **41,** 704 (1955).

194. A. W. Kozinski, *Virology,* **13,** 124 (1961).

195. A. Roller, *J. Mol. Biol.,* **9,** 260 (1964).

196. E. Shahn and A. W. Kozinski, *Virology,* **30,** 455 (1966).

197. F. R. Frankel, *J. Mol. Biol.,* **18,** 127 (1966).

198. A. W. Kozinski and P. B. Kozinski, *Proc. Natl. Acad. Sci. U.S.,* **54,** 634 (1965).

199. A. W. Kozinski, P. B. Kozinski, and R. James, *J. Virol.,* **1,** 758 (1967).

200. N. Anraku and J. Tomizawa, *J. Mol. Biol.,* **11,** 501 (1965).

201. S. S. Cohen, J. G. Flaks, H. D. Barner, M. R. Loeb, and J. Lichtenstein, *Proc. Natl. Acad. Sci. U.S.,* **44,** 1004 (1958).

202. E. H. Simon, *Science,* **150,** 760 (1965).

203. J. Tomizawa, N. Anraku, and Y. Iwama, *J. Mol. Biol.,* **21,** 247 (1966).

204. F. R. Frankel, *J. Mol. Biol.,* **18,** 144 (1966).

205. R. E. Murray, Ph.D. thesis, Yale University, 1969.

206. S. Benzer, *J. Bacteriol.,* **63,** 59 (1952).

207. A. W. Kozinski and M. J. Bessman, *J. Mol. Biol.,* **3,** 746 (1961).

208. A. W. Kozinski and T. H. Lin, *Proc. Natl. Acad. Sci. U.S.,* **54,** 273 (1965).

209. D. P. Snustad, *Genetics,* **54,** 923 (1966).

210. D. P. Snustad, *Genetics,* **54,** 937 (1966).

211. K. G. Lark, *Bacteriol. Revs.,* **30,** 3 (1966).

212. M.-L. Dirksen, J. S. Wiberg, J. F. Koerner, and J. M. Buchanan, *Proc. Natl. Acad. Sci. U.S.,* **46,** 1425 (1960).

213. K. Ebisuzaki, *J. Mol. Biol.,* **20,** 545 (1966).

214. M. Sekiguchi and S. S. Cohen, *J. Mol. Biol.,* **8,** 638 (1964).

215. E. Sercarz, *Virology,* **28,** 339 (1966).

216. K. Ebisuzaki, *J. Mol. Biol.,* **7,** 379 (1963).

217. T. Kano-Sueoka and S. Spiegelman, *Proc. Natl. Acad. Sci. U.S.,* **48,** 1942 (1962).

218. B. D. Hall, A. P. Nygaard, and M. H. Green, *J. Mol. Biol.,* **9,** 143 (1964).

219. E. K. F. Bautz and E. Reilly, *Science,* **151,** 328 (1966).

220. E. K. F. Bautz, T. Kasai, E. Reilly, and F. A. Bautz, *Proc. Natl. Acad. Sci. U.S.,* **55**, 1081 (1966).

221. T. Kasai and E. K. F. Bautz, *Federation Proc.,* **26**, 350 (1967).

222. G. Edlin, *J. Mol. Biol.,* **12**, 363 (1965).

223. S. P. Champe and S. Benzer, *Proc. Natl. Acad. Sci. U.S.,* **48**, 532 (1962).

224. M. I. Baldi and R. Haselkorn, *J. Mol. Biol.,* **27**, 193 (1967).

225. E. P. Geiduschek, L. Snyder, A. J. E. Colvill, and M. Sarnat, *J. Mol. Biol.,* **19**, 541 (1966).

226. M. Green, *Proc. Natl. Acad. Sci. U.S.,* **52**, 1388 (1964).

227. S. E. Luria, *Biochem. Biophys. Res. Commun.,* **18**, 735 (1965).

228. N. Sueoka and T. Kano-Sueoka, *Proc. Natl. Acad. Sci. U.S.,* **52**, 1535 (1964).

229. T. Kano-Sueoka and N. Sueoka, *J. Mol. Biol.,* **20**, 183 (1966).

230. L. C. Waters and G. D. Novelli, *Proc. Natl. Acad. Sci. U.S.,* **57**, 979 (1967).

231. C. Shalitin and S. Sarid, *J. Virol.,* **1**, 559 (1967).

232. C. Shalitin, *J. Virol.,* **1**, 569 (1967).

233. S. S. Cohen, in *The Molecular Biology of Viruses* (J. S. Colter and W. Paranchych, eds.), Academic Press, New York, 1967.

234. K. Ebisuzaki, *J. Mol. Biol.,* **5**, 506 (1962).

235. C. Levinthal, J. Hosoda, and D. Shub, in *The Molecular Biology of Viruses* (J. S. Colter and W. Paranchych, eds.), Academic Press, New York, 1967.

236. D. J. McCorquodale, A. E. Oleson, and J. M. Buchanan, in *The Molecular Biology of Viruses* (J. S. Colter and W. Paranchych, eds.), Academic Press, New York, 1967.

237. E. Kellenberger and J. Séchaud, *Virology,* **3**, 256 (1957).

238. E. Kellenberger, J. Séchaud, and A. Ryter, *Virology,* **8**, 478 (1959).

239. G. Kellenberger and E. Kellenberger, *Virology,* **3**, 275 (1957).

240. W. B. Wood and R. S. Edgar, *Sci. Am.,* **217**, 61 (1967).

241. M. M. Piechowski and M. Susman, *Virology,* **28**, 386 (1966).

242. S. Brenner and L. Barnett, *Brookhaven Symp. Biol.,* **12**, 86 (1959).

243. M. Susman, M. M. Piechowski, and D. A. Ritchie, *Virology,* **26**, 163 (1965).

244. M. M. Piechowski and M. Susman, *Virology,* **28**, 396 (1966).

245. D. Korn, J. J. Protass, and L. Leive, *Biochem. Biophys. Res. Commun.,* **19**, 473 (1965).

246. D. Korn, *J. Biol. Chem.,* **242**, 160 (1967).

247. G. Streisinger, F. Mukai, W. J. Dreyer, B. Miller, and S. Horiuchi, *Cold Spring Harbor Symp. Quant. Biol.,* **26**, 25 (1961).

248. A. H. Doermann, *J. Bacteriol.,* **55**, 257 (1948).

249. F. Mukai, G. Streisinger, and B. Miller, *Virology,* **33**, 398 (1967).

250. E. Terzaghi, Y. Okada, G. Streisinger, J. Emrich, M. Inouye, and A. Tsugita, *Proc. Natl. Acad. Sci. U.S.,* **56**, 500 (1966).

251. Y. Okada, E. Terzaghi, G. Streisinger, J. Emrich, M. Inouye, and A. Tsugita, *Proc. Natl. Acad. Sci. U.S.,* **56**, 1692 (1966).

252. F. H. C. Crick, L. Barnett, S. Brenner, and R. J. Watts-Tobin, *Nature,* **192**, 27 (1961).

253. S. Benzer, in *Symposium on the Chemical Basis of Heredity* (W. D. Elroy and B. Glass, eds.), Johns Hopkins Press, Baltimore, Md., 1957.

254. B. Rutberg and L. Rutberg, *J. Bacteriol.,* **91**, 76 (1966).

255. W. H. McClain and S. P. Champe, *Proc. Natl. Acad. Sci. U.S.,* **58,** 1182 (1967).

256. A. Garen, *Virology,* **14,** 151 (1961).

257. G. FerroLuzzi-Ames and B. N. Ames, *Biochem. Biophys. Res. Commun.,* **18,** 639 (1965).

258. M. L. Brock, *Virology,* **26,** 221 (1965).

259. M. Sekiguchi, *J. Mol. Biol.,* **16,** 503 (1966).

260. C. W. M. Orr, S. T. Herriott, and M. J. Bessman, *J. Biol. Chem.,* **240,** 4652 (1965).

261. A. V. Paul and I. R. Lehman, *Federation Proc.,* **24,** 287 (1965).

262. M. J. Bessman, S. T. Herriott, and M. J. V. Orr, *J. Biol. Chem.,* **240,** 439 (1965).

263. C. K. Mathews, unpublished data.

264. R. Hausmann, *J. Virol.,* **1,** 57 (1967).

265. R. Hausmann and B. Gomez, *J. Virol.,* **1,** 779 (1967).

266. C. A. Thomas, D. A. Ritchie, and L. A. MacHattie, in *The Molecular Biology of Viruses* (J. S. Colter and W. Paranchych, eds.), Academic Press, New York, 1967.

267. D. A. Ritchie, C. A. Thomas, L. A. MacHattie, and P. C. Wensink, *J. Mol. Biol.,* **23,** 365 (1967).

268. J. Abelson and C. A. Thomas, *J. Mol. Biol.,* **18,** 262 (1966).

269. Y. T. Lanni, *Virology,* **10,** 514 (1960).

270. Y. T. Lanni, D. J. McCorquodale, and C. M. Wilson, *J. Mol. Biol.,* **10,** 19 (1964).

271. Y. T. Lanni, *Proc. Natl. Acad. Sci. U.S.,* **53,** 969 (1965).

272. Y. T. Lanni and D. J. McCorquodale, *Virology,* **19,** 72 (1963).

273. E. B. McKinley, *Compt. Rend. Soc. Biol.,* **93,** 1050 (1925).

274. F. M. Burnet and M. McKie, *Australian J. Exptl. Biol. Med. Sci.,* **6,** 277 (1929).

275. A. Lwoff, *Science,* **152,** 1216 (1966).

276. A. Lwoff and A. Gutmann, *Ann. Inst. Pasteur,* **78,** 711 (1950).

277. A. Lwoff, L. Siminovitch, and N. Kjeldgaard, *Compt. Rend.,* **231,** 190 (1950).

278. A. D. Kaiser, *J. Gen. Physiol.,* **49,** 171 (1966).

279. E. Burgi and A. D. Hershey, *Biophys. J.,* **3,** 309 (1963).

280. L. A. MacHattie and C. A. Thomas, *Science,* **144,** 1142 (1964).

281. D. S. Hogness, W. Doerfler, J. B. Egan, and L. W. Black, in *The Molecular Biology of Viruses* (J. S. Colter and W. Paranchych, eds.), Academic Press, New York, 1967.

282. A. D. Kaiser and D. S. Hogness, *J. Mol. Biol.,* **2,** 392 (1960).

283. A. D. Kaiser, *J. Mol. Biol.,* **4,** 275 (1962).

284. A. D. Hershey, E. Burgi, and C. I. Davern, *Biochem. Biophys. Res. Commun.* **18,** 675 (1965).

285. A. D. Hershey, E. Burgi, and L. Ingraham, *Proc. Natl. Acad. Sci. U.S.,* **49,** 748 (1963).

286. H. B. Strack and A. D. Kaiser, *J. Mol. Biol.,* **12,** 36 (1965).

287. A. D. Kaiser and R. B. Inman, *J. Mol. Biol.,* **13,** 78 (1965).

288. M. Meselson and J. Weigle, *Proc. Natl. Acad. Sci. U.S.,* **47,** 857 (1961).

289. M. Meselson, *J. Mol. Biol.,* **9,** 734 (1964).

290. N. D. Zinder and J. Lederberg, *J. Bacteriol.*, **64**, 679 (1952).
291. W. Hayes, *The Genetics of Bacteria and their Viruses*, Wiley, New York, 1964.
292. H. Ikeda and J. Tomizawa, *J. Mol. Biol.*, **14**, 85 (1965).
293. M. L. Morse, E. M. Lederberg, and J. Lederberg, *Genetics*, **41**, 42 (1956).
294. E. L. Wollman, *Compt. Rend.*, **257**, 4225 (1963).
295. A. M. Campbell, *Virology*, **14**, 22 (1961).
296. A. M. Campbell and A. del Campillo-Campbell, *J. Bacteriol.*, **85**, 1202, (1963).
297. K. Brooks, *Virology*, **26**, 489 (1965).
298. W. F. Dove, *J. Mol. Biol.*, **19**, 187 (1966).
299. A. Joyner, L. N. Isaacs, H. Echols, and W. S. Sly, *J. Mol. Biol.*, **19**, 174 (1966).
300. J. J. Weigle, *Proc. Natl. Acad. Sci. U.S.*, **55**, 1462 (1966).
301. H. Eisen, C. R. Fuerst, L. Siminovitch, R. Thomas, L. Lambert, L. Pereira da Silva, and F. Jacob, *Virology*, **30**, 224 (1966).
302. C. M. Radding, *Proc. Natl. Acad. Sci. U.S.*, **52**, 965 (1964).
303. D. Korn and A. Weissbach, *J. Biol. Chem.*, **238**, 3390 (1963).
304. H. Echols, B. Butler, A. Joyner, M. Willard, and L. Pilarski, in *The Molecular Biology of Viruses* (J. S. Colter and W. Paranchych, eds.), Academic Press, New York, 1967.
305. C. M. Radding and D. Shreffler, *J. Mol. Biol.*, **18**, 251 (1966).
306. W. E. Pricer and A. Weissbach, *J. Biol. Chem.*, **242**, 1701 (1967).
307. R. C. Schuster, T. R. Breitman, and A. Weissbach, *J. Biol. Chem.*, **242**, 3723 (1967).
308. C. M. Radding, J. Szpirer, and R. Thomas, *Proc. Natl. Acad. Sci. U.S.*, **57**, 277 (1967).
309. A. D. Kaiser, *Virology*, **3**, 42 (1957).
310. A. D. Kaiser and F. Jacob, *Virology*, **4**, 509 (1957).
311. G. Kellenberger, M. L. Zichichi, and J. J. Weigle, *J. Mol. Biol.*, **3**, 399 (1961).
312. G. S. Stent, C. R. Fuerst, and F. Jacob, *Compt. Rend.*, **244**, 1840 (1957).
313. F. Jacob and E. L. Wollman, *Sexuality and the Genetics of Bacteria*, Academic Press, New York, 1961.
314. A. Campbell, *Advan. Genetics*, **11**, 101 (1962).
315. E. Calef and G. Licciardello, *Virology*, **12**, 81 (1961).
316. J. R. Rothman, *J. Mol. Biol.*, **12**, 892 (1965).
317. E. Calef, C. Marchelli, and F. Guerrini, *Virology*, **27**, 1 (1965).
318. E. T. Young and R. L. Sinsheimer, *J. Mol. Biol.*, **10**, 562 (1964).
319. V. C. Bode and A. D. Kaiser, *J. Mol. Biol.*, **14**, 399 (1965).
320. A. Lipton and A. Weissbach, *J. Mol. Biol.*, **21**, 517 (1966).
321. E. Signer and J. R. Beckwith, *J. Mol. Biol.*, **22**, 33 (1966).
322. L. Fischer-Fantuzzi, *Virology*, **32**, 18 (1967).
323. J. Zissler, *Virology*, **31**, 189 (1967).
324. H. O. Smith and M. Levine, *Virology*, **31**, 207 (1967).
325. R. Sussman and F. Jacob, *Compt. Rend.*, **254**, 1517 (1962).
326. M. Ptashne, *Proc. Natl. Acad. Sci. U.S.*, **57**, 306 (1967).
327. M. Ptashne, *Nature*, **214**, 232 (1967).
328. M. Levine and H. O. Smith, *Science*, **146**, 1581 (1964).
329. H. O. Smith and M. Levine, *Virology*, **25**, 585 (1965).

330. A. L. Lisio and A. Weissbach, *J. Bacteriol.,* **90,** 661 (1965).

331. T. Ogawa and J. Tomizawa, *J. Mol. Biol.,* **23,** 225 (1967).

332. J. Tomizawa and T. Ogawa, *J. Mol. Biol.,* **23,** 247 (1967).

333. J. Tomizawa and T. Ogawa, *J. Mol. Biol.,* **23,** 277 (1967).

334. D. Goldthwait and F. Jacob, *Compt. Rend.,* **259,** 661 (1964).

335. K. Brooks and A. J. Clark, *J. Virol.,* **1,** 283 (1967).

336. W. Dove and J. J. Weigle, *J. Mol. Biol.,* **12,** 620 (1965).

337. L. A. Salzman and A. Weissbach, *Virology,* **30,** 579 (1967).

338. L. A. Salzman and A. Weissbach, *Virology,* **31,** 70 (1967).

339. M. G. Smith and A. Skalka, *J. Gen. Physiol.,* **49,** 127 (1966).

340. L. A. Salzman and A. Weissbach, *J. Mol. Biol.,* **28,** 53 (1967).

341. A. Weissbach and L. A. Salzman, *Proc. Natl. Acad. Sci. U.S.,* **58,** 1096 (1967).

342. Z. Hradecna and W. Szybalski, *Virology,* **32,** 633 (1967).

343. A. Skalka, *Proc. Natl. Acad. Sci. U.S.,* **55,** 1190 (1966).

344. S. Naono and F. Gros, *J. Mol. Biol.,* **25,** 517 (1967).

345. S. N. Cohen, *Federation Proc.,* **26,** 449 (1967).

346. K. Taylor, Z. Hradecna, and W. Szybalski, *Proc. Natl. Acad. Sci. U.S.,* **57,** 1618 (1967).

347. J. Marmur and C. M. Greenspan, *Science,* **142,** 387 (1963).

348. G. P. Tocchini-Valentini, M. Stodolsky, A. Aurisicchio, F. Grasiosi, M. Sarnat, S. B. Weiss, and E. P. Geiduschek, *Proc. Natl. Acad. Sci. U.S.,* **50,** 935 (1963).

349. M. Hayashi, M. N. Hayashi, and S. Spiegelman, *Proc. Natl. Acad. Sci. U.S.,* **50,** 664 (1963).

350. S. N. Cohen, U. Maitra, and J. Hurwitz, *J. Mol. Biol.,* **26,** 19 (1967).

351. A. Skalka, B. Butler, and H. Echols, *Proc. Natl. Acad. Sci. U.S.,* **58,** 576 (1967).

352. R. L. Sinsheimer, *J. Mol. Biol.,* **1,** 43 (1959).

353. W. Fiers, and R. L. Sinsheimer, *J. Mol. Biol.,* **5,** 424 (1962).

354. R. L. Sinsheimer, C. A. Hutchinson, and B. H. Lindqvist, in *The Molecular Biology of Viruses* (J. S. Colter and W. Paranchych, eds.), Academic Press, New York, 1967.

355. D. A. Marvin and H. Hoffman-Berling, *Z. Naturforsch.,* **18b,** 844 (1963).

356. W. O. Salivar, H. Tzagoloff, and D. Pratt, *Virology,* **24,** 359 (1964).

357. E. S. Tessman, in *The Molecular Biology of Viruses* (J. S. Colter and W. Paranchych, eds.), Academic Press, New York, 1967.

358. I. Tessman, S. Kumar, and E. S. Tessman, *Science,* **158,** 267 (1967).

359. R. L. Sinsheimer, B. Starman, C. Nagler, and S. Guthrie, *J. Mol. Biol.,* **4,** 142 (1962).

360. A. K. Kleinschmidt, Burton, and R. L. Sinsheimer, *Science,* **142,** 961 (1963).

361. A. Burton, and R. L. Sinsheimer, *Science,* **142,** 962 (1963).

362. T. F. Roth and M. Hayashi, *Science,* **154,** 658 (1966).

363. D. T. Denhardt and R. L. Sinsheimer, *J. Mol. Biol.,* **12,** 647 (1965).

364. R. R. Rueckert and W. Zillig, *J. Mol. Biol.,* **5,** 1 (1962).

365. B. H. Lindqvist and R. L. Sinsheimer, *J. Mol. Biol.,* **28,** 87 (1967).

366. D. T. Denhardt and R. L. Sinsheimer, *J. Mol. Biol.,* **12,** 663 (1965).

367. M. J. Yarus and R. L. Sinsheimer, *J. Virol.,* **1,** 135 (1967).

368. D. T. Denhardt, D. H. Dressler, and A. Hathaway, *Proc. Natl. Acad. Sci. U.S.,* **57,** 813 (1967).

369. J. Spizizen, *Proc. Natl. Acad. Sci. U.S.*, **44**, 1072 (1958).

370. A. M. Brodetsky and W. R. Romig, *J. Bacteriol.*, **90**, 1655 (1965).

371. W. R. Romig, *Virology*, **16**, 452 (1962).

372. P. Abel and T. A. Trautner, *Z. Vererbungslehre*, **95**, 66 (1964).

373. K. E. Bayreuther and W. R. Romig, *Science*, **146**, 778 (1964).

374. S. Okubo and W. R. Romig, *J. Mol. Biol.*, **14**, 130 (1965).

375. D. M. Green, *J. Mol. Biol.*, **10**, 438 (1964).

376. D. M. Green, *J. Mol. Biol.*, **22**, 1 (1966).

377. H. T. Epstein, *Biochem. Biophys. Res. Commun.*, **27**, 258 (1967).

378. I. Takahashi and J. Marmur, *Nature*, **197**, 794 (1963).

379. F. M. Kahan, *Federation Proc.*, **22**, 406 (1963).

380. I. Takahashi and J. Marmur, *Biochem. Biophys. Res. Commun.*, **10**, 289 (1963).

381. R. G. Kallen, M. Simon and J. Marmur, *J. Mol. Biol.*, **5**, 248 (1962).

382. E. Rosenberg, *Proc. Natl. Acad. Sci. U.S.*, **53**, 836 (1965).

383. F. Kahan, E. Kahan, and B. Riddle, *Federation Proc.*, **23**, 318 (1964).

384. D. H. Roscoe and R. G. Tucker, *Virology*, **29**, 157 (1966).

385. E. A. Haslam, D. H. Roscoe and R. G. Tucker, *Biochim. Biophys. Acta*, **134**, 312 (1967).

386. H. V. Aposhian and G. Y. Tremblay, *J. Biol. Chem.*, **241**, 5095 (1966).

387. D. M. Trilling and H. V. Aposhian, *Proc. Natl. Acad. Sci. U.S.*, **54**, 622 (1965).

388. J. J. Pène and J. Marmur, *J. Virol.*, **1**, 86 (1967).

389. L. Snyder and E. P. Geiduschek, *Proc. Natl. Acad. Sci. U.S.*, **59**, 459 (1968).

390. J. F. Pulitzer, L. Snyder and E. P. Geiduschek, *Federation Proc.*, **27**, 592 (1968).

391. A. Bolle, R. H. Epstein, W. Salser, and E. P. Geiduschek, *J. Mol. Biol.*, **31**, 325 (1968).

392. R. Greene and D. Korn, *J. Mol. Biol.*, **28**, 435 (1967).

393. A. Guha, W. Salser, and W. Szybalski, *Federation Proc.*, **27**, 646 (1968).

394. W. Salser, R. F. Gesteland, and A. Bolle, *Nature*, **215**, 588 (1967).

395. R. F. Gesteland, W. Salser, and A. Bolle, *Proc. Natl. Acad. Sci. U.S.*, **58**, 2036 (1967).

396. J. E. Celis and T. W. Conway, *Proc. Natl. Acad. Sci. U.S.*, **59**, 923 (1968).

397. F. C. Neidhardt and C. F. Earhart, *Cold Spring Harbor Symp. Quant. Biol.*, **31**, 557 (1966).

398. C. F. Earhart and F. C. Neidhardt, *Virology*, **33**, 694 (1967).

399. W-T. Hsu, J. W. Foft, and S. B. Weiss, *Proc. Natl. Acad. Sci. U.S.*, **58**, 2028 (1967).

400. J. W. Foft, W-T. Hsu, and S. B. Weiss, *Federation Proc.*, **27**, 341 (1968).

401. F. R. Frankel, *Proc. Natl. Acad. Sci. U.S.*, **59**, 131 (1968).

402. W. B. Wood, R. S. Edgar, J. King, I. Lielausis, and M. Henninger, *Federation Proc.*, **27**, 1160 (1968).

403. C. Verses, M. Lute, L. Crosby, and L. M. Kozloff, *Bacteriol. Proc.*, p. 153 (1968).

404. R. Gingery and H. Echols, *Proc. Natl. Acad. Sci. U.S.*, **58**, 1507 (1967).

405. M. E. Gottesman and M. B. Yarmolinsky, *J. Mol. Biol.*, **31**, 487 (1968).

406. H. Echols, L. Pilarski, and P. Y. Cheng, *Proc. Natl. Acad. Sci. U.S.*, **59**, 1016 (1968).

407. M. Green, B. Gotchel, J. Hendershott, and S. Kennel, *Proc. Natl. Acad. Sci. U.S.*, **58**, 2343 (1967).

408. M. W. Konrad, *Proc. Natl. Acad. Sci. U.S.*, **59**, 171 (1968).

409. B. D. Howard, *Science*, **158**, 1588 (1967).

410. M. Goulian, A. Kornberg, and R. L. Sinsheimer, *Proc. Natl. Acad. Sci. U.S.*, **58**, 2321 (1967).

411. S. Silver, E. Levine, P. M. Spielman, *J. Virol.*, **2**, 763 (1968).

412. C. S. Buller and L. Astrachan, *J. Virol.*, **2**, 298 (1968).

413. M. H. Furrow and L. I. Pizer, *J. Viol.*, **2**, 594 (1968).

414. R. Werner, *J. Mol. Biol.*, **33**, 679 (1968).

415. K. Sugimoto, T. Okazaki, and R. Okazaki, *Proc. Natl. Acad. Sci. U.S.*, **60**, 1356 (1968).

416. R. Okazaki, T. Okazaki, K. Sakabe, K. Sugimoto, and A. Sugino, *Proc. Natl. Acad. Sci. U.S.*, **59**, 598 (1968).

417. J. R. Menninger, M. Wright, L. Menninger, and M. Meselson, *J. Mol. Biol.*, **32**, 631 (1968).

418. M. Ptashne and N. Hopkins, *Proc. Natl. Acad. Sci. U.S.*, **60**, 1282 (1968).

419. E. T. Young, II, and R. L. Sinsheimer, *J. Mol. Biol.*, **33**, 49 (1968).

420. A. Bolle, R. H. Epstein, W. Salser, and E. P. Geiduschek, *J. Mol. Biol.*, **33**, 339 (1968).

421. C. K. Mathews, *J. Biol. Chem.*, **243**, 5610 (1968).

422. D. M. Trilling and H. V. Aposhian, *Proc. Natl. Acad. Sci. U.S.*, **60**, 214 (1968).

423. T. Kasai, E. K. F. Bautz, A. Guha, and W. Szybalski, *J. Mol. Biol.* **34**, 709 (1968).

424. H. Echols, R. Gingery, and L. Moore, *J. Mol. Biol.*, **34**, 251 (1968).

425. E. R. Singer and J. Weil, *J. Mol. Biol.*, **34**, 361 (1968).

426. J. Weil and E. R. Signer, *J. Mol. Biol.*, **34**, 273 (1968).

427. C. M. Radding and H. Echols, *Proc. Natl. Acad. Sci. U.S.*, **60**, 707 (1968).

428. W. Doerfler and D. S. Hogness, *J. Mol. Biol.*, **33**, 661 (1968).

429. W. Doerfler and D. S. Hogness, *J. Mol. Biol.*, **33**, 635 (1968).

430. R. W. Davis, and N. Davidson, *Proc. Natl. Acad. Sci. U.S.*, **60**, 243 (1968).

431. M. Meselson and R. Yuan, *Nature*, **217**, 1110 (1968).

432. S. Linn, and W. Arber, *Proc. Natl. Acad. Sci. U.S.*, **59**, 1300 (1968).

433. K. Okamoto, J. A. Mudd, J. Mangan, W. M. Huang, T. V. Subbaiah, and J. Marmur, *J. Mol. Biol.*, **34**, 413 (1968).

434. K. Okamoto, J. A. Mudd, and J. Marmur, *J. Mol. Biol.*, **34**, 429 (1968).

435. J. Emrich and G. Streisinger, *Virology*, **36**, 387 (1968).

436. D. H. Duckworth, *J. Virol.*, **3**, 85 (1969).

437. R. E. Murray and C. K. Mathews, *Federation Proc.*, in press, 1969.

438. A. Tsugita and M. Inouye, *J. Mol. Biol.*, **37**, 201 (1968).

439. M. Inouye and A. Tsugita, *J. Mol. Biol.*, **37**, 213 (1968).

440. T. Ogawa, J. Tomizawa, and M. Fuke, *Proc. Natl. Acad. Sci. U.S.*, **60**, 861 (1968).

12

BIOCHEMISTRY OF INTERFERON

HILTON B. LEVY, SAMUEL BARON, and
CHARLES E. BUCKLER

U. S. DEPARTMENT OF HEALTH, EDUCATION, AND WELFARE
NATIONAL INSTITUTE OF ALLERGY AND INFECTIOUS DISEASES
LABORATORY OF VIRAL DISEASES
BETHESDA, MARYLAND

12-1. Introduction

This chapter reviews recent biochemical developments in the field of interferon research. Older information, beginning with the discovery by Isaacs and Lindenmann, is mentioned to provide continuity, but for a more complete review of the older literature, the reader is referred to the book edited by Finter (1).

Interferon is a protein elaborated by most cells as a response to infection by most viruses (2). The amount of interferon synthesized varies greatly with the cells synthesizing the interferon and with the virus used to stimulate its induction. When solutions containing this protein are incubated with other cells for several hours, under conditions where the cells can synthesize RNA and protein, these cells develop resistance to subsequent virus infection. That is to say, they synthesize less virus when infected than do cells not previously exposed to interferon. These findings are interpreted to mean that the antiviral activity requires the mediation of a new interferon-induced protein. The amount of resistance developed varies greatly, both with the type of cell and the virus used to test the resistance. Interferon shows a high degree of specificity for the animal species—that is chicken interferon does not protect mouse cells, etc. It shows no specificity for the virus; for example, interferon produced by stimulation with influenza virus is indistinguishable from that produced upon stimulation by Newcastle disease virus.

12-2. Properties of the Interferon Protein

A. Induction of the Cell to Produce the Interferon Protein

Important clues to the mechanism by which cells are induced to synthesize interferon has come from several investigations: observations that various stimuli may lead to initiation of such synthesis; study of conditions which prevent induction; and attempts to identify the active component of crude inducers. Although these approaches have led to the proposal of several reasonable mechanisms for induction of the cell, the absence of a general biochemical theory for the mechanism of expression and repression of cistrons, coupled with the technical difficulties of performing certain definitive experiments with interferon, has to date prevented identification of the correct mechanism for induction of interferon (1, 2). This section considers proposed mechanisms for induction in the light of recent studies.

Isaacs originally predicted that the foreign nucleic acid was the essential stimulus for induction of interferon (2, 3). This hypothesis was based upon the knowledge that simple viruses which were composed of only protein and nucleic acid could induce interferon, whereas incomplete virus which contains much less nucleic acid failed to induce. Supporting evidence came from the finding that nonviral, heterologous nucleic acids (e.g., yeast RNA) stimulated the development of interferon-like resistance in cells and trace amounts of a soluble viral inhibitor which was never fully characterized as interferon (4–6). Additional studies of induction of resistance to viruses by heterologous nucleic acids or by some of their bases indicated that certain forms of the resistance were unrelated to interferon in that they lacked cell species specificity (7) or acted through noninterferon mechanisms (8, 9). Although many laboratories were unable to repeat the induction by nucleic acids, there was at least one confirmative study (10). As a result of these several negative findings the foreign nucleic acid hypothesis had been considered unproven for a period of time (11).

Recent findings have reawakened interest in nucleic acid as the stimulus for induction of interferon (12, 13). Treatment of Semliki Forest virus with hydroxylamine, which is thought to preferentially inactivate viral RNA and not viral protein, led to inactivation of the interferon-stimulating capacity of the virus in primary chick embryo cell cultures. Similarly, hydroxylamine was found to inactivate the infectivity and interfering properties of influenza virus without affecting antigenicity which is determined by the viral protein (14). These findings have been interpreted to support the view that viral nucleic acid is necessary for induction of interferon by the virus.

To define further whether the stimulus for induction of interferon is due to the infecting viral RNA itself or due to substances which are specified by the viral RNA, the effect of inhibition of the early stages of viral replication on the induction of interferon was determined (12, 13, 15). Specifically, chick embryo cell cultures were infected with Semliki Forest virus and incubated at 36°C for varying times before transfer to 42°C. At 42°C, but not at 36°C, the production of virus and virus-induced RNA polymerase were markedly or completely inhibited, as was also the production of other virus-specified proteins. Similarly no viral RNA synthesis could be detected at 42°C.

When infected cells were incubated at 42°C throughout the experiment no interferon was produced. For induction of full yields of interferon it was found necessary to incubate infected cells at 36°C for 2 hr before transfer to 42°C. During the 2 hr at 36°C, viral polymerase and both single-stranded and double-stranded viral RNA were formed in small amounts (13).

Similar studies have been performed in which conditional lethal mutants of Sindbis virus (arbovirus group A) were substituted for the conditional lethal mutants of Semliki Forest virus (also arbovirus group A) (16). At elevated temperatures where Sindbis virus penetrates and uncoats but does not initiate replication there was no induction of interferon. As in the studies with Semliki Forest virus, induction of interferon required a minimum of 2 hr incubation at the permissive temperature before the shift to the elevated lethal temperature. These findings also suggest that an early viral replicative event is required for induction of interferon.

Further evidence bearing on the viral-specified inducer of interferon comes from the findings that double-stranded RNA derived from MS 2 phage, reovirus, or several synthetic double-stranded RNA's can induce interferon (20a, 161). In one particular study, single-stranded synthetic RNA's were not active (17–20) nor were single-stranded ribosomal RNA from yeast and mouse cells; whole cell RNA from yeast and bovine cells; yeast core RNA; and myxovirus and tobacco mosaic virus RNA. (As is discussed later, single-stranded RNAs are active under proper conditions.) Calf thymus double-stranded DNA was also inactive. These findings have been interpreted to suggest that the double-stranded replicative form of single-stranded RNA virions is the stimulus for induction of interferon during infection. Also the short time for induction of interferon by double-stranded reovirus or synthetic RNA's *in vivo* in comparison with the longer induction time by intact reovirus was interpreted to be consistant with the idea that whole reovirus does not become an interferon inducer until its replicated double-stranded RNA has formed. In addition to inducing more quickly, double-stranded reovirus RNA induced far more interferon in the rabbit than did whole reovirus (19).

Although the foregoing evidence indicates that viral replication must be initiated to induce interferon, other forms of evidence suggest that under some conditions the input viral RNA itself may induce interferon.

In studies where cell cultures were stimulated to produce interferon with an RNA virus and simultaneously treated with inhibitors of protein synthesis, interferon was produced after reversal of the protein inhibitor and replacement with actinomycin D (*21, 22*). The interpretation placed upon these findings is that the inhibitor of protein synthesis prevented replication of the input virus but nevertheless the messenger RNA for the interferon was synthesized. Therefore, parental viral RNA or protein was probably the stimulus of the cell for induction of the messenger RNA.

The hypothesis that double-stranded RNA is the inducer does not fully explain the mechanism by which DNA viruses, which do not produce double-stranded RNA, act as stimuli. It is possible that certain forms of double-stranded viral DNA–RNA complexes act as inducers (*20*). Some suport comes from the finding that a DNA–RNA complex can induce cellular resistance to virus infection (*23*).

A major difficulty with the concept that double–stranded viral RNA is the principal inducer is the recent finding that synthetic single-stranded RNA's can induce large amounts of interferon under certain conditions (*24*). Certain preparations of polyinosinic or polycytidylic RNA homopolymers were 1/10 to 1/100 as active as double-stranded polyinosinic–polycytidylic acid. both in rabbits and in rabbit or human tissue cultures. More recent studies have shown that virtually all preparations of single-stranded polyinosinic or polycytidylic acid act as inducers in many cell systems when combined with the polybasic substances such as methylated albumin, protamine sulfate, or neomycin. Even yeast RNA becomes active under these conditions (*25*). It is not yet clear how these findings relate to earlier reports of induction of interferon-like substances by naturally occuring single-stranded RNA's (*2–5, 26*).

Some of the findings obtained with double-stranded RNA as inducers of interferon (*17–20*) may not clearly support an interpretation that the double-stranded replicative form of viral RNA is the stimulus of the cell to produce interferon. For example, the short time for induction of interferon *in vivo* by double-stranded RNA as compared with intact viruses does not hold true in all systems. Several intact viruses initiate interferon synthesis *in vivo* in equally short time (*27*). Furthermore certain viruses can induce interferon *in vitro* more rapidly than can double-stranded RNA's (*28*).

Similarly the finding that more interferon is induced in rabbits by double-stranded reovirus RNA than by intact reovirus may not mean that the double-stranded RNA is the primary structure. Other viruses such as Sindbis virus (29) can induce as much interferon in the rabbit as does double-stranded reovirus RNA.

Guanidine, which is thought to inhibit virus replication after formation of the double-stranded replicative RNA (30), blocks interferon formation by human amnion cells infected with poliovirus (31). This result is inconsistant with the concepts that input poliovirus RNA or its replicative form are sufficient for induction of interferon. It may suggest that full accumulation of poliovirus RNA, which is prevented by guanidine, may be necessary for induction. Alternately, the complexed viral RNA in the guanidine treated cell may not be available to act as an inducer.

Thus the available information does not allow final resolution of the apparent conflict between the hypothesis that infecting viral RNA (single stranded) or protein is the inducer and the hypothesis that the replicated viral RNA (perhaps double stranded) or protein is the inducer. The evidence does favor RNA as the stimulus. A possible reconciliation of the two views could come from the interpretation that sufficient quantities of either single-stranded or double-stranded viral RNA may stimulate. Isolated double-stranded RNA may appear more efficient because it is better able to reach intracellular sites intact due to its relative resistance to ribonuclease. Unexcluded is the possibility that the double-stranded RNA must dissociate into its single-stranded components intracellularly in order to induce. The requirement of early replicative events by some viruses for induction of interferon (13, 16) could be explained by the need for increased amounts of these types of viral RNA for induction over that provided by parental material.

An observation which may be difficult to interpret in terms of nucleic acid induction of interferon is that treatment of adenovirus with trypsin eliminates its interferon-inducing capacity in primary chick embryo cell cultures but does not decrease its infectivity for human KB cells (32, 33). The possibility that interferon was induced by a soluble protein in the viral preparation was ruled out because the interferon-stimulating capacity followed adenovirus migration in a cesium chloride gradient (33). One possible interpretation is that the component of the adenovirus preparation which induces interferon is a preformed protein which is linked to the virion. An alternative explanation is that although trypsin treatment does not eliminate the adenovirus infectivity for human KB cells, trypsin does impair viral functions in chick cells.

Another discrepant finding is that mutants of Sindbis virus which induce full production of viral RNA but not complete virus fail to induce interferon (*16*). It is not yet clear whether absence of induction is due to absence of the required stimulus or whether the mutant virus blocks the ability of the cell to synthesize interferon.

The concept of a single inducer substance or a single inducer mechanism for interferon is appealing for its simplicity and may hold true for most cell types. However, the variety of inducers under certain conditions suggests that the depression mechanism may vary with the cell being induced. There is increasing evidence to suggest that a few cell types are capable of being induced by stimuli which are insufficient to induce most cell types. Examples include phytohemaglutin stimulation of leukocytes (*34*), polkweed mitogen stimulation of leukocytes (*35*), endotoxin stimulation of explanted macrophages (*36*), antigen stimulation of sensitized lymphocytes (*37*), mannan stimulation of macrophages (*37a*), and spontaneous production of interferon by macrophages and leukocytes (*38, 38a*). It is not yet clear whether pyran copolymers induced cell types other than leukocytes and macrophages (*39, 40*).

It may be hoped that the stimuli which can induce interferon will be well defined in the near future. The mechanisms by which they act will be equally interesting but perhaps even more difficult to define.

B. SYNTHESIS OF THE INTERFERON PROTEIN

1. *Control by Host-Cell Genome*

Five observations suggest that the information for the synthesis of interferon resides in host-cell DNA: (a) the high degree of cell specificity of the antiviral action of interferon, (b) the similarity of the interferon produced regardless of the inducing virus, (c) the lack of immunological relationship between viral proteins and interferon, (d) species-specific antibody against interferon (*41*), (e) evidence of a more biochemical nature obtained through the use of actinomycin D, which inhibits DNA-dependent RNA synthesis (*42, 43*). This inhibitor blocks the synthesis of interferon both in tissue culture and *in vivo* (*44–49*), even when the inducing virus is insensitive to actinomycin D. However, host DNA synthesis is not needed for interferon production (50, 51).

The fact that blocking of cell RNA synthesis inhibits interferon production has been used to determine when the messenger RNA for interferon is synthesized (46–48). If groups of cell cultures, treated with actinomycin at different times after infection, are allowed to incubate further and then tested for the presence of interferon, one finds that after a certain time of infection the production of interferon is not inhibited by actinomycin. It can be assumed that the messenger RNA has been synthesized at this time. For Newcastle disease virus in mouse tumor cells this time was 6 hr postinfection (47), while for chikungunya virus in chick embryo cells the time was 2 hr (48).

Studies using the protein inhibitors puromycin and fluorophenylalanine showed that active synthesis of protein was needed to produce interferon (52–55). This evidence is supported by the previously mentioned fact that actinomycin blocks the production of interferon. Presumably actinomycin blocks the formation of the messenger RNA for interferon, and puromycin and fluorophenylalanine inhibit the synthesis of functional protein.

The finding that inhibition of cellular RNA synthesis blocks interferon production may explain why certain viruses are poor inducers of interferon production. For example, virulent poliovirus, which ordinarily inhibits cellular RNA synthesis, is a poor inducer of interferon, but the attenuated RMC strain of poliovirus which is less cytocidal does lead to its production (56). The herpes simplex strain dk+, which grows in dog kidney cells, is not an interferon inducer and does cut off cellular RNA, while the dk− strain does not cut off RNA synthesis rapidly, does induce interferon, and fails to grow (57). Vesicular stomatitis virus in chick embryo cells is also able to turn off cellular RNA synthesis (58) and interferon production. Apparently, in this case, the input virus itself contains something that inhibits cell RNA synthesis, as contrasted with other agents, such as Mengo virus or poliovirus, that require the early synthesis of a virus-coded protein for this inhibition. There are, however, cytocidal viruses that do induce interferon formation. It would be desirable to examine their effect on cell RNA synthesis before generalizing on this point. Possibly one factor that determines virulence of a virus is this ability to switch off cellular RNA synthesis and thus prevent interferon formation. Since both RNA and protein synthesis are necessary for the production of interferon and the antiviral state, comparable considerations might very well apply to cell systems in which protein synthesis is blocked by virus infection.

2. Suppression and Enhancement of Interferon Production

The experimental conditions under which interferon production is altered, help to understand the regulation of its synthesis. As referred to above, interferon production is blocked by those metabolic inhibitors which suppress cellular RNA or protein synthesis. Similarly those virus infections which cause early inhibition of cellular RNA and protein syntheses lead to production of little interferon (57, 58). Other viral infections suppress production of interferon although they do not severely inhibit cellular synthetic mechanisms, as evidenced by only minor cytotoxicity in infected cells (58–66). The mechanism of suppression of interferon production is unknown in these instances. It is possible that successful replication of many of these viruses is dependent on their ability to suppress interferon production or action.

When uninfected cells are treated with interferon before being stimulated by virus to produce new interferon, one of two effects may be observed. If the pretreatment is with low to moderate doses of interferon, subsequent stimulation with virus leads to the formation of increased amounts of new interferon and an earlier synthesis (28, 67–69). The mRNA for interferon also is synthesized earlier than in nonprimed cells, within ½ hr of infection in the chikungunya virus–chick embryo cell system mentioned above (28). When large amounts of interferon are used to prime the cells there is a decreased amount of new interferon synthesized (68–71). The nature of these enhancing and inhibiting effects is not understood; however, it resembles in certain ways the induction by interferon of the virus-resistant state in cells. That is, both RNA and protein synthesis are needed subsequent to interferon treatment in order for cells to convert to the "enhanced" state (69).

It is not known if the pretreatment with interferon exerts these effects because (1) somehow virus growth is modified leading to altered ability to induce or because (2) the inducibility of the cell genome is changed. Since there is a refractory state in reinduction of interferon by nonreplicating inducers such as statolon or synthetic RNA (72, 73), it seems that replicative events are not necessary for the altered inducibility. Although these findings favor the second possibility, final interpretation must await clearer understanding of viral and polynucleotide induction of interferon and effects of distinct inhibitors of interferon in some preparations of interferon (73a, 73b).

It is evident that the "priming" effect may play a role in recovery from disease. If an infected cell releases interferon to surrounding cells before the surrounding cells become infected, these cells would be enabled to respond to infection by providing interferon earlier and in larger quantities than cells not pretreated with interferon and so should be able more rapidly to shorten the spread of virus to other cells.

It has been established that several chemical classes of carcinogens as well as X-ray and ultraviolet radiation act to suppress interferon production (74, 74a). Related but noncarcinogenic compounds do not suppress interferon. These findings have been interpreted to suggest that suppression of interferon may be related to chemical carcinogenesis.

Steroid hormones have been observed to suppress, enhance, or to have no effect on interferon production depending on the type of hormone used, the dose, the cell or animal system, and the inducing virus (cf. 1). This is not surprising considering that the varied biochemical and physiological effects of steroids are extremely dependent on experimental conditions.

An interesting but as yet unconfirmed observation is the report that colchicine can markedly depress both interferon production and the antiviral action of interferon in mouse cells (75). Similarly, heavy water has been reported to inhibit the induction of interferon in infected cells (76).

Alteration of cellular protein synthetic rates are reported to be inversely correlated with the yield of interferon (77). Conditions leading to increased protein synthesis inhibits interferon production and, conversely, decreased protein synthesis is associated with increased yields of interferon.

Other factors which influence interferon production include temperature (78) and immune status of macrophages and lymphocytes (79).

12-3. Properties of the Interferon Protein

As noted in the introduction the word interferon was conceived originally because of the viral interfering activity of a substance in fluids that had been in contact with virus-infected cells. It soon became apparent that the responsible factor was proteinaceous. It later developed that this biological activity was associated with entities of different molecular weights. The variation in molecular weight depended on (a) the means of inducing synthesis and (b) the cells in which the interferon originated. The discussion below emphasizes the properties of interferon recognized since an earlier review (80).

A. BIOLOGICAL PROPERTIES

Several properties are shared by all of the substances which have been called interferon: (a) Interferons do not directly inactivate virus but induce cellular metabolic alterations which inhibit viral replication (81). (b) There is relative species specificity, in that an interferon produced by cells of a given animal species is most active in cells of the same species (2, 82). Diminished antiviral activity is frequently found in the cells of closely related species (83). Mouse interferon, for example, has been shown to have activity in rat and hamster cells but not in chicken cells. An exception is the antiviral activity of one type of human interferon in rabbit cells (84). (c) Antiviral activity is exerted against a wide variety of viruses. (d) There is no immunological relationship between interferon, which is a product of a cell gene, and the virus or any virus-specified product (85). The specific activity of interferon is among the highest of biologically active substances and greatly exceeds 10^6 units mg protein (86). A unit of biological activity has as yet no universal meaning but, in general, it is related to the minimum amount of interferon needed to induce significant resistance to some measurable effect of virus replication.

B. PURIFICATION OF INTERFERONS

One of the first observed chemical properties of interferon preparations was the destruction of the antiviral activity by proteolytic enzymes such as trypsin (85) and not by enzymes which hydrolyze other classes of compounds. Thus, the interferon molecule is either a protein or requires a protein component for its activity. Procedures used for purification of proteins have been applied, with moderate success, in attempts to obtain purified interferon (80, 80a, 81, 87, 88). Purifications on the order of 4000–20,000-fold increase in specific activity have been obtained (89).

In general, methods used for purification of interferon from a variety of species have included a preliminary elimination of some impurities by precipitation with perchloric acid, followed by concentration with a protein precipitant, such as Zn^{2+} or ammonium sulfate. Final purification involved chromatography on one or more cellulose or sephadex derivatives such as DEAE-, CM-, (2), or sulfomethylcellulose columns and subsequent electrophoresis (80).

The most successful purification has been obtained with chicken interferon (80, 89). It was the first interferon described, it was easy to produce in high titer, and it is relatively stable. The following is a summary of the series of steps that have been most successful. The interferon activity is adsorbed from the starting material by sodium aluminum silicate, followed by elution of activity in a smaller volume of 0.5 M potassium thiocyanate at pH 7.5. Acidification, first to pH 3.5 and then to pH 2 removes further inactive protein by precipitation, as does the subsequent addition of five volumes of methanol. The interferon activity is then precipitated by neutralizing the acidic aqueous methanol solution. It is then extracted from this precipitate with 0.01 M phosphate buffer at pH 7.5 and passed through DEAE-cellulose in the same buffer. The final step is to subject the active fraction eluted from DEAE-cellulose to pH gradient chromatography on CM-sephadex-G-50.

Using these methods a product with a specific activity of 1.6×10^6 units/mg protein was obtained. This represented nearly a 20,000-fold purification. The purified material was then subjected to acrylimide gel electrophoresis at an alkaline pH with a recovery of interferon activity which coincided with a double band of protein. However, when electrophoresis was carried out at acid pH in similar gels the activity was found to coincide with an area in the gel between protein bands and was probably not associated with any of the detectable protein. The interpretation of these findings is that the specific activity of interferon is much greater than 1.6×10^6 units/mg protein.

Different methods have been used to obtain partially purified human interferon. Crude interferon was prepared by infecting human foreskin fibroblast cultures with NDV virus (90) and treating the medium at pH 2 to destroy virus infectivity. This material was concentrated, with high yields by precipitation with zinc acetate (0.02 M) or by lyophilization. Only 30% yields were obtained using ammonium sulfate precipitation. The residual 70% was lost through either absorption or denaturation. The concentrated material, adjusted to pH 5, was then purified by chromatography on a XE-64 ion exchanger; the active material was eluted with a 0.3 M sodium succinate buffer, pH 7.0;

dialyzed against $0.2\,M$ ammonium bicarbonate; and further concentrated by lyophilization. These procedures give only a 5% final yield with a 10-fold increase in specific activity. A problem with all attempts to obtain purified interferon is that it seems to become less stable as it is purified with resultant low yields. In this regard the order of stability of interferons is chick > mouse > human.

C. PHYSICAL PROPERTIES

The physical and chemical properties ascribed to various interferons have been obtained by studying impure preparations. The extent to which impurities in the preparations have contributed to the observed properties is as yet unknown. Interferons have been examined by various means to determine the molecular weight of the active protein.

TABLE 12-1
Selected Molecular Weights Observed for Various Interferons

Molecular weight	Species	Inducer	Interferon from	Reference
110,000	Chicken	Statolon	Serum	91
50,000	Chicken	WS influenza virus	Allantoic fluid	93
38,000	Chicken	WS influenza virus	Allantoic fluid	93
30,000–40,000	Chicken	Statolon	Serum	91
26,000	Chicken	WS influenza virus	Allantoic fluid	163
75,000–90,000	Mouse	Gram negative bacteria	Serum	72
75,000–90,000	Mouse	Newcastle disease virus (NDV)	Serum	94
85,000	Mouse	Statolon	Tissue culture (T. C.)	91
54,000	Mouse	*Brucella abortus*	Serum	72
50,000	Mouse	TRIC agent	L-cells	163
50,000	Mouse	TRIC agent	Serum	163
38,000	Mouse	NDV	Serum	94, 161
34,000	Mouse	Statolon	T. C.	91
33,000	Mouse	*Salmonella*	Serum	72
30,000	Mouse	Statolon	Spleen extract	91
26,000	Mouse	Chikunguna virus	T. C.	91
26,000	Mouse	NDV	Serum	91
25,000	Mouse	NDV	Serum	161
25,000	Mouse	NDV	Spleen cell T. C.	161
25,000	Mouse	NDV	L-cell T. C.	161
>134,000	Rabbit	NDV	Macrophage culture	36
>134,000	Rabbit	NDV	Kidney cell T. C.	36

TABLE 12-1 (*Continued*)

Molecular weight	Species	Inducer	Interferon from	Reference
>134,000	Rabbit	NDV	Serum	*36*
>134,000	Rabbit	Endotoxin	Macrophage culture	*36*
>100,000	Rabbit	NDV	Serum and urine	*164*
>100,000	Rabbit	Endotoxin	Serum	*164*
75,000	Rabbit	Endotoxin	Macrophage	*36*
54,000	Rabbit	Endotoxin	Serum	*164*
50,000	Rabbit	NDV	Serum	*36*
44,000	Rabbit	NDV	Macrophage	*36*
44,000	Rabbit	NDV	Kidney cell T. C.	*36*
44,000	Rabbit	NDV	Serum	*36*
44,000	Rabbit	Endotoxin	Macrophage	*36*
46,000	Rabbit	NDV	Serum and urine	*164*
37,000	Rabbit	NDV	Macrophage	*36*
37,000	Rabbit	Endotoxin	Macrophage	*36*
35,000	Rabbit	Endotoxin	Urine	*164*
30,000	Rabbit	Endotoxin	Macrophage	*36*
30,000	Rabbit	NDV	Spleen cell T. C.	*161*
160,000	Human	Sendai	Human amnion	*160*
25,000	Human	Sendai	White blood cells	*160*
26,000	Human	NDV	Skin fibroblast T. C.	*90*
60,000	Rat	Sindbis virus	Tumor cell T. C.	*87*
100,000	Rat	Sindbis virus	Tumor cell T. C.	*87*
30,000	Rat	Sindbis virus	Tumor cell T. C.	*87*

The most popular recent methods have been molecular sieving on sephadex gels or centrifugation in sucrose gradients (*91, 93,* cf. *88*). A very wide range of molecular weights have been observed (Table 12-1). Many factors seem to influence the molecular weight obtained, such as the inducing agent, the cell or animal species, and the time of production after induction. Most of the work has been done with chicken and mouse interferon, but determinations have been made of the molecular weight of human, monkey, rabbit, and calf interferons (*87, 88, 91, 93, 94*).

From the studies with gel filtration three major molecular classes > 90,000; 40,000–50,000; and 25,000–30,000 have been found. There is as yet no adequate explanation for this wide range but several possibilities have been suggested. Polymerization of a basic monomer of lower molecular weight could give rise to the higher molecular weight material as could complexing with other cell products during synthesis or extraction (*36, 89, 92*).

The finding that many preparations of interferon contain more than one molecular species may be related to the heterogeneity of the cell types responsible for the synthesis *in vivo* and *in vitro* (see human amnion and leukocyte interferon in Table 12-1).

Many of the early appearing interferons, which tend to have higher molecular weights, may be "preformed." That is, their production is not inhibited under conditions of strong inhibition of cell RNA and protein synthesis *in vivo* and *in vitro* (*95*). For example, endotoxin stimulation of animals leads to appearance of a high molecular weight interferon in the serum within 2 hr. The appearance of this interferon is not prevented by inhibitors of RNA or protein synthesis. A similar phenomenon has been observed in bovine macrophages in culture. Virus stimulation of animals leads to early appearance of this high molecular weight interferon and to later appearance of much larger amounts of low (30,000) or intermediate (68,000) molecular weight interferons (*94*), the synthesis of which, in contrast, is blocked by inhibition of RNA and protein synthesis. These observations have been interpreted as indicating that the early-appearing high molecular weight interferon was already present in some cells and endotoxin acted to release it into the serum. The light interferon was newly synthesized after stimulation.

Early studies of molecular charge probably were influenced by effects of contaminating protein (*80, 80a*). The most recent work with chicken interferon, with specific activity in the range of 10^{-9} g protein/ unit of activity, gives results indicating an isoelectric point between pH 6.5 and 7.0 (*81, 89, 93*). Human and mouse interferon seem to have similar isoelectric points (*90, 93*).

Most interferons have been found to be stable over a wide pH range. In general, exposure of interferon to pH 2–10 has no effect upon biological activity (*85*). This stability makes it possible to destroy most residual viruses by acidifying interferon preparations to pH 2. Most species of interferon studied seem to retain this pH stability even after extensive purification. Some forms of human interferon, however, have been reported to be unstable at pH 3 or less (*34, 96, 97*).

The biological activity of interferons from different animal species vary in heat stability. Chicken interferon, the most stable, resists heating at 66°C (pH 7) for at least 1 hr even when purified (*80*). Human interferon the least stable type, lost activity when heated above 30°C. The rate of inactivation was the same for crude or partially purified tissue culture interferon. Human interferon in serum, on the other hand, was stable up to 40°C suggesting a protective effect of contaminating proteins (90). Other interferons studied are of intermediate stability, being more or less stable at 56°C.

The few studies on the effect of repeated freeze–thaw cycles or uv irradiation suggest that crude preparations seem more stable to these treatments than do more purified preparations (80).

D. Chemical Properties

Early determinations of the chemical composition of interferon were invalid because of the gross contamination of the preparations. Studies with proteolytic enzymes indicate that interferon either is a protein or requires intimate association with a protein for its activity. Disulfide bonds (81, 86, 87) but not free-sulphhydryl groups (86) are required for activity. Some, but not all, interferons are inactivated by urea (81, 87). The unavailability of pure interferon has precluded meaningful studies of amino acid analysis (86).

12-4. Mechanisms of the Antiviral Activity of Interferon

A. Inductive Action of Interferon on Cells

Early studies reported a variable extent of disappearance of interferon from fluids in contact with cell cultures (2, 98–100). However, the interferon that disappeared from the fluid could not be recovered from the cells (99). Under the experimental conditions of other laboratories, the loss of interferon from culture fluids was not observed even though the treated cells developed full antiviral activity (101–103).

Under conditions where interferon was lost from fluids the degree of induced virus resistance was related to both volume (total amount) and concentration of the interferon preparation; where interferon was not lost the degree of resistance depended only on concentration. These findings are consistent with the interpretation that interferon is continually decreasing in the former situation and not in the latter. Since under certain conditions antiviral activity can develop in the absence of detectable loss of interferon, it necessarily follows that disappearance of measurable quantities of interferon is not a prerequisite for development of antiviral activity. The loss of interferon may be due to nonspecific inactivation or nonspecific binding to cells.

Such inactivation may have been observed with a partially purified interferon which lacks stabilizers (*81*).

The lack of detectable uptake does not clarify the mechanism of the initial interferon–cell reaction but does show that only an undetectably small fraction of the applied interferon could be bound to cells at any one time. Certainly at least one molecule of interferon must react with each cell at reasonable intervals of time in order to account for the maintenance of a constant antiviral activity in the continued presence of interferon (*104*).

If adsorption of a few molecules of interferon by cells is a necessary prerequisite for the induction of the state of resistance to virus multiplication then conditions which enhance or favor the uptake of proteins by cells should produce an apparent increase in the titer of a given interferon preparation. Poly-L-ornithine, which is capable of enhancing the uptake by cells of both particulate materials and smaller molecules, has been shown to enhance the antiviral titer of an interferon preparation up to 10-fold (*105, 106*). If one assumes that poly-L-ornithine acts exclusively to increase uptake of interferon, this evidence suggests: (a) uptake of some interferon into cells is necessary to induce some antiviral activity; (b) under ordinary conditions (in the absence of poly-L-ornithine) that uptake is a small fraction of the interferon applied. Unfortunately these conclusions are tenuous because poly-L-ornithine exhibits considerable cell toxicity. For this reason the use of these conditions to enhance standard interferon assay procedures has not yet been employed. Dimethyl sulfoxide (DMSO) which can also enhance uptake of macromolecules, has been studied for its effect on interferon action (*107*). Mixed with interferon at concentrations of 5–10% it not only failed to increase the effect of interferon but also markedly inhibited the action of interferon. Additional studies with other compounds known to enhance cell uptake of macromolecules are needed before an interpretation of the results observed with poly-L-ornithine and DMSO can be made.

It has been suggested that the initial reaction of interferon with cells is attachment to the cell surface. This is based upon the observation that cells incubated at 1°C with interferon, washed, and then incubated at 37°C develop antiviral activity. Treatment with trypsin subsequent to reaction at 1°C and washing prevented the development of viral resistance, presumably because trypsin destroys bound interferon (*108–110*). It was also found that the initial interferon–cell

interaction proceeded rapidly at 1°C so that a maximum reaction had
occurred by 10–20 min. The amount bound was too small to measure
as loss from the fluid although perhaps 1% could be recovered from
the cells when the level of applied interferon was high. It should be
mentioned that many proteins will be found nonspecifically associated with cells at about this level.

B. CONDITIONS REQUIRED FOR INDUCTION

Several factors have been observed to affect the development of
cellular resistance to viruses during exposure to interferon. This section utilizes the working interpretation that resistance which is induced
by interferon is a reflection of the intracellular production and accumulation of an antiviral protein (52). The proposal that interferon
induced an antiviral protein was preceded by the observation that
cells treated with interferon had to be incubated at 37°C for several
hours before developing full resistance to virus growth. (85, 111). It
was generally agreed that this meant that metabolic activity was
needed for development of interferon's antiviral action. Taylor (52)
noted that actinomycin D blocked the development of antiviral activity in cells treated with interferon and concluded that DNA-directed
RNA synthesis was needed. She also said, but did not document the
statement, that fluorophenylalanine (FPA), an inhibitor of functional
protein synthesis, also prevented the antiviral action of interferon.
The documentation which was presented subsequently (53) showed
that FPA inhibited the development of interferon action when the
interferon was added to cells together with FPA, but not when interferon was added 5 hr before the FPA. These findings are in accord
with those reported independently (32, 54, 112). An unexpected observation was that mouse embryo cultures treated with interferon
in the presence of cycloheximide (a potent inhibitor of protein synthesis) developed almost the same level of resistance as did the control cultures treated with interferon alone (113). Cultures treated with
a combination of interferon, cycloheximide, and FPA and then washed
free of these substances developed resistance in the subsequent presence of medium plus actinomycin D but not in the presence of FPA.
These findings have been interpreted to indicate that in the presence
of interferon and cycloheximide, mRNA for the antiviral protein may
be transcribed, and it is then rapidly translated after removal of cycloheximide. It also suggests that the hypothesized mRNA is unstable

in the presence of FPA alone. Actinomycin D, added at the time of removal of interferon and cycloheximide, did not prevent the development of antiviral resistance. This finding that the proposed mRNA for interferon is synthesized in the absence of protein synthesis and it is translated when production of additional RNA is inhibited by actinomycin D indicates that the formation of proposed antiviral protein (AVP) is directly induced by interferon without synthesis of newly induced, intermediary proteins. Taken together these findings support the working model that interferon exerts its antiviral activity not directly but by inducing the cell to make a new protein.

This need for protein synthesis to induce antiviral activity is related to the synthesis of modified ribosomal subunits and ribosomes (interferon-type ribosomes) in which reside the ability to block viral replication (see Sec. E).

C. SOME DETERMINANTS OF THE DEVELOPMENT OF THE INTERFERON-INDUCED ANTIVIRAL STATE

Using the concept that modified ribosomal subunits and ribosomes account for interferon's antiviral action, the available evidence is probably best explained by the interpretation that the finally established level of viral resistance is maintained by a continual induction by interferon of modified subunits and ribosomes which serve to replace decaying subunits and ribosomes (114). The supporting evidence and some of the variables involved are now considered.

The time course of development of antiviral activity of cultured mouse embryo cells with varying amounts of interferon is shown in Fig. 12-1. The level of antiviral activity is expressed as $\log_{10}$ inhibition of vesicular stomatitis virus yield by interferon-treated cultures in comparison with untreated cultures, both infected at an input multiplicity of 20 or more. It may be seen in Fig. 12-1 that the time of onset of antiviral activity is directly correlated with the units of interferon applied. There is an early rise in antiviral activity at each dose level and the antiviral activity continues to rise for 5–7 hr. These findings are similar to those reported for mouse L-cells exposed to interferon (115). After this time, the degree of antiviral activity at each dose level reaches a fixed plateau which is related to the concentration of interferon in the extracellular fluid. Under the experimental conditions used, the extracellular concentration of interferon did not de-

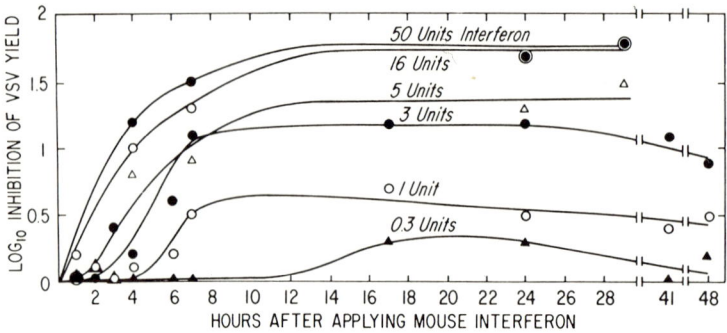

FIG. 12-1. Time course of development of resistance to multiplication of vesicular stomatitis virus, by cultures of mouse embryo cells treated with varying concentrations of interferon.

tectably decrease during the development and maintenance of the antiviral state (106, 116).

Evidence that the final established level of resistance is continually dependent on the extracellular concentration of interferon comes from the observation that the established level of resistance can be altered by increasing or decreasing the concentratiton of extracellular interferon (114, 115, 117). These findings also tend to rule out the possibility that the leveling off of resistance in the presence of interferon could be due to an acquired refractoriness of cells to the action of interferon. Evidence favoring the interpretation that the extracellular interferon acts to continually induce messenger RNA's and dependent protein(s) comes from decrease of the established level of resistance by application of metabolic inhibitors of RNA protein and synthesis (114). This interpretation applies to primary mouse embryo tissues reacting with interferon. It also probably applies to primary chick embryo cultures reacting with interferon since preestablished resistance in chick embryo cultures was entirely abolished by treatment for 6 hr with FPA. It is not yet clear whether this interpretation could explain the resistance level induced by interferon in all types of cell culture systems.

D. Factors Influencing the Initial Rise of Resistance

As was seen in Fig. 12-1, reaction of interferon with primary mouse embryo cultures was followed by a rising resistance to virus infection over the first 6–8 hr. This section considers factors which help determine the increasing resistance in this initial period before the final development of a fixed level of resistance. Timed addition of actinomycin D has been used to estimate the rate of formation of the hypothesized mRNA for the antiviral protein in cultures of primary mouse embryo cells (113). Earliest formation is detectable by 1–2 hr and the amount of mRNA required for the production of the maximum resistance at each dose level is synthesized by 5–7 hr.

Studies involving addition of actinomycin D with challenge virus during the period of rising resistance indicated that upon removal of interferon from mouse embryo cultures there is a rapid cessation of production of antiviral protein from mRNA transcribed after virus challenge (113). Further, when viral challenge was delayed up to 6 hr after removal of interferon and addition of actinomycin D, the level of resistance remained unchanged from the time interferon was removed. This finding suggests that translation of the hypothesized mRNA into antiviral protein is completed very rapidly and that the mRNA is not repeatedly translated and is therefore unstable relative to antiviral protein. Removal of interferon from cultures in the presence or absence of inhibition of protein synthesis was also followed by an unchanged level of resistance. These results together are consistent with the idea that removal of interferon from mouse embryo cultures rapidly results in the cessation of production of the AVP.

The interplay of factors which lead to the antiviral state may differ depending on the experimental conditions. For example, simple removal of interferon during the initial rise of resistance in chick embryo cultures and in mouse L-cell cultures (108, 111, 116, 118) is followed by a continued rise in the level of resistance. The continued rise of resistance after removal of interferon is prevented by actinomycin D (108, 118). This finding implies that under some conditions synthesis of the messenger RNA for the antiviral protein may continue at least in part after removal of interferon. A possible reconciliation of these apparently opposing observations is that the continued rise of resistance after removal of interferon is relatively small and therefore can be detected only during the very early stage of development of resistance.

There is a small but growing list of conditions under which interferon's antiviral action is altered. As referred to elsewhere in this chapter, interferon's action is inhibited by those metabolic inhibitors which suppress cellular RNA or protein synthesis. Similarly, interruption of cellular RNA or protein synthesis, which is caused by lowering the pH of cell culture medium, leads to inhibition (*119*). Steroid hormones have been reported to inhibit the antiviral action of interferon *in vitro* and *in vivo* under certain experimental conditions (*1*). Colchicine and heavy water have been reported to markedly inhibit the action of interferon *in vitro* (*75*). A particularly interesting group of substances, termed stimulons, which are derived from normal or infected cells, are reported to stimulate virus replication by inhibiting interferon's action (*120*). The reported augmentation of interferon's antiviral action by poly-L-ornithine has been discussed in Sec. 13-4A. Finally, a substance derived from *E. coli* augments the antiviral action of interferon (*121*).

E. The Molecular Basis of Interferon Action

The locus of action of interferon has been shown to be on a step in virus replication that occurs after the penetration and uncoating of virus, since interferon exerts its antiviral action even when infectious viral RNA is used as the challenge (*122, 123*). Also, the synthesis of new viral RNA (or DNA) is inhibited in cells treated with interferon. This has been shown for several virus–cell systems (*122, 124, 125*). The synthesis of replicative forms of viral RNA has been shown to be blocked in interferon-treated cells (*126, 127*). An earlier effect of interferon can be seen in cell systems where virus infection quickly leads to an inhibition of a high endogenous cellular RNA synthesis. Infection of exponentially growing L-cells by Mengo virus has been shown (*128*) to lead, within 30 min, to a decrease in cellular RNA synthesis. About 3–4 hr postinfection, a secondary rise in RNA synthesis is seen, which is attributable to the synthesis of viral RNA. In cells pretreated with interferon, the cutoff of cellular RNA synthesis is delayed by an hour (*125*); the secondary rise in RNA synthesis is not seen (*123, 125*). One can, therefore, conclude that interferon is affecting an event that occurs with 30 min after infection.

These as well as other consequences of interferon on virus-specified events, such as synthesis of infectious RNA, virus-induced enzymes, and proteins, are all explainable by the fact that interferon has an effect on the earliest interaction of the RNA of the infecting particle with a cellular component (*129*). During the course of a normal infection of L-cells with Mengo virus, the earliest detected event involving the viral genome is its association with the 40S ribosomal subunit (*130, 130a*). This association is detectable 30 min after infection.

As virus replication proceeds, the RNA of the parental virus is found on the viral polysome where viral proteins are made. Some viral RNA synthetase is found in association with this polysome. In cells pre-treated with interferon, the association of the viral RNA with 40S subunit is greatly inhibited. Some virus polysome-like material is found late in infection in these interferon-treated cells, but it is not functional, in that it does not synthesize viral protein (130). As would be expected, the viral RNA synthetase is found in greatly decreased amounts in interferon-treated cells (130–132).

Interferon treatment of vaccinia virus-infected L-cells also pre-vents the incorporation of virus mRNA into a virus polysome (133). Analysis of this reaction is complicated by the fact that initial associa-tion of the mRNA with the 40S subunit is not always measurable in L-cells. To examine the effects of interferon on vaccinia virus mRNA-40S subunit association it would be necessary to utilize HeLa cells in which this association is more readily observable (133).

Evidence of a different type to support the idea that interferon-type ribosomes can distinguish between cell and viral mRNA comes from work with the T antigen associated with SV40 virus (134). Inter-feron inhibits the synthesis of T antigen in cells productively infected with SV40 virus, where the virus genome and its mRNA exist as free entities in the cell. In cells transformed by SV40 virus, in which the virus genome may be integrated with the host genome, synthesis of the SV40 T antigen is not blocked by interferon.

Of interest in this connection are observations that interferon in-hibits the growth of *Toxoplasma* in tissue culture, and may inhibit the growth of *Plasmodium* berghei in animals. Both of these organisms are obligate intracellular parasites. Their sensitivity to the action of inter-feron may indicate that they need host-cell ribosomes for their replica-tions or that their ribosomes may become modified by the host's anti-viral protein.

These observations made in whole cells are supported and ampli-fied by experiments made with isolated ribosome from interferon-treated and control cells. Interferon-type and control ribosomes from mouse L-cells bind cell RNA and bind and translate synthetic poly-nucleotides equally well, but only the control ribosomes bind and translate Mengo virus RNA (135, 136). Also, the interferon-type ribo-somes from chick cells do not bind Sindbis virus RNA as well as con-trol ribosomes and translate it into protein even more poorly (137). A short treatment of the interferon-type chick ribosomes with trypsin augments somewhat its ability to bind and translate the viral RNA, suggesting that the interferon-type ribosomes contain additional proteins that enable it to distinguish between cell and viral RNA (137).

The *in vivo* and *in vitro* data indicate that the action of interferon is to cause the modification of the 40S ribosomal subunits and the ribosomes. These modified structures are able to carry out the synthesis of normal cell proteins and so support normal cell growth. They are able to reject viral RNA to a large extent and so virus growth is inhibited. Since the incorporation of viral mRNA into a ribosomal subunit and a polysome is the point of convergence in the replication of DNA and RNA viruses, it is possible to explain the action of interferon on both types of virus with a single hypothesis. By this mechanism the observed inhibition of synthesis of all virus-directed proteins, not just coat proteins, would be expected. This type of selective association with different types of RNA represents a type of control of protein synthesis at the ribosomal level.

In this regard, interferon's action resembles that of certain hormones. Isolated ribosomes from hypophysectomized or thyroidectomized rats do not translate added mRNA as well as do ribosomes from normal rats (138). Ribosomes from diabetic animals also have a low endogenous rate of protein synthesis *in vitro* (139). Other similarities to hormonal systems have been pointed out (140).

It is apparent that the basic effect of interferon is on normal cells and not on the infecting virus. It can be recalled from Sec. 12-4B that the establishment of the virus-resistant state in interferon-treated cells requires the synthesis of cellular RNA and protein subsequent to the exposure to interferon and prior to exposure to the challenge virus. The next section discusses the effect of interferon on normal cells.

F. EFFECTS ON NORMAL CELLS

It was originally suggested that the action of interferon was to inhibit some major cell function on which the virus was critically dependent. There were reports of uncoupling of oxidative phosphorylation, inhibition of RNA synthesis, and inhibition of cell growth. However, it was shown that interferon could act without any effect on oxidative phosphorylation and that uncoupling of oxidative phosphorylation did not necessarily affect virus growth (141–143). The reported effects on RNA synthesis and cell growth were probably attributable to impurities in the interferon preparations used (144) since some preparations of interferon, both crude and purified, were able to induce strong resistance to virus without affecting cell growth or cell macromolecule synthesis (*145–148*). Recent observation by Johnson et al. (*149*), while agreeing with the conclusion that moderate concentrations of interferon do not grossly affect cell metabolism, do show that high concentrations of partially purified interferon do affect protein synthesis. They characterized the effect as being probably attributable to the interferon in their preparation. Confirmation of this potentially important observation would be highly desirable.

This discussion indicates that the antiviral action of interferon is

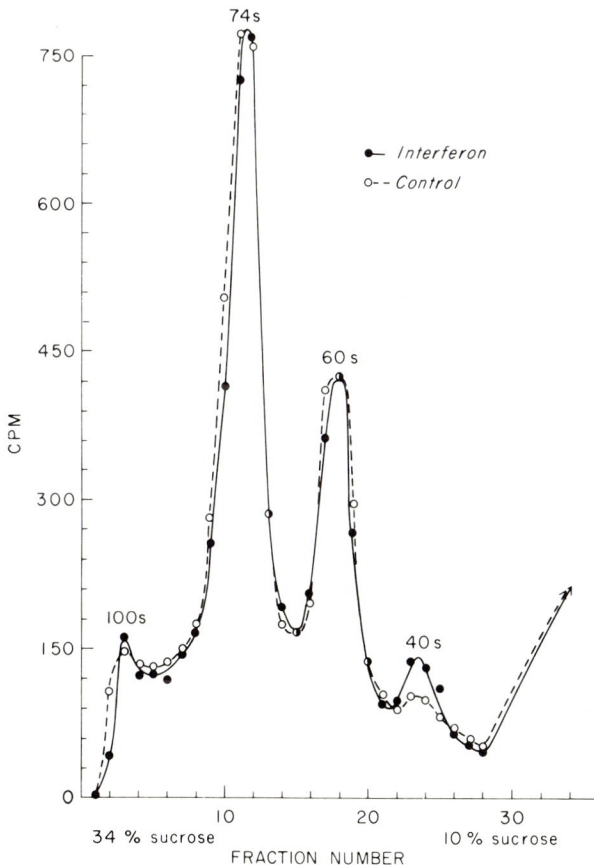

FIG. 12-2. Effect of interferon on incorporation of ^{3}H-valine into ribosome proteins. Suspension cultures of mouse L-cells were exposed to 100 units of interferon/ml for 16 hr, or to an equal volume of control fluid. Following a 1-hr pulse with ^{3}H-valine the ribosomes were isolated and analyzed by velocity sedimentation in sucrose gradients.

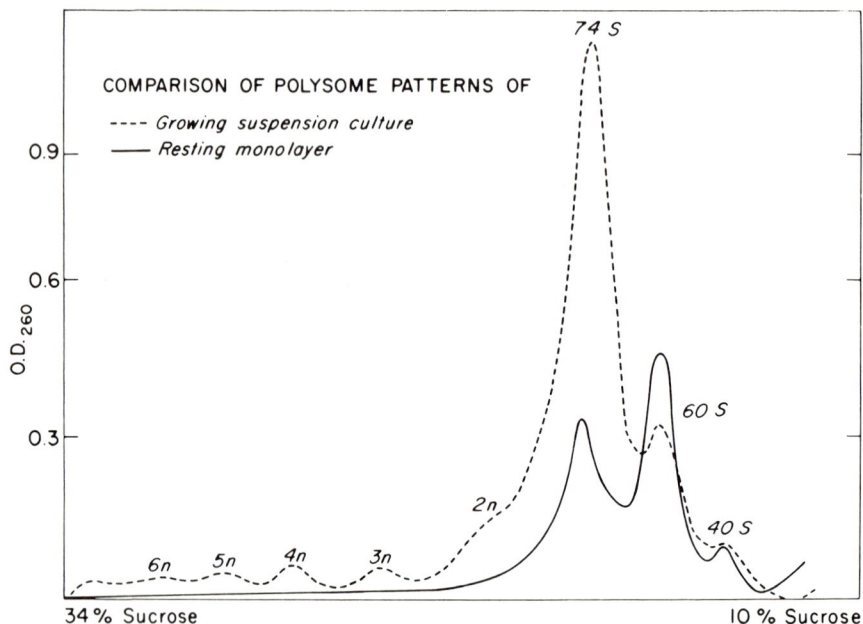

Fig. 12-3. Comparison of a ribosome pattern from suspension cultures of L-cells with that from a heavily sheeted-out monolayer culture.

not attributable to a gross disruption of cell metabolism. It is also self-evident that interferon must be causing some more subtle change in uninfected cells. The preceding section indicates that the change is in the ribosome and in particular, the 40S ribosomal subunit. These alterations in ribosomal subunit enable it and the ribosome to distinguish between cellular RNA and viral RNA but the nature of the alteration is not clear. The following observations, however, indicate that the changes are substantial.

Mouse L-cells reacting with a high concentration of interferon were exposed briefly to a radioactive amino acid and then their ribosomes were examined by sucrose gradient centrifugation. The proteins on the 40S ribosomal subunit became more heavily labeled in the preparation from interferon-treated cells than in the control preparation (Fig. 12-2). These results were obtained when the amino acid was valine, phenylalanine, histidine, or leucine (Ref. *165*). In an independent study, gel electrophoresis of radioactive proteins of chick embryo cells indicated the presence of increased amounts of radioactivity among the fast moving proteins of interferon-type ribosomes and 40S subunits (*150*). The authors were not sure this represents an interferon-induced effect. The increased radioactivity could be a new protein peculiar to the interferon ribosomal subunit or it could be due to the increased amounts of a normal protein.

With the L-cells, increased labeling of the 40S subunit was also observed when a radioactive RNA precursor, uridine, was used. This augmented uptake of macromolecule precursors is observed with both rapidly growing cells in suspension culture and fully sheeted out contact-inhibited monolayers. Comparison of the optical-density profiles of polysome gradients from rapidly growing and heavily grown out contact-inhibited cells is seen in Fig. 12-3. The low rate of protein synthesis in the contact-inhibited cells is consistent with the low level of polysomes. Interferon treatment of the logarithmically growing culture has an effect only on the labeling rates of the 40S subunit and not on the optical-density profiles. However, interferon treatment of the contact-inhibited cells does increase both the labeling rates and the amount of polysomes and the 40S subunit as shown by the altered optical-density profiles seen in Fig. 12-4.

One further effect of interferon on nongrowing contact-inhibited cells is that it induces an increase in the ratio of 16S ribosomal RNA to 28S ribosomal RNA, which is consistent with the increased amount of 40S ribosomal subunit seen in these cells. That these effects are attributable to interferon and not to impurities is attested to by the following. (1) It is detectable only with homologous, not heterologous

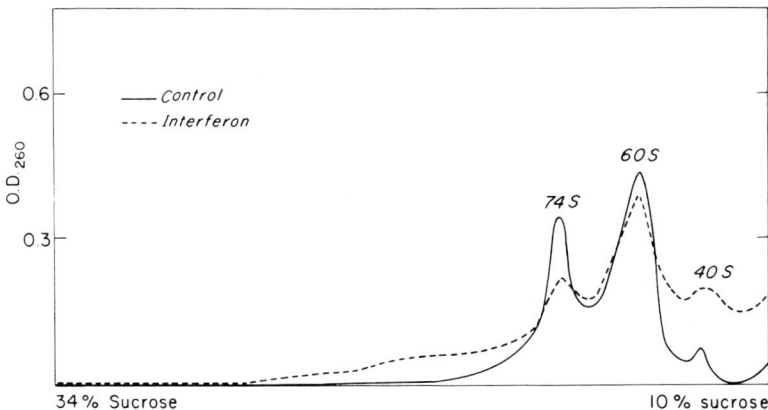

FIG. 12-4. Effect of interferon on polysomes from resting monolayers.

interferons. (2) It is correlated in time with the development of resistance to virus. (3) It is stable to pH 2. (4) It is not rapidly lost on washing the interferon preparation off the cells.

A reasonable interpretation of all the available data is that interferon induces qualitative changes in the 40S ribosomal subunit and also increases its rate of synthesis. In cells where the endogenous rate of synthesis is extremely small this result in a detectable increase in the quantity of 40S subunit. The observation that there are more polysomes in interferon-treated contact-inhibited cells would suggest that these cells are more actively engaged in protein synthesis than control cells. That this increased protein synthesis is attributable to new proteins that are to become associated with newly made 40S subunits can only be suggested.

12-5. Discussion and Conclusion

Future research with interferon may prove useful in understanding a wide range of biochemical and biological problems, some of which have been suggested by topics considered in this chapter. Studies of the two derepression steps involved in the interferon system may improve understanding the activation of gene function. Also, since the action of interferon involves the ability of ribosomes to make distinctions among different RNA's, future investigations should throw light on the control of protein synthesis at the ribosome level. Closely related to the RNA recognition mechanisms is the interpretation that the RNA for the oncogenic virus T antigen is inhibited in interferon-treated cells during acute infection but not in transformed cells where the viral genome is somehow integrated with the host genome (151). Extension of the recent findings that interferon and interferon inducers inhibit the growth of two obligate intracellular protozoal parasites (152–154) may lead to better understanding of the relationship of these organisms to their hosts.

The interferon system is one that has been studied at three levels, molecular, physiological, and medical. We have covered only the first of these here. The biological studies that deal with the sites of interferon production in animals, the types of cells involved, the role of leucocytes, and the transport of interferon from its initial place of synthesis to the other organs of the body are equally significant (1, 2). These studies have supported the concept that the interferon system is an important defense mechanism against virus infection (*2, 156, 157*). Of great importance from the public health viewpoint is the possible application of the interferon system in the treatment of human viral disease. Ever since the original suggestion of Isaacs that interferon production is the cells defense against the invasion of a foreign nucleic acid (*3*), there have been attempts to develop noninfectious agents that would stimulate animals to synthesize large quantities of interferon as a means of therapy against viral disease. The discovery of several natural and synthetic inducers of interferon (*16a–20a, 158, 159*) appears to promise practical and effective antiviral therapy in the very near future. Some of these compounds, the synthetic RNA's, have shown strong inhibitory effects on the growth of a variety of tumors in animals and the antitumor effect is currently under study (*155*).

REFERENCES

1. N. D. Finter (ed.), *Interferons,* North Holland, Amsterdam, 1966, pp. 1–340.

2. A Isaacs, *Advan. Virus Res.,* **10,** 1–38 (1963).

3. A. Isaacs, *Sci. Am.,* **204,** 51 (1961).

4. A. Isaacs, R. A. Cox, and Z. Rotem, *Lancet,* **1963-II,** 113.

5. K. E. Jensen, A. L. Neal, and R. E. Owens, *Nature,* **200,** 433 (1963).

6. Z. Rotem, *Israel J. Exptl. Med.,* **11,** 174 (1964).

7. K. Takano, J. Warren, K. E. Jensen, and A. L. Neal, *J. Bacteriol.,* **90,** 1542 (1965).

8. E. Mecs, *Acta Virol.,* **8,** 475 (1964).

9. G. E. Gifford, *Proc. Soc. Exptl. Biol. Med.,* **119,** 9 (1965).

10. H. Kohlhage and D. Falke, *Arch. Ges. Virusforsch.,* **14,** 404 (1964).

11. A. Isaacs, *Australian J. Exptl. Biol. Med. Sci.,* **43,** 405 (1965).

12. D. C. Burke, J. J. Skehel, and M. Low, *J. Gen. Virol.,* **1,** 235 (1967).

13. D. C. Burke, J. J. Skehel, A. J. Hay, and S. Walters, in *Interferon,* Ciba Foundation Symposium (G. E. W. Wolstenholme and M. O'Connor, eds.), Churchill, London, 1968, p. 4.

14. K. Grossgebauer, *Z. Naturforsch.,* **21b,** 1063 (1966).

15. J. J. Skehel and D. C. Burke, *J. Gen. Virol.,* **3,** 191 (1968).

16. R. Z. Lockhart, Jr., L. N. Bayliss, S. T. Joy, and F. H. Yin, *J. Virol.,* **2,** 962 (1968).

16a. M. Rytel, R. E. Shope, and E. D. Kilbourne, *J. Exptl. Med.,* **123,** 577 (1966).

17. G. P. Lampson, A. A. Tytell, A. K. Field, M. M. Nemes, and M. R. Hilleman, *Proc. Natl. Acad. Sci. U.S.,* **58,** 782 (1967).

18. A. K. Field, A. A. Tytell, G. P. Lampson, and M. R. Hilleman, *Proc. Natl. Acad. Sci. U.S.,* **58,** 1004 (1967).

19. A. A. Tytell, G. P. Lampson, A. K. Field, and M. R. Hilleman, *Proc. Natl. Acad. Sci. U.S.,* **58,** 1719 (1967).

20. A. K. Field, G. P. Lampson, A. A. Tytell, M. M. Nemes, and M. R. Hilleman, *Proc. Natl. Acad. Sci. U.S.*, **58**, 2102 (1967).
20a. W. J. Kleinschmidt, L. F. Ellis, R. M. VanFrank, and E. B. Murphy, *Nature*, **220**, 167 (1968).
21. R. R. Wagner and A. S. Huang, *Proc. Natl. Acad. Sci. U.S.*, **54**, 1112 (1965).
22. F. Dianzani, C. E. Buckler, S. Baron, and G. Rita, personal communication, 1968.
23. J. Vilcek, M. H. Ng, A. E. Friedman-Kien, and T. Krawciw, personal communication, 1968.
24. S. Baron, N. Bogomolova, H. B. Levy, A. Billiau, and C. E. Buckler, personal communication, 1968.
25. A. Billiau, C. E. Buckler, F. Dianzani, and S. Baron, personal communication, 1968.
26. A. Isaacs, *Sci. Am.*, **209**, 46 (1963).
27. S. Baron and C. E. Buckler, *Science*, **141**, 1061 (1963).
28. H. B. Levy, C. E. Buckler, and S. Baron, *Science*, **152**, 1274 (1966).
29. M. Ho, in *Interferons* (N. B. Finter, ed.), North Holland, Amsterdam, 1966, pp. 21–54.
30. D. Baltimore, *Virus Growth and Cell Metabolism* (H. B. Levy, ed.), Dekker, New York, 1969, Chap. 3 (this volume).
31. T. C. Johnson and L. C. McLaren, *J. Bacteriol.*, **90**, 565 (1965).
32. I. Beladi and R. Pusztai, *Z. Naturforsch.*, **22B**, 165 (1967).
33. M. Ho and K. Kohler, *Arch. Ges. Virusforsch.*, **22**, 69 (1967).
34. E. F. Wheelock, *Science*, **149**, 310–311 (1965).
35. R. M. Friedman and H. L. Cooper, *Proc. Soc. Exptl. Biol. Med.*, **125**, 901 (1967).
36. T. J. Smith and R. R. Wagner, *J. Exptl. Med.*, **125**, 559 (1967).
37. J. A. Green and S. Kibrick, *Federation Proc.*, 561 (1968).
37a. L. Borecky, V. Lakovic, D. Blaskovic, L. Masler, and D. Sikl, *Acta. Virol.*, **11**, 264 (1967).
38. R. M. McCombs and M. Benyish-Melnick, *J. Natl. Cancer Inst.*, **39**, 1187 (1967).
38a. Y. Nagano, Y. Kozima, J. Arakowa, and R. S. Kanashiro, *Japan J. Exptl. Med.*, **36**, 481 (1966).
39. T. C. Merigan and M. S. Finkelstein, *Virology*, **35**, 363 (1968).
40. P. DeSomer, E. DeClercq, A. Billiau, E. Schonne, and M. Claesen, *J. Virol.*, **2**, 878 (1968).
41. K. Paucker, *J. Immunol.*, **94**, 371 (1965).
42. E. Reich, R. M. Franklin, A. J. Shatkin, and E. L. Tatum, *Science*, **134**, 556 (1961).
43. R. M. Franklin, *Biochem. Biophys. Acta.*, **72**, 555 (1963).
44. E. Heller, *Virology*, **21**, 652 (1963).
45. M. Ho, *Bacteriol. Rev.*, **28**, 367 (1964).
46. M. Ho, *Science*, **146**, 1472 (1964).
47. R. R. Wagner, *Nature*, **204**, 49 (1964).
48. H. B. Levy, D. Axelrod, and S. Baron, *Proc. Soc. Exptl. Biol. Med.*, **118**, 384 (1965).
49. J. S. Youngner, W. R. Stinebring, and E. S. Traube, *Virology*, **27**, 541 (1965).

50. H. B. Levy, D. Axelrod, and S. Baron, *Proc. Soc. Exptl. Biol. Med.,* **118,** 1013 (1965).

51. D. C. Burke and J. M. Morrison, *Virology,* **28,** 108 (1966).

52. J. Taylor, *Biochem. Biophys. Res. Commun.,* **14,** 447 (1964).

53. R. M. Friedman and J. A. Sonnabend, *Nature,* **203,** 366 (1964).

54. R. Z. Lockart, Jr., *Biochem. Biophys. Res. Commun.,* **15,** 513 (1964).

55. S. Levine, *Virology,* **24,** 229 (1964).

56. M. Ho and J. Enders, *Proc. Natl. Acad. Sci. U.S.,* **45,** 385 (1959).

57. L. Aurelian and B. Roizman, *J. Mol. Biol.,* **11,** 539 (1965).

58. R. R. Wagner and A. S. Huang, *Virology,* **28,** 1 (1966).

59. J. Lindenmann, *Z. Hyg. Infecktionskrankh.,* **146,** 287 (1960).

60. A. Isaacs, *Cold Spring Harbor Symp. Quant. Biol.,* **27,** 343 (1962).

61. S. Hermodsson, *Virology,* **20,** 333 (1963).

62. S. Hermodsson, *Acta Pathol. Microbiol. Scand.,* **62,** 224 (1964).

63. K. Maeno, S. Yoshii, I. Nogato, and T. Matsumoto, *Virology,* **29,** 255 (1966).

64. E. F. Wheelock, *Proc. Natl. Acad. Sci. U.S.,* **55,** 774 (1966).

65. H. Diderholm and Z. Dinter, *Proc. Soc. Exptl. Biol. Med.,* **121,** 976 (1966).

66. J. E. Osborn and D. N. Medearis, Jr., *Proc. Soc. Exptl. Biol. Med.,* **124,** 347 (1967).

67. A. Isaacs and D. C. Burke, *Nature,* **182,** 1073 (1958).

68. R. Z. Lockart, Jr., *J. Bacteriol.,* **85,** 556 (1963).

69. R. M. Friedman, *J. Immunol.,* **96,** 872 (1966).

70. K. Cantell and K. Paucker, *Virology,* **21,** 11 (1963).

71. D. Stancek and J. Vilcek, *Acta. Virol.,* **9,** 1 (1965).

72. J. S. Youngner, in *Medicine and Applied Virology* (M. Sanders and E. H. Lennette, eds.), W. H. Green, St. Louis, 1968, p. 210.

73. F. Dianzani, G. Rita, P. Cantagalli, and S. Gagnoni, personal communication, 1968.

73a. A. Isaacs, Z. Rotem, and K. H. Fantes, *Virology,* **29,** 248 (1966).

73b. K. Paucker and M. Boxaca, *Bacteriol. Rev.,* **31,** 145 (1967).

74. J. DeMaeyer-Guignard and E. DeMaeyer, *J. Natl. Cancer Inst.,* **34,** 265 (1965).

74a. J. DeMaeyer-Guignard and E. DeMaeyer, *Science,* **155,** 482 (1967).

75. V. D. Soloviev and L. M. Mentkevich, *Acta Virol.,* **9,** 308 (1965).

76. V. Mayer and E. Dobrocka, *Acta Virol.,* **12,** 247 (1968).

77. R. M. Friedman, *J. Bacteriol.,* **91,** 1224 (1966).

78. J. Ruiz-Gomez and A. Isaacs, *Virology,* **19,** 8 (1963).

79. L. A. Glasgow, *J. Bacteriol.,* **91,** 2185 (1966).

80. K. H. Fantes, in *Interferons* (N. B. Finter, ed.), North Holland, Amsterdam, 1966, p. 119.

80a. K. H. Fantes, in *The Interferons, An International Symposium* (G. Rita, ed.), Academic Press, New York, 1968, p. 213–222.

81. T. C. Merigan, C. A. Winget, and C. B. Dixon, *J. Mol. Biol.,* **13,** 679–691 (1965).

82. D. A. J. Tyrrell, *Nature,* **184,** 452–453 (1959).

83. C. E. Buckler and S. Baron, *J. Bacteriol.,* **91,** 231–235 (1966).

84. J. Desmyter, W. E. Rawls, and J. L. Melnick, *Proc. Natl. Acad. Sci. U.S.,* **59,** 69 (1968).

85. J. Lindenmann, D. Burke, and A. Isaacs, *Brit. J. Exptl. Pathol.*, 38, 551–562 (1957).

86. K. H. Fantes, in *Interferon*, Ciba Foundation Symposium (G. E. W. Wolstenholm and M. O'Connor, eds.), Churchill, London, 1968, p. 78.

87. E. Schonne, *Biochem. Biophys. Acta*, 114, 429–439 (1966).

88. J. Vilcek, in *Interferon*, Springer, Berlin, 1968.

89. K. H. Fantes, in *Medical and Applied Virology. Proceedings of the Second International Symposium* (M. Sanders and E. H. Lennette, eds.), W. H. Green, St. Louis, 1968, pp. 223–229.

90. T. C. Merigan, D. F. Gregory, and J. K. Petralli, *Virology*, 29, 515–522 (1966).

91. T. C. Merigan, *Bacteriol. Rev.*, 31, 138–144 (1967).

92. R. R. Wagner and T. J. Smith, in *Interferon*, Ciba Foundation Symposium (G. E. W. Wolstenholme and M. O'Connor, eds.), Churchill, London, 1968, pp. 95–106.

93. G. P. Lampson, A. A. Tytell, M. M. Nemes, and M. R. Hilleman, *Proc. Soc. Exptl. Biol. Med.*, 118, 441–448 (1965).

94. J. V. Hallum, J. S. Youngner, and T. C. Merigan, *Virology*, 34, 802–804 (1968).

95. J. S. Youngner and W. R. Stinebring, *Virology*, 29, 310 (1966).

96. I. Gresser and K. Naficy, *Proc. Soc. Exptl. Biol.*, 117, 285–289 (1964).

97. M. M. Freshman, T. C. Merigan, J. S. Remington, and J. E. Brownlee, *Proc. Soc. Exptl. Biol. Med.*, 123, 862–866 (1966).

98. D. C. Burke and A. Buchan, *Virology*, 26, 28–35 (1965).

99. R. R. Wagner, *Bacteriol. Rev.*, 24, 151–166 (1960).

100. K. H. Ke, R. Armstrong, and M. Ho, personal communication, 1967.

101. C. E. Buckler, S. Baron, and H. B. Levy, *Science*, 152, 80–82 (1966).

102. J. S. Youngner, S. E. Taube, and W. R. Stinebring, *Proc. Soc. Exptl. Biol. Med.*, 123, 795–797 (1966).

103. L. A. Glasgow, personal communication, 1966.

104. S. Baron, C. E. Buckler, H. B. Levy, and K. Wong, in *Medical and Applied Virology* (M. Sanders and E. H. Lennette, eds.), W. H. Green, St. Louis, 1968, p. 146.

105. W. J. Kleinschmidt and L. F. Ellis, in *Interferon*, Ciba Foundation Symposium (G. E. W. Wolstenholme and M. O'Connor, eds.), Churchill, London, 1968, pp. 39–46.

106. J. Tilles, *Proc. Soc. Exptl. Biol. Med.*, 125, 996–999 (1967).

107. J. Vilcek and D. R. Lowy, *Arch. Ges. Virusforsch.*, 21, 254–264 (1967).

108. S. Levine, *Proc. Soc. Exptl. Biol. Med.*, 121, 1041–1045 (1966).

109. R. M. Friedman, *Science*, 156, 1760–1761 (1967).

110. R. A. Goldsby, *Experientia*, 23, 1073 (1967).

111. J. Vilcek and B. Rada, *Acta Virol.*, 6, 9 (1962).

112. G. Rita, M. Russi, F. Dianzani, and A. Quercioli, *Proc. Ital. Soc. Micro-Symp. Arboviruses* (1965).

113. F. Dianzani, S. Baron, and C. E. Buckler, in *The Interferons* (G. Rita, ed.), Academic Press, New York, 1968, p. 147.

114. S. Baron, C. E. Buckler, H. B. Levy, and R. M. Friedman, *Proc. Soc. Exptl. Biol. Med.*, 125, 1320 (1967).

114a. S. Baron, C. E. Buckler, and F. Dianzani, in *Interferon,* Ciba Foundation Symposium (G. E. W. Wolstenholme and M. O'Connor, eds.), Churchill, London, 1968, pp. 186–200.

115. R. Z. Lockart, Jr., and B. Horn, *J. Bacteriol.,* **85,** 996 (1963).

116. J. S. Youngner, A. W. Scott, J. V. Hallum, and W. R. Stinebring, *J. Bacteriol.,* **92,** 862 (1966).

117. K. Pauker and K. Cantell, *Virology,* **31,** 22 (1963).

118. R. Z. Lockart, Jr., *J. Virol.,* **1,** 1158 (1967).

119. J. V. Hallum, J. S. Youngner, and N. J. Arnold, *J. Virol.,* **2,** 772 (1968).

120. C. Chaney and C. Brailovsky, *Proc. Natl. Acad. Sci. U.S.,* **57,** 87 (1967).

121. J. Vilcek and M. H. Ng, *Virology,* **31,** 552 (1967).

122. S. E. Grossberg and J. J. Holland, *J. Immunol.,* **88,** 708 (1962).

123. P. DeSomer, A. Prinzie, P. Denys, Jr., and E. Schonne, *Virology,* **16,** 63 (1962).

124. M. Ho, *Proc. Soc. Exptl. Biol. Med.,* **112,** 511 (1963).

125. H. B. Levy, *Virology,* **22,** 575 (1964).

126. R. M. Friedman and J. Sonnabend, *Nature,* **206,** 532 (1965).

127. I. Gordon, S. Chenault, D. Stevenson, and J. Acton, *J. Bacteriol.,* **91,** 1230 (1966).

128. R. Franklin and D. Baltimore, in *Viruses, Nucleic Acids and Cancer,* Williams & Wilkins, Baltimore, Md., 1964, p. 310.

129. S. Baron and H. B. Levy, *Ann. Rev. Microbiol.,* **20,** 291 (1966).

130. H. B. Levy and W. A. Carter, "Abstracts 1st International Symposium on Interferon, National Institutes of Health, December, 1955," *Interferon Information Exchange,* January, 1966.

130a. H. B. Levy and W. A. Carter, *J. Mol. Biol.,* **31,** 561–577 (1968).

131. N. Miner, W. J. Ray, Jr., and E. H. Simon, *Biochem. Biophys. Res. Commun.,* **24,** 264 (1966).

132. J. A. Sonnabend, E. M. Martin, E. Mecs, and K. H. Fantes, *J. Gen. Virol.,* **1,** 41 (1967).

133. W. K. Joklik and T. C. Merigan, *Proc. Soc. Natl. Acad. Sci. U.S.,* **56,** 558 (1966).

134. M. N. Oxman, S. Baron, P. H. Black, K. K. Takemoto, K. Habel, and W. P. Rowe, *Virology,* **32,** 122 (1967).

135. W. A. Carter and H. B. Levy, *Science,* **155,** 1254 (1967).

136. W. A. Carter and H. B. Levy, *Biochem. Biophys. Acta.,* **155,** 437 (1968).

137. P. I. Marcus and J. M. Salb, in *The Interferons* (G. Rita, ed.), Academic Press, New York, 1968, p. 111.

138. L. D. Garren, R. L. Ney, and W. W. Davis, *Proc. Natl. Acad. Sci. U.S.,* **53,** 1443 (1965).

139. I. G. Wool and P. Cavicchi, *Biochemistry,* **6,** 1230 (1967).

140. S. Baron, in *Interferons* (N. B. Finter, ed.), North-Holland, Amsterdam, 1966, p. 268.

141. H. B. Levy, L. F. Snellbaker, and S. Baron, *Virology,* **21,** 48 (1963).

142. G. P. Lampson, A. A. Tytell, M. M. Nemes, and M. R. Hilleman, *Proc. Soc. Exptl. Biol. Med.,* **112,** 468 (1963).

143. A. Beloff-Chain, R. Cantazaro, F. Pocchiari, M. Balducci, and D. Balducci, *Biochim. Biophys. Acta,* **90,** 228 (1964).

144. C. Cocito, E. Schonne, and P. Desomer, *Life Sci.,* **4,** 1253 (1965).

145. H. B. Levy and T. C. Merigan, *Proc. Soc. Exptl. Biol. Med.,* **121,** 531 (1966).

146. S. Baron, T. C. Merigan, and M. L. McKerlie, *Proc. Soc. Exptl. Biol. Med.,* **121,** 50 (1966).

147. T. C. Johnson, M. P. Lerner, and G. J. Lancz, *J. Cell. Biol.,* **36,** 617 (1968).

148. R. R. Wagner, *Am. J. Med.,* **38,** 726 (1965).

149. T. C. Johnson, M. P. Lerner, and G. J. Lancz, *J. Cell. Biol.,* **36,** 617 (1968).

150. J. A. Sonnabend, E. M. Martin, and E. Mecs, in *Interferon* (G. E. W. Wolstenholme and M. O'Connor, eds.), Churchill, London, 1968, p. 143.

151. M. N. Oxman, W. P. Rowe, and P. H. Black, *Proc. Natl. Acad. Sci. U.S.,* **57,** 941 (1967).

152. R. I. Jahiel, J. Vilcek, R. Nussenzweig, and J. Vanderberg, *Science,* **161,** 802 (1968).

153. J. S. Remington and T. C. Merigan, *Science,* **161,** 804 (1968).

154. W. Schultz, K. Y. Huang, and F. B. Gordon, *Nature,* **220,** 1707 (1968).

155. H. B. Levy, A. H. Rabson, and L. Law, *Proc. Natl. Acad. Sci. U.S.,* in press (1969).

156. S. Baron, *Advan. Virus Res.,* **10,** 39 (1963).

157. S. Baron, in *Modern Trends in Medical Virology* (R. B. Heath and A. P. Waterson, eds.), Butterworths, London, 1967, p. 77.

158. W. Regelson, *Proc. Intern. Symp. Atheros. & Reticuloendothelial System, Advan. Exptl. Med. Biol.,* **1** (1966).

159. W. Regelson and T. Merigan, *Proc. Am. Soc. Cancer Res.,* March, abstract No. 219 (1967).

160. E. Falcoff, F. Fournier, and C. Chaney, *Ann. Inst. Pasteur,* **111,** 241 (1966).

161. G. P. Lampson, A. A. Tytell, M. M. Nemes, and M. R. Hilleman, *Proc. Soc. Exptl. Biol. Med.,* **121,** 377 (1966).

162. L. E. Kreuz and A. H. Levy, *J. Bacteriol.,* **89,** 462 (1965).

163. T. C. Merigan and L. Hanna, *Proc. Soc. Exptl. Biol. Med.,* **122,** 421 (1966).

164. Y. H. Ke and M. Ho, *J. Virol.,* **1,** 883 (1967).

165. H. B. Levy and W. A. Carter, unpublished observations.

AUTHOR INDEX

Numbers in parentheses are reference numbers and indicate that an author's work is referred to although his name is not cited in the text. Numbers in italics show the page on which the complete reference is listed.

A

Abbot, A., 41(169), *53*

Abe, C., 338–339(73), *357*

Abel, P., 371(29), *409*, 558(372), *576*

Abelson, J., 535(268), *573*

Acheson, N. H., 41(173), *53*

Achong, B., 29(101), *51*, 418(6, 14), *477*

Ackermann, W. W., 241(141), *256*, 391 (100), *411*, 430(76), *478*

Acton, J., 600(127), *611*

Ada, G. L., 32(112), 36(130), 41(169), 42(112), *52, 53*, 223(198, 200), 227 (45), 232(74), 236(45), *253, 254, 257, 258*

Adams, M. H., 486(4), *566*

Adams, W. A., 314(128), *326*

Agol, V. I., 156(171), 188(97, 98), *175*, 197(68), *217*

Agrawal, H. O., 34(146), 42(146), *52*, 221(59), 229(59), *254*

Ainbender, E., 188(123), *218*

Alcott, J., 353(127), *359*

Alexandrova, G. I., 252(224), *258*

Algranati, I., 121(91), *172*

Allen, B. V., 296(18), *323*

Allen, D. W., 18(195), 48(195), *54*

Allison, A. C., 32(114), 42(114), 44 (114), *52*, 77(123), *98*, 223(75), 232 (75), *254*, 302(55), 308(55), *324*, 341 (91), 342(97), *358*

Almeida, J. D., 77(1), 78(178), 83(2, 184), 86(41, 150), 88(178), 93(82), *95, 96, 97, 100*, 221(196), 224(22), *253, 257*

Amako, K., 88(130, 166), *99*, 162(189), *175*

Ames, B. N., 489(24), 534, *566, 573*

Amies, C. R., 303(58), *324*

Amos, H., 430(63), *478*

Anderer, V. F. A., 13(26), 14(26), 15 (53), 18(53), 20(26), 21(26), 22(53, 71), 34(117), *49, 50, 52*, 57, 74(4), *95*, 109–110(22), *170*, 180(6), *215*

Anderson, E. C., 117(80), *172*

Anderson, J., 243(149), *256*

Anderson, N., 260(16), 262(16), 276(16), *288*

Anderson, S. G., 41(169), *53*, 232–233 (83), *255*

Anderson, T. F., 128(110), *173*, 331(11), 342(11), *356*, 488(9, 12, 14), 492, 493 (48, 49), 494, 495, *566*

Andrewes, C. H., 83(5), 90(5), *95*, 220 (1), 222(1, 10, 11), *252*, 417(5), *477*

Anken, M., 332(31), 337(31), *356*

Anraku, N., 522, 523, *571*

Aoyagi, T., 44(177), *53*, 224(26), *253*

Aposhian, H. V., 374(94), 386(88), *411*, 513, 560(387), *570 576*

Apostolov, K., 81(6), *95, 97*, 224(24), *253*

Appleman, M. M., 149(159), *174*

Appleyard, G., 365(14), 390(95, 96), 391 (105, 106), 393, *409, 411*

Arakowa, J., 585(38a), *608*

Arangio-Ruiz, G., 224(23), *253*

Arber, W., 515(167), 565, *570, 577*

Archetti, J., 224(23), *253*

Arhelger, R. B., 382(71), *410*

Arlinghaus, R. B., 144(149), *174*

Armstrong, D., 300(41), 305(41), *324*

Armstrong, R., 594(100), *610*

Arnold, N. J., 600(19), *611*

Arnott, S., 269(80, 123), *290, 291*

Asano, K., 497(65), *567*

Astrachan, L., 499, 500, *568, 577*

Atanasiu, P., 86(7), *95*

Figueroa, M. E., 29(102a), *51*
Finch, J. T., 56, 57(44), 63, 64(47), 67 (44), 72(48, 49, 99, 100), 73(48), 76, 77(48, 50, 52, 99, 100), 79(50), 80(53), 84, 91(49), 92(50, 53), 93 (49), *96, 97, 98*, 110(27), 122(99), *171, 173*, 182(19), *215*
Fink, M. A., 318(141, 145), *326, 327*
Finkelstein, M. S., 585(39), *608*
Finland, M., 262(40), 264–265(40), *289*
Finter, N. D., 580, 581(1), 600(1), 606 (1), *607*
Fischer, H., 444(133), 446(133), *480*
Fischer, S., 112(50), *171*
Fischer-Fantuzzi, L., 547, *574*
Flaks, J. G., 392(110a), *411*, 502(81), 503, 512(82), 552, *568, 571*
Flanagan, J. F., 162(193), *175*, 336(44), 341(44, 88), 344(88), 350(88), *357, 358*, 434(101), 435, 441(119), 465 (101), *479, 480*
Fleming, W. H., 511(128), *569*
Fletcher, E. W. L., 221(196), *257*
Flewett, T. H., 81(6), *95, 97*, 224(24), *253*
Foft, J. W., 563, *576*
Follett, E. A. C., 15(34a), 16(34a), *49*
Forget, B. G., 337(53), *357*
Forsyth, B. R., 474(195), *481*
Force, E. E., 311(115), *326*
Fouad, M. T. A., 261(33, 37), 263(37), 264(33, 37), *288, 289*
Fournier, F., 592(160), *612*
Fowcell, O. W., 418(16), *477*
Fowler, C. B., 501, *568*
Fox, H. H. 222(11), *253*
Fox, J. P., 264(60), *289*
Francis, T., Jr., 232(82), *255*
Francki, R. I. B., 86(28, 183), *96, 100*
Frank, H., 13(26), 14(26), 15(53), 18 (53), 20(26), 21(26), 22(53), *49, 50*, 77(4), 95(4), 251(209), *258*, 331 (16), *356*
Frankel, F., 376, *409*
Frankel, F. R., 521, 523, 530, 563, 565 (197), *571, 576*
Franklin, R. E., 57(58), 63, *97, 98*, 112, *171*, 600(128), *611*

Franklin, R. M., 111(41, 43, 44), 113 (44), 114(66), 115(44), 116(41, 44, 66, 79), 117–118(44), 143(138, 141), 150(161, 163, 166), 152(141), 153(151), 162(191), 165(191), *171, 174, 175*, 209(89), *217*, 231(68a), 234(93), 236(100), 243(100), *254, 255*, 260–263(12), 272(75), 274(12), 275(12), 278(12), 279(12), *288, 290*, 379, *410*, 585(42, 43), *608*
Fraser, D., 493(47), 496(59), 511(126), *567, 569*
Fraser, D. K., 501(76), *568*
Fraser, K. B., 243(146, 149), *256*, 418 (13), *477*
Frearson, P. M., 441(113), *479*
Freeman, A. E., 23, 24(79), *51*
Freeman, N. K., 32(113), 42(113), 44 (113), *52*
Freiman, M. E., 113, *172*
French, R. C., 498(67), *567*
Freshman, M. M., 593(97), *610*
Frey, S., 78(112, 113), 79(113), 84 (112), 93(112), *98*
Friedewald, W. T., 474(195), *481*
Friedkin, M., 502(83), 510(119), *568, 569*
Friedman, M., 352–353(112), *358*
Friedman, R. M., 34(172), 41(172), *53*, 585(35), 586(53), 587(69), 588 (77), 595(109), 596(53), 597(114), 598(114), 600(126), *608–611*
Friedman-Kien, A. E., 583(23), *608*
Friend, C., 297(24), 303(24), *323*
Friesen, J. D., 402, *412*
Frisch-Niggemeyer, W., 233(87), *255*
Frist, R. H., 86(161), *99*
Frommhagen, L. H., 32(113), 42(113), 44(113), *52*, 109(21), *170*, 223 (199), *257*
Fuerst, C. R., 543(301), 545(312), *574*
Fujimura, R., 497(64), *567*
Fujinaga, K., 13(20b), 23(80–83), 24 (80–83), 26(82, 83), 27(80–83), *49, 51*, 336(51), *357*
Fujinami, A., 295, *323*
Fujiwara, S., 444(128), *480*

SUBJECT INDEX

recombination tests, 192–197

growth cycle, 111

  inhibition, 112, 154–160

  proteins involved in, 165–167

mutants characteristics, 200–201

  complementation tests on, 197–198

  physiological tests with, 199–207

  problems of, 190–192

  temperature-shift experiments, 198–199

phenotypic mixing, 129

proteins, 178–186

  functions, 185–186

  synthesis, 122

replication, 103ff.

  general view of, 165–167

RNA, 37–38

  base ratios, 131

  preparation, 105–106

  synthesis, 122–129, 167, 202

    defects in, 204–205

  structure, 110

  virion formation, 122–129

  X-ray diffraction studies, 56, 64–65, 110

Polykaryocytosis, induction by myxoviruses, 248

Polynucleotide kinase, in phage-infected cells, 516

Polynucleotide ligase, in phage-infected cells, 517–518

5′-Polynucleotide phosphatase, in phage-infected cells, 516

Polyoma viruses, DNA properties, 21

  structure, 95

Potato avcuba mosaic virus, electron microscopy, 85

Potato virus X, electron microscopy, 85

Potato virus Y, electron microscopy, 85

Potato yellow dwarf virus, electron microscopy, 86

Poxviruses, Appleyard's "protective antigen," 391–392

  chemical composition, 13

  chemical and physical properties, 30–31

  classification and properties, 4, 362–364

composition and structure, 364–368

core, 372

DNA, 374–382

  composition, 374–376

  control, 380–381

  properties, 15

  replication, 376–380, 394

electron microscopy, 88–89

enzymes, 394–399

genome, expression of, 383–392

infection, enzymes of, 383–390

  initiation and uncoating, 368–374

"M" form, 368

major groups, 6

messenger RNA synthesis, 399–400

particle weight, 14

polypeptides, 18

proteins, 390–392

regulation of gene expression, 392–399

replication, 361–413

Proteins (viral), manufacture, 117–122

Puromycin, as viral growth inhibitor, 149–150

R

Rabbitpox virus, electron microscopy, 88

Rabies virus, chemical and physical properties, 45–46

  electron microscopy, 86

Rauscher murine leukemia virus, (MLV), chemical and physical properties, 46–48

  electron microscopy, 90

  RNA, properties, 46–47

Red clover vein mosaic virus, electron microscopy, 85

Reoviruses, 259–291

  assay, 260–261

  chemical composition, 32, 263–264

  chemical and physical properties, 38–40

  classification and properties, 4

  hemagglutinin, 264–265

  members, 7

  morphology, 262–263

  particle weight, 33

  physical properties, 261–262